PLANT
BREEDING

PLANT BREEDING
Theory and Practice

Neal C. Stoskopf
UNIVERSITY OF GUELPH

with Dwight T. Tomes
PIONEER HI-BRED INTERNATIONAL

and B. R. Christie
AGRICULTURE CANADA

Routledge
Taylor & Francis Group

LONDON AND NEW YORK

First published 1993 by Westview Press, Inc.

Published 2019 by Routledge
52 Vanderbilt Avenue, New York, NY 10017
2 Park Square, Milton Park, Abingdon, Oxon OX14 4RN

Routledge is an imprint of the Taylor & Francis Group, an informa business

Copyright © 1993 Taylor & Francis

Library of Congress Cataloging-in-Publication Data
Stoskopf, Neal C., 1934–
    Plant breeding : theory and practice / by Neal C. Stoskopf, with Dwight T. Tomes and B. R. Christie.
        p.   cm.
    Includes bibliographical references and index.
    ISBN 0-8133-1764-9
    1. Plant breeding.   I. Tomes, D. T. (Dwight Thomas), 1946–
II. Christie, B. R. (Bertram R.), 1933–   III. Title.
SB123.S89   1993
631.5'23—dc20
                                                                    92-38294
                                                                          CIP

ISBN 13: 978-0-367-28301-8 (hbk)
ISBN 13: 978-0-367-29847-0 (pbk)

Contents

Tables and Figures

Figures

Preface

This book attempts to present a readable format on plant breeding principles and their application, based on the collective experience of the three authors, but with a heavy dependence on the scientific literature. Modern pedagogy recognizes that teaching can occur when students are motivated to learn. Subject matter must be communicated in an interesting, appealing, and understandable fashion. In preparing the text, every effort has been made to translate pertinent plant breeding references into a clear, logical, and comprehensible format for those studying the challenging and dynamic field of plant breeding.

Plant breeding is a huge subject on which many sciences impact, and this in itself can present a formidable challenge to clear, logical prose. In addition, plant breeding is a biological science, which means that no single best method of producing superior crop cultivars can be presented. There is no assured recipe for success. Instead, each situation must be analyzed, and the system of plant improvement adopted will depend on objectives, resources, the time frame, and the information available.

The three authors, representing separate disciplines, developed the book from a course for senior undergraduate students. Presented without bias toward a favorite crop, breeding system, or geographic location, the book has application for students in horticulture, field crops, and forestry.

In attempting to make the principles of breeding plants pedagogically palatable, the authors have not "written down" to the student. A key objective has been one of adequacy of presentation without popularization or condensation. There is an assumption that students using this book have a biological background that includes genetics. A basic review of genetics is included, however, for those whose recall of genetics has slipped.

Two important aspects of teaching are concepts and content. Concepts, or development goals, make up the main part of the text and are aimed at developing problem-solving strategies, process skills, and attitudes about learning and self. Concepts and principles are presented on the philosophy that when principles are understood, practices based on sound reasoning and understanding follow logically and with a degree of confidence.

With regard to content, it is impossible to cover specific breeding applications for the rich diversity of species in the plant kingdom. Attempts to cover the major species would be repetitious, wearisome, and laborious. On the other hand, it would be unfortunate if in covering concepts and principles the applications were missed. To avoid this situation, figures are included, each with a detailed caption that ties theory to practice.

An additional reason for including figures is to liberate the reader from a continuous flow of principles and concepts. The detailed caption has the effect of broadening the scope of the text and provides a practical example of the principle involved. Figures and photographs add a dimension of appeal and readability. Considerable knowledge may be gleaned by a student who browses through the text and, being attracted by a photograph, reads the caption. The figures, photographs, and tables should be regarded as an integral part of the script, not as a supplement, and as such provide an exemplary teaching technique to tie theory to practice.

We trust that the references that have been integrated in the text will serve their intended role by providing substance to statements and giving the reader an opportunity to pursue a topic in greater detail. The names of researchers are included, rather than references by number, so that the reader will become familiar with the names of leading authorities, both past and present.

Gratitude is extended to those who contributed to the preparation of this book beyond the call of friendship or financial return. The authors acknowledge a debt of gratitude to the University of Guelph, especially the library facility, for providing an environment that was conducive to preparing a manuscript. The individual encouragement and cooperation of faculty in securing information and photographs is also acknowledged. Photos without a credit line were taken by the senior author.

All the word processing, typesetting, and computer production of diagrams was undertaken by individuals at Clavis Software Inc., Guelph. The authors are truly grateful for their perseverance and constructive editorial comments.

The following individuals contributed in various ways, and their time and effort is sincerely appreciated: Gerald Stoskopf, Mitchell; Duane Falk, Art Davis, Jack Tanner, Dave Hume, Peter Pauls, Ken Kasha, Tissa Senaratna, Ernie Reinbergs, Peter Kevan, Ross Nazar, Jay Patel, Carol German, John Baker, Josef Papadopoulos, Mohamed Elhalwagy, Jenny Gusso, Randy Gadawshi, and Norman McCollum, University of Guelph; Randall Olson, Nova Scotia Agricultural College; Marvin Dudas and Keith Briggs, University of Alberta, Edmonton; Bob Wolfe and Daphne Fairey, Beaverlodge, Alberta; H. W. Anderson, Cliff Holland, and David Bates, Ontario Tree Improvement and Forest Biomass Institute, Maple; W. R. Meredith, Jr., United States Department of Agriculture Delta States Research Center, Stoneville, Mississippi; Brad Fraleigh, Plant Gene Resources of Canada, Agriculture Canada; personnel from the International Maize and Wheat Improvement Center, El Batan, Mexico, including Willy Villena, Benjamin Burke, Jorge Bolanos, Greg Edmeades, and Gonzalo Tsuzuki, El Batan and Tlaltizapan, Mexico; D. A. Fuccillo, International Crops Research Institute for the Semi-Arid Tropics, India; Edwin Tout, International Rice Research Institute, Philippines; Peter Schmiediche, International Potato Centre, Lima, Peru; Ken Barton, Daryl Marriette, Dan Vautour, and Brian Watkin, Ontario Ministry of Agriculture and Food, Audio Visual Productions; and Herb Rauscher, Illustration Services, University of Guelph.

The authors wish to thank Max Epp, Huttonville; Peter Hannam and Jim Timmings, Woodrill Farms and First Line Seeds; Peter and Hank Vander Pol, Rol-land Farms, Blenheim; Fred Rauch, University of Hawaii at Manoa; Bryan Harvey, University of Saskatchewan, Saskatoon; Brian Anderson, Doug Bailey, Lynne Mitchell, Mike Nuttal, and Bill Parks, Pioneer Hi-Bred Corn Co., Ridgetown and Chatham, Ontario; Daniel Pfister, Pfister Hybrid Corn Co.,

El Paso, Illinois; John de Witt, Northern Clonal Forestry Centre, Moonbeam; Allan Farquharson, Jane Ventry, Kelly Hunter, Gerry Roberts, Glenn Meatherall, and Tom Madrid, University of Guelph technical assistants; Rodale Press; the Mennonite Library and Archives, Bethel College, North Newton, Kansas; and the Canadian Seed Growers' Association.

To all of those who helped and who may have been overlooked, we extend our gratitude.

Neal C. Stoskopf
Dwight T. Tomes
B. R. Christie

CHAPTER 1

Plant Breeding:
Historical Perspective

Plant breeding can be defined as the application of techniques for exploiting the genetic potential of plants. Success in plant breeding depends on the ability to identify promising parents, to combine desirable attributes through hybridization, and to recognize and select effectively among segregating populations. It has taken humans many centuries to discover these basic concepts and to develop systems of plant breeding. Plant breeding as a science is new and is still under development, with plant biotechnology and molecular biology recently added as new tools.

Even before any of the basic concepts of plant breeding were understood, before anything was known about genetics and the **gene**,[1] and before sex and hybridization in plants were recognized, considerable progress in plant improvement occurred. This progress can, by definition, be said to be plant breeding, and by transforming wild plants into economically desirable crops, early humans achieved the first major accomplishment of civilization, namely, **agriculture**. It was plant **domestication** through plant breeding that allowed agriculture to develop.

This chapter provides the historical background that led to the enormous advances in plant breeding about A.D. 1900. That history reveals a painstaking process that began 8,000 to 10,000 years ago and was accomplished as an art without the benefit of science, which was not applied to plant breeding until the discovery of the laws of genetics in 1865. How success was achieved is largely speculation.

The normal reaction to a study of the historical development of plant breeding, especially a study covering the period when plant breeding was more an art form than a science, is that it will serve no useful purpose. On the contrary, its historical development is an integral part of understanding modern plant breeding. To ignore the past is like moving blindfolded into the future. History provides vision, and "where there is no vision, the people perish" (Proverbs 29:18, King James Version). History can enrich the imagination of the plant breeder, for it is with imagination that a dream can be turned into reality.

---

1. Words or terms that are defined or explained in the Glossary are set in boldface the first time they appear in the text and at strategic points throughout the text.

THE EARLIEST STEPS IN PLANT BREEDING

Wild plants usually have small, unappealing seeds that shatter at maturity to ensure seed dispersal and perpetuation of the species in the wild. Considerable improvement was needed to transform such plants into economically desirable forms with large, edible seeds or fruit that were retained on the plant after maturity. Variability undoubtedly existed, but whether humans recognized this variation and deliberately selected for it or whether progress was indirect and unconsciously conducted is open to debate. Arnold (1985) suggested that all crops were developed by humans through conscious selection. Jenkins (1966) argued that very little if any conscious selection occurred.

The opportunity for speculation about the early forms of plant breeding is considerable, the key point being that progress did occur. **Neolithic** farmers may have adopted a series of agricultural practices that embodied selection for improved plants. Harvesting the seed of wild plants for food purposes has a negligible effect, if any, on the genetic structure of the population. It is only when such seed subsequently is sown that a genetic shift can be expected (Harlan, de Wet, and Price, 1973) (Figure 1.1).

Exactly when humans unraveled the great secret of managed growth and changed from collecting seed of wild plants for food to placing seeds in the soil and tending their growth cannot be determined accurately. However, such controlled production should not be viewed as a single act or discovery but rather as a process that began in several areas (Harlan, 1975). The **transformation** to nonshattering types, that is, those that retained their seeds at maturity, was neither a simple nor a rapid process but one that occurred over a considerable time span.

Other features such as seed dormancy needed modification before plants were agriculturally acceptable. Dormancy is a biological mechanism that ensures the survival of seed until conditions occur that are favorable for establishment and growth. Seeds that were dormant might be lost to a subsequent population seeded by humans. Other seeds having specific requirements may have been influenced to germinate by cultural practices, such as varying the seeding depth, that in turn modified light, temperature, and moisture requirements. Again, selection occurred under human influence but not as an intentional act.

Perhaps it was in the area of seed size that deliberate human action exerted conscious selection and direct plant breeding. Large seeds would be favored over small seeds, and humans might select the large seeds for planting. If seed size was genetically controlled, a large-seeded population eventually would emerge (Figure 1.2).

Our knowledge of when and where such early developments in plant breeding took place is based on archaeological evidence. The first efforts to grow wild plants probably occurred in Southeast Asia and culminated in the domestication of some of the many plants with which humans experimented in 13,000 B.C. (Solheim, 1972). By 10,000 B.C. advanced knowledge existed of such plants as rice (*Oryza sativa*), several legumes such as the pea (*Pisum* spp), either the bean or the broad bean (*Phaseolus* or *Vicia* spp), and possibly the soybean (*Glycine* spp). It was at this point that crops became domesticated. Possibly ten separate genera of horticultural crops were domesticated in this region. Although the beginnings of civilization in China and Southeast Asia are obscure and fragmented (Treistman, 1967; Chang, 1981), the number of successful species domesticated indicates considerable success in plant breeding.

Braidwood (1960) suggested that the first successful experiment in food production in

Figure 1.1 It is still possible to find plants that are undomesticated and which, if successfully domesticated, may have considerable economic value. One such crop is wild rice (*Zizania palustris* L.), a grasslike plant found in shallow lakes and rivers in North America and highly prized by gourmets for its distinctive, nutty flavor and chewy texture.

The feature that distinguishes a wild plant from its domesticated counterpart is the ability to disperse seed at maturity. Because wild rice shatters its seed, the traditional method of harvesting is to press the panicles against the gunwales of a canoe or boat and dislodge the seed into the bow of the boat. This nondestructive method of gathering seed allows harvesting to occur several times per season. Wild rice is an annual, and because so much seed is dropped, the stand for the following year is assured.

This traditional method of harvesting has been going on for centuries, but if nonshattering was genetically controlled, no progress toward domestication could be expected. If harvesting was delayed so as to favor those plants that retained their seed, and if this seed was subsequently planted, a shift toward domestication might occur.

Adding to the complexity of domesticating wild rice is the fact that seed cannot be stored dry for an extended period of time. Normally wild rice exhibits seed dormancy that may last three months or longer when stored wet.

Deliberate selection for nonshattering seed habit led to the discovery in 1963 of a cultivar with reduced shattering habit. Although shattering losses still occur, the rise of a more shatter-resistant cultivar has permitted controlled production and the use of harvesting equipment similar to that for cultivated rice (*Oryza sativa* L.).

Learning to domesticate the wild rice crop is analogous to the situation of crop development by early agriculturists.

Photo courtesy Ontario Ministry of Agriculture and Food.

Figure 1.2   Early agriculturists may have consciously selected large seeds from among a mass of seeds and if seed size was genetically controlled, progress through plant breeding would occur.

The small wheat seeds on the left are those of einkorn (*Triticum monococcum*), one of three kinds of wheat originally domesticated in southeastern Turkey. It is a diploid with seven chromosome pairs. The wheat on the right is common wheat (*Triticum aestivum* L. em. Thell.). The seed of common wheat in this photograph is 3.6 times larger by weight than the seed of einkorn wheat. It is possible that a fivefold increase in weight per grain was achieved in the evolution of wheat from a wild grass.

Other wheats that were domesticated are tetraploid forms (14 chromosome pairs) called emmer (*Triticum dicoccum*) and *Triticum timopheevi*, which has no common name. As humans manipulated these wheats, combinations were formed that produce the hexaploid common wheat with 21 chromosome pairs.

The degree of transformation, as evident from this photo, is substantial, and although it took centuries to achieve, the plant breeding accomplishments of early agriculturists are significant.

The sequel to controlled agricultural production was to free humans from the task of obtaining their daily food supply. Those so freed were able to achieve advances in art and technology, the ingredients of a civilization.

Humans were able to domesticate wild species only if natural variability existed. A limited number of genera dominate the agricultural scene, possibly because of limited variation in other species. The *Gramineae* must have possessed a range of genetic variability, for eight genera make up a major share of important world crops, namely, *Avena* (oat), *Hordeum* (barley), *Oryza* (rice), *Pennisetum* (millet), *Secale* (rye), *Sorghum* (sorghum), *Triticum* (wheat), and *Zea* (maize).

Photo courtesy Ontario Ministry of Agriculture and Food.

Eurasia took place on the forest-free areas along the Tigris and Euphrates rivers in present-day Iraq, extending into Iran and southeast Turkey on the north and south into Lebanon, Israel, and the southern Jordan highlands. This horseshoe-shaped area is referred to as the Fertile Crescent and is an area where cereals and pulses show the rich diversity so essential for selection, a component of plant breeding. Imprints on clay vessels suggest that wheat, barley, rye, oat, and millet were produced between 9000 and 7000 B.C.

The third site of independent agricultural development and plant breeding was in south central Mexico between 6700 and 5000 B.C. Squash and avocados were domesticated first

(MacNeish, 1964), and in time a remarkable number of species were domesticated, with maize (*Zea mays*) as the foremost example.

The evidence of independent domestication of diverse crops in three separate areas suggests that advances in plant breeding were not isolated accidents or that such advances occurred without conscious human effort. Domestication in the Middle East and China included the self-fertilized cereals and soybeans. These species have relatively small and inconspicuous reproductive organs, and it is highly doubtful that humans practiced any hybridization. Self-fertilization leads to homozygous plants, but opportunity for selection could arise from within the heterogeneous population that had developed by natural hybridization and spontaneous mutation.

Maize is a cross-fertilized crop that was in an advanced state of development and extensively grown when the Europeans came to the Americas. Adapted maize **cultivars**[2] extended from the southern part of South America to the north shore of the St. Lawrence River; from sea level to elevations of 3,355 m (11,000 ft.). Types included flint, flour, pod, and popcorn as well as red, blue, black, yellow, white, and variegated kernels. There is no doubt about the competence in plant breeding of the American Indians, as it took the development of F_1 hybrid maize of modern agriculture to exceed the performance capability of Indian maize. It is understandable that European settlers grew to respect the maize crop and its developers, for maize is said to be the greatest gift from the Indians.

Exactly how the American Indian accomplished the monumental task of transforming the maize plant from an unresolved but probably primitive and unproductive grass (Figure 1.3) to a major food crop remains a mystery, for the details of how it was accomplished were not passed to European settlers. Weatherwax (1954) suggested the task was accomplished without the realization of what was being done, and hence how it was achieved was not understood. The credibility of this view is enhanced by the fact that the modification occurred over 4,000 years.

Some features of the maize plant may have facilitated breeding advances compared to the small grain cereals in Europe, but this should not discredit the plant breeding accomplishments of the American Indian. A large ear, for example (Figure 1.3), provided a major asset in that selection pressure was directed toward individual plants that produced 200 to 1,000 progeny. In addition, the maize plant is cross-fertilized, which assures heterozygosity, yet large separate male and female sex organs provide for manual removal of the staminate parts of some plants to control pollination. Moreover, maize is a full-season crop, and as the crop was moved to short-season areas, natural selection eliminated plants that did not reach maturity. If humans practiced selection, it was directed to individual ears, in contrast to small cereals that are mass harvested and sown in broadcast stands. The combination of natural selection coupled with deliberate and either conscious or unwitting selection resulted in a highly desirable domesticated crop. For agriculturists, there was no turning back, for once a crop was domesticated it became dependent on humans for survival (Figure 1.4).

Evidence exists that the American Indians did not understand sex and **hybridization** in

---

2. In addition to the term cultivar, plant breeders use a number of terms to designate plants at various stages of development in a plant breeding program. Terms include line, clone, hybrid, type, cultigen, stock, biotype, selection, strain, race, breed, brand, accession, segregate, phenotype, and genotype. All are defined in the Glossary.

A.

B.

Figure 1.3 Considerable speculation exists about the ancestry of cultivated maize. Three principal theories suggest that (1) the ancestor was a wild pod-popcorn, possibly extinct but the existence of which is based on archaeological evidence found in Mexican caves; (2) cultivated maize represents a domesticated, annual teosinte that is the closest relative of corn (the seed-bearing section is shown in Photo B); and (3) annual teosinte is not an ancestor but rather the product of the hybridization of *Zea diploperennis* with a maize plant in the early stages of domestication.

The third theory was developed by Wilkes (1977, 1985), based on the discovery in 1977 in Mexico of a previously unknown wild relative of maize, a perennial teosinte, *Zea diploperennis* (Photo A), having the same chromosome number as corn and cross-fertile with it. Wilkes's hypothesis has gained support and has compelled almost all students of the ancestry of maize to modify their views (Nault and Findley, 1981-1982).

Crosses between *Zea diploperennis* and a primitive Mexican popcorn of *Zea mays* produced annual teosinte plants. Evidence is strong, therefore, that the ancestry of cultivated corn is biparental, with *Zea mays* and *Zea diploperennis* serving as coequal ancestors (Mangelsdorf, 1986).

The accomplishments of the American Indian in developing modern corn cultivars from the first crosses 4,000 years ago in Jalisco, Mexico, are indeed major plant breeding milestones. The extent of breeding progress can be appreciated by comparing annual teosinte ears in Photo A with yellow dent corn ears in Photo B.

Photo A courtesy National Gardening Association, Vermont.

A.   *B.*

Figure 1.4   In the development of primitive crop plants, wild, weedy, and domesticated forms can be found. Wild forms can survive without the help of humans; weedy forms survive because agriculture provides a suitable environment despite efforts to get rid of them; and domesticated forms require care and cultivation for survival. Domestication is associated with the loss of the brittle rachis in cereals or the brittle cob in maize, the structures on which the seeds are carried.

The clumps of maize plants in Photos A and B are striking evidence of the fate of modern-day maize if left on its own for survival. The massive ear of modern maize cultivars is characterized by many rows of seed adhering tenaciously to a rigid cob that is enclosed by husks and adheres to the stock by a sturdy shank. When an ear of maize reaches the soil, it is usually intact and numerous seedlings emerge that are so crowded that they either choke each other or develop into numerous spindly plants that fail to reproduce seed. Photo B shows tassels but no productive ears. Domesticated plants are dependent on humans for survival.

In contrast, teosinte, an ancient form of maize, is a self-sowing wild plant that disperses its seed as the rachis segments from the disarticulating and slender cob. Teosinte seeds are arranged in two vertical rows on opposite sides of the rachis (distichons), as compared to polystichons in modern maize ears.

Domestication of a crop should not be confused with crop cultivation. Domestication involves genetic changes that make the plant better suited to human needs and the environment associated with agriculture. Domestication is an evolutionary change, whereas cultivation refers to human efforts to care for plants. Wild plants may be cultivated without being domesticated.

plants and believed the mixing of different kernel colors of maize when grown adjacent to each other resulted from underground fusion of small rootlets. In an effort to keep ceremonial maize uncontaminated, isolation was practiced and this may have aided the development of specific types.

PLANT BREEDING FOLLOWING HYBRIDIZATION

Hybridization to increase genetic variability for subsequent selection is an important component in plant breeding. The extent to which sex in plants was understood by early agriculturists appears to have been very limited. Without a knowledge of genetics, there could be no understanding of the fact that pollen carried half the hereditary features from the male parent. Positive evidence that sex was recognized and hybridization practiced is associated with the modern era.

Pollination of crops was practiced in the ancient world, but as an **empirical process** having no association with fertilization and probably with no appreciation of the fact that characteristics of the parents are transferred to offspring. The Assyrians and Babylonians in ancient times recognized that date palms were of two sorts, male and female. As early as 2400 B.C., and possibly earlier, people in these civilizations hand-pollinated date palms, which represent some of the oldest known domesticated tree crops (Figure 1.5). The ancient Arabs regarded the palm tree as an animal plant because like animals, it needs male fertilization to bear fruit (Zirkle, 1935), but curiously they appear not to have recognized sex in other plants.

Centuries later, Greek and Roman writers on natural history, including Aristotle (384-323 B.C.), Herodotus (484-425 B.C.), Pliny (A.D. 23-79), and Theophrastus (371-285 B.C.), commented on the supposed existence of the sexes in plants. Pollination in the date palm and fig tree was referred to as caprification, named after the wild fig *Caprificius*, a tree in which the fruit prematurely falls to the ground if not pollinated (Zirkle, 1935).

Unfortunately, many of the botanical writings of Aristotle and his student Theophrastus, among the greatest biologists of the ancient world, have been lost. Evidently these ancient workers failed to recognize that a single ovary developed when it was fertilized by a pollen grain. Pollen was thought to act in mass, the more the better. The whole pollen mass rather than a single pollen grain was regarded as the fertilizing substance. Although selection of superior seed during Roman times is mentioned (Jenkins, 1966), no firm evidence exists that hybridization was conducted for plant improvement purposes.

Evidence of active plant breeding in China can be gained from the philosopher and scholar Shunce (300-225 B.C.), who is quoted as saying, "Rather than allow things to propagate randomly, one should exercise one's skills and transform" (Braks, 1984, p. 5). Although active plant breeding was rare in the seventeenth century, a rice selection made by the Emperor Khang-Hi (1662-1723) made history (Reed, 1942). The emperor noticed a rice plant in a peasant's field that matured early and was of a desirable quality. The precocious plant was gathered, propagated, and named Imperial rice. It was the only rice of its time that matured north of the Great Wall. In longer-season areas, two crops of Imperial rice could be produced per year. This example illustrated the importance of progress in plant breeding by selection.

Figure 1.5 The photo shows the reproductive organs of the coconut palm (*Cocos nucifera* L.). Both coconut and date palms have been a part of cultivated fields and gardens since the dawn of civilization. Hand-pollination over thousands of years has genetically modified these crops to such an extent that neither the date nor the coconut palm have ever been found in modern times in a truly wild state.

Although most palm trees (*Phoenix dactylifera*) are **monoecious** and bear male and female flowers on the same tree, several species are **dioecious**, bearing male and female flowers on different trees. The date and palmyra palms exhibit this dioecious characteristic. To assure maximum fruit yield when cultivating dioecious palms, it is necessary to provide an adequate number of male trees in a field. Alternatively, only female trees could be planted if they are pollinated with pollen collected from superior male trees. This is an ancient practice of date growers and one that is standard procedure in modern date plantations. The advantage is that a single male tree can be used to pollinate an entire grove of female trees, thereby allowing for considerably more fruit-bearing female trees to be grown than the normal 50:50 ratio of male to female trees.

Despite the evidence of pollination in the ancient world, it is clear that these people did not understand that a single pollen grain united with an egg to form a fruit or that heritable traits were associated. Rather, pollination was a mechanical act, perhaps considered to be of religious significance.

PLANT BREEDING UNTIL 1900

Plant hybridization for crop improvement was scarcely used until after 1900. In cereals and pseudocereals with small inconspicuous flowers, such as amaranth (*Amaranth* spp), it may not be surprising that little hybridization occurred (Figure 1.6). On the other hand, early agriculturists recognized human and animal sexuality and that mating superior animals was the basis for improvement. Some plants have large conspicuous sexual organs, and it is curious therefore that the first real description of sex and plant hybridization was delayed until 1694 when the German worker Rudolph Camerer (1665-1721), better known by the Latinized name of Camerarius, described his experimental work on the question of sex in plants (Roberts, 1929). Camerarius appears to have been the first botanist to discover by actual experimentation that pollen produced on male flowers is indispensable to fertilization and seed development on female plants. Of interest is the fact that the experiments of Camerarius were conducted on the **dioecious** plants of hemp, spinach, and maize and not on plants with both sex organs contained in a single floret.

A second step toward hybridization in plant breeding was the establishment in England of the science of plant anatomy by Nehemiah Grew (1628-1711). He described flowers as organs of reproduction: the pistil as the female organ, the stamens as the male organ, and the pollen as seed (Durant and Durant, 1963; Roberts, 1929).

The first systematic investigations into plant hybridization in a number of species were conducted in Germany by Joseph Koelreuter (1733-1806), who used the large, obvious floral parts of the tobacco plant (*Nicotiana*). Between 1760 and 1766 Koelreuter systematically hybridized many plants and is regarded as one of the founding fathers of plant hybridization.

But knowledge of hybridization techniques appeared to have little impact. In England in 1717, Thomas Fairchild crossed the Sweet William (*Dianthus barbatus* L.) with pollen of the carnation (*Dianthus caryophyllis* L.), an **interspecific** cross, which is normally much more difficult to complete successfully than between a cross of the same species. The product of this cross was Fairchild's Sweet William (Roberts, 1929). Some examples of hybridization are described by Zirkle (1932, 1935).

Fairchild's work appears to have failed to motivate others or to have been utilized by science. Possibly the reason was related to attempts and probable failures to make **intergeneric** and interspecific crosses. Koelreuter reported attempts to hybridize 54 species belonging to 13 genera. His fame rests with the finding that when different species are crossed, they generally result in sterile progeny, so that hybridization in general is only possible between closely related plants.

Sterile plants did not produce offspring and were named mules after a sterile animal produced by crossing a male donkey with a mare. Sterile plants did not provide encouragement for continued hybridization, even though the enormous possibilities of hybridization were sensed by workers such as Camerarius. It was a century after Koelreuter before there was a revival of interest in hybridization within a species and before progress in plant breeding was resumed.

The Swedish botanist Carolus Linnaeus (1707-1778) helped the cause of hybridization by bringing a semblance of organization to the confusing world of nature through plant classification. Within the plant kingdom, Linnaeus formed a broad grouping called classes; each class

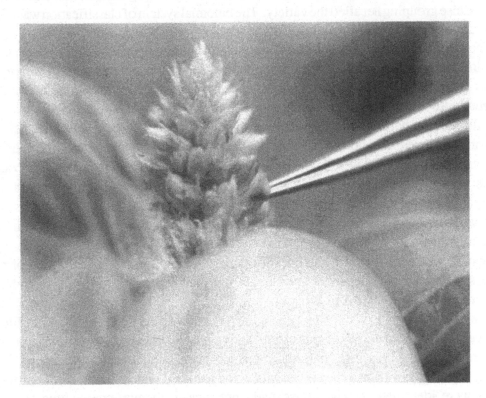

Figure 1.6 The inconspicuousness of the flowers of cereals, forages, and amaranth meant that hybridization was of minor importance until the second half of the nineteenth century.

Amaranth is of interest because it is an ancient crop that the Aztecs had developed to a high level of productivity. The grain amaranth (*Amaranth* spp) is a pseudocereal, but like true cereals has small inconspicuous flowers and like most cereals is self-fertilized. The amaranth plant being emasculated in this photo shows the small size of florets carried in clusters. Hundreds of such florets are arranged on a leafless axis on a complex panicle. Each floret consists of one staminate (male) flower surrounded by an indefinite number of pistillate (female) flowers.

To effect controlled hybridization, emasculation or removal of the stamens takes place before **anthesis** occurs to avoid selfing of the female parent. Male and female flowers can be distinguished before anthesis. The staminate flower swells just before the anthesis and stands out among the more demure female flowers surrounding it. Anthers are removed with the aid of fine surgical forceps. A magnifying glass is almost essential to remove all the staminate flowers.

Pollination of the female flower is accomplished the morning following the emasculation process by gathering pollen from a selected male parent and rubbing it gently onto the stigmatic surface of the female flowers.

Only astute ancient observers could successfully hybridize selections of amaranth, yet it was once cultivated extensively in Mexico and Central America. Production ceased in the early sixteenth century when the Spanish conquistadors banned the crop because of its sacred role in Aztec religious ceremonies (Tucker, 1986). Just how the crop was developed is a mystery.

Photo courtesy Rodale Press, Inc.

was subdivided into narrower groupings called orders; then genus, and within each genus, groupings were defined that would normally breed freely among themselves; and finally the most concise grouping he called the variety. The binomial system of classification was adopted, which consisted of a generic name and a species name, e.g., *Oryza sativa* (rice).

The skillful but flawed classification Linnaeus created is called the "sexual system" because it employed the stamens and pistils as critical characteristics for nomenclature. Its shortcoming was its dependence upon one set of organs instead of aggregate characters, but it made a contribution by serving as a fundamental system upon which others might build.

A further step that aided plant breeding was taken by Christian Konrad Sprengel (1705-1816). His work on flower morphology resulted in a much clearer understanding of the meaning of flowers. Sprengel identified self- and cross-fertilized plants, considered cross-fertilization to be most common, and concluded that, as far as possible, nature avoided self-fertilization.

Sprengel associated nectar-producing flowers with insect pollination and flowers lacking nectar with wind pollination. He observed that wind-pollinated flowers produced light and abundant pollen with the reverse found in insect-pollinated flowers.

Despite advances toward techniques for purposeful and routine hybridization of plants, progress in plant breeding depended upon the ability to handle progeny following hybridization. A major problem was that all F_1 progeny were identical, and breeders did not appreciate the importance of segregation in the F_2 generation. Nevertheless, Thomas Knight, in England, effectively selected superior cultivars of fruit and vegetable crops from crosses he produced in the years 1811 to 1838. Knight hybridized many species, including currants, grapes, apples, pears, and peaches, with the objective of producing hardier plants. When superior progeny of a cross were identified, they were asexually propagated, thereby overcoming subsequent segregation and reselection problems.

Knight discovered that crosses between red and white currants produced more progeny with red fruit, demonstrating the dominance of red color (Roberts, 1929). Unfortunately, instead of checking the segregating F_2 populations, he refertilized each F_1 plant, thereby confounding the segregating ratios that eventually led Gregor Mendel to determine dominant and recessive gene action and segregating ratios. Knight published this work in 1799, but advances in genetics had to await the publication of Mendel's work in 1865.

A major stumbling block to the development of genetics was the consideration of the whole parental complement rather than attention to the inheritance of specific traits. Without an understanding that an individual gene may influence a specific trait, purposeful breeding and selection was not possible. Consideration of the plant as a whole was analogous to seeing the proverbial forest but not the individual trees.

In species with an eye-catching feature that directed attention away from the whole plant, some progress was achieved. Attention was directed to ornamental plants with brilliant flowers (Figure 1.7). About A.D. 1070, the Chinese horticulturist Wang Kuan, for example, described 39 cultivars of peonies. In Japan, many new cultivars of flowering plants, including iris, were developed during the 1600s. The tulip craze in Holland undoubtedly arose from Dutch gardeners hybridizing cultivars and selecting progeny with appealing flowers. Asexual propagation aided progress.

Patrick Shirreff began work on wheat and oat crops in 1819, hybridized those crops, and selected superior cultivars. Shirreff was among the first cereal breeders of any consequence and

Figure 1.7　In contrast to grass plants with enclosed and inconspicuous floral parts, some plants such as the hibiscus (*Hibiscus* spp) flower in this photo have large, distinct, and exposed sexual organs that most surely attracted the attention of amateur hybridizers. As early as 1870, 12 cultivars of ornamental hibiscus were known.

Note the reproductive organs carefully. The flower is large and colorful and bears both male and female parts. The sexual column is brightly colored, as is the five-parted style, culminating in knoblike stigmas. An abundance of easily observable yellow pollen is produced from a cluster of stamens united for most of their variable length. Emasculation of these stamens could easily be accomplished by cutting away the folded petals of an advance flower to expose the anthers for complete removal. Since hibiscus flowers drop within 24 hours if fertilization has not occurred, emasculation and immediate pollination is necessary. The entire bud can be protected from stray pollen by bagging. Successfully fertilized flowers will develop seed pods that will mature in five to six weeks and produce 1 to 20 seeds per pod.

Evidence exists that the desire to produce new and commercially attractive colors led to hybridization, at least in plants with large sexual organs. Asexual propagation of F_1 plants eliminated selection of segregating progeny in the F_2 and subsequent generations.

Photo courtesy Ontario Ministry of Agriculture and Food.

developed the principle of selecting pure lines, a noteworthy plant breeding achievement in itself. Superior individuals were isolated and maintained as a line. Most of his work, however, was based on selecting individual plants from heterogeneous populations where **pure line** selection had never been practiced. The names of his cultivars reflect the phenotypic selections he emphasized and include bearded red, bearded white, and early yellow.

Shirreff took the important and previously unrecognized step of growing his selections among a collection of named cultivars for comparative purposes. These plots were not sampled for grain yield but were used to determine comparative height, maturity, and other agronomic features. Shirreff observed that new cereal cultivars could be obtained from three sources — hybridization, natural mutations that he called sports, and introduction. He stressed the need for carefully selected parents that best met the objectives of the cross. But without an understanding of segregation, he selected plants with the assumption they could breed true following hybridization.

In the New World, recognition of sexuality in the maize plant and the phenomenon of cross-fertilization (Figure 1.8) were reported independently by three observers in the early eighteenth century. Many of the steps leading to improvements of maize probably occurred by

Figure 1.8   When maize with colored kernels is planted adjacent to plants with yellow kernels, cross-fertilization will occur because of the wind-borne pollen grains. The results can readily be detected in the multicolored kernels on ears, as shown in this photo. This observation led workers in the early eighteenth century to conclude that sexuality did exist in the maize crop and that cross-fertilization occurred.

In 1716, the famed puritan clergyman Cotton Mather observed the transfer of kernel color to the progeny when contrastingly marked maize was grown in close proximity (Wallace and Brown, 1956). A similar and independent observation was reported by Paul Dudley in 1724, and in 1729 James Logan conducted what was probably the first truly scientific experiment with maize in America. By removing silks and tassels, he proved the precise function of each structure and the importance of wind as a pollinating agent (Reed, 1942; Roberts, 1929; Zirkle, 1935).

Much of the work of plant breeding prior to 1900 was conducted by nonscientists and amateurs. No plant breeders were recognized. Cotton Mather is best known as a witch-hunter; Paul Dudley was attorney general in Massachusetts and later a judge of the Superior Court; James Logan was a colonial governor of Pennsylvania. The father of genetics, Gregor Mendel, was an Austrian monk.

Photo courtesy Ontario Ministry of Agriculture and Food.

chance. There is evidence, however, that some crosses were directed and purposeful, as evidenced by the report of Lorain (1814). Lorain not only developed breeding methods but produced a high-yielding maize with wide adaptation that influenced production until advances associated with hybrids were introduced commercially in the 1930s (Hardeman, 1981) (Figure 1.9).

Yellow dent maize became so widespread in North America that its very origin probably was forgotten. There is not the slightest doubt that yellow dent was developed by a deliberate plant breeding process in the nineteenth century and not by the American Indians. The Indians provided the parental materials, however. Modern maize breeders should recognize that the vigor on which they capitalize in F_1 hybrids is largely the dispersed hybrid vigor of the first dent crosses (Anderson, 1944).

Despite the commendable advances of Lorain, much of the credit for maize improvement is given to Robert Reid in Illinois for the cultivar Reid's Yellow Dent. In 1846, Reid took a

A.   B.

Figure 1.9 In 1813, John Lorain of Phillipsburg, Pennsylvania, reported on the effect of crossing dent and flint maize and produced a superior genetic combination that was recognized quickly for its superiority. The cross involved Northern flint and Southern gourdseed maize. The progeny from crossing these two distinctly different maize types was a yellow dent maize, as shown at the bottom center of Photo A. Yellow dent maize is characterized by large cylindrical ears, as shown in Photo B. This maize formed the foundation of the agricultural economy of North America (Anderson and Brown, 1952).

Northern flint corn is characterized by small, yellow kernels produced on a long slender ear with eight to ten rows (upper left of Photo A). Almost all the endosperm is hard and translucent. The entire tip and sides of the endosperm consist of a thick cap of corneous tissue, and the grain does not change its shape on drying.

Southern gourdseed corn, shown in the upper right of Photo A, is characterized by large, soft, white kernels carried on ears having 18 to 24 rows. Gourdseed maize has a thin layer of corneous endosperm at the outer end of the kernel, which contracts on drying to produce a dent structure.

John Lorain (1814) must have recognized the potential of flint x dent crosses, for he advocated their production and gave directions for the production of first generation seed by detasseling one parent. Lorain noted that there was no blending of parental characteristics following hybridization; various properties did not mix like wine and water, but rather individual traits were transmitted to individual plants.

Photo courtesy Ontario Ministry of Agriculture and Food.

A.

B.

C.

Figure 1.10   In recognition of a major step forward in plant breeding, a historical cairn (Photo A) was erected to tell the story of Red Fife spring wheat and to honor David Fife, an Ontario farmer who made the selection. Fife lived before the establishment of any experimental research on plant breeding programs in Canada, and his efforts demonstrated the need for such a program. The development of

cultivar known as Gordon Hopkins Red from Ohio to Illinois. A poor stand developed, and an early-maturing yellow flint was used to replant the empty hills. The seed produced from this accidental cross-fertilization produced a cultivar of outstanding performance.

During this era, the state of plant breeding is shown by the use of bags to cover tassels and silks and the development of an early-maturing white sweet maize by Noyes Darling in 1844 (Anderson and Brown, 1952; Singleton, 1944). The shriveled kernels in the account verifies it as high-sugar sweet maize.

In self-fertilized crops, individual plants are homozygous but populations heterogeneous, and rather than hybridize plants to add to this confusing mixture, observant farmers and amateur plant breeders made individual plant selections. Henry Zimmerman, for example, in 1837 in Maryland was one of the earliest individuals in North America to select a pure line wheat cultivar (Ball, 1930).

The earliest recorded barley produced from hybridization in America was made in 1879 or 1880 by F. H. Horsford in Vermont. He crossed a hooded, hulless and an awned, hulled cultivar that resulted in a hooded, hulled cultivar (Pope, 1944). The fact that it was distinguishable may have overshadowed its agronomic performance.

Progress in breeding improved cultivars was slow. Although many selections were made following hybridization, performance was not reliable either because of further segregation in self-fertilized crops or because of outcrossing in cross-fertilized crops. Breeding methods based on identifiable phenotypic features seldom produced positive advances. Claims of superiority were not substantiated because no comparative tests were conducted. A standard of performance was lacking.

Occasionally, real progress did occur due more to good luck than good management, as illustrated by Red Fife wheat, a selection made in 1842 by Ontario farmer David Fife (Figure 1.10). This famous spring wheat was not selected for superior quality or agronomic

---

Figure 1.10 continued

Red Fife wheat was a significant event in the economic history of North America for four reasons. First, it was among the first pure line selections in North America and contributed to the development of the fledgling science of plant breeding. Second, Red Fife became the predominant parent in the development of improved cultivars, including Marquis spring wheat that became so prominent during the first half of the twentieth century in North America. In 1988, a stamp commemorating Marquis wheat was released (Photo B). Third, Red Fife set a standard of excellence for agronomic performance and bread-making quality and established a benchmark, or **check cultivar** system, vital to the developing wheat program. Finally, Red Fife was rust resistant.

The plaque on the cairn (Photo C) is entitled "The Story of Red Fife Wheat" and reads as follows: "David Fife, desiring to improve the quality of Canadian wheat, requested and received from a friend some new wheat samples from a Baltic ship docked in Glasgow, Scotland. In the spring of 1841 the samples obtained by Mr. Fife were sown . . . [and the one plant that reached reproductive development was saved and subsequently] proved superior to any other kind grown. The rust resistant Red Fife wheat was grown in the West and for 50 years remained the undisputed strain until it was crossed with another strain of wheat to produce the early maturing Marquis which became the basic wheat to be grown in North America."

Photos A and C courtesy Ontario Ministry of Agriculture and Food. Photo B courtesy Canada Post Corporation.

traits but because it was the only plant that reached reproductive development in a spring sowing of winter wheat. Red Fife was the first truly hard red spring wheat known in North America, and it set a standard of performance for yield and quality.

Other pure line selections were developed and, although less renowned than Red Fife, made a substantial contribution to the agriculture of the day. Fultz, a beardless soft red winter wheat, was developed in 1862 and Rudy in 1871 (Ball, 1930).

Pure line development in maize was difficult because of cross-fertilization, but named cultivars were developed in the latter part of the nineteenth century by judicious hybridization of preferred open-pollinated cultivars. Cultivars were often named after their originators or geographic origin and include such famous names as Krug Yellow Dent, Lancaster Surecrop, Iowa Stiff-Stalk, Synthetic, Boone County White, and Minnesota 13 (Mangelsdorf, 1974). These cultivars were grown extensively from 1890 to 1920 and were the source of breeding lines for hybrid maize development.

A 1979 survey of the source and breadth of maize germplasm revealed that 39% of the germplasm in the United States was related to the Lancaster source, 42% to the Reid source, and 18% to other sources (Zuber and Darrah, 1980). In 1984, 44% of the hybrid germplasm available was of Reid origin and 88% of the hybrids in the United States contained Reid in their parentage (Darrah and Zuber, 1986). This has caused concern about a narrow base and genetic vulnerability to disease. The detailed study of U.S. maize indicated a broader germplasm base than would have been expected solely on the basis of the number of races involved in its origins (Smith, Goodman, and Stuber, 1985). This fact is based on the considerable genetic diversity found in Reid Yellow Dent and Iowa Stiff-Stalk Synthetic.

The process of blending maize cultivars through crossing and recrossing continued up to the advent of commercial hybrid maize production. Superiority of parental selections was based on phenotype, and superior types were judged as exhibits at state fairs without regard for the environmental conditions under which the crop was produced or for its ability to produce a subsequent crop.

Progress in plant breeding is contingent upon the recognition and acceptance of **evolution**, for plant breeding based on selection, combination, and recombination of genetic traits is controlled evolution. The concept of evolutionary changes was introduced by Charles Darwin in the 1860s in his epochal work on the origin of species. His work lacked the clarity it might have enjoyed had he a knowledge of genetics, for he proposed the principle of pangenesis, or Lamarckism, the concept of inheritance of acquired characters.

Darwin was the first to inbreed maize deliberately, but a decrease in vigor and performance discouraged continuation of the experiment. His observations were published in 1876 in a work entitled "The Effects of Cross- and Self-Fertilization in the Vegetable Kingdom" and were prophetically described by Wallace and Brown (1956, p. 69) as "containing ideas which were destined to reach into the heart of the Corn Belt and change the nature of the corn plant for all time."

The major step in the progression toward the science of plant breeding was provided by Gregor Mendel (1822-1884) when he formulated the fundamental laws of inheritance by hybridizing the common garden pea in a series of experiments starting in 1858. The results were published in an obscure journal in 1865 and escaped the notice of scientists until 1900. No previous contribution to biological knowledge has had a greater influence on the development of plant breeding than Mendel's work. He demonstrated that when distinctly different cultivars are hybridized, contrasting characters segregate and recombine in definite, predictable proportions and do not blend together.

Mendel succeeded in his work because he had testable hypotheses and because he considered plant characteristics as specific, single, and separate units rather than as a group. In addition, he systematically followed through in subsequent segregating generations following hybridization, kept exact numerical records, carefully selected parents with contrasting characters, and presented his results in precise numerical fashion.

In one experiment, Mendel crossed tall and short cultivars. The first generation (F_1) progeny were all tall and not intermediate in height. From this observation, he developed the law of uniformity of the F_1, which states that when two different but homozygous individuals are hybridized, the F_1 progeny will be uniform because each is genetically identical.

Unlike many of his predecessors who hybridized plants, Mendel allowed the F_1 plants to self-fertilize, with the result that the second filial generation (F_2) contained both tall and short plants in a phenotypic ratio of 3 tall to 1 short. Mendel concluded that factors such as tallness or shortness occur in pairs. When contrasting factors occur together, one is **dominant** and masks the expression of the **recessive** factor. He reasoned that a short plant is homozygous for the recessive **gene** (tt) and tall plants could be either homozygous (TT) or heterozygous (Tt).

Mendel had no knowledge of **meiosis** but reasoned that some process must take place whereby the egg and pollen grain (**gametes**) receive one of a pair of genes and that members of a pair of genes must **segregate** from each other during gametic formation. A given gamete can carry T or t but not both, and hence Mendel's law of segregation (Figure 1.11).

The era of plant breeding purely as an art was closing, to be replaced by scientific principles based on genetics. Mendel's contribution goes beyond his discovery of the basic principles of genetics. His approach initiated the transition from empirical observations to hypotheses based on **inductive** and **deductive reasoning**.

Mendel's treatise "Experiments on Plant Hybrids," his letters to Carl Nägeli, the famous botanist and discoverer of **chromosomes**, and the rediscovery of his findings by Correns, de Vries, and Tschermak are included in a source book edited by Stern and Sherwood (1966).

OVERVIEW OF PLANT BREEDING DEVELOPMENTS BEFORE 1900

The enormous achievements of developing plants from a wild to a commercially acceptable state stand in sharp contrast to the painfully slow progress in plant breeding prior to 1900. Why did it take so long for the science of genetics to be developed? No simple, single explanation is adequate, but the following are presented for consideration:

1. Hybridization in plants was discouraged by religious and social forces. The word *hybrid* means to cross two unlike races and **breeds**. The Greek meaning of the word is to insult or outrage, especially when connected with sex. Progeny produced by hybridization were viewed as mongrels and considered an indignity, an affront, and an outrage to nature and the gods. So strong was this feeling that plant breeders felt obligated to justify their attempts at hybridization (Zirkle, 1935). Fear that a newly created species would be an impropriety and a tacit criticism of the deity of creation discouraged all but the boldest

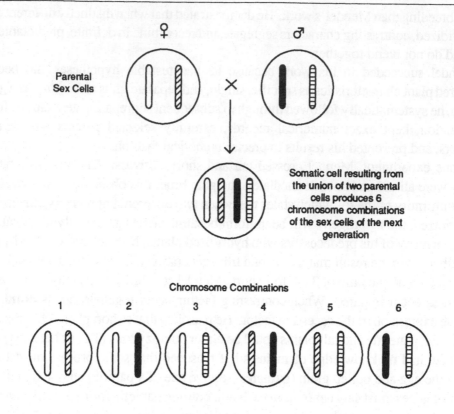

Parental Sex Cells

♀   ×   ♂

Somatic cell resulting from the union of two parental cells produces 6 chromosome combinations of the sex cells of the next generation

Chromosome Combinations

1   2   3   4   5   6

Figure 1.11   Mendel's discovery of genetics established a firm basis for the science of genetics and raised questions in selected fields that attracted the attention of biologists.   Fundamental to an understanding of flexibility in inheritance is the interchange of genes on a chromosome.   Random assortment of the genes received from two parents assures such flexibility.

The diagram shows six ways of combining four chromosomes in pairs.   Combinations 2 and 5 are between homologous chromosomes, and only combinations 1, 3, 4, and 6 are possible in the formation of sex cells.   Combinations 1 and 6 are the same as the parents, and combinations 3 and 4 are new ones. In species with a larger number of chromosomes, such as 7 in barley, 10 in maize, and 21 in wheat, a larger number of new combinations can occur.

This example illustrates a limited possibility of varying the genetic combinations because the genes are carried on chromosomes.   Crossing over can occur between contrasting genes, called **alleles**, of homologous chromosomes.   Crossing over provides a mechanism whereby entirely new genetic combinations can occur.   Superior recombinations of factors are always the hope of the plant breeder.

from attempting crosses. The impact of religious beliefs on agriculture is illustrated in Figure 1.12.

2. Despite the potential wrath of the gods, agnostics may have been tempted to produce a new species by hybridization. Their imagination was piqued by the successful development of the mule and by mythological reports of fantastic animal combinations (Zirkle, 1935). Bestiality was at least a sporadic practice among ancient herdsmen, and failure in the pasture field may have diverted attention to the garden, where dreams of producing commercially desirable and exotic new plants prompted the most unlikely

A.   *B.*

Figure 1.12 Evidence of plant breeding in ancient civilizations may be found in religious ceremonies. Agriculture was believed to be of divine origin and under divine guidance. In Egypt, the goddess Isis was the source of agriculture; in Greece, Demeter; in Rome, Ceres; in China, the ox-headed god Shen-nung; in Mexico, Quetzalcoatl; and in Peru, Verococha.

In Chinese history, Shen-nung is said to have instituted a custom of ritually sowing five kinds of grain at spring planting, but the purpose is unclear today. Was it to facilitate chance cross-fertilization to promote genetic variation either through natural or human means?

Fertility rites associated with pollination in crops implies hybridization was recognized as an important aspect in crop production. Mayan epic suggests that Quetzalcoatl taught the Indians tasks such as "selecting seeds" (Harlan, 1975, p. 40), a hint that plant breeding procedures were encouraged. Plant breeding came under divine guidance.

Deification of agriculture reached its most refined forms in objects of worship such as the ancient Aztec maize goddess, similar to the pottery replicas shown in these photographs. Note the large maize ears with straight rows that embellish the figures and attest to the success of ancient maize breeders.

Some of the religious ceremonies in Mayan culture recognized aspects of plant breeding. The goddess Xilomen, the guardian of the immature ears of corn, was honored by festivities starting with the first emergence of the silks on the ear, the beginning of pollination. Women with long, flowing, loose hair spread over their shoulders, backs, and bare breasts danced for eight days, the duration of the pollination period, as an admonition to the maize ears to produce long and abundant silks for pollination and grain production.

Modern society is inclined to look askance at ancient ceremonies as pagan cults. Nevertheless, these customs do reveal recognition of and respect for the events and processes associated with plant breeding.

Photo courtesy Ontario Ministry of Agriculture and Food.

crosses. Failure to effect all but a few such crosses and subsequent sterility discouraged further attempts at hybridization.

3. The heterogeneous nature of existing species discouraged crosses between the plants of the same species, which merely resulted in segregating progeny that resembled parental populations. Successful hybridization within a species was not noteworthy.

4. In the absence of science, empirical experiments conducted without an attempt to verify a theory or hypothesis generally were unproductive. Progress occurs when crosses and subsequent selection are directed and purposeful. Without the realization that individual traits were genetically influenced, selection for the plant as a whole was too complex to contribute to the advance of knowledge.

5. Success did occur when aided by natural selection or when a new and desirable segregate from a fruit tree, potato, or ornamental plant could be propagated asexually (Hedrick, 1950). Among the outstanding contributors was Luther Burbank (1849-1926), who released cultivars of numerous asexually propagated species (Howard, 1945) and popularized crop breeding before it was a recognized branch of science.

6. Despite considerable advances during Greek and Roman times, the period from the third to the end of the sixteenth century was a retrogressive era in agriculture and science, and associated advances were negligible.

7. Philosophical views, apart from religious implications, contributed nothing to plant breeding. The idea of a combination of both sexes in a single plant was not in harmony with the philosophical concepts of the Babylonians or the Greeks, for it might be concluded that plants were superior to animals, which was unthinkable. The controlling philosophy connected with botanical thought adhered closely to the doctrine of immutability of plant species, which included a belief in special creation that could not be improved upon by human efforts.

8. Advances in plant breeding are often small and subtle but important and may go unnoticed without a proper system of testing or standard cultivars. There is no evidence of the early establishment of research stations to check new cultivars systematically, let alone conduct plant breeding. The Rothamsted Experiment Station, considered the oldest agricultural research station in the world, was established in England in 1843 for the purpose of studying crop nutrition and management systems. The Morrow plots, the oldest in the United States, were established in 1876. Leadership was lacking in plant breeding until late in the nineteenth century.

PLANT BREEDING ABOUT 1900

The rediscovery of Mendel's laws in 1900 prompted a surge of interest in plant breeding, but systems of handling segregating populations had to be developed. In Sweden at the end of the nineteenth century, W. H. Nilsson was possibly the first to realize that a pure line is a population descended from the self-fertilization of an individual homozygous plant. Examples of individual plant selection had been shown before Nilsson, but the reasons for success were not clearly understood.

One of the earliest workers to inspire the modern movement in systematic plant breeding was Wilhelm Rimpau (1842-1903) (Anonymous, 1910). Rimpau's plant breeding methods were developed on the cross-fertilized rye crop and were termed "broad breeding," in contrast to American and Swedish approaches using self-fertilized crops that developed the method of "narrow breeding." The **centgener method** of testing was developed based on narrow breeding systems.

The centgener method is no longer a common term or plant breeding system, but it offered breeders the first systematic steps in handling segregating populations. It was the forerunner of the **pedigree method**. Willet M. Hays, at the Minnesota Agricultural Experiment Station, is credited with developing the centgener system (Ball, 1930), but the implications were not appreciated fully, for it was viewed as a testing rather than a breeding procedure. Hays (1901) probably was the first to publish a treatise on plant breeding methodology including the centgener method.

The centgener method, which was widely used (Buller, 1919; Beaven, 1947), is based on individual plant selections from segregating generations of self-fertilized crops. The system involves seed obtained from a single plant, spaced in nursery trials. The name developed from the fact that 100 plants were grown from each plant selected. Ten F_3 plants were selected based on careful observation and records. Selected plants were threshed and weighed and the five best were bulked and used to establish a centgener plot the following year. The criterion to measure yield was the number of grains harvested per grain sown rather than total grain weight. This measurement was an integral part of the centgener method, where 100 plants were equally spaced in a square of a specified size.

The development of pure lines that exhibited uniformity in appearance, performance, and reliability was a giant step forward, for it revealed a major deficiency — the inability to identify superior selections. Yield per unit area replaced number of grains harvested per grain sown as the measure of yield. From the centgener system evolved test plots 1/100 acre in size that were mechanically harvested and threshed, and the results were summarized for farmer recommendations over a five-year period. This proved too cumbersome for the large number of entries, which in Ontario grew to over 300 cereal plots between 1890 and 1919 (Zavitz, 1919). The time was right for statistical methods to be developed, and small replicated plots soon emerged.

Although yield was a dominant objective of early breeding programs, quality was addressed in Canada in 1892 by William Saunders, who crossed Red Fife and Hard Red Calcutta to obtain early maturity and bread-making quality. In 1903, William's son, Charles, systematically evaluated the progeny for quality by chewing samples and in 1909 released Marquis, which set a standard of quality (Figure 1.10).

W. J. Farrer is Australia's most famous pioneer wheat breeder and one of the world's leaders in plant breeding. Much of his work starting in 1886 was conducted prior to the understanding of genetics (Farrer, 1898). R. H. Biffen was a post-Mendelian geneticist and wheat breeder and was the first to describe **emasculation** and pollination techniques in barley (Biffen, 1907).

In the cross-fertilized maize crop, efforts to control pollination in maize cultivars by self-fertilization were undertaken by G. H. Shull at Cold Spring Harbor and E. M. East at the Connecticut Agricultural Experiment Station. They inbred maize starting in 1904, securing stable homozygous lines that were combined to produce F_1 hybrids. The advantages of hybrid vigor and reproducible progeny subsequently were applied to forest, horticultural, and other field crops.

The early advances in plant breeding led to a series of principles, breeding methods, and testing and licensing procedures. New tools have been added to make modern-day plant breeding a dynamic science, and these are spelled out in this book.

REFERENCES

Anderson, E. 1944. The Source of Effective Germplasm in Hybrid Maize. *Missouri Botanical Garden Annals* 31:355-361.

Anderson, E., and W. L. Brown. 1952. The History of the Common Maize Varieties of the United States Corn Belt. *Agricultural History* 26:2-8.

Anonymous. 1910. Koelreuter, Bakewell, and Rimpau. *American Breeders Magazine* 1(3):159-164.

Arnold, M. H. 1985. Sugar Beet. *Outlook on Agriculture* 14:109-114.

Ball, C. R. 1930. The History of American Wheat Improvement. *Agricultural History* 4(2):48-71.

Beaven, E. S. 1947. *Barley, Fifty Years of Observation and Experiment.* Duckworth, London. 394 pp.

Biffen, R. H. 1907. The Hybridization of Barleys. *Journal of Agricultural Science* 2:183-206.

Braidwood, R. J. 1960. The Agricultural Revolution. *Scientific American* 203(3):130-138, 143-148.

Braks, G. 1984. *Efficiency in Plant Breeding.* Proceedings of the 10th Congress of the European Association for Research on Plant Breeding. EUCARPIA Wageningen. The Netherlands, 19-24 June 1983. W. Lange, A. C. Zeven, and N. G. Hogenboom (eds.). Purdoc, Wageningen, p. 5.

Buller, A.H.R. 1919. *Essays on Wheat.* Macmillan Company, New York. 339 pp.

Chang, K. C. 1981. In Search of China's Beginnings: New Light on an Old Civilization. *American Scientist* 69:148-160.

Darrah, L. L. and M. S. Zuber. 1986. 1985 United States Farm Maize Germplasm Base and Commercial Breeding Strategies. *Crop Science* 26:1109-1113.

Durant, W., and A. Durant. 1963. *The Story of Civilization VIII: The Age of Louis XIV.* Simon and Schuster, New York, p. 519.

Farrer, W. J. 1898. The Making and Improvements of Wheats for Australian Conditions. *Agricultural Gazette New South Wales* 9:131-168, 241-250.

Hardeman, N. P. 1981. *Shucks, Shocks and Hominy Blocks: Corn as a Way of Life in Pioneer America.* Louisiana State University Press, Baton Rouge. 271 pp.

Harlan, J. R. 1975. *Crops and Man.* American Society of Agronomy Publication, Madison, Wisconsin. 295 pp.

Harlan, J. R., J.M.J. de Wet, and E. G. Price. 1973. Comparative Evolution of Cereals. *Evolution* 27:311-325.

Hays, W. M. 1901. *Plant Breeding.* Bulletin 29. United States Department of Agriculture, Division of Vegetables Physiology and Pathology, Washington, D.C., pp. 5-72.

Hedrick, U. P. 1950. *A History of Horticulture in America to 1860.* Oxford University Press, New York. 551 pp.

Howard, W. L. 1945. Luther Burbank. *Chronica Botanica* 9:299-506.

Jenkins, J. A. 1966. The Origin of Cultivated Wheat. *Canadian Journal of Genetics and Cytology* 8:220-232.

Lorain, J. 1814. Observations on Indian Corn and Potatoes. *Philadelphia Society for Promoting Agriculture, Memoirs* 3:303-325.

MacNeish, R. S. 1964. The Origins of New World Civilization. *Scientific American* 211(2):29-37.

Mangelsdorf, P. C. 1974. *Corn: Its Origin, Evolution and Improvement.* Harvard University Press, Cambridge, Massachusetts. 262 pp.

———. 1986. *The Origin of Corn.* Scientific American 255(2):80-86.

Nault, L. R., and W. R. Findley. 1981-1982. Corn (*Zea mays*) in Relation to its Wild Relatives. *Desert Plants* 3(4):193-202.

Pope, M. N. 1944. Some Notes on Technique in Barley Breeding. *Journal of Heredity* 35:99-111.

Reed, H. S. 1942. *A Short History of the Plant Sciences*. Ronald Press Company, New York. 320 pp.

Roberts, H. F. 1929. *Plant Hybridization Before Mendel*. Princeton University Press, Princeton. 374 pp.

Singleton, W. R. 1944. Noyes Darling, First Maize Breeder. *Journal of Heredity* 35:265-267.

Solheim, W. G., II. 1972. An Earlier Agricultural Revolution. *Scientific American* 226(4):34-41.

Smith, J.S.C., M. M. Goodman, and C. W. Stuber. 1985. Genetic Variability Within U.S. Maize Germplasm. I: Historically Important Lines. *Crop Science* 25:550-555.

Stern, C., and E. R. Sherwood (eds.). 1966. *The Origin of Genetics: A Mendel Source Book*. W. H. Freeman and Company, San Francisco. 179 pp.

Treistman, J. M. 1967. China and 1000 B.C.: A Cultural Mosaic. *Science* 160:853-856.

Tucker, J. B. 1986. Amaranth: The Once and Future Crop. *BioScience* 36(1):9-13.

Wallace, H. A., and W. L. Brown. 1956. *Corn and Its Early Fathers*. Michigan State University Press, East Lansing. 134 pp.

Weatherwax, P. 1954. *Indian Corn in Old America*. Macmillan Company, New York. 253 pp.

Wilkes, H. G. 1977. Hybridization of Maize and Teosinte in Mexico and Guatemala and the Improvement of Maize. *Economic Botany* 31:254-293.

———. 1985. Teosinte: The Closest Relative of Maize Revisited. *Maydica* 30:209-223.

Zavitz, C. A. 1919. *Farm Crops*. Ontario Department of Agriculture Bulletin 268, Guelph. 79 pp.

Zirkle, C. 1932. Some Forgotten Records of Hybridization and Sex in Plants. *Journal of Heredity* 23:443-448.

———. 1935. *The Beginnings of Plant Hybridization*. University of Pennsylvania Press, Philadelphia. 231 pp.

Zuber, M. S., and L. L. Darrah. 1980. 1979 U.S. Corn Germplasm Base. *Proceedings of the Corn and Sorghum Research Conference* 35: 234-249.

CHAPTER 2

Reproduction in Plants

SELF- AND CROSS-FERTILIZATION IN PLANTS

The transfer of pollen from anther to stigma is known as pollination. The union of male and female gametes is known as fertilization. Pollination is often used as a synonym of fertilization, but the two events are distinctly separate. Pollination does not ensure fertilization.

Fertilization resulting from the union of gametes produced on the same plant is **autogamy** and that from different plants is **allogamy**. Pollination and subsequent fertilization may occur within the same flower, between flowers on the same plant, or between flowers of different plants. The former two are both self-pollination, and the latter is cross-pollination.

Correct breeding procedures require a knowledge of the mode of pollination and fertilization, of whether a plant is self- or cross-fertilized, and of whether it is capable of being either crossed or selfed, respectively. If the latter occur, the extent of each, as well as the behavior of the crop when inbred or crossbred, is important. The type of variation found within and among cultivars and the breeding strategy used with a given pollination-fertilization system may differ. Initiating a breeding program in a new species would require an early determination of the type of pollination-fertilization system that is prevalent in the species.

The classification of crops as self- and cross-fertilized is not absolute because a range from none to complete cross-fertilization can occur in most crop species. Nevertheless, the classification is useful from a plant breeding standpoint. Soybeans, for example, are almost 100% self-fertilized, whereas tobacco or wheat may have as much as 4 to 8% natural cross-fertilization under field conditions. Other cross-fertilized crops have some degree of self-fertilization, which varies depending on the floral mechanisms present to ensure cross-fertilization and the availability of other genotypes to furnish pollen for cross-fertilization. Plants generally self- or cross-fertilized are listed in Table 2.1.

Self- and cross-fertilization lead to distinct attributes in plants or plant populations. Repeated self-fertilization results in homozygous plants; cross-fertilization leads to heterozygous plants. To describe a plant population, the terms **homogeneous** and **heterogeneous** are appropriate.

In an F_1 hybrid crop, the population is homogeneous, but the individuals within the population are heterozygous. Likewise, an unselected population of self-fertilized plants may have homozygous plants in a heterogeneous population. Cultivars of self-fertilized crops may be homozygous and homogeneous or homozygous and heterogeneous, depending on how the breeder constructed the cultivar, as they are composed of a large population of individuals each with identical genotypes.

If the alleles on the chromosome inherited from the female and from the male parent are the same, the plant is homozygous. Such plants would always produce sex cells with the same

Table 2.1   Abbreviated List of Predominantly Self- and Cross-fertilized Species

Predominantly Self		Predominantly Cross	
Barley	*Hordeum vulgare*	Alfalfa	*Medicago sativa*
Chickpea	*Cicer arietinum*	Amaranth	*Amaranthus*
Clover	*Trifolium spp*	Asparagus	*Asparagus officinalis*
	(about 70% of 240 species)	Banana	*Musa*
Coffee	*Coffea arabica*	Birdsfoot trefoil	*Lotus corniculatus*
Common bean	*Phaseolus vulgaris*	Buckwheat	*Fagopyrum esculentum*
Cotton	*Gossypium spp*	Cabbage	*Brassica oleracea*
Cowpea	*Vigna unguiculata*	Carrot	*Daucus carota*
Crambe	*Crambe abyssinica*	Cassava	*Manihot esculentum*
Eggplant	*Solanum melongena*	Castor	*Rinincus communis*
Flax	*Linum usitatissimum*	Clover	*Trifolium spp*
Guar	*Cynanopsis tetragono*		(about 30% of 240 species)
Jute	*Corchorus espularis*	Coconut	*Cocos nucifera*
Lentil	*Lens culinaris*	Cool season	*Agropyron*
Lettuce	*Lactuca sativa*	forages	*Bromus*
Lupine	*Lupinus spp*		*Dactylis*
Oat	*Avena sativa*		*Elymus*
Pea	*Pisum sativum*		*Festuca*
Peach	*Prunus persica*		*Lolium*
Peanut	*Arachis hypogaea*		*Phalaris*
Rice	*Oryza sativa*		*Phleum*
Safflower	*Carthamus tinctorius*	Cucumber	*Cucumis sativa*
Sorghum	*Sorghum bicolor*	Hop	*Humulus lupulus*
Soybean	*Glycine max*	Maize	*Zea mays*
Tobacco	*Nicotiana tabacum*	Mango	*Mangifera indica*
Tomato	*Lycopersicon esculentum*	Muskmelon	*Cucumis melo*
Trees		Onion	*Allium*
Eucalyptus	*Eucalyptus*	Pearl millet	*Pennisetum americanum*
Teak	*Tectona grandis*	Pepper	*Capsicum spp*
Triticale	*X Triticosecale*	Pineapple	*Ananas bracteatus*
Wheat	*Triticum aestivum*	Potato	*Solanum tuberosum*
		Pumpkins &	
		squash	*Curcurbita*
		Radish	*Raphanus sativus*
		Rape	*Brassica*
		Rye	*Secale cereale*
		Sesame	*Sesanum indicum*
		Sugarbeet	*Beta vulgaris*
		Sunflower	*Helianthus annuus*
		Sweet potato	*Ipomoea batatas*
		Trees	*Acer*
			Larix
			Picea
		Watermelon	*Citrullus lanatas*
		Wild rice	*Zizania palustris*

alleles, and in spite of crossing over, self-fertilization would lead to plants with identical, true-breeding genotypes. Plants that have different alleles in their chromosome pairs are heterozygous and do not breed true. The terms homozygous and heterozygous may be used to describe a few specific genes or a single allele instead of the entire gene complex. The characteristics of individual plants and populations associated with mode of fertilization are shown in Table 2.2.

Many self-pollinating plants have inconspicuous flowers that are small in size, such as wheat and barley. Pea and tobacco plants, however, are strongly self-pollinating species that have showy flowers, possibly an **evolutionary development**.

From an evolutionary standpoint, it is possible that cross-fertilization is favored over self-fertilization because genetic diversity is increased. As new gene combinations and mutations occur, they are passed from one plant to another instead of being confined. Self-fertilization should not be condemned, however, because it has the advantage of exposing recessive deleterious genes. Recessive mutations, desirable or undesirable, are uncovered by self-fertilization.

Cross-pollination is achieved mainly by wind or insects. Cross-fertilized species that rely on air currents or wind to distribute pollen are characterized by inconspicuous flowers, large, feathery stigmas, and an abundance of light, buoyant pollen (Figure 2.1). Such adaptations are found in maize and forage grasses. Maize tassels, for example, bear a thousand or more flowers, each containing three anthers. Each anther produces about 2,500 pollen grains, so that a single plant can produce millions of pollen grains. Possibly 30 to 60 million pollen grains are produced per plant, and in a field as many as 25,000 to 50,000 are produced for each kernel fertilized (Mangelsdorf, 1974).

Species that are insect-pollinated often have showy, brightly colored flowers, may have an odor, and contain nectar that attracts insects or other pollinating agents such as birds. Some flowers have mechanical adaptations that prevent or promote a specific form of pollination. Sticky pollen, for example, facilitates cross-pollination by insects.

Table 2.2 Impact of Self- and Cross-fertilization on Plant Characteristics

Characteristic	Self-fertilized	Cross-fertilized
Natural population	Homogeneous	Heterogeneous
Single plants from natural population	Homogeneous and homozygous	Heterogeneous and heterozygous
Individual plant from natural population	Homozygous	Heterozygous
2n genotype	Homozygous	Heterozygous
Genotype of haploid gametes	All the same	All different
Inbreeding depression	None	Expected
Self-incompatibility	None	Common

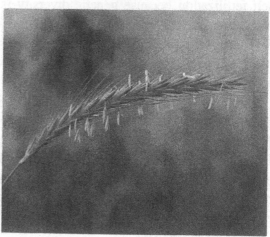

A.

B.

Figure 2.1   The inflorescences in these photos are at full anthesis, the period when pollen is being shed from the dehiscing anthers. The plants shown are bromegrass (*Bromus inermis* Leyss.) (Photo A) and wild barley (*Hordeum bulbosum*) (Photo B). Note the extrusion of anthers from the florets caused by a rapid extension of the filament.

Cross-fertilization in these species is achieved when the light and buoyant pollen is carried by air currents or wind to adjacent plants. Flowers are small, compact, and inconspicuous, capable of producing relatively large amounts of pollen, and not attractive to insects. Note that in the bromegrass panicle (Photo A), individual florets are open to allow pollen to penetrate into each of them. The feathery stigmas are exposed and are effective in trapping pollen. In normally cross-fertilized species, self-pollination may occur as a last resort.

Individual cross-fertilized plants in a wild or natural population are heterozygous at many loci and the population of such plants heterogeneous. When breeding cross-fertilized plants, caution must be exercised to control the source of pollen by bagging seed-producing plants or by isolation.

Photo courtesy Ontario Ministry of Agriculture and Food.

MECHANISMS INFLUENCING POLLINATION

Morphological Mechanisms

Self-fertilization. Complete flowers containing both male and female reproductive organs, are by their very nature subject to self-pollination. If **anthesis** occurs before the anthers or styles are exposed, then self-pollination occurs—a situation known as **cleistogamy**. Cleistogamy is common in soybeans and wheat, barley, and oat cultivars, and when it occurs it is effective in ensuring almost exclusive self-pollination. Cleistogamous flowers may eventually open, but not until the stigma has been pollinated. If fertilization does not occur, perhaps because of a lack of viable pollen, cross-pollination may take place. Cleistogamy must be overcome to enforce cross-fertilization in the production of hybrid crops.

A second mechanism to promote self-fertilization is the development of a staminal sheath, in which the stamens are joined and completely surround the pistil. The stigmas of most cotton cultivars are not exserted clear of the uppermost anthers, so that selfing usually occurs immediately after anther **dehiscence**. Shortly after the cotton flower opens, the stigma becomes receptive, the anthers split and discharge pollen grains, which adhere to the sticky stigmatic surface, and self-pollination normally takes place. Cotton pollen grains retain their viability for about 12 hours, and although very little, if any, pollen is disseminated by air movement, some cross-pollination and fertilization may occur as a result of insect visits to the flowers.

In the potato flower, five anthers surround the pistil, but the stigma generally protrudes beyond the ring of anthers. Potatoes have a high percentage of self-pollination achieved when a bee lands on the anther cone, causing the anther cone to hang down and to spill pollen onto its abdomen and onto the stigma (Glendinning, 1976). The same insect may travel to another plant and cause cross-pollination. This is of interest mainly to plant breeders, as the potato is generally a clonally produced crop.

A fourth mechanism that promotes autogamy is the presence of a sticky exudate on the stigmatic surface (Figure 2.2). Pollen readily adheres to this exudate.

Cross-fertilization. Separation of the male and female organs is a morphological mechanism promoting cross-pollination, but it does not preclude self-fertilization. A plant with male and female sexual organs occurring at separate locations on the same plant is described as **monoecious**. Maize is a common example of a monoecious crop, with staminate flowers on the top of the plant and pistillate flowers midway on the stalk. In addition to maize, walnut and wild rice are monoecious (Figure 2.3).

Dioecious plants bear flowers of only one sex on a plant; in such plants cross-pollination is obligatory. Dioecious species include the hop (*Humulus lupulus* L.), hemp (*Cannabis sativa* L.), ginkgo (*Ginkgo biloba*), pistachio (*Pistachio vera* L.), asparagus (*Asparagus officinalis*), and the oil palm (*Elaeis quineensis* Jacq.). The understanding of fertilization procedures in the oil palm has been a tremendous asset to that industry (Figure 2.4).

Another mechanism to promote cross-fertilization is the unequal maturation of either the male or female reproductive organ. **Protandry** is the maturation of anthers and pollen release before the stigma is receptive; **protogyny** is the maturation of the stigma before pollen from the same plant is released, as evidenced in pearl millet (Figure 2.5).

It is difficult, if not impossible, to make crosses by hand of such small-flowered crops as

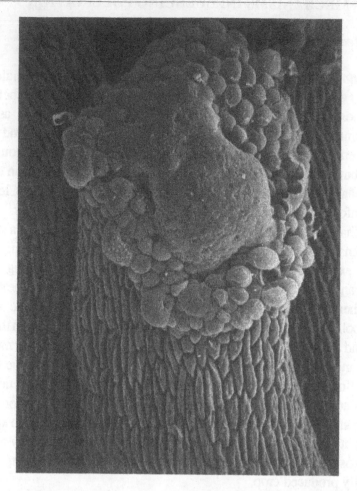

Figure 2.2   The saskatoon berry (*Amelanchier alnifolia* Nutt.), a species native to western Canada, produces a secretion from the stigma that forms a sticky exudate on the papillate surface at full bloom. In this photo, the exudate covering the papillae on a stigma is at a magnification of 275x. Just prior to anthesis, the flow of exudate is sufficient to inundate the stigmatic surface. Exudates are generally believed to contribute to the pollen-stigma recognition response and the nutritional well-being of any subsequent pollen tube growth.

The saskatoon berry secretes nectar, and the bomb-shaped perianth (a collective term for the calyx and corolla) of the flower facilitates pollination by foraging insects, both of which suggest this plant is a cross-fertilized species. Saskatoon berry plants generally are believed, however, to be self-fertilized, a situation enhanced by the sticky, stigmatic exudate.

Photo courtesy A. R. Olson.

Figure 2.3   Monoecious plants have separate male and female organs on the same plant, which can either have perfect or imperfect flowers. In maize, the staminate flowers occur on the tassel at the top of the plant (Photo A), and the pistillate flowers form at a mid-plant internode (Photo B). Each silk/style is attached to a single ovary on the cob. Silks from the upper portion of the ear appear first. A pollen tube must grow down each silk to effect double fertilization in the ovary.

Figure 2.3 caption continues on p. 34

A.

B.

C.

D.

carrot (*Daucus carota* L.) and celery (*Aspium graveolens* L.), but the protandrous nature of these crops makes hand-emasculation unnecessary. In both of these crops, the male organs develop and wither before the stigmas become receptive, so bagging of the inflorescence causes such a plant to function as a female parent only.

Both protandry and protogyny can occur in maize under certain environmental conditions. Under growth room conditions, pollen will be shed two to three days or more before the stigma is receptive to pollen. This protandrous condition appears to be associated with rapid plant development. Under a different environmental regime involving slow growth, plants may be protogynous.

Under field conditions, the maturation of the anthers and the silks generally coincides in a homogeneous population. In a heterogeneous population, maturation of reproductive organs may differ by several days. In crossing programs, it is essential that pollen release on the male plant coincide with the receptivity of the stigma on the female parent, a situation known as **nicking**. Proper synchronization of the male and female reproductive development may be achieved by delayed planting of one parent of a cross.

A number of floral architectural features, along with the arrangement of stigmas and anthers, have evolved to facilitate cross-pollination by insects. Specific modifications influence the number of visits by insect pollinators. The valvular arrangement is an example in which the stamens and the pistil project from the keel petal as long as the pressure of a visiting bee is present. The pump arrangement consists of thickened ends of the filament that press pollen out of the keel tip in successive visits by insect pollinators. This modification found in birdsfoot trefoil (*Lotus*) allows for multiple insect visits. Birdsfoot trefoil flowers secrete nectar, but this mechanism is also present in onions, which do not have nectar.

The explosive floral modification has stamens and a pistil that abruptly springs out of the keel upon rupture by an insect pollinator or mechanical means. This structure is found in alfalfa (*Medicago*), and although nectar is attractive to bees, many honeybees (*Apis mellifera* L.) are reluctant to pollinate alfalfa because of the sound thumping received when the stamen and pistil spring out of the keel.

The brush arrangement consists of pubescence or hair on the style, which sweeps the pollen

---

Figure 2.3 continued

Another monoecious plant, the walnut (*Juglans* spp), is shown in Photo C. The hanging catkins contain the staminate flowers, and the structure at the top left is the pistillate flower.

A panicle of wild rice (*Zizania palustris* L.), shown in Photo D, is a third example of a monoecious plant. It bears male and female flowers on the same panicle. Pistillate flowers arise in the upper portion of the panicle, with staminate flowers at the base. Pistillate flowers arise on short, stiff pedicels and are held closely to the panicle rachis; staminate florets are suspended on threadlike pedicels from the spreading panicle branches. Cross-fertilization is promoted by the unisexual nature of each floret as well as by the fact that as the panicle pushes up above the leaf sheath, the two white, feathery stigmas of the pistillate flowers extrude and are usually pollinated by adjacent plants before the staminate flowers on the same tiller are exposed. Following anthesis, staminate flowers normally fall away from the panicle.

In all three of the plants illustrated, the monoecious nature of the plant promotes, but does not mandate, cross-fertilization.

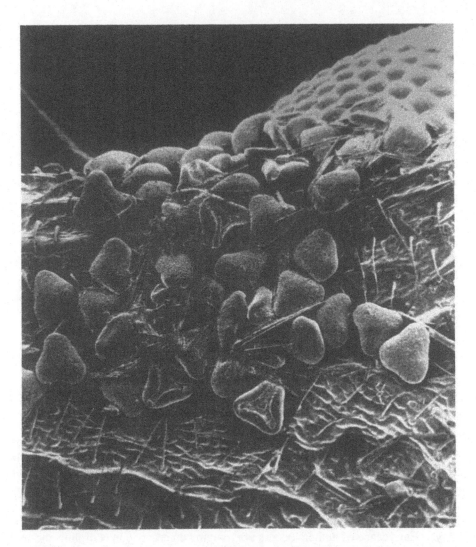

Figure 2.4 The oil palm (*Elaeis quineensis* Jacq.) bears male and female flowers on separate trees, is termed dioecious, and is therefore an obligate cross-fertilized plant. To assure maximum fruit set, it is necessary to provide an adequate number of male trees in a field. Hand pollination of palm trees is an ancient practice of growers. The advantage is that a single male tree can be used to pollinate an entire grove of female trees, thereby allowing for considerably more fruit-bearing female trees to be grown in an area of land than the normal 1:1 ratio of male to female trees.

In the early 1980s, the weevil *Elaeidobius kamerunicus* was introduced into Malaysia from West Africa to serve as a pollinating agent. As a result, fruit set in Malaysia oil palms increased from an average of 34% to nearly double that value with substantial economic benefit.

This photograph shows the heart-shaped oil palm pollen, grains magnified 182x, between the eyes of the pollinating weevil. The insect's compound eye can be seen at the top right.

Three years of careful research established that members of the genus *Elaeidobius* were safe to import into Southeast Asia to perform the pollinating role. The action is viewed as one of the most foresighted and beneficial events in the history of the oil palm industry in Malaysia (Barlow, 1985).

Photo courtesy P. G. Kevan.

Figure 2.5   A member of the ICRISAT staff examines a high-tillering, dwarf inbred line of pearl millet (*Pennisetum* spp), a crop that exhibits a mechanism known as protogyny to reduce self-pollination.

Pearl millet is an annual, normally cross-pollinated crop that has both male and female reproductive organs contained in each flower of the inflorescence borne on the top of each plant. Protogyny occurs when the stigma develops and matures prior to anther development and pollen release. Stigmas are exserted first in the top third of the head, proceeding downward until stigma extrusion is completed in about three days. Emergence of anthers occurs after stigma emergence is complete and in the same order. Stigmas remain receptive for two or three days.

About 80% of pearl millet flowers are naturally cross-pollinated, but complete cross-pollination is rare because stigmas of late tillers receive pollen from flowers of earlier tillers and occasionally anthers may pollinate late-emerging stigmas on the same spike. Variations in the degree of protogyny may influence the extent of natural cross-pollination.

For controlled self-fertilization, heads are enclosed in bags prior to the emergence of the stigmas, when most of the spike is in the boot. If protogyny results in an extended time lag between stigma emergence and pollen release, poor seed set results.

For controlled cross-fertilization, heads of the female parent are enclosed in glassine bags before stigmas emerge. When most of the stigmas are extruded, pollen obtained from the male parent is dusted on the spike.

Photo courtesy International Crops Research Institute for the Semi-Arid Tropics (ICRISAT), India.

from the keel tip. With this mechanism, multiple visits by insect pollinators are necessary to achieve cross-pollination. The brush is among the least effective of the floral mechanisms to promote cross-pollination and occurs in some largely self-pollinated species such as pea (*Pisum*), bean (*Phaseolus*), and broad beans (*Vicia*).

A protective film over the stigmatic surface in alfalfa and narrow-leafed trefoil (*Lotus tenuis*) helps ensure cross-fertilization. Pollen will not germinate on the stigma as long as this film is intact. The practical result is that self-pollination is prevented and cross-pollination encouraged when the film is ruptured through activity by an insect pollinator.

Lodicules, two small glands that lie at the base of the ovary and opposite the palea of grass flowers, expand and force open the flower, allowing pollen from adjacent plants to enter. Lodicules are found in the normally self-fertilized cereal crops and become instrumental in opening the floret to allow cross-pollination to occur (Figure 2.6) if self-fertilization has not already taken place.

Physiological Mechanisms

A highly evolved and effective mechanism controlling fertilization is **self-incompatibility**, which was defined by de Nettancourt (1977) as "the inability of a fertile **hermaphrodite** seed plant to produce zygotes after self-pollination" (p. 2). Pollen may be normal, viable, and fully functional on another genotype, but self-fertilization is prevented, thereby denying the plant breeder the gene combination planned. Because self-incompatibility systems are based upon the inherited capacity of a flower to reject its own male gamete, the system includes postfertilization failures due to zygotic inviability, embryo abortion, and endosperm abnormalities (Hayman, 1956).

Various incompatibility systems have evolved, as inbreeding is usually disadvantageous: It frequently leads to reduced vigor and other undesirable characteristics that weaken the genotype. Plant breeders, however, may wish to impose inbreeding temporarily on an otherwise naturally cross-fertilized species to improve genetic uniformity and to reveal undesirable recessive genes.

Apple and pear trees exhibit self-incompatibility mechanisms but are dependent on seed set to produce fruit. Monoculture production in an asexually propagated homogeneous orchard is not possible; to resolve this problem a compatible cultivar to serve as a pollen source is required. A sound knowledge of the physiological self-incompatibility barriers is necessary if the gene pool is to be enlarged successfully.

Homomorphic Incompatibility. **Homomorphy** is a form of incompatibility that regulates pollen tube growth in certain pollen-pistil combinations and that is a gene-controlled physiological system. Both the gametophytic and sporophytic incompatibility systems are controlled by genes at different alleles at the S locus. If the pollen and the pistil contain the same S allele, the combination is incompatible. In diploids there is one S gene in the pollen and two in the somatic or sporophytic style through which the pollen tube must develop. Under such conditions, the pollen tube is either excessively slow growing or grows in styles containing all alleles in common with that of the pollen (Figure 2.7). This form of homomorphic incompatibility is sporophytic because it is regulated by the S alleles in the sporophyte.

The genetic basis control of self-incompatibility was established by Lundqvist (1956) and

Figure 2.6 Photo of the male and female reproductive organs of barley (*Hordeum vulgare* L.), showing an ovary with a two-branched feathery stigma; three anthers with pollen sacs that split longitudinally to release the pollen, filaments arising from the base of the ovary and bearing the pollen sacs (the anther on the left has been repositioned for clarity); and two lodicules at the base of the ovary that have been pulled away from the ovary.

Although barley is normally self-fertilized, its features are common to most wind-pollinated, cross-fertilized crops. Barley florets lack petals, nectar, and odor and do not attract insects; they have enlarged or feathery stigmas adapted to the interception of airborne pollen.

Lodicules are small glands situated at the base of the ovary palea of grass flowers, which when stimulated expand and force the floret open, allowing the feathery stigmas to extrude and allowing cross-fertilization if self-fertilization does not occur.

The commercial success of F_1 hybrids in normally self-fertilized crops requires that these plants be cross-fertilized. With the incorporation of male sterility and elimination of self-pollination, the lodicules become important organs in the female plant. To achieve satisfactory cross-fertilization, breeders may select for large lodicules, increased size of the feathery stigma, filaments that extend to push the anthers out of the floret, and large anthers that shed abundant amounts of pollen.

Photo courtesy Ontario Ministry of Agriculture and Food.

Hayman (1956). The suggestion that two polyallelic loci are involved was challenged by Spoor (1976), who suggested that in ryegrass (*Lolium perenne*) at least three loci are involved. Interpretations may be clouded by a third category of partially compatible crosses, in addition to compatible and incompatible categories (Cornish, Hayward, and Lawrence, 1979).

Self incompatibility in tall buttercup (*Ranunculus acris* L.) and sugarbeet (*Beta vulgaris* L.) was reported by Lundqvist *et al.* (1973). These workers suggested that incompatibility genes are ancient constituents of the breeding systems of angiosperms, appear to be widespread within both monocots and dicots, and occur more and more frequently than hitherto expected. With gametophytic incompatibility, the genotype of the haploid pollen grain determines its

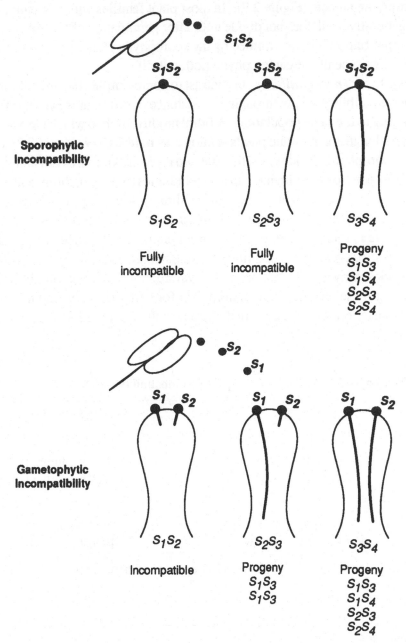

S_1, S_2, S_3 & $S_x...S_n$ denote incompatibility alleles

Figure 2.7 Diagram of sporophytic and gametophytic incompatibility systems.

Double fertilization is dependent upon two generative nuclei produced by the pollen grain. Two generative nuclei may be present when the pollen is shed from the anther or may be formed just prior to fusion with the egg or polar nuclei.

Sporophytic incompatibility is associated with pollen grains having two generative nuclei when the pollen is shed from the anther. This situation occurs in *Compositae*, *Cruciferae*, and *Rubiaceae* and is important in the breeding of cabbages and other forms of *Brassica oleracea*, radish, cosmos, iberis, and

Figure 2.7 continues overleaf

incompatibility reaction and occurs in pollen grains having one generative nucleus at the time of release from the anthers (Figure 2.7). In most plant families with gametophytic incompatibility, the generative cell does not divide until after germination of the pollen tube. Gametophytic incompatibility is not determined by the sporophytic plant that produces the pollen but rather by the genotype of the gametophytic pollen itself.

Heteromorphic Incompatibility. In contrast to homomorphic incompatibility, heteromorphic self-incompatibility occurs through the existence of different flower forms in individual plants belonging to the same population. A floral modification in which the anthers and pistils are of different lengths in separate plants is known as **heterostyly** (Figure 2.8) and is believed to be always sporophytic (Pandey, 1970). The two types are known as **pin** and **thrum flowers**. All crosses between flowers with identical morphologies (pin x pin; thrum x thrum) are sterile, whereas crosses between morphologies (pin x thrum; thrum x pin) are fertile (Table 2.3). Incompatibility is governed by a single gene that segregates as a simple Mendelian factor. One dominant S allele is found in heterozygous thrum plants and it is theoretically not possible to obtain in the homozygous (SS) condition (de Nettancourt, 1977).

Plant species may exhibit a number of variations of heteromorphic incompatibility. Included among these variations are dimorphism for both the pollen and the stigma, with or without the association of heterostyly, or dimorphic pollen (Figure 2.9) and monomorphism for the stigma.

Table 2.3   Mating Types of the Heteromorphic Incompatibility System

	Type	*Progeny*
Compatible		
	Pin (ss) x thrum (Ss)	1 pin (ss); 1 thrum (Ss)
	Thrum (Ss) x pin (ss)	1 pin (ss); 1 thrum (Ss)
Incompatible		
	Pin (ss) x pin (ss)	No progeny
	Thrum (Ss) x thrum (Ss)	No progeny

Source: Briggs and Knowles, 1967, p. 184. Reprinted by permission.

Figure 2.7 continued

primula. Although pollen tube growth is retarded in sporophytic systems, apparently incompatible crosses can sometimes result in fertilization because blockage is incomplete.

Gametophytic incompatibility is associated with only one generative nucleus in the pollen tube. Division occurs just prior to fertilization if a pollen tube develops. Gametophytic incompatibility is found in red clover, white clover, alsike clover, and yellow sweet clover.

When identical s genes are involved in self-incompatibility, the pollen and pistil become incompatible with one another and self-sterility results. Cross-incompatibility can restrict plant breeders from making selected crosses and hence enlarging gene pools.

A detailed account of pollen-stigma interaction in cross-incompatible crosses was provided by Heslop-Harrison, 1982.

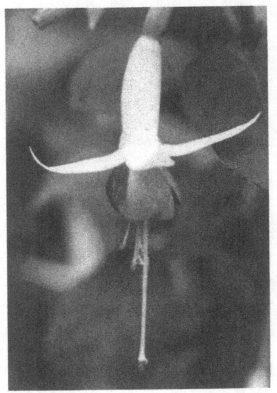

A.   B.

Figure 2.8   The fuchsia (*Fuchsia* spp) flowers in these photos are heterostyled, a heteromorphic form of sporophytic incompatibility. The flower in Photo A, which has a long style and short stamens, is termed a **pin flower**, and in Photo B the flower with the short style and long stamens is termed a **thrum flower**. Pin has the genotype *ss* and thrum *Ss*, with *S* completely dominant to *s*. The incompatibility reaction is not based on the difference in relative lengths of the style or stamens but rather on the genotype of the plant and the ability of the pollen to grow on the stigma.

Buckwheat (*Fagopyrum* spp) is the most common crop plant displaying heteromorphic incompatibility and is a normally cross-fertilized species. Self-fertile flowers of buckwheat with styles and stamens the same length have been reported (Marshall, 1969).

Other plants exhibiting pin and thrum flowers are *Primula, Forsythia, Lenium, Oxalis, Lythrum, Mitchella, Silia*, and other genera. Lewis (1947, 1949, and 1954) and Mather (1950) have studied heterostyly (*Primula sinensis*) and have reported on the inheritance of this feature. Heterostyly was reported to be a complex character involving a number of variations, including long, medium, and short styles along with long, medium, and short stamens. Full fertility is contingent upon matings of flowers having stamens and styles of the same length.

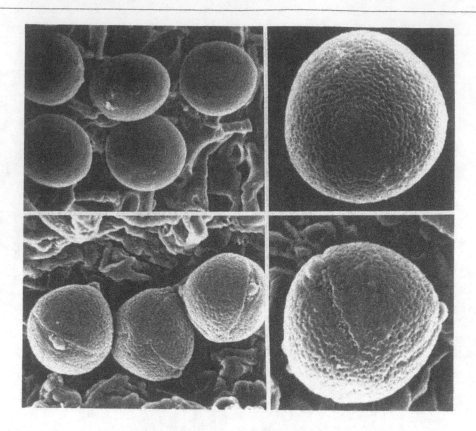

Figure 2.9   The pollen grains in this photo are those of the wild grape (*Vitis riparia*), showing pollen dimorphism, a form of heteromorphic incompatibility. The wild grape is a dioecious plant, but both male and female plants produce pollen. Pollen produced from female plants (top row), however, is incapable of germinating. Only pollen from the male plant (bottom row) is capable of germinating, producing a pollen tube, and fertilizing the egg.

Like heterostyly, the name pin (upper row) and thrum (lower row) are applied. Pollen grains from both male and female are well formed, and nonfunctional pollen on fruit-bearing female plants may serve to attract pollinating insects bearing functional pollen from a male plant.

A pollen tube can be seen emerging through the outer pollen layer in the functional pollen in the bottom row, but no evidence of germination exists. The sculpturing on the nonviable pollen appears more dense, but pollen size is similar. The large single grains on the right are at a magnification of 1,900x, the grains on the upper left at 930x, and those on the lower left at 1,000x.

In domesticated grapes, breeding and selection have eliminated nonbearing male plants. Breeders selected either consciously or unconsciously against dioecy, presumably to increase vineyard efficiency by selecting for fruit-bearing plants. This was done to ensure production by favoring plants capable of producing by both cross- and self-fertilization. In the muscadine grape (*Vitis rotundifolia*), pollination difficulties may be experienced unless interplantings of pollinator plants are used. An understanding of grape pollination may allow breeders to eliminate the need for a pollinator in muscadine grape cultivars by increased and continued selection pressure against dioecy (Kevan, Longair, and Gadowski, 1985).

Dimorphism for pollen differs from floral modifications because it requires genetic differentiation for expression of incompatibility and represents the most highly evolved form of pollination control.
Photo courtesy P. G. Kevan and R. Gadowski.

Other Genetic and Cytological Mechanisms

Sterility. The failure of anthers to be differentiated, to develop fully, or to dehisce is known as male sterility and is a morphological mechanism that prevents self-pollination. Female sterility occurs when the stigma, style, or ovary does not develop normally. Both male and female sterility are determined by genetic, cytoplasmic, genetic-cytoplasmic factors or by chemical induction (Johnson and Brown, 1978; Miller and Lucken, 1977). The expression of sterility may be influenced by the environment.

Genetic male sterility is inherited according to Mendelian principles in which the alleles for sterility generally are recessive to the alleles for pollen fertility. Male-sterile genotype plants (*ms ms*) must be obtained by mating them with male-fertile (*Ms ms* or *Ms Ms*) individuals. In a male-sterile (*ms ms*) x male-fertile (*Ms ms*) cross, half the progeny would be sterile (*ms ms*), but the remaining plants would be fertile (*Ms ms*) and capable of producing viable pollen, thereby reducing the effectiveness of the desired male parent. Since fertile plants with the *Ms ms* genotype cannot be phenotypically identified, such plants cannot be removed from a population to prevent viable pollen from being released.

Cytoplasmic male sterility is governed by factors in the maternal cytoplasm that induce sterility. This system allows for reproduction of the female by fertilization with fertile pollen.

Genetic-cytoplasmic male sterility is conditioned by an interaction between a **nuclear gene** (recessive in the case of onion [*Allium*] and dominant in most grasses) and a cytoplasmic factor. The cytoplasmic factor is referred to as N for normal fertile cytoplasm and S for the sterile condition. The nuclear inherited sterile condition is *ms ms*, and *Ms* for the fertile. Cytoplasmic male sterility can be maintained by having two lines. The female line with cytoplasmic male sterility must be homozygous for *ms ms*, and the male line must have the normal, or nonsterile, cytoplasm and be homozygous for *ms ms*. When crossed, the fertile cytoplasmic factor does not transfer to the female during the cross, and the N cytoplasm provides for normal fertility in the male maintainer or **B-line**, yet does not restore fertility in the progeny when crossed to the female **A-line**. Any other combination of genetic and cytoplasmic factors in the male line will give either fertile or segregating progenies.

The major use of male sterility is to facilitate commercial-scale cross-fertilization in the production of F_1 hybrid seed in such crops as corn and sunflowers. Commercial success in utilizing male sterility depends upon genes that will restore fertility in commercial progeny.

Combination of Mechanisms Ensuring Cross-fertilization. Many cross-fertilized species have a combination of factors that facilitate pollination by insects along with a highly developed incompatibility system, that together ensure cross-fertilization. As already noted in maize, two criteria may ensure cross-fertilization: the monoecious nature and unequal maturation of male and female flowers.

ASEXUAL REPRODUCTION

Two techniques may be used to achieve **asexual reproduction**: **vegetative propagation**, in which plant tissue other than a seed is used to produce a clone, and **apomixis** involving plant embryos developed without fertilization.

Asexual reproduction achieved by apomixis is a result of seed development in the ovary in which the embryo is formed without union of the sperm and egg cells. Generally, the embryo originates by mitotic division of a nucleus in a somatic cell of the ovule, producing a **zygote** or seed, with exactly the same chromosome number as the mother plant. The resulting offspring is a true clone identical to the female parent. Apomixis has been reported in more than 300 plant species, involving at least 35 families (Nygren, 1954; Hanna and Bashaw, 1987) and including numerous perennial forage grasses, grain sorghum, pearl millet, and many citrus species.

Some species, including sorghum, are **facultative apomicts,** which means that the same ovule is capable of either sexual or apomictic reproduction. **Obligate apomicts** reproduce only by means of apomictic seed set, producing uniform progeny identical to the mother plant, since it concerns only the gametophytic generation. Apomixis may be recurrent, that is it may occur generation after generation, or may be nonrecurrent with apomictic seed occurring sporadically.

Four mechanisms of apomictic seed development are recognized that collectively are called **agamospermy.** They are adventitious embryony, apospory, diplospory, and parthenogenesis. The mechanism distinguishing each of the four forms of agamospermy is based on the site of origin and subsequent developmental pattern of the cell that gives rise to the embryo. Cytological examination of developing ovaries during megasporogenesis and embryo sac development is required to distinguish among the four mechanisms.

Adventitious embryony occurs when a cell in the integuments divides into an embryo in the presence of the haploid embryo sac. This phenomenon occurs in *Citrus* species.

Apospory is characterized by a cell in the integuments growing into an embryo in the absence of the haploid embryo sac; the sexual process is bypassed completely. This occurs in some species of *Rubus* (raspberries, blackberries), *Allium* (onion, leek), and *Opuntia* (prickly pear).

Diplospory occurs when the diploid mother cell develops into an embryo sac without meiosis or reduction division. This is the most frequent type of apomixis. In many cases, pollination is necessary before seed is produced, and this type of apomixis is referred to as **pseudogamy.** Nonviable pollen may cause pseudogamy.

Parthenogenesis, the development of an individual from a gamete without fertilization, is found in alfalfa and tobacco. A notable example is haploid barley production in *Hordeum bulbosum* (Kasha and Kao, 1970). The long English forcing cucumbers develop parthenocarpic fruit freely without benefit of a male parent. If the egg cell is unreduced, the embryo will have the same genotype as the parent plant and the process is synonymous with apomixis. The distinction between apomixis and parthenogenesis occurs when the reduced egg cell develops without fertilization to produce a haploid plant. Parthenogenesis cannot be identified cytologically but is detected by the presence of haploid offspring.

Apomixis is of interest to plant breeders as a system of fixing heterosis since it provides a means for producing uniform heterozygous progeny that breed true. In buffelgrass (*Cenchrus ciliaris* L.), an obligate apomictic species, three apomictic hybrids have been released as commercial cultivars (Bashaw, 1962, 1980). The ability to induce apomixis could provide the plant breeder with a system of vegetative reproduction but with the convenience of seed. Apomixis is covered more thoroughly in Chapter 20.

DETERMINATION OF POLLINATION SYSTEMS

When a plant breeder begins work with a species, one of the first tasks is to determine the type of reproductive system and the type of fertilization involved. Such information will suggest the amount of variability that can be expected in introductions or among cultivars of a species and, more importantly, will determine how much of the existing variability is available for exploitation.

In practical terms, it is more important to determine the major type of reproduction or fertilization system rather than the details of how much self- or cross-fertilization occurs. An indication of the reproductive or mating system likely to exist can be obtained from preliminary observations of flower morphology, size, and brightness or coloring. The fertilization system can be verified by isolating or bagging a single plant to protect it from foreign pollen and observing the amount of seed obtained. If no seed is produced, the crop may be cross-fertilized; if seed is produced, it may be self-fertilized, although many normally cross-fertilized species will form seed upon self-pollination.

To reach conclusive results on the mode of fertilization of a given species, additional observations may be required. For example, among most cross-fertilized species that form seed on a single plant as a result of self-fertilization, progeny will be less vigorous than the parent plant as a result of inbreeding.

Further experiments to determine the degree of self- or cross-fertilization would require the use of genetic markers. A plant with a known genetic marker could be planted among plants with a deficient genetic marker. Progeny grown from plants with the known genetic marker should be observed for segregation or the lack of it. Plants with a recessive marker, interplanted among plants with a dominant marker in the homozygous condition, would help reveal the degree of cross- and self-fertilization.

If apomixis or complete self-fertilization is the normal form of reproduction, then all of the progeny would appear phenotypically recessive. If some cross-fertilization has occurred, then the progeny of recessive plants would exhibit the dominant **phenotype**. Determination of the number of dominant phenotypes among the progeny of recessive **genotypes** would provide a quantitative estimate of cross-fertilization.

In the case where apomixis or complete self-fertilization cannot be distinguished, controlled crosses with a known homozygous, dominant genotype would produce progeny that deviates from the recessive parental genotype by showing all dominant progeny. If apomictic seed was produced, progeny that retained the phenotype of the female parent would prevail.

REFERENCES

Barlow, C. 1985. The Oil Palm Industry. *Outlook on Agriculture* 14:204-212.

Bashaw, E. C. 1962. Apomixis and Sexuality in Buffelgrass. *Crop Science* 2:412-415.

———. 1980. Apomixis and Its Application in Crop Improvement. In *Hybridization of Crop Plants*, edited by W. R. Fehr and H. H. Hadley. American Society of Agronomy and Crop Science Society of America, Madison, Wisconsin, pp. 45-63.

Briggs, F. N., and P. F. Knowles. 1967. Introduction to Plant Breeding. Reinhold Publishing Corporation. Davis, California. 426 pp.

Cornish, M. A., M. D. Hayward, and M. J. Lawrence. 1979. Self Incompatibility in Ryegrass I. Genetic Control in Diploid *Lolium perenne* L. *Heredity* 43:95-106.

de Nettancourt, D. 1977. *Incompatibility in Angiosperms*, Springer-Verlag. Berlin. 230 pp.

Glendinning, D. R. 1976. Neo-tuberosum: New Potato Breeding Material. 4: The Breeding System of Neo-tuberosum and the Structure and Compositon of the Neo-tuberosum Gene-Pool. *Potato Research* 19:27-36.

Hanna, W. W., and E. C. Bashaw. 1987. Apomixis: Its Identification and Use in Plant Breeding. *Crop Science* 27:1136-1139.

Hayman, D. L. 1956. The Genetical Control of Incompatibility in *Phalaris coerulescens* Desf. *Australian Journal of Biological Sciences* 9:321-331.

Heslop-Harrison, J. 1982. Pollen-Stigma Interaction and Cross-Incompatibility in the Grasses. *Science* 215:1358-1364.

Johnson, R. R., and C. M. Brown. 1978. Use of DPX 3778 to Produce Hybrid Wheat Seed. *Crop Science* 18:1026-1028.

Kasha, K. J., and K. N. Kao. 1970. High Frequency Haploid Production in Barley (*Hordeum vulgare* L.). *Nature*, London 225:874-875.

Kevan, P. G., R. W. Longair, and R. M. Gadowski. 1985. Dioecy and Pollen Dimorphism in *Vitis riparia* (Vitaceae). *Canadian Journal of Botany* 63:2263-2267.

Lewis, D. 1947. Competition and Dominance of Incompatibility Alleles in Diploid Pollen. *Heredity* 1:85-108.

———. 1949. Incompatibility in Flowering Plants. *Biological Review*, Cambridge Phil. Society 24:472-496.

———. 1954. Comparative Incompatibility in Angiosperms and Fungi. *Advances in Genetics* 6:235-285.

Lundqvist, A. 1956. Self-Incompatibility in Rye I. Genetic Control in the Diploid. *Hereditas* 42:293-348.

Lundqvist, A., U. Osterbye, K. Larsen, and I. Linde-Laursen. 1973. Complex Self-Incompatibility Systems in *Ranunculus acris* L. and *Beta vulgaris* L. *Hereditas* 74:161-168.

Mangelsdorf, P. C. 1974. *Corn: Its Origin, Evolution and Improvement*. Harvard University Press, Cambridge, Massachusetts. 262 pp.

Marshall, H. G. 1969. Isolation of Self-fertile, Homomorphic Forms in Buckwheat, *Fagopyrum sagittatum* Gilib. *Crop Science* 9:651-653.

Mather, K. 1950. The Genetical Architecture of Heterostyly in *Primula sinensis*. *Evolution* 4:340-352.

Miller, J. F., and K. A. Lucken. 1977. Gametocidal Properties of RH-531, RH-532, RH-2956, and RH-4667 on Spring Wheat (*Triticum aestivum* L). *Euphytica* 26:103-112.

Nygren, A. 1954. Apomixis in Angiosperms. *Botanical Review* 20:577-649.

Pandey, K. K. 1970. Time and Site of the S Gene Action: Breeding Systems and Relationships in Incompatability. *Euphytica* 19:369-372.

Spoor, W. 1976. Self-Incompatibility in *Lolium perenne* L. *Heredity* 37:417-421.

CHAPTER 3

Genetic and Cytogenetic Structure of Plants

As evident from the painstaking progress in plant breeding prior to the development of the science of genetics, modern plant breeding techniques are largely unintelligible without a working knowledge of genetics. This chapter is designed to be a refresher course on fundamental genetics and cytogenetics.

SEXUAL REPRODUCTION

Variation is essential for success in plant breeding and may be achieved in higher plants called **angiosperms** through sexual union of male and female **gametes** to form a **zygote**. The single-celled zygote divides numerous times to produce an adult plant, which then produces male and female gametes to start the cycle again. Two distinct phases can thus be identified in angiosperms, namely, the **gametophytic** and **sporophytic generation**. The two processes associated with this **alternation of generations** are **fertilization** and **meiosis**.

During the sporophytic stage, cell division requires that each daughter cell has exactly the same genetic makeup as the parent cell to produce a new plant such as a spruce tree, a cotton plant, or a tomato plant. The process of cell division is **mitosis**, whereby resulting **somatic** or **diploid** cells all contain the normal, or 2n, number of **homologous chromosomes**.

To initiate the gametophytic stage, meiosis occurs so that the male and female sex cells have precisely half the genetic makeup of the parent. In this way the normal genetic constitution of the progeny can be restored upon sexual union to begin the sporophytic generation once again. The zygote is diploid (2n) and each of the gametes is **haploid** (n). In sexually reproducing plants, all of the haploid structures or gametes produced during the life cycle belong to the gametophytic generation, and the diploid structures that produce the cells that initiate meiosis belong to the sporophytic generation.

Plant breeders traditionally **screen** and select on the sporophytic generation. Screening and selecting pollen *in vitro* may allow plant breeders an opportunity to use the gametophytic generation in breeding programs. Pollen grows quickly *in vitro*, is haploid, varies in genetic content, and 60% of the genes expressed in the sporophyte can be expressed in the gametophyte (Tanksley, Zamir, and Rick, 1981). The overlap of genetic expression allows a breeder to expose pollen to selective pressures that will be expressed in the sporophytic generation.

Screening of the gametophytic generation has the advantage of allowing plant breeders to screen thousands and perhaps millions of individuals in a culture flask in just a few hours, as compared to hundreds of hectares in perhaps years by traditional methods with the sporophytic generation. The extent to which selection of traits during the gametophytic stage will lead to desired improvements of the whole plant remains somewhat unclear, but tomato pollen has been used to screen successfully for salt tolerance (Sacher and Mulcahy, 1981).

Asexual, or vegetative, reproduction provides an alternative form of reproduction. Both sexual and asexual reproduction play important roles in plant breeding that will be considered in this chapter.

GENETIC REPRODUCTION

Whether a plant develops from a single, sexually produced zygote or from an asexually propagated cell, a plant similar to the parent plants invariably is produced. Since Carl Nägeli discovered strands, later named chromosomes, in 1842 and Gregor Mendel described the units of heredity, later named genes, geneticists have described in precise detail the physical structure of the genetic material. To comprehend how genetic information is carried, copied, decoded, and put into practical use and how the genetic information may be changed by mutation, it is necessary to understand some basic biochemical genetics.

Genetic materials consist of three basic ingredients: deoxyribonucleic acid (**DNA**), ribonucleic acid (**RNA**), and protein. A chromosome consists of DNA and other components in a molecule that is double stranded, helical, long, and slender, and which can be described as a circular stairway. The railings of the stairway consist of sugar-phosphate, and the steps that hold the two railings together consist of four kinds of nitrogenous bases, namely, adenine (A), cytosine (C), guanine (G), and thymine (T). The four nitrogenous bases represent four **nucleotide** building blocks of DNA, each consisting of about 30 atoms of carbon, oxygen, hydrogen, phosphorus, and nitrogen. The nitrogenous bases are arranged at random, and although only four in number, they combine to produce thousands of variations for the steps of the stairway. For example, the steps could be arranged as

GGTAAATCGGATTCTGACGCGTAC.

Another gene might have the arrangement

TCGTTAAAACCGGCGCCGGACTGAT.

A second critical aspect of the double helix is base pairing. Adenine always pairs with thymine, guanine with cytosine as well as the reciprocal T with A and C with G. The physical, three-dimensional structure of DNA is shown in Figure 3.1.

Each gene consists of possibly 700 to 1,000 base pairs, and its particular configuration forms a blueprint for the action of that gene. With any of the four possible bases at each of perhaps 800 sites, there are many different kinds of genes possible. Flowering plants have a very large number of genes per plant, and the potential exists for the 300,000 plant species in the world each to have a unique genetic makeup.

Replication of DNA is facilitated through specific base pairing. At the proper time, the bonds holding the two halves together release and the double strands separate. By means of base pairing, adenine, cytosine, guanine, and thymine manufactured in the cell are used to form a new copy of each half of the original chromosome. The process is facilitated by enzymes.

For genes to be expressed, the information contained must be carried from within the

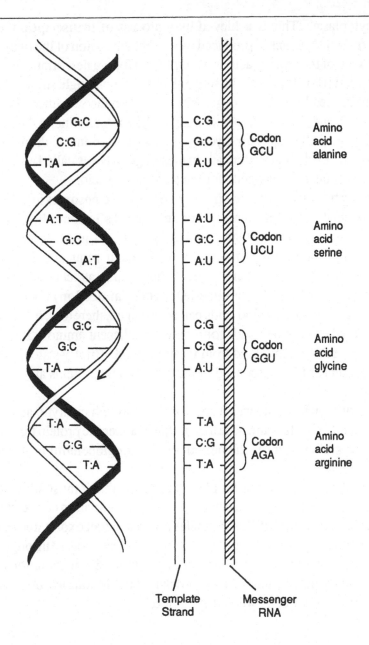

Figure 3.1   The twisted, **double helix** DNA (deoxyribonucleic acid) molecule on the left is the famous Watson-Crick model (Watson and Crick, 1953) of a chromosome that is contained in the nucleus of every living cell. Each step in the spiral-staircase-like structure is composed of a compatible pair of two of four possible nucleotides held by hydrogen bonds. Release of the bonding allows for the double structure to separate like a zipper, and each side then serves as a template against which nucleotide building blocks are properly ordered to construct a new chromosome half.

Each gene consists of a number of base pairs that serve as a blueprint for a particular expression of gene action. Since chromosomes are found in the nucleus and protein synthesis occurs in the cytoplasm, instructions for each gene are carried to the cytoplasm by transcription, the first step in protein synthesis. Transcription results in the formation of RNA copied from a template strand of the

Figure 3.1 continues overleaf

nucleus to the cytoplasm. This is achieved by a process of **transcription** associated with messenger RNA (m RNA), which is produced by a slight separation of the two complementary strands of DNA. One of the strands acts as a template to form a new strand, except that where thymine occurs, uracil (U) is formed. Messenger RNA remains single stranded. Discrete units of information are carried by the messenger RNA based on the sequence of three nucleotide bases (letters A, U, G, and C). ACC, for example, codes for threonine, AAA codes for lysine, CCU codes for proline, GGU codes for glycine, so that all 20 amino acids, the building blocks of protein, are formed. Each three-letter base in messenger RNA is called a **codon**. More than one three-letter base sequence may code for the same amino acid.

Protein assembly from a pool of 20 amino acids occurs on **ribosomes** in the cell. Messenger RNA becomes attached to the ribosome. Protein synthesis is controlled by transfer RNA molecules found in a cell and is regulated by base pairing. A given protein is formed by a particular amino acid sequence determined by a particular sequence of three nucleotides (codons) in a messenger RNA molecule. Each amino acid is attached initially to a specific transfer RNA molecule that also has a three-letter code (**anticodon**) complementary to the codons of the messenger RNA. Amino acids can be arranged, therefore, in a designated order. Proteins are formed when peptide bonds form between adjacent amino acids. The process is repeated when transfer RNA is released from its amino acid. Briefly stated, protein synthesis is controlled by messenger RNA produced in the nucleus, and proteins are synthesized in the cytoplasm.

The formation of protein influences a specific trait through catalytic activities. A dwarf cereal plant, for instance, may lack the gene that produces a specific protein that in turn produces the growth hormone **gibberellin**. If the protein producing gibberellin is absent or its action blocked, dwarf plants result.

A requirement of genetic material for plant breeders is that it must allow for variation to occur; this flexibility can then be used to meet human needs. Variation can be obtained through hybridization programs that bring different **alleles** and hence trait expressions together and by means of **mutation**, both key elements in plant breeding. Genes evolved through modifications in their nucleotides over time, and this is known as mutation. Some mutations are minor and may go unnoticed; others may be undesirable or deleterious. Mutations are errors in the DNA that are passed on to daughter cells when they divide.

Figure 3.1 continued

double helix when a slight separation occurs, as shown in the center of the figure. To form RNA, thymine (T) is replaced by uracil (U) to form a single-stranded messenger RNA.

A specific sequence of three bases in a messenger RNA chain is called a **codon**, which dictates or codes for the 20 amino acids, the subunits of protein shown on the right side of the figure. UUA, for example, codes for leucine, GUU for valine, GGU for glycine. More than one three-base sequence will code for the same amino acid. Protein molecules are completed with the aid of transfer RNA, which exists in the cell. Refer to Weinberg (1985), Darnell, Jr. (1985), and Felsenfeld (1985) for details of DNA, RNA, and protein.

CELL DIVISION: MITOSIS AND MEIOSIS

Before cell division can occur, the components of a cell are doubled through biosynthesis of chromosomes, followed by a division that distributes the components to the daughter cells. The replication of the molecules that carry the genetic code in the nucleus is the most fundamental aspect of the process.

The genetic material packaged in the chromosomes goes through a complex but intelligible series of steps during the process of division called mitosis. Before mitosis begins, a cell is in a state called interphase, the time when RNA and proteins are synthesized. During this period, chromosomes are stretched out, cannot be distinguished individually, and appear as a granular mass. Four stages of mitosis—**prophase, metaphase, anaphase,** and **telophase**—have been distinguished to facilitate description, but the process is a continuous one lasting about 60 minutes.

During prophase, the chromosomes condense into short, thick threads and can be distinguished individually. The **nucleolus** disappears, and threadlike **spindle fibers** from **microtubules** are formed from opposite poles of the nucleus. As prophase proceeds, the chromosomes with their conspicuous **chromatids** move toward the central region of the nucleus, or equatorial plane. The final aspect of prophase is the disappearance of the nuclear envelope. During metaphase, the **centromere** of each chromosome becomes positioned along the equatorial plane, and spindle fibers attach themselves to the centromere of each chromosome. At anaphase, the centromere divides, thereby allowing the two chromatids to be pulled one to each pole by the contracting spindle fiber. Each chromatid is a new daughter chromosome. In the final stage, telophase, the daughter chromosomes at each pole gradually lose their identity as they uncoil and revert to **chromatin**. Nucleoli are reformed, and **cytokinesis** completes the process.

Any chemical or physical factor that disrupts the microtubules (spindle fibers) can prevent the successful division of the chromosomes and results in a doubling of the chromosomes in a single cell. Application of **colchicine** provides a mechanism for chromosome doubling and **polyploid** development.

Mitosis occurs in the sporophyte and results in cells having the 2n chromosome number. In the gametophytic stage, sexual union results in a zygote containing a maternal and a paternal set of chromosomes. In order for the zygote to contain the 2n number, gametes must be formed that are haploid, and this is achieved by meiosis.

Meiosis, which is fundamental to the sexual reproduction of plants, occurs in two stages, involves two nuclear and two cell divisions, and results in four haploid nuclei from one diploid nucleus. All four haploid gametes are functional in the male, whereas in the female, only one out of four generally survives. The first stage in meiosis is termed the **reductional division**, and the second, the **equational division**. During reductional division, the chromosome number is halved but the centromeres of homologous chromosomes do not separate. The second, equational, division is similar to mitosis except that the number of chromosomes is half that of a usual sporophytic cell. The pairing of homologous chromosomes during the first meiotic division results in a close association between the chromosomes, which often exchange chromosomal material. The exchange is of equivalent segments and is referred to cytologically as **crossing over**, and genetically as **recombination**. The four stages of mitosis—prophase, metaphase, anaphase, and telophase—are repeated twice in meiosis. In Figure 3.2 the various stages are presented schematically, and mitosis and meiosis are compared.

In prophase I of meiosis, five stages—leptotene, zygotene, pachytene, diplotene, and diakinesis—have been recognized. During leptotene, each chromosome resembles a thin thread as it emerges from interphase or transition period from leptotene to zypotene stage. Maternal and paternal chromosomes pair or synapse to form homologs during the zygotene stage. Paired chromosomes are known as **bivalents**. Failure to pair along their entire lengths or along sections indicates that chromosomes are structurally nonhomologous and as such are incapable of synapsing. Because of the intimate pairing between chromosomes, only half as many chromosomes appear to be present. During pachytene, chromosomes appear as thick strands because of contraction. The paired chromosomes then double to form four chromatids at the diplotene stage. The paired chromosomes relax, and crossing over of doubled chromosomes occurs with only two of the four strands involved; each point of contact known as **chiasma**. The number of chiasmata between chromosome pairs varies with chromosome length. At diakinesis, the chromosomes become more contracted. The paired chromosome bivalents tend to separate from each other.

The positioning of the chromosomes in the center of the cell indicates metaphase I. Positions of bivalents are such that homologous centromeres face opposite poles. Spindle fibers are anchored to the centromeres. In anaphase I, the centromeres of the two paired chromosomes separate, but unlike in mitosis, each centromere of a chromosome remains undivided, and the two chromatids remain secured by the centromere. Reduction occurs when the double-stranded homologous chromosomes go to opposite poles. In telophase I, the nuclear membrane is reformed. The chromosome number at the end of the first division is half that observed in earlier stages.

During the second part of meiosis, the reduction is completed when each chromatid is separated into a full-fledged chromosome, which requires the splitting of the centromeres. During this second stage, prophase II, metaphase II, anaphase II, and telophase II are completed.

Preceding meiosis, the 2n chromosome number exists. At the end of the first meiotic division, each daughter cell has n chromosomes, consisting of one bipartite chromosome of each pair. At the end of meiosis II, each of the four daughter cells has only single-stranded chromosomes with the haploid chromosome number. Photomicrographs of meiotic stages are shown in Figures 3.3 to 3.6.

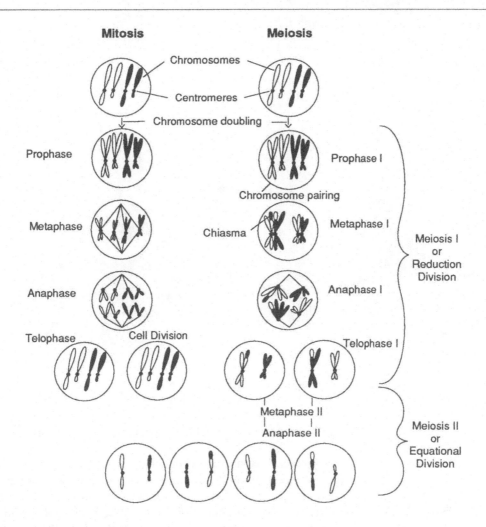

Figure 3.2   Diagram comparing the major steps in mitosis and meiosis in a diploid cell containing four chromosomes, or two homologous pairs. White chromosomes are from one parent and the black chromosomes from the other parent.

Mitosis involves one fully equational division, whereas meiosis involves one numerically reductional and one numerically equational division. In mitosis no pairing of chromosomes occurs, just a doubling; In meiosis synapsis occurs at prophase I and continues to metaphase I.

No chiasmata occur in mitosis, whereas chiasmata and associated crossing over occur during prophase I of meiosis. Centromeres split in metaphase of mitosis and separate in anaphase, whereas in meiosis centromeres do not split at metaphase I and separate undivided in anaphase I.

The essential difference between the two processes is that the chromosome number is maintained in each daughter cell produced by mitosis, whereas the chromosome number is reduced to half after meiosis I. Two diploid gametes are produced from mitosis, and four haploid gametes are produced from meiosis. Note that the four haploid gametes are of two genetic types because of the random distribution of the homologous pairs during metaphase I. Chromosomes that originate from one parent normally segregate randomly to the four haploid cells during reduction division. Crossing over, which normally occurs during prophase I, can result in entirely new genetic combinations.

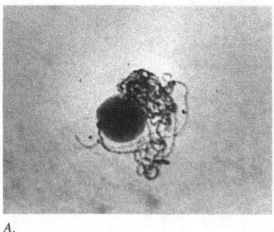
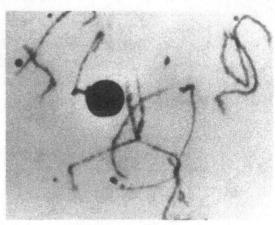

A.   B.

Figure 3.3 The structures in these photomicrographs illustrate the different stages of meiosis. These microphotographs, at a magnification of 2,200x, and are of maize, which has 20 chromosomes as its 2n number. In Photo A, the zygotene stage, the chromosomes appear as intimately paired structures called homologues, representing a maternal and a paternal chromosome. The large dark body is the nucleolus.

The pachytene stage, in which chromosome pairing is complete, is shown in Photo B. Only half as many chromosomes appear to be present as in Photo A because of the intimate and appressed pairing. The large, dark object is the nucleolus, and the conspicuous dark stains on the chromosomes are known as knobs. These are found at characteristic positions along the chromosome arms. Some cultivars of maize have chromosomes lacking in knobs, while others have many knobs that aid in the identification of the different chromosomes. The ten pachytene chromosomes of maize in Photo B are individually recognizable by their relative lengths, by other dark stains known as chromomeres, by the deeply stained knobs at characteristic positions, and by the position of the centromeres. Maize chromosomes have been numbered according to their ten different lengths, with chromosome 1 as the longest and chromosome 10 as the shortest member of the haploid complement.

Pachytene pairs are designated as tetrads. The centromeres of nonhomologous chromosomes and the heterochromatic knobs on different chromosomes may be stuck together, presumably because of the nonspecificity of these regions.

Photos courtesy K. J. Kasha.

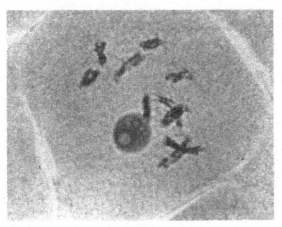
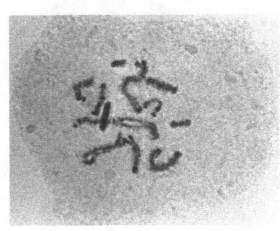

A.   B.

Figure 3.4   The chromosomes in photomicrograph A are those of maize and in B of grasshopper at the diplotene stage of meiosis. Both are at a magnification of 2,200x. It is evident that the two synapsed chromosomes have doubled. A somewhat relaxed pairing is apparent, and the chromosomes are held together by points of overlap known as chiasmata. The diplotene stage is characterized by an opening out of the doubled chromosomes to form loops and nodes. Twisting of the paired homologues is evident. Most of the nodes represent chiasmata making an exchange of pairing partners possible.

Meiosis is basically the same in all organisms with a nucleus, yet certain stages can be seen better in some species than in others. The diplotene stage shows up well in the grasshopper, which has eight pairs of chromosomes and one sex chromosome.

A complete account of meiosis in maize, a crop for which more extensive cytogenetical studies have been conducted than for any other crop, along with photomicrographs and diagrammatic sketches, has been presented by Rhoades (1950).

Photos courtesy K. J. Kasha.

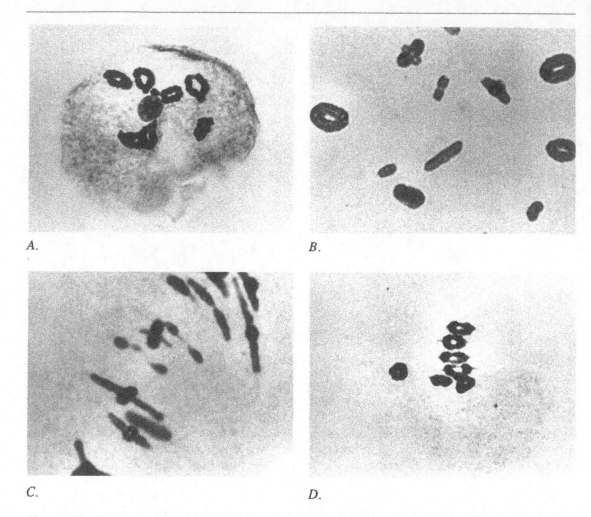

A.

B.

C.

D.

Figure 3.5 Photomicrographs of barley chromosomes (A) at a magnification of 2,200x and grasshopper chromosomes at a magnification of 3,300x, both at late stages of diakinesis in meiosis, a stage difficult to distinguish from the diplotene stage. During diakinesis, the chromosomes become more contracted and chiasmata appear to move to the ends of the chromosome, resulting in what appear as ringlike structures.

Photos C and D are at metaphase I of meiosis of grasshopper (3,300x) and barley (2,200x), respectively. Eleven bivalents and one univalent can be seen in C. The seven bivalents of the 14-chromosome barley are seen in D.

At metaphase I, the nucleolus has disappeared (at late diakinesis), the nuclear membrane breaks down, and the tetrads (paired bivalents) are oriented at the central region of the cell such that the two homologous centromeres of each tetrad lie on opposite sides of the equatorial plate. Spindle fibers, which cannot be seen, are attached to the centromere, and as they retract, draw the bivalents to their respective poles. Chiasmata present at metaphase I become unraveled in preparation for unimpeded movement to the poles in anaphase.

Photos courtesy K. J. Kasha.

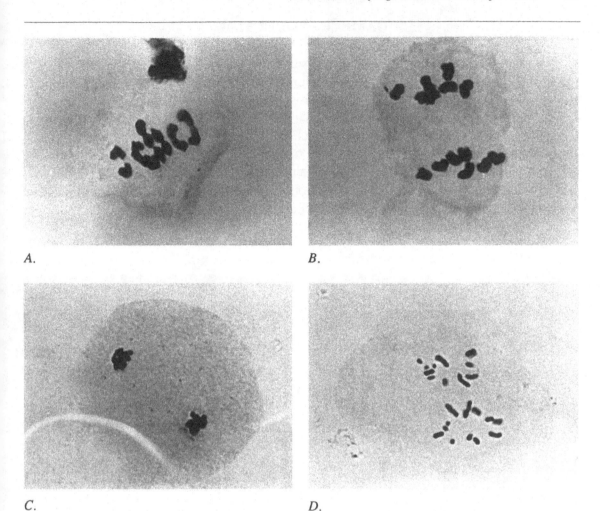

A.

B.

C.

D.

Figure 3.6 Separation of the tetrad into two pairs of chromatids begins at anaphase of meiosis. An early stage for barley chromosomes (B) is shown in Photo A and a late stage in Photo B at a magnification of 2,200x. At anaphase I, the centromeres begin to separate. The centromere does not divide, so that the two chromatids are held together. The bipartite homologous chromosomes move to opposite poles, with the arms of the paired chromatids passively pulled along. The arms appear as a double V if the centromere is near the middle and as a double J if the centromere is more terminally placed. The four arms of the paired chromatids do not lie close to their counterpart as they move poleward but diverge as if mutually repelling each other.

The first division in meiosis is reductional. Early telophase I and late telophase I stages for maize are shown in microphotographs C and D at a magnification of 2,200x. The four chromatids are much contracted and appear as a spherical mass of chromatin as they form a compact group at the pole. The cell plate that arises at telophase divides the cell into two daughter cells, each with half the chromosome number.

In meiosis II, the doubled chromosomes are split; it is completed when a cell plate is formed. Each of the four chromatids composing the pachytene tetrad have been moved to one of the four haploid cells formed as a consequence of two meiotic divisions.

Photos courtesy K. J. Kasha.

GAMETE PRODUCTION

After meiosis is completed, gametes, or sex cells, are formed. Gamete production is a specialized process, with male gamete production (pollen) referred to as microsporogenesis and female gamete production (eggs) as megasporogenesis.

In microsporogenesis, each of the four haploid cells resulting from meiotic division is capable of development into a pollen grain. Each of the four cells undergoes mitosis, producing a two-celled pollen grain consisting of a large tube, or vegetative cell, and a small generative cell. It is at this stage that the pollen grain contained in pollen sacs within the anther is released and carried to the stigma to complete pollination. The tube, or vegetative nucleus, directs the germination and growth of the pollen tube (Figure 3.7).

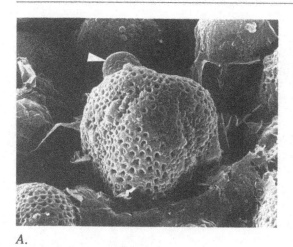
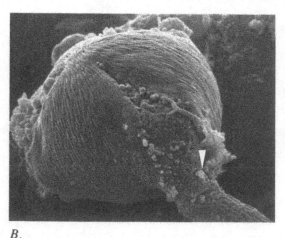

A.   B.

Figure 3.7   These scanning electron photomicrographs show pollen grains in the early stages of germination. The pollen grain in Photo A is that of rape (*Brassica napus* L.) cultivar Regent and in Photo B is that of the Saskatoon berry (*Amelanchier alnifolia* Nutt.).

Immediately after pollination, a pollen tube develops and grows through the style toward the ovary. Each pollen tube contains a generative nucleus and a tube, or vegetative, nucleus, which directs pollen tube development. The newly developing pollen tube is indicated by the arrows. During pollen tube development, the generative nucleus divides once by mitosis to form two sperm cells.

The pollen grain is a highly efficient structure for the transfer of the male gametes. Each pollen grain contains sufficient stored energy to allow the pollen tube to reach its destination and fertilize the ova of the female organ. The pollen tube reaches its destination without any need for water.

Pollen grains have an intine, or inner, and an exine, or outer, wall. The exine wall varies according to species and may exhibit distinct patterns of ridges, mounds, geometrical designs, and large or small spines, which, along with size and shape, can be used for taxonomic identification. The exine wall is composed of sporopollenin, a chemical substance that is highly resistant to degradation, with the result that pollen grain fossils are commonly found. Substances in the pollen wall may cause allergenic reactions in some people.

The exine wall normally fails to develop in one or more areas, thus exposing the intine layer. At the time of germination, the intine layer bulges out to become the pollen tube walls.

Photo A courtesy A. Davis and P. G. Kevan; Photo B courtesy A. R. Olson.

Megasporogenesis, which occurs in the embryo sac, is more complex than microsporogenesis. Only one of the four nuclei resulting from meiosis ultimately gives rise to a gamete. The other three disintegrate. The remaining megaspore undergoes three mitotic divisions, resulting in eight haploid nuclei lying free in the cytoplasm of the embryo sac. The eight cells usually consist of three antipodals, three synergids, and two polar nuclei, but grasses may have many antipodals.

Depending on the species, one or more ovules can be produced within an ovary. A small opening, the **micropyle**, occurs at one end of the ovule. The three antipodals lie opposite the micropyle, the three synergids lie adjacent to the micropyle, and the two polar nuclei move to the center of the embryo sac and fuse, so that seven cells remain. The structure of a flower and the cells of microsporogenesis and megasporogenesis are shown in Figure 3.8.

Fertilization of the egg cell with one sperm nuclei results in the development of the embryo, the first cell of the next sporophytic generation. Fertilization of the fused diploid polar nuclei by the other pollen nucleus forms the endosperm, the nutritive tissue immediately accessible to the embryo when germination of the seed eventually occurs. The process is known as **double fertilization**. Double fertilization results in a diploid embryo (egg plus pollen nucleus) and a **triploid** endosperm (fused polar nuclei plus pollen nucleus). Fertilization is viewed as a specialized version of protoplast fusion and replaces the long-held view that only a male nucleus was delivered to the egg and polar nuclei (Chapman, 1986).

Double fertilization introduces genetic material from the pollen parent into both the embryo and the endosperm. The influence of the genes from the pollen parent on the endosperm is called **xenia** and may affect such characteristics as the endosperm color (Figure 3.9). In addition to the 2n embryo and 3n endosperm tissue in the seed, unfertilized diploid cells from the ovule also divide by mitosis to form part of the seed called the testa. Such tissue is unaffected by genes from the pollen parent.

The function of the synergids is unclear. Chapman (1986) suggested that one synergid collapses and in some instances appears to provide a chemical signal to guide the advancing pollen tube.

MENDELIAN INHERITANCE

Independent Assortment

Perhaps Gregor Mendel, considered the father of genetics, was surprised to find that when he crossed a pea plant with red flowers to one with white flowers, he did not get a blending of factors to produce pink or some other flower color, but rather the F_2 progeny produced distinctly red and white flowers. From this observation, he established the concept of dominant and recessive factors and the rule that each trait segregated independently.

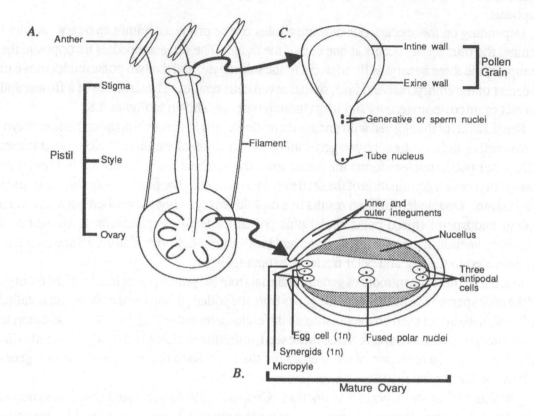

Figure 3.8   Diagrams of a pistil of an angiosperm, showing an embryo sac with multiple ovaries (A), an ovule (B), and a germinating pollen grain (C).

The pollen grain produces three nuclei, two generative, or sperm, nuclei and a tube nucleus, which guides the pollen tube through the style and into the micropyle of an ovary. Light micrographs D and E show a longitudinal section through a pollinated stigma at a magnification of 130x. The arrows indicate pollen tubes. Note the pollen tube growth in between the fusiform of the transmitting tissue of the stigma.

The mature embryo sac, known as the megagametophyte, has seven cells, except in grasses where many are found: the 3 antipodals, which degenerate at fertilization to nourish the embryo; the fused polar nuclei, considered as one cell; the egg cell; and two synergids, which disintegrate.

Two fertilizations occur in the megagametophyte. One of the generative, or sperm, nuclei from

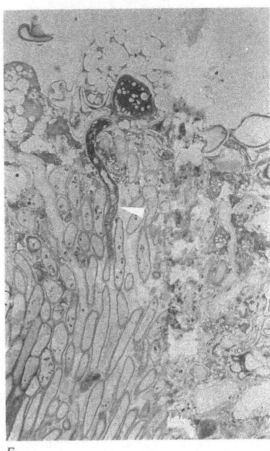

D.   E.

Figure 3.8 continued

the pollen grain unites with the egg to produce the diploid zygote. This develops into the embryo of a seed and is the first cell of the sporophytic generation. A second fertilization occurs between the other generative nucleus and the fused, and hence diploid, polar nuclei to form the 3n, or triploid, nucleus, which produces endosperm tissue. Following fertilization, both tissues develop by mitosis. In monocotyledons, the endosperm often forms the bulk of the seed. Since there are two fusion events, the whole phenomenon, unique to flowering plants, is termed double fertilization.

Solid styles are common in dicots and hollow styles in monocots, although some grasses do have a solid style. In solid styles, the pollen tube pushes its way among the cells, as shown in light micrographs D and E; in hollow stems, the pollen tube grows in a nutrient fluid secreted by cells lining the inner surface.

Photos D and E courtesy A. R. Olson.

62

Figure 3.9 Double fertilization in angiosperms results in the phenomenon of xenia. Xenia is the immediate effect of the genotype of foreign pollen on characteristics of the endosperm in the female parent. Seeds with a large endosperm, such as maize, may exhibit character differentiation traceable to genes contributed by the pollen parent. The endosperm receives two sets of chromosomes from the mother plant and one from the pollen parent, and is triploid. Yellow endosperm color is conditioned by a dominant gene (*Y*). Recessive alleles (*yy*) produce a white endosperm. The endosperm thus receives two genes, either *YY*, *Yy*, or *yy* depending on the genetic background of the mother plant, and either a *Y* or *y* gene from the pollen parent. The genetic composition of the endosperm and its corresponding xenia color could be *YYY* (deep yellow), *YYy* (medium yellow), *Yyy* (light yellow), or *yyy* (white). Other endosperm characteristics that may exhibit xenia in maize are starchy vs. sugary, waxy vs. nonwaxy, and shrunken vs. nonshrunken.

The protective coat surrounding a seed, which is called the testa, is maternal tissue, usually colorless and unaffected by genes from the pollen parent. Some of the kernels in this photo exhibit the brown, yellow, cherry, and variegated patterns that are found in maize in Peru and some parts of Mexico and may exhibit coloration of the testa.

Xenia is of concern to producers of sweet maize if pollen from adjacent starchy maize plants serve as the pollinator, since the triploid endosperm will contain one third starchy genes and two thirds sugary genes. Likewise, popcorn plants, which normally have a hard, flinty endosperm, may be partially dented and soft if a soft, dented pollen parent is used for fertilization.

Xenia can be used to demonstrate segregation of the parental cultivars since the kernels can be classified in the same year of production, whereas most sporophytic responses are not apparent until the seed is planted and the character becomes visible in the progeny.

Photo courtesy Ontario Ministry of Agriculture and Food.

Linked Genes

Mendel observed dihybrid crosses involving two contrasting characters. To determine more easily the genetic combinations, the Punnett square can be used (Figure 3.10). Dihybrid crosses led Mendel to the law of independent assortment, which states that one gene pair segregates independently of another gene pair.

Mendel was either very fortunate or had tremendous foresight into the genetic factors he selected, because the characteristics he studied were transmitted independently from generation to generation. When genes are located on the same chromosome, **linkage** may exist and a

---

Figure 3.10 In addition to studying the inheritance of only one trait, or a monohybrid cross, Mendel also studied the inheritance of two traits, or a dihybrid cross. To avoid the confusion of determining the combinations of the F_2 genotypes, the Punnett square was devised to illustrate clearly the 16 possible combinations of F_1 gametes. In this example, pea plants with yellow and round seed (*YYRR*), both dominant features, are crossed with a pea plant with green and wrinkled seed (*yyrr*), both recessive features. All the possible gametes that could be produced by one of the parents are written along one side of the square and all of the possible gamete types of the other parent along the other side. The genetic contribution of the zygotes in the theoretically expected proportion are shown.

Sorting of the resulting individuals on the basis of phenotype produces a 9:3:3:1 ratio of 9 yellow and round (*Y-R-*), 3 yellow and wrinkled (*Y-rr*), 3 green and round (*yyR-*), and 1 green and wrinkled. These phenotypes can be expressed as follows:

1	*YYRR*	
2	*YYRr*	
2	*YyRR*	
4	*YyRr*	9 yellow round seeds
1	*yyRR*	
2	*yyRr*	3 green round seeds
1	*YYrr*	
2	*Yyrr*	3 yellow wrinkled seeds
1	*yyrr*	1 green wrinkled seed

		Pollen from the male F_1 parent			
		YR	*Yr*	*yR*	*yr*
Egg cells	*YR*	*YYRR*	*YYRr*	*YyRR*	*YyRr*
from the	*Yr*	*YYRr*	*YYrr*	*YyRr*	*Yyrr*
female	*yR*	*YyRR*	*YyRr*	*yyRR*	*yyRr*
F_1 parent	*yr*	*YyRr*	*Yyrr*	*yyRr*	*yyrr*

Y represents a gene for dominant yellow color.
y represents a gene for recessive green color.
R represents a gene for dominant round seed.
r represents a gene for recessive wrinkled seed.

number of factors may be inherited as a unit. For genes on the same chromosome, the law of independent assortment does not apply because linked genes may not segregate, depending on how tightly they are linked, but tend to stay together during metaphase I and anaphase I. The expected 9:3:3:1 dihybrid ratio does not occur, and instead of nine different genotypes, only three appear in a 1:2:1 ratio.

The number of linkage groups corresponds to the haploid chromosome number of the organism. Linked genes may remain together from generation to generation. Plant breeders might wish to separate linked genes, especially if a desired trait is linked with an undesirable trait. Gene linkages may be broken by crossing over, depending on how closely one gene is linked to another.

Crossing Over

Crossing over increases genetic diversity by **recombination**. The frequency of crossing over depends on the distance between two gene **loci**. Genes that are close together on the chromosome will probably not be separated by crossing over, whereas genes at opposite ends of a chromosome are likely to be separated most of the time.

When genes are closely linked, large populations are required to find the genetic combination in a progeny in which crossing over and separation of linked genes has occurred. To identify a plant in which the linkage has been broken, a clear image of what to expect is needed. Complete linkage infers that there is no recombination between genes on a chromosome. Crossing over is less frequent near the centromere. Crossing over occurs during prophase of meiosis I when two homologous chromosomes, each consisting of two chromatids, lie next to each other (Figure 3.11).

Chromosome Mapping

Information gained from crosses involving linked genes has allowed for the position of genes on a chromosome to be determined and a gene map to be constructed. Map units, or

---

Figure 3.11   Diagram illustrating crossing over of chromatids with one chiasma (A) and two chiasmata (B).

Crossing over in A is the result of breaks in sister chromatids. The broken ends are rejoined, but chromatid segments are exchanged. If the broken ends are not rejoined, the result is an **acentric fragment** without a centromere and a **centric fragment** with a centromere. The centric fragment will be pulled by the spindle fiber during division, but the acentric fragment is lost and excluded from the nucleus.

In the sequence listed as B, two breaks in the same chromatid occur and can lead to either an exchange of a chromosome segment or an inversion where the middle section is reattached in the inverted configuration. An outcome of this new arrangement is **position effect,** resulting from a changed linkage arrangement.

When an inversion heterozygote synapses at meiosis, a loop is formed to accommodate the precise pairing. Inversions may result in zygotes that fail to develop.

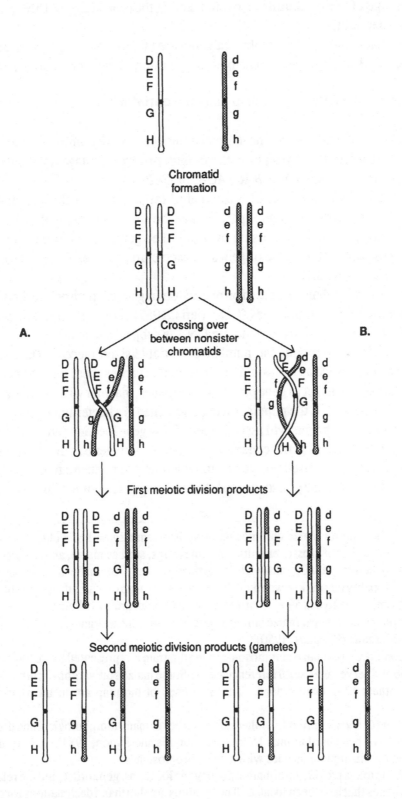

Chromatid formation

A.

Crossing over between nonsister chromatids

B.

First meiotic division products

Second meiotic division products (gametes)

crossover units, express the distance between any two genes on a chromosome and correspond to the percentage of **recombinants** recovered among the progeny, e.g., 1.0% recombination is equal to one map unit.

Suppose that there are 40 recombinant genotypes for two linked genes A and B identified among a total of 360 F_2 progeny. The percentage of crossing over would be

$$\frac{40}{360} \times 100 = 11.1\% \text{ or } 11.1 \text{ map units or } \textbf{centiMorgans}$$

The order of the genes on the chromosome can be determined from crossing-over percentages. In a three-gene sequence with a crossing-over percentage for *A* and *C* of 4.9%, and for *B* and *C* of 16.0%, the order is *BAC*, *ACB*, or *CAB*.

Genes with cross-over values of 1 to 6 centiMorgans are closely linked; genes that are situated 50 or more cross-over units apart behave the same as nonlinked genes.

Information obtained from crossing over of linked genes can be used to construct a gene map showing the order of distance of genes on each chromosome, as shown for chromosome five for maize in Figure 3.12.

Genetic mapping techniques, based on restriction fragment length polymorphisms (RFLPs), offer superior mapping opportunities (Helentjaris, 1987) using a **restriction enzyme** such as restriction endonuclease(s). When genomic DNA is subjected to a restriction enzyme, it is cut at specific sites into a reproducible number of variable-sized pieces. These fragments are inserted into a vector such as a plasmid or virus, which carries them into the cell of a selected organism. The fragments act as gene probes, or templates, that pair with similar gene sequences in the selected organism. If the population to be screened is a bacterium, plating and screening colonies for specific fragments will assist in identifying gene locations and in mapping. RFLPs can provide a rich source of genetic markers for species with genetic diversity, such as tomato and maize. RFLPs can be used on species in which no prior genetic map is known.

Chromosome maps may also be constructed using **isozyme** markers. In this case,

---

Figure 3.12   The map for maize chromosome five is essentially part of a chromosome atlas and is constructed from studies of crossover units. In some crops, such as maize, all ten chromosomes have been mapped (Rhoades, 1950), although modifications may be made as increased evidence is accumulated. Probably more is known about the chromosome map of maize than any other plant species.

Maize chromosome five is 59.82 centiMorgans long, with arms of almost equal length. Chromosome number one is the longest maize chromosome at 82.40 centiMorgans; number ten is the shortest at 36.93 centiMorgans (Rhoades, 1950).

The centromere on the maize chromosome in the diagram is represented by a clear circle. The solid circle is a knob and serves as a useful reference point. Some maize cultivars have chromosomes lacking knobs, while others have many knobs. The zero ends of the map are in the short arms of each chromosome.

In cucumber (*Cucumis sativus* L.), inheritance studies of many morphological and disease-resistant characteristics have been used to map 15 loci (Fanourakis and Simon, 1987). Eighty-three loci have been described on this diploid species with seven chromosomes.

Chromosome maps provide information about gene locations, gene order, and the relative positions of the various genes that have been located. The symbols are shorthand designations for each gene, and the numbers to the left of the chromosomes are linkage units expressed in centiMorgans.

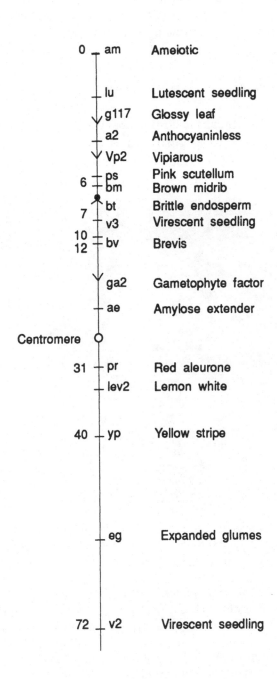

0	am	Ameiotic
	lu	Lutescent seedling
	g117	Glossy leaf
	a2	Anthocyaninless
	Vp2	Vipiarous
	ps	Pink scutellum
6	bm	Brown midrib
	bt	Brittle endosperm
7	v3	Virescent seedling
10		
12	bv	Brevis
	ga2	Gametophyte factor
	ae	Amylose extender
Centromere		
31	pr	Red aleurone
	lev2	Lemon white
40	yp	Yellow stripe
	eg	Expanded glumes
72	v2	Virescent seedling

morphological features on the chromosome serve in mapping. Isozyme methods are less powerful than RFLP because the number of isozymes is less than that of RFLP probes.

Geneticists and cytologists have dreamed of finding large chromosomes to facilitate the visual detection of gene locations and intricate morphological features instead of relying on other techniques. The giant salivary gland chromosomes of the fruit fly (Figure 3.13) have provided a unique initial research tool by allowing qualitative differences to be seen. Such chromosomes have aided gene mapping.

Changes in Chromosome Structure

Crossing over is associated with breaks in chromatids. If for some reason the broken ends are not rejoined as expected, structural changes can occur. Chromosome changes are of interest and value to plant breeders because they provide a means of chromosome engineering. Not all

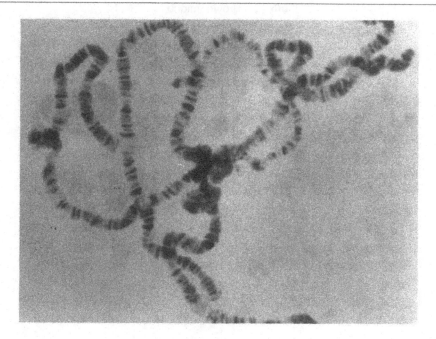

Figure 3.13   Photomicrograph of the giant salivary gland chromosomes of the fruit fly (*Drosophila melanogaster*), which are approximately 200 times as large as the chromosomes in the reproductive cells. These chromosomes show very conspicuous bands, or disks, of deeply stained material alternating with clearer areas.

The light and dark bands represent a definite pattern, which can be used to identify the chromosome. Genes can be traced to definite regions of the chromosomes and are useful in chromosome mapping. Gene maps can be used advantageously for the study of all types of chromosome arrangements resulting from crossing over.

Gene maps take the guesswork out of a genetic study of chromosome aberrations and provide a more accurate picture of what has occurred than that which could be obtained by any but the most exhaustive breeding tests. Simple mutual translocations, inversions, complex rearrangements of chromosome segments, or missing chromosome segments can all be identified through cytological examination.

Photo courtesy K. J. Kasha.

chromosome changes are desirable as they may lead to somatic abnormalities or failure of the zygote, resulting in sterility. Structural changes are illustrated in Figure 3.14.

Gene Interaction

Segregation ratios reported by Mendel are found when the genes involved do not interact. A number of situations arise if the expected ratio is modified when two genes influence the same character. Complementary gene effects occur when at least one dominant allele is present at two different loci, i.e., both genes *A* and *B* must be present and the character is not expressed with the genotype *AAbb*, *Aabb*, *aaBb*, or *aabb*, resulting in a 9:7 dihybrid F_2 ratio.

Additive gene action occurs when the intensity of a character is increased, as in the situation where both genes affecting a character are present in the dominant state. The genotype aabb gives no expression at all, and AaBb produces a more intense expression than Aabb. Additive gene action is recognized by a 9:6:1 ratio in a dihybrid cross.

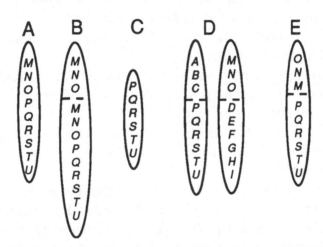

Figure 3.14 Diagrams of four structural changes in a chromosome. The chromosome in A has nine genes and is considered normal. Any modifications in chromosome structure are known as chromosome mutations and may have a harmful effect since the zygote may fail to develop and sterility may result.

Chromosome B has twelve genes instead of nine, three of which are duplicated, and hence this structural mutation is known as duplication. The resulting nucleus cell, tissue, or individual is **hyperploid** for that segment. The chromosome segment that is gained may be inserted into various positions in a homologous or nonhomologous chromosome.

In C, only six of the original nine genes are present, a situation known as a **deficiency**. The loss of a chromosome segment may be acentric (without a centromere) occurring at the end of the chromosome, or in some cases in interstitial regions of the chromosome.

In D, a segment of one chromosome has been interchanged with a segment of a nonhomologous chromosome; this is known as a **translocation**.

In E, the segment bearing genes *M N O* has a reversed position relative to the remainder of the chromosome. Such a situation is called an **inversion**. This results from two breaks in the chromosome, followed by rejoining of ends in a manner opposite from the original chromosome.

Duplicate gene action is found when both genes have similar expression, as exemplified by red kernel color in wheat. The same phenotype is expressed with the *AAbb*, *aaBb*, *Aabb*, and *aaBb* genotypes. Only the full recessive *aabb* has no expression, and hence the dihybrid ratio of an *AaBb* cross is a 15:1 ratio.

Suppressor or inhibiting genes produce a 13:3 dihybrid ratio because gene *A* is dominant, but its expression is inhibited when *B* is also dominant. Only the *AAbb* or *Aabb* genotypes are expressed. In the other thirteen genotypes produced in a dihybrid ratio, the homozygous recessive *aabb* genotype is not expressed because the homozygous *aa* is recessive and the expression of *A* in all the other combinations is suppressed by the *B* allele.

The dominance of one factor over another occurs when the character is controlled by two genes, rather than by alleles of a single gene, and is known as **epistasis** or masking genes. Epistatic dominant genes produce a 12:3 dihybrid ratio. Gene-*A* effect dominates *B* effect, which is expressed only when no dominant *A* allele is present. The *B* gene can be expressed only in the one *aaBB* and two *aaBb* combinations. In the *aabb* combination, the *B* gene is not expressed because all genes are in a homozygous recessive state.

In contrast to epistatic dominant genes, epistatic recessive genes occur when gene-*A* effect dominates *B*, and *B* intensifies *A*. Epistatic recessive genes are known also as modifying genes. There is no expression in the *aaBB*, *aaBb*, or *aabb* genotypes. In the remaining 12 combinations, expression is intensified in *AABB*, *AABb*, *AaBB*, and *AaBb* but not in the other combinations, resulting in a 9:3:4 ratio.

The term epistasis was originally was used to refer to gene interactions in which the genes at one locus show dominance over the genes at another locus. Currently, epistasis refers to all types of nonallelic gene interaction.

Genes that influence more than one character are called **pleiotropic** genes. For example, gene hl in sorghum results in both an increase in the lysine content of storage protein in the seed and a shrunken endosperm.

Pleiotropy may be confused with gene linkage. A single pleiotropic gene will not produce a recombinant genotype, whereas linked genes can be separated. A large number of segregates may be required before a recombinant genotype involving linked genes can be demonstrated. The presence of a single recombinant genotype is sufficient to establish that linked genes are involved.

The ratios listed in this section serve as a useful tool for plant breeders in identifying types of gene interactions. Whether a dihybrid cross is made to determine gene interactions or to recover a desired genotype, a sufficiently large population must be produced in order to provide a reasonable assurance of success. In a dihybrid F_2, it would be necessary to grow 71 plants for a 99% chance of obtaining a triple recessive, and 296 plants would have to be grown to recover the desired genotype in a trihybrid cross.

Heterosis

The increased vigor that frequently is observed in F_1 progeny following a cross of inbred lines, cultivars, or clonal lines is known as hybrid vigor or **heterosis**. Heterosis may be expressed in a single attribute, such as economic yield, or it may be expressed in a number of traits, such as increased vigor and hardiness, increased plant size, pest resistance, adaptability, or plant uniformity.

In contrast to the heterosis that results following a cross, a depression in vigor results from inbreeding. Jones (1939) observed that maize plants exhibited height reductions for five years and yield reductions for 20 years as a result of inbreeding conducted over 30 generations. Inbreeding leads to homozygosity, and deleterious genes can be identified and eliminated when they occur in the homozygous recessive state. Jones (1939) noted that one maize line failed to survive continued inbreeding. Inbred line performance may be at a level of about 50% of that of the cross-fertilized cultivar from which they were derived. Hybridizing two unrelated homozygous inbreds, both lacking in vigor can result in progeny with a high level of hybrid vigor and agronomic performance.

The genetic reasons for the vigor of F_1 hybrids and for the loss of vigor resulting from inbreeding have never been adequately elaborated. A number of nineteenth-century plant breeders reported on hybrid vigor as a result of plant hybridization. Despite advances in genetics, heterosis could not be explained adequately in terms of Mendelian gene action. Several theories of gene action have been developed within the framework of classical Mendelian genetics and include the theories of dominance relationships, of overdominance, and of epistatic interactions. These theories still provide the most satisfactory explanations of hybrid vigor yet devised, but the question is whether they account for all of the manifestations of heterosis.

1. *Dominance Theory*. This theory suggests that favorable dominant genes are accumulated in the F_1 from each parent and that homozygous recessive alleles are masked by dominant genes. Linkage and a large number of genes, however, prevent plant breeders from breeding and selecting inbred lines containing all homozygous dominant alleles.

2. *Overdominance Theory*. According to the overdominance theory, heterosis results from the interactions of alleles at a single locus. The heterozygous genotype *Aa* is superior to the homozygous condition *AA* or *aa*. If there are more than two alleles at a locus, however, then AA^1 may be superior to either *AA* or A^1A^1. Overdominance may provide one explanation of why all polyploids, such as wheat (*Triticum aestivum*), have vigor in the homozygous state.

3. *Epistasis*. The broad sense of epistasis referring to all types of nonallelic gene interaction has been proposed as an explanation of heterosis. This theory is based on a view that every character of an organism is influenced by all genes, and that every gene participates in controlling all characters. Epistasis may provide a second explanation of why hexaploid bread wheat with six sets of chromosomes exhibits vigor.

Geneticists and plant breeders have come to realize that theories of dominance, overdominance and epistasis, all set in the framework of classical Mendelian genetics, do not provide a complete explanation of heterosis. Possibly heterosis can best be explained not by one or several theories but rather as the result of the interaction of several genetic factors.

A contemporary explanation suggests that several genetic causes of heterosis can be identified. Although all are involved in heterosis, their appearance and expression may be variable (Konarev, 1976). The hypothesis of genetic balance and genetic complex completeness is suggested as a concept capable of binding all available data into a general theory of heterosis.

In addition to Mendelian factors, cytoplasmic factors may be involved in the expression of heterosis. The possible role of **mitochondria** in the cell has received attention in recent years (McDaniel, 1986; Sage and Hobson, 1973; Sarkissian and Srivastava, 1967), but experimental problems have raised questions about such a role.

The progress toward an understanding of heterosis is extensive, and reviews can be obtained from the following sources: Gowen (1964), Janossy and Lupton (1976), Frankel (1983), Sinha and Khanna (1975), Srivastava (1981).

Gene Changes

Gene mutations are changes in individual genes that may occur during the production of gametes or in the formation of somatic cells. Mutations may be dominant, recessive, beneficial, or harmful and occur in various combinations.

Most mutations result in recessive genes, but a recessive gene rarely mutates to a dominant form. Mutations may be reversible. Minor mutations are a common occurrence and many may go undetected. Many mutations are agronomically undesirable, resulting in sterility, morphological malformations, or lack of adaptation.

Mutations that occur during the production of gametes are carried into the progeny. A recessive mutation cannot show up in the first generation, called the M_1 or the first mutation generation, but may appear when in the homozygous state in the M_2 or subsequent generations.

Mutation rates are low, but a mutation rate is found for a particular gene. In maize, the *Sh* gene for shrunken grain mutates at a rate of about 1 per million gametes; the *I* gene inhibiting color mutates at a rate of about 100 per million gametes; and the *R* gene for red color in maize mutates at a rate of about 500 per million gametes.

Plant breeders wishing to increase the mutation rate may subject seeds or seedlings to **x-rays**, gamma rays, or mutagenic agents such as ethyl methyl sulphonate. X-rays and gamma rays produce a high level of mutation, but a high level of harmful chromosome mutations also occurs. In contrast, chemical **mutagens** produce a high mutation rate with relatively few chromosome changes.

Changing Chromosome Number

Doubling Chromosome Number. The discovery in the 1930s that colchicine can be extracted from the autumn crocus plant (*Colchicum autumnale* L.) (Kostoff, 1938) attracted considerable attention as a breeding tool (Blakeslee and Avery, 1937) because colchicine prevents the spindle fibers from drawing the chromosomes to their respective poles during cell division. The result is a doubling of the chromosome number. Methods for colchicine treatment were developed quickly and applied to a number of crops, such as red clover, alsike clover, turnip, lettuce, spinach, and rye (Bragdo, 1955). Colchicine appears to be effective on the cells of all plants. Chromosome doubling may be achieved by a number of treatments, including soaking seeds, immersing roots, or placing decapitated **coleoptiles** in a 0.1 to 0.05% colchicine solution (Bremer-Reinders and Bremer, 1952). Various techniques for application to selected crops were reported in detail by Jensen (1974).

The ability to double chromosomes with colchicine can be regarded as another plant breeding tool. Chromosome doubling has found useful applications in restoring bivalent pairing, thereby allowing genomes of different species to be successfully combined. The use of colchicine to induce polyploidy indiscriminately has not led to a great deal of commercial success and is regarded by some plant breeders as the result of the misapplication of colchicine.

Halving the Number of Chromosome Sets. When the gametophytic chromosome number of higher plants is halved, sporophytes called haploids are produced. The utilization of haploids has become a rapidly expanding field since the early 1960s. Haploids can be produced by culturing anthers (Sunderland, 1974) or pollen (Nitsch, 1974). In barley, haploids can be produced by chromosome elimination following interspecific hybridization of *Hordeum vulgare* x *Hordeum bulbosum*, a common and a wild tetraploid barley, respectively (Kasha, 1974). The bulbosum method is an effective means of producing large numbers of barley haploids from any genotype of cultivated barley and has led to the haploid breeding method in barley (Figure 3.15).

POLYPLOIDY

Dreams of producing commercially desirable, exotic new plants may have prompted ancient breeders to attempt the most unlikely crosses. Lack of success discouraged work in this area; if a cross was successful, seed sterility prevented its exploitation unless it was propagated asexually. A knowledge of genetics revealed the reason for sterility. Fertility depends upon regular meiosis, which itself depends upon regular two-by-two pairing in bivalents. Diploid plants indicated by the 2n number of chromosomes are usually fertile, whereas crosses that introduce a different chromosome number are sterile. Barley, for example, has a 2n chromosome number of 14, 7 from the female and 7 from the male. The 2n number of maize is 20; tomato, 24; pea, 14; and sugarcane, 80.

An entire set of chromosomes, the n number, is known as a **genome**. The basic or haploid set of chromosomes in the genome is represented as x. The term 2n designates the sporophytic or somatic tissue and n the chromosome number of microspores, megaspores, and gametophytic tissue. The somatic chromosome number is always represented as 2n, so that for a diploid species, 2n = 2x.

About 40% of crop plants, and indeed of all plants, have three or more sets of chromosomes (genomes) and are polyploid. Over 20% of forage grasses and possibly 33% of all domesticated species are polyploid.

Polyploids have more than two copies of a genome or genomes. **Autoploids** have three or more copies of the same genome; others have three or more different genomes and are called **alloploids**. Cultivated alfalfa, for example, is an autotetraploid with 2n = 32 = 4x; that is, there are 8 chromosomes in each genome. Most potatoes are autotetraploid with 2n = 48 = 4x; bread wheat is an allohexaploid with 2n = 42 = 6x; triticale plants may be either hexaploid or octoploid with 2n = 42 = 6x or 2n = 56 = 8x, respectively; and many strawberries are octoploid with 2n = 56 = 8x. Some sugarcane and sisal plants have five to ten genomes.

With the exception of mules, virtually all animals have two sets of chromosomes. Mules are sterile animals simply because regular two-by-two, or bivalent, pairing is impossible. Likewise, many polyploid plants are sterile.

Figure 3.15   The elimination of chromosomes following an interspecific cross of *Hordeum bulbosum* (plant on left) with normal barley, *Hordeum vulgare* (plant on right), has proven to be a practical method of deriving haploid plants (plant in center). Interspecific crosses between *H. vulgare* and *H. bulbosum* result in the formation of a hybrid zygote, which during embryogenesis and endosperm development undergoes chromosome elimination. In this case, the elimination process is preferential for the loss of chromosomes in the *H. bulbosum* genome. Excision of the embryo is necessary to avoid the embryos from aborting, with subsequent culturing *in vitro*. The resulting plants are haploid. By treating these plants with colchicine, a doubled haploid plant (2n) is produced. Such plants are homozygous for all gene loci in the haploid gamete of the *H. vulgare* parent used in the initial interspecific cross.

   Hordeum bulbosum is a wild, cross-pollinated barley species with no commercial value. Haploid plants generally are small and lacking in vigor. They may exhibit undesirable abnormalities and in many cases are completely sterile.

   Photo courtesy Ontario Ministry of Agriculture and Food.

   Polyploids with an uneven number of genomes (3x, 5x, 7x) are usually sterile, whereas polyploids with an even number of genomes (4x, 6x, 8x) may be fertile or sterile, depending on whether their constituent genomes are all alike or not (Figure 3.16). If the genomes are from the same source and are alike, as occurs in autoploids, reduced fertility can be expected because chromosomes may associate in sets other than the normal bivalent pairing. If there are balanced sets of unlike genomes in alloploids, there is an opportunity for bivalent pairing and high fertility.

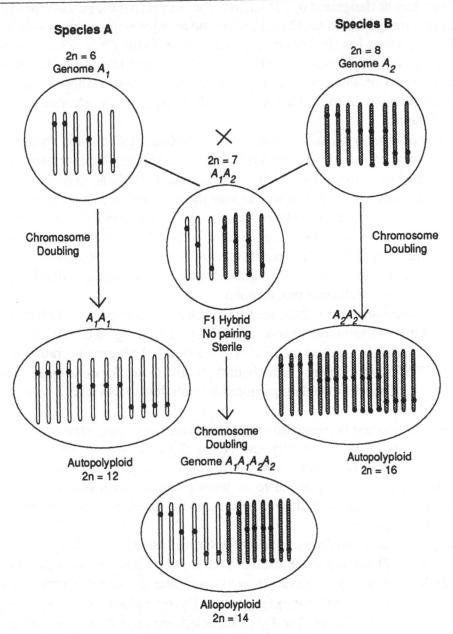

Figure 3.16 The origin of autoploids and alloploids is shown in this schematic diagram. Autoploids occur through a doubling of the complete genome containing the n chromosomes. Doubling species A and B, with six and eight chromosomes respectively, results in autoploids with 2n = 4x = 12 and 16, respectively.

Autoploids usually have reduced fertility because there are four homologous chromosomes, which may all synapse to form a quadrivalent instead of the normal bivalent formation during meiosis. When the chromosomes separate toward the opposite poles during anaphase I, the division of chromosomes may be unequal. All four may go to one end, or three may go to one end and one to the other, all resulting in nonfunctional growth.

Alloploids are produced by hybridization of two different species, in this case A and B, followed by a doubling of the chromosome number, as shown in the center sequence of the diagram. The F₁ plant

Figure 3.16 continues overleaf

Genomes may be designated by a letter. Rye for example has the genomic designation R. A diploid rye is designated as RR (2n = 14 = 2x) and an autotetraploid rye as $RRRR$ (2n = 28 = 4x). Durum wheat has the genomic designation of $AABB$ (2n = 28 = 4x) and is an allotetraploid; triticale is designated $AABBRR$ (2n = 42 = 6x) or $AABBDDRR$ (2n = 56 = 8x) and as an allohexaploid or allooctoploid, respectively; bread wheat is allohexaploid with the genomic designation of $AABBDD$ (2n = 42 = 6x). The letter symbols indicate the origin of the constituent genomes.

The nature of allopolyploids can be illustrated with tobacco and common wheat. Cultivated tobacco (*Nicotiana tabacum*) is a tetraploid (2n = 48 = 4x), which probably originated from two diploid species, possibly *Nicotiana sylvestris* with an n number of 12 and a genome designation of SS and *Nicotiana tormentosiformis* with an n number of 12 and a genomic designation of TT. The species or their ancestors hybridized, and although the two species are similar enough to share common genes between them, chromosome pairs did not occur and sterility resulted. Chance doubling to form the $SSTT$ genomes restored fertility. This situation is referred to as duplicate genetic loci, which means that a gene conditioning a particular trait is found at more than a single locus in the chromosome complex.

Cultivated wheat originated from three component progenitor species. The first is *Triticum monococcum* (2n = 14), known as einkorn, with the A genome designation. Hybridization with a grass species, possibly *Aegilops speltoides* (2n = 14) with the B genome, produced a fertile durum wheat (*Triticum turgidum*). Further hybridization with *Aegilops squarrosa*, carrying the D genome, and subsequent doubling produced a fertile hybrid, $AABBDD$, common wheat (*Triticum aestivum*).

Common bread wheat is composed of three separate, but closely related, basic genomes, which, however, behave cytologically and genetically as diploids. The seven pairs of chromosomes from each of the three genomes have triplicate segments. Chromosomes that share common segments are referred to as **homoeologous chromosomes**. In contrast, homologous chromosomes have identical genetic and cytogenetic characteristics along the entire length of the chromosome.

The origin of alfalfa (*Medicago sativa*) will serve to illustrate the principles of autoploidy. Alfalfa (2n = 4x = 32) probably is the result of a fertile hybrid between two diploid progenitor species, possibly *Medicago coerules* (2n = 18) with the genome designation A_1A_1, and *Medicago falcata* (2n = 16), with the A_2A_2 genomes. Hybridization produced a partially fertile A_1A_2 plant with 16 chromosomes. The A_1A_2 terminology suggests the genomes are similar but with possibly enough structural differentiation that a doubling of chromosomes might have resulted in higher fertility. When this happened, a tetraploid $A_1A_1A_2A_2$ plant was formed that was more vigorous and competitive.

Figure 3.16 continued

is sterile because pairing cannot take place between chromosomes of different genomes. When the chromosomes are doubled, normal pairing can occur between homologous chromosomes. The new allopolyploid produced is designated as 2n = 4x = 14.

Note the knobs at the ends of some chromosomes in species B, which serve to identify these specific chromosomes. Genomes are designated as A_1 and A_2. The genome, or x, numbers for species A and B are 3 and 4 respectively.

Autoploids differ from alloploids in that chromosome pairing can occur at random between any of the homologous or nearly homologous chromosomes of the component progenitor species. When a particular chromosome is present four times, it is called a **tetrasomic**. Because the genomes are closely related genetically and cytologically, multiple allelic series for particular genes commonly are found in these species.

EFFECT OF POLYPLOIDY ON PHENOTYPE

Duplicate genetic loci are commonly observed in alloploids. The inheritance differs from that of diploids in several aspects. For example, the yellow burley character in tobacco that results in light-colored leaves has two separate genes that condition the trait, one from the S genome and another from the T genome. In addition, the alkaloid or nicotine level in the cured leaf is conditioned by two separate genes, one from each of the component genomes, but with unequal effects. One gene in the recessive condition reduces nicotine content in the leaf from approximately 4.0 to 1.5% dry weight. A second gene from the other genome lowers the nicotine content from 1.5 to about 0.5% or less when in the recessive state.

When a gene is located that affects a particular trait in one genome, it can be assumed that a similar gene trait exists in the other component genome(s) that conditions a similar trait. If, on the other hand, the two genomes are very different, alloploids may behave like diploids.

Red kernel color in hexaploid wheat is an example of genes from each of the three genomes that affect a particular trait. Red color is dominant to white kernel color, but the degree of coloring is determined at three gene loci from the three genomes referred to as R_1R_1, R_2R_2 and R_3R_3. The only genotype that gives rise to white kernels is r_1r_1, r_2r_2, r_3r_3, whereas all of the other combinations give some combination of red color. Depending on the genotypes of parent plants, genetic ratios of 3:1, 15:1, or 63:1 can be observed in F_2 segregates. The cross $R_1R_1R_2R_2r_3r_3$ x $r_1r_1r_2r_2r_3r_3$ gives a ratio of 15:1. Each genome apparently conditions kernel color to the same degree such that it is possible for two parents with red kernels to give rise to individuals with white kernels. This is called **additive gene effects.**

In autoploid plants, the increased number of alleles can produce phenotypes that are not possible in diploids. For example, in an autotetraploid that has the genotype $A_1A_2A_3A_4$ and that has four alleles at each locus, there are five possible genotypes for the A_1 and A_2 genes alone— $A_1A_1A_1A_1$, $A_1A_1A_1A_2$, $A_1A_1A_2A_2$, $A_1A_2A_2A_2$, and $A_2A_2A_2A_2$—as compared with three genotypes— A_1A_1, A_1A_2, and A_2A_2 —in a diploid. When all possible gene combinations are produced, many phenotypes are possible.

Homozygosity is approached more slowly in autoploids than in diploids. In an autotetraploid, 3.8 generations of selfing are equivalent to one generation of selfing in a diploid, and inbreeding depression is more severe than in diploids. Response to selection is usually much slower in autoploids than in diploids.

Polyploids produced from chromosome doubling are often a little bigger, lusher, and more luxuriant, with thicker leaves and larger flowers, than corresponding diploids. Such characteristics are called **gigas features** and are a consequence of larger cells. Most alloploids were developed over a long time period, and gigantism has been reduced by selection.

The chief advantages offered by alloploids are genetic flexibility and permanent hybridity.

This feature must be recognized when hybrid wheat work is undertaken. Newly produced alloploids must be made as stable and as reliable as diploids, a process called **diploidization**.

In newly developed alloploids, partial sterility is a problem, which Gaul and Friedt (1975) suggested occurs on average 40% of the time because of one of three phenomena: (1) irregular chromosome distribution caused by unequal separation of multivalents, (2) irregular distribution of chromosomes as a result of either genetic control or meiotic abnormalities of a physiological nature, or (3) genetic-physiological sterility of an unexplained nature, which is not associated with meiotic irregularity. In addition to sterility and a complete lack of seed set, the formation of **aneuploid** seeds produces aneuploid plants lacking in vigor, which have a reduced yield. The occurrence of unaborted aneuploid seeds can be attributed mainly to the first two causes.

Based on the relative frequency of quadrivalent and bivalent association of autotetraploid maize chromosomes, Gilles and Randolph (1951) concluded that autoploids that form multivalents with a relatively high frequency at the time of their origin may become diploidized and shift to the bivalent type of synapsis during their evolutionary development.

Deliberate selection for regular distribution of chromosomes at meiosis, however, has had only a slight effect, if any, on seed fertility. Attempts to diploidize barley following polyploidization by selection, hybridization, and reselection and repeated for a number of consecutive generations generally did not increase fertility (Kao, Reinbergs, and Harvey, 1970; Reinbergs, 1963; Reinbergs and Shebeski, 1959; Smith, 1960).

The diploidization method suggested by Gaul and Friedt (1975) consists of three steps: (1) inducing chromosome and gene mutations in as many autotetraploid cultivars as possible by means of a mutagenic treatment in each reproductive generation, (2) hybridizing the mutagenically treated population at random to bring together as many factors influencing fertility as possible, and (3) selecting plants for fertility and agronomic performance.

The lessons that students should learn from a discussion of polyploidy are as follows:

1. Allopolyploids are commonly found, and plant breeders must learn to deal with all levels of ploidy.
2. Most successful long-term alloploids are functionally diploids, but recessive mutants are rarely expressed because of the number of genes affecting a specific character. Mutation breeding programs are therefore less successful in polyploid plants, although examples of wheat cultivars produced by mutations are found.
3. Induced polyploids have many uses including
   - facilitation of genetic transfer in genetic complexes
   - the synthesis of new crop species
   - broadening of the genetic base, and,
   - development of commercial cultivars.
4. Systematic efforts to exploit new allopolyploids have been few, perhaps because the work tends to be laborious with no guarantee of success. The mere production of a polyploid population does not ensure practical results. Induced polyploids should be considered as new and sometimes valuable raw material for further breeding work requiring a long-term commitment.
5. The realization that polyploids are possible through chromosome doubling or through

interspecific and intergeneric crosses has allowed plant breeders to fulfill an age old desire to produce new species.

6. Following a great deal of effort with polyploids, the practical outcome has been small. Where seed is the economic component, autotetraploids have, in effect, failed because of sterility. Several tetraploid rye cultivars have been released and are marginally acceptable. Tetraploid barley has failed to date. Work on tetraploid sorghum continues, but prospects for success are uncertain. Asexually propagated crops offer promise of success in the polyploid state. They may also be useful where reduced seed fertility is not so severe as in the ryegrasses and clovers. Success has been enjoyed in ornamental crops and in triploid banana, hops, pyrethrums, and watermelons, notably where seed set is undesirable.

7. The gigas characteristic generally has not proven to be agriculturally useful because larger cells are associated with increased moisture content rather than with increased **biomass** potential. In ornamental plants, larger flowers are desired. However, larger flowers in tetraploid red clover have made it difficult for bees to get nectar from the flowers, effect cross-pollination, and promote seed set in a crop already suffering from chromosome irregularities.

AUTOPOLYPLOIDY

Two kinds of polyploids exist—alloploids and autoploids. Allopolyploids are characterized by two or more component **genomes,** which are derived from different species. Meiosis in allopolyploids usually is regular, with diploid segregation ratios, and for this reason breeding methods are similar to those in diploids.

Autopolyploids have two or more genomes, which have their origin from the same or similar species. An autotetraploid, such as potato, alfalfa, coffee, or orchardgrass, has four sets of the same basic genome. Examples of cross-fertilized autohexaploids are sweet potato and timothy. Modifications to breeding programs for autopolyploids to take into account polyploid inheritance are often necessary because of different patterns of chromosome pairing and gene segregation. Because autotetraploids form the majority of autopolyploids and because more information is available on autotetraploids, the remainder of this section will concentrate on autotetraploid inheritance.

Gene Segregation in Autotetraploids

Possible Genotypes in Autotetraploids. In a diploid population in which genes *A* and *a* are segregating, three possible genotypes for one genetic locus exist: *AA*, *Aa* and *aa*. For an autotetraploid in which alleles *A* and *a* are segregating, five different genotypes are possible:

AAAA	quadruplex
AAAa	triplex
AAaa	duplex
Aaaa	simplex
aaaa	nulliplex

Types of Gametes. In a diploid species in which two genes are segregating in genotypes *AA*, *Aa*, and *aa*, the only gametes possible are either *A* or *a*. For an autotetraploid in which two alleles *A* and *a* are segregating, the gametic frequencies are much more complex. Gametes in autotetraploids have two copies of the gene, and therefore gametes, *AA*, *Aa*, and *aa* are possible. The gamete frequencies in an autotetraploid with chromosome segregation are shown in Table 3.1.

The phenotypic ratios observed following crosses of specific genotypes for the *A* locus are shown in Table 3.2. The segregation ratios for duplex x duplex, duplex x simplex, and duplex x nulliplex are different than would be encountered in a similar cross in a diploid. Observation of such phenotypic ratios in a species is indicative of the possible autotetraploid nature as autotetraploid inheritance can be distinguished by certain segregation ratios. However, as shown in Table 3.3, a phenotypic ratio of 3:1 for a completely dominant trait could be obtained by crossing either *Aa* x *Aa* or *Aaaa* x *Aaaa*. Determination of whether such an individual possessed autotetraploid segregation ratios would require an additional generation of controlled matings or self-fertilizations. Selfed diploid-dominant phenotypes would either produce all dominant progeny or segregate in a 3:1 ratio. A further generation of selfing or controlled mating in an autotetraploid would rarely result in individuals that produced all dominant genotypes and would result also in some progeny segregating in a 35:1 ratio.

Alternative Types of Segregation in Autotetraploids

In autotetraploids, chromosome segregation occurs more frequently than chromatid segregation, which can result in an altered segregation pattern. With chromatid segregation, a triplex *AAAa* can give rise to homozygous *aa* gametes under certain conditions. Similarly, the

Table 3.1   Gametic Frequencies Produced by Autotetraploid Genotypes with Chromosome Segregation

Genotype	Gametic Frequency		
	AA	*Aa*	*aa*
AAAA	*1*	*0*	*0*
AAAa	*1/2*	*1/2*	*0*
AAaa	*1/6*	*4/6*	*1/6*
Aaaa	*0*	*1/2*	*1/2*
aaaa	*0*	*0*	*1*

Note: Using these gametic frequencies, self-fertilization of genotype *AAaa* or cross-fertilization to the genotype *AAaa* would result in the following proportion of progeny:

1/36	*AAAA*
8/36	*AAAa*
18/36	*AAaa*
8/36	*Aaaa*
1/36	*aaaa*

If *A* is completely dominant over allele *a*, then the phenotypic ratio would be 35:1.

Table 3.2 Phenotypic Segregation Observed After Crossing Genotypes Assuming Complete Dominance and Chromosome Segregation

Parents	Progeny (dominant : recessive)
AAAA x AAAA	1:0
AAAa x AAAa	1:0
AAaa x AAaa	35:1
AAaa x Aaaa	11:1
AAaa x aaaa	5:1
Aaaa x Aaaa	3:1
Aaaa x aaaa	1:1
aaaa x aaaa	0:1

Table 3.3 Gametes Produced by Autotetraploid Genotypes with Chromatid Segregation

Genotype	AA	Aa	aa
AAAA	1	0	0
AAAa	$\frac{2+\alpha}{4}$	$\frac{2-2\alpha}{4}$	$\frac{\alpha}{4}$
AAaa	$\frac{1+2\alpha}{6}$	$\frac{4-4\alpha}{6}$	$\frac{1-2\alpha}{6}$
Aaaa	$\frac{\alpha}{4}$	$\frac{2-2\alpha}{4}$	$\frac{2+\alpha}{4}$
aaaa	0	0	1

Note: Double reduction does not occur at high frequencies. It can result in unusual segregation patterns that deviate from the chromosome type of segregation.

simplex *Aaaa* can, using the same segregation pattern, give rise to gametes that are homozygous *AA*. **Double reduction** is the term used to describe homozygous gamete formation from either triplex or simplex genotypes.

A number of conditions are necessary for double reduction to occur. The first is that quadrivalents must be formed during meiosis. Second, a crossover must occur between the centromere and the allele that undergoes double reduction so that sister chromatids are attached to two different centromeres. Third, the centromeres with sister chromatids must go to the same pole at second anaphase.

The probability that double reduction will occur can be calculated. Where double reduction is given the designation α, then the probability of α occurring is $a \times e$ where a equals the proportion of quadrivalents formed in meiosis and e equals the product of the probabilities that a crossover will occur between the centromere and the allele. Centromeres with sister chromatids must go to the same pole at second anaphase. The proportion of quadrivalents can vary from 0 to 1 in which quadrivalent formation occurs 100% of the time.

The probability of having a crossover between the centromere and the allele so that sister chromatids are attached to two centromeres has a maximum value of 6/7. The maximum probability that the centromeres with sister chromatids will go to the same pole at first anaphase is 1/3 because of the way in which quadrivalents are distributed at anaphase I. Last, the probability that the same sister chromatids will go to the same pole is 1/2. Therefore, the maximum value of a in the formula $\alpha = a \times e$ is 1, and the maximum value of e is the product of the three probabilities mentioned earlier—6/7 x 1/3 x 1/2—or 1/7. The maximum value of α is 1/7. When double reduction is occurring, the gametic frequencies shown in Table 3.1 will be altered. The gametic frequencies observed with chromatid segregation are shown in Table 3.2.

CHANGES IN CHROMOSOME NUMBER

Changes in the chromosome number characteristic of a species may occur. **Euploids** have an exact multiple of the haploid number and may be triploid, tetraploid, hexaploid, and so on. **Aneuploids** have a chromosome number other than an exact multiple. An important contribution to wheat genetics and breeding investigations was made by Sears (1939, 1944). He developed **monosomic** and **multisomic** series in Chinese Spring wheat, which permitted the association of genes with specific chromosomes and the expression of a number of genes through modification of the character controlled. The most exciting aspect, however, involves the various manipulations with aneuploids. Chromosomes and parts of chromosomes can be studied for the genes they carry, and chromosomes from various sources can be systematically and deliberately combined.

Common aneuploids are not only monosomic (2n-1) but may be double monosomic (2n-1-1), nullisomic (2n-2), trisomic (2n+1), double trisomic (2n+1+1), or tetrasomic (2n+2, the addition of a complete chromosome pair).

Chinese Spring wheat is generally accepted as the standard cultivar for cytogenetic research with wheat because it is the cultivar in which almost all of the more than 250 available aneuploids were isolated (Sears and Miller, 1985) (Figure 3.17). Chinese Spring is awnless and has a tendency to shatter, which makes threshing easy. It is productive when grown under favorable conditions, and its susceptibility to disease avoids complications when hybridized with cultivars carrying genes for disease resistance. However, Chinese Spring wheat, which may have originated in Szechuan province in China, is less than satisfactory for commercial use.

In polyploid species such as wheat, the identification of chromosomes has been difficult. The development of monosomic stocks for each of the 21 chromosomes of hexaploid wheat (Sears, 1954) has made it possible to observe the structural features of each chromosome because of the unique behavior of the univalent (Morris and Sears, 1967).

The effects of each chromosome on character expression in trisomic sugarbeet plants (*Beta vulgaris* L) in eight of the nine trisomic types was reported by Romagosa *et al.* (1986, 1987). These trisomic genetic stocks provide an opportunity for systematic genetic linkage studies, leading to the establishment of linkage groups and chromosome maps in sugarbeet.

Chromosomes may be substituted into a species from the same or another alien species.

Figure 3.17   Photo of a spike of Chinese Spring wheat (n=21), a cultivar of *Triticum aestivum* that has been used to study the effect of each chromosome through deficient chromosome pairs, or **nullisomics**. Chinese Spring offers distinct advantages for cytogenetic work and is generally accepted as the standard cultivar for cytogenetic research on wheat. One advantage is that Chinese Spring crosses readily with rye (*Secale cereale*) (n=7), and from such a cross E. R. Sears of the USDA at the University of Missouri obtained two wheat haploids. When fertilized by Chinese Spring, a number of monosomes, trisomes and reciprocal translocations were produced (Sears, 1939). By selfing monosomic plants (2n-1, i.e., 20 chromosome pairs plus one), a low frequency of nullisomic offspring (20 chromosome pairs minus a homologous pair) were produced. One of these nullisomic plants proved to be chromosome 3B, which is partially asynaptic and was therefore a good source of additional monosomes and trisomes (three chromosomes missing) (Sears, 1944). Subsequently, more than 250 aneuploids (having a chromosome number other than an exact multiple of the **monoploid**, or basic, number) were produced.

Monosomes are useful in the genetic analysis of a species since they greatly facilitate the locating of genes on the chromosomes. The transferring of whole chromosomes from one cultivar to another can be achieved with monosomes. Monosomic analysis is possible only in polyploid organisms because the loss of an entire chromosome from a strictly diploid organism is too deleterious to be tolerated.

If they can be obtained, nullisomes are of greater value than monosomes for genetic analysis. Nullisomes are viable only in higher polyploids, such as allohexaploid wheat. If nullisomes in wheat can be selfed, they breed true for basic chromosome numbers. Some nullisomes are male or female sterile.

In addition to monosomes, nullisomes, and trisomes, tetrasomes (with either four missing or four extra chromosomes) have been produced by chance.

Photo courtesy E. R. Sears, University of Missouri-Columbia.

When chromosome substitution involves another species, the progeny is called an alien chromosome substitution line.

CYTOPLASMIC (NON-MENDELIAN) INHERITANCE

Cells consist of two major components of inheritance - the nucleus, containing DNA encoded with the genetic information, and the cytoplasm that makes up the remainder of the cell. Factors influencing inheritance may be carried in the cytoplasm and are known as cytoplasmic, nonchromosomal, non-Mendelian, or maternal inheritance.

When fertilization occurs to produce a zygote, the male contributes, through the pollen, a haploid nucleus with almost no cytoplasm. The female contributes a haploid nucleus and virtually all the cytoplasm through the egg (instead of 50% of the cytoplasm being inherited from each parent, as in Mendelian genetics), with the result that cytoplasmically inherited factors are passed from generation to generation through the female but not the male. Reciprocal crosses demonstrate this concept quite clearly. As long as the individual possessing a specific cytoplasmic factor is used as the female, the factor will be passed from generation to generation.

Cytoplasmic male sterility (cms) is inherited in an extrachromosomal, or non-Mendelian, fashion in approximately 80 different plant species (Edwardson, 1970). Levings (1983) presented evidence that implicates cytoplasmic male sterility in the mitochondria of maize. Other forms of male sterility are contained in the **chloroplasts**. In addition to male sterility, a number of chlorophyll deficiencies and some other agronomically ill-defined traits are cytoplasmically inherited (Harvey, Levings, and Wernsman, 1972).

Male sterility does not affect female fertility. Male sterility can be restored by genes in the nucleus. The discovery of cytoplasmically inherited male sterility in wheat precipitated extensive research into finding male sterility restorer gene systems that could facilitate hybrid wheat production (Sage, 1976). Cytoplasmic factors determining male sterility do, on the whole, appear to be very stable; the interaction of fertility-restoring genes with the environment may mean that restoration sometimes is incomplete.

REFERENCES

Blakeslee, A. F., and A. G. Avery. 1937. Methods of Inducing Doubling of Chromosomes in Plants by Treatment with Colchicine. *Journal of Heredity* 28:393-411.

Bragdo, M. 1955. Production of Polyploids by Colchicine. *Euphytica* 4:76-82.

Bremer-Reinders, D. E., and G. Bremer. 1952. Methods Used for Producing Polyploid Agricultural Plants. *Euphytica* 1:87-94.

Chapman, G. P. 1986. Modified Fertilisation in Plants: A Strategy for Directed Genetic Change. *Outlook on Agriculture* 15(1):27-32.

Darnell, J. E., Jr. 1985. RNA. *Scientific American* 253(4)68-78.

Edwardson, J. R. 1970. Cytoplasmic Male Sterility. *Botanical Review* 36:341-420.

Fanourakis, N. E., and P. W. Simon, 1987. Analysis of Genetic Linkage in the Cucumber. *Journal of Heredity* 78:238-242.

Felsenfeld, G. 1985. DNA. *Scientific American* 253(4):58-67.

Frankel, R., ed. 1983. *Heterosis: Reappraisal of Theory and Practice*. Monographs on Theoretical and Applied Genetics, 6. Springer-Verlag, Berlin. 290 pp.

Gaul, H., and W. Friedt. 1975. Progress in the Diploidization of Autotetraploid Barley. In *Barley Genetics III*. Proceedings of the Third International Barley Genetics Symposium, Garching. Verlag Karl Thiemmig, Munchen, pp. 378-387.

Gilles, A., and L. F. Randolph. 1951. Reduction of Quadrivalent Frequency in Autotetraploid Maize During a Period of Ten Years. *American Journal of Botany* 38:12-17.

Gowen, J. W., ed. 1964. *Heterosis*. Hafner Publishing, New York. 552 pp.

Harvey, P. H., C. S. Levings III, and E. A. Wernsman. 1972. The Role of Extrachromosomal Inheritance in Plant Breeding. *Advances in Agronomy* 24:1-27.

Helentjaris, T. 1987. A Genetic Linkage Map for Maize Based on RFLPs. *Trends in Genetics* 3:217-221.

Janossy, A., and F. G. H. Lupton, eds. 1976. *Heterosis in Plant Breeding*. Proceedings of the Seventh Congress of Eucarpia. Elsevier Publishing Co., Amsterdam. 366 pp.

Jensen, C. J. 1974. Chromosome Doubling Techniques in Haploids. In *Haploids in Higher Plants: Advances and Potential*, edited by K. J. Kasha. University of Guelph, pp. 153-190.

Jones, D. F. 1939. Continued Inbreeding in Maize. *Genetics* 24:462-473.

Kao, K. N., E. Reinbergs, and B. L. Harvey. 1970. Selection for Seed Setting in Hybrid Populations of Autotetraploid Barley, *Hordeum vulgare* L. emend Lam. *Crop Science* 10:491-492.

Kasha, K. J. 1974. Haploids from Somatic Cells. In *Haploids in Higher Plants: Advances and Potential*, edited by K. J. Kasha. University of Guelph. pp. 67-87.

Konarev, V. G. 1976. Physiological and Biochemical Aspects of Heterosis. In *Heterosis in Plant Breeding*, edited by A. Janosy and F.G.H. Lupton. Elsevier Publishing Co., Amsterdam, pp. 81-93.

Kostoff, D. 1938. Colchicine and Acenophthlene as Polyploidizing Agents. *Nature* 142:753.

Levings, C. S. III. 1983. Cytoplasmic Male Sterility. In *Genetic Engineering in Plants. An Agricultural Perspective*. Edited by T. Kosuge, C. Meredith, and A. Hollaender. Plenum Press, New York. pp. 81-92.

McDaniel, R. G. 1986. Biochemical and Physiological Basis of Heterosis. *CRC Critical Reviews in Plant Sciences* 4:227-246.

Morris, R., and E. R. Sears. 1967. The Cytogenetics of Wheat and Its Relatives. In *Wheat and Wheat Improvement*, edited by K. S. Quisenberry and L. R. Reitz. American Society of Agronomy, Madison, Wisconsin, pp. 19-87.

Nitsch, C. 1974. Pollen Culture: A New Technique for Mass Production of Haploid and Homozygous Plants. In *Haploids in Higher Plants: Advances and Potential*, edited by K. J. Kasha. University of Guelph, pp. 123-135.

Reinbergs, E. 1963. Autotetraploids in Barley. *Barley Genetics I*. Proceedings First International Barley Genetics Symposium, Wagenengen. pp. 151-154.

Reinbergs, E., and L. H. Shebeski. 1959. Fertility of Barley Autotetraploids. I: Fertility in Successive Generations of Four Autotetraploid Barley Varieties and the Effect of Selection for Fertility in the O.A.C. 21 Autotetraploid. *Canadian Journal of Plant Science* 39:98-107.

Rhoades, M. M. 1950. Meiosis in Maize. *Journal of Heredity* 41:59-67.

Romagosa, I., L. Cistue, T. Tsuchiya, J. M. Lasa, and R. J. Hecker. 1987. Primary Trisomics in Sugarbeet. II: Cytological Identification. *Crop Science* 27:435-439.

Romagosa, I., R. J. Hecker, T. Tsuchiya, and J. M. Lasa. 1986. Primary Trisomics in Sugarbeet. I: Isolation and Morphology Characterization. *Crop Science* 26:243-249.

Sacher, R. F., and D. L. Mulcahy. 1981. Gametophytic Selection for Salt Tolerance in Tomato and Related Species. Supplement to *Plant Physiology* 67:19 (abstract no. 96).

Sage, G.C.M. 1976. Nucleo-Cytoplasmic Relationships in Wheat. *Advances in Agronomy* 28:267-300.

Sage, G.C.M., and G. E. Hobson. 1973. The Possible Use of Mitochondrial Complementation as an Indicator of Yield Heterosis in Breeding Hybrid Wheat. *Euphytica* 22:61-69.

Sarkissian, I. V., and H. K. Srivastava. 1967. Mitochondrial Polymorphism in Maize, II. Further Evidence of Correlation of Mitochondrial Complementation and Heterosis. *Genetics* 57:843-850.

Sears, E. R. 1939. Cytogenetic Studies with Polyploid Species of Wheat. I: Chromosomal Aberration in the Progeny of a Haploid of *Triticum vulgare*. *Genetics* 24:509-523.

———— 1944. Cytogenetic Studies with Polyploid Species of Wheat. II: Additional Chromosomal Aberrations in *Triticum vulgare*. *Genetics* 29:232-246.

———— 1954. *The Aneuploids of Common Wheat*. Missouri Agricultural Experiment Station Research Bulletin 572. 59 pp.

Sears, E. R., and T. E. Miller. 1985. The History of Chinese Spring Wheat. *Cereal Research Communications* 13:261-263.

Sinha, S. K., and R. Khanna. 1975. Physiological, Biochemical and Genetic Basis of Heterosis. *Advances in Agronomy* 27:123-174.

Smith, W. E. 1960. Fertility of Autotetraploid Varieties and Hybrids of Barley. *Canadian Journal of Plant Science* 40:434-442.

Srivastava, H. K. 1981. Intergenomic Interaction, Heterosis, and Improvement of Crop Yield. *Advances in Agronomy* 34:117-195.

Sunderland, N. 1974. Anther Culture as a Means of Haploid Induction. In *Haploids in Higher Plants: Advances and Potential*, edited by K. J. Kasha. University of Guelph, pp. 91-122.

Tanksley, S. D., D. Zamir and C. M. Rick. 1981. Evidence for Extensive Overlap of Sporophytic and Gametophytic Gene Expression in *Lycopersicon esculentum*. *Science* 213:453-455.

Watson, J. D., and F.H.C. Crick. 1953. Molecular Structure of Nucleic Acids: A Structure for Deoxyribose Nucleic Acid. *Nature* 171 (4356):737-738.

Weinberg, R. A. 1985. The Molecules of Life. *Scientific American* 253(4):48-57.

CHAPTER 4

Genetic Variation in Plants

The first step of any plant breeding program is to identify plants that exhibit variation for the trait or traits of interest to the plant breeder. Desirable traits or combinations of traits should be sought among plants in existing populations, such as recommended cultivars, breeding lines, and **landrace** cultivars. If traits are not found in any of these **germplasm resources**, the breeder should evaluate exotic or related genotypes, wild relatives, and even alien species. When the desired traits cannot be found in any of the existing germplasm sources, the breeder creates variation by hybridization or induced mutation.

GERMPLASM RESOURCES

Germplasm is an essential resource for successful plant breeding. Germplasm resources were recognized and collected in the mid-eighteenth century when Sir Joseph Banks, the director of Kew Gardens, the famous Royal Botanic Gardens in London, England, accompanied Captain Cook on a plant-collecting voyage. Centuries later, N. I. Vavilov (1949-1950) pointed out that certain areas of the world exhibit a particularly high level of plant genetic variability. Such centers of genetic diversity were regarded by Vavilov as indicators of centers of origin. Eight centers were identified, each associated with a particular complex of crops—six in the Old World and two in the New World.

Subsequent research suggests that crop domestication and the spread of agriculture resulted in a greater dispersion of crop genetic diversity than Vavilov realized. The concept of a center of origin was questioned (Harlan, 1971, 1976), as a single crop may have originated in more than one center. In addition, many crops did not originate in the centers identified by Vavilov. The term *regions of diversity* generally is used to account for the variability observed in regions of the world.

Figure 4.1 shows nine regions of genetic diversity and their associated crops, based on an empirical distribution of the world's plant genetic resources. Such maps are of value in plant collection since new and valuable germplasm resources may be found in these regions.

Replacement of landrace cultivars by modern, pure-line cultivars (and widespread production of a single, superior cultivar) has reduced the genetic variation in cropping programs. There is also a danger that valuable genetic resources may be lost to future breeding programs as the areas of genetic diversity are developed industrially and agriculture becomes more intensified. Consideration of who owns these valuable resources must recognize that the economies of the world have a growing reliance on each other and that the nations of the world are linked in a network of plant genetic interdependence (Shulman, 1986; Kloppenburg and Kleinman, 1987).

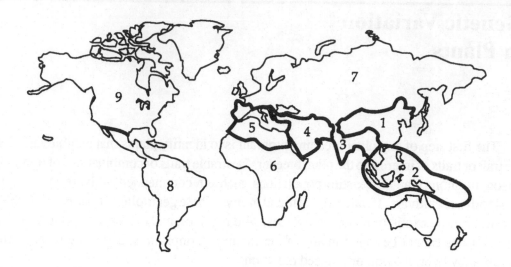

Figure 4.1   The nine centers shown on this map are regions of diversity for 36 leading food and industrial plant products (rice is common to three regions). Genetic diversity is critical to success in plant breeding, and this map illustrates the global distribution of the world's plant genetic resources of the economically most important food and industrial crops. The nine regions and their crops, from a report by Kloppenburg and Kleinman (1987), are largely based on the original work of Harlan (1971) and are as follows:

Sino-Japanese: soybean, orange, rice, tea.
Indo-Chinese: banana, coconut, yam, rice, sugarcane.
Hindustani: rice, jute.
West Central Asiatic: wheat, barley, grape, apple, flax (oil and fiber), sesame.
Mediterranean: cabbage, sugarbeet, olive, rapeseed.
African: sorghum, millet, oil palm, coffee.
Euro-Siberian: oat, rye.
Latin American: maize, potato, sweet potato, cassava, tomato, cotton, tobacco, rubber, cocoa.
North American: sunflower.

Source: Kloppenburg and Kleinman, 1987, p. 193. Reprinted by permission.

The realization that genetic resources may be the most precious asset of the world led to the establishment of **gene banks** for purposes of conserving genetic resources. Exploration and conservation programs are based on one or both of the following objectives (Hawkes, 1985):

1.   To meet the needs of plant breeders. Genetic resource materials, particularly of wild species, weed races, primitive forms, and landraces, may be needed urgently by breeders because they know or hope that desirable genetic features of pest resistance or adaptation can be obtained for continued progress in plant breeding.

2.   To reduce the threat or actual loss of genetic diversity in the field in the form of landraces and primitive forms of crop species as they are replaced by high-yielding, modern cultivars. Wild species or weed races may be lost through changes in land use resulting from human population pressures, war, or other factors.

Over the years, many plant introductions have been made, but many early introductions have been lost because of inadequate storage facilities. Concern about this loss led to the concept of national seed storage units. In 1958, a milestone was achieved in the conservation of germplasm when the United States opened the National Seed Storage Laboratory at Colorado State University, Fort Collins, Colorado. Other nations followed this example and constructed controlled storage rooms designed to preserve the viability of seeds during intervals between their periodic growth outdoors. Inventories of stocks are published, and seed is made available to plant breeders (Figure 4.2).

A leading role in establishing a European network of germplasm conservation was assumed by **EUCARPIA** starting in 1960. To serve southern Europe and the Mediterranean region, a gene bank was established at Bari, Italy; a gene bank at Lund, Sweden, was constructed for Scandinavia; and another was set up at Braunschweig, Germany, for northwestern Europe (Hawkes and Lambert, 1977).

In 1974, the International Board for Plant Genetic Resources (IBPGR), headquartered in Rome, Italy, was formed to assist and encourage the collecting, preservation, and exchange of plant genetic material on a global scale. This program is part of a larger global experiment in agricultural research and development sponsored by International Research Centers (Table 4.1). The primary purpose of the IBPGR is to conserve genetic resource materials both for long-term storage (base collections) and for short- to medium-term storage (active collections) for distribution to plant breeders and others.

International Research Centers are concerned with basic food or starch-producing crops. Gene banks for these crops are needed because most are grown in monoculture production over large areas and are often based on narrow genetic bases. These two factors have heightened concerns about the outbreak of pests or diseases in potential epidemic proportions. Such problems may be alleviated by exotic genes obtained from various sources (Figure 4.3).

Commercial cultivars of vegetable crops have become genetically more uniform as once over machine harvesting, extended storage, long-distance hauling, and marketing systems have developed. Only cultivars with satisfactory yield and quality are acceptable, and the price of this has been a loss of genetic variation, the plant breeder's basic resource. To minimize genetic erosion in vegetables, the Vegetable Gene Bank at the National Vegetable Research Station in the United Kingdom was established in 1981 (Astley, 1985, 1987).

Likewise, the large number of forest, fruit, and nut-bearing trees makes conserving genetic tree resources a challenging task. To manage genetic resources more objectively, Crisp and Astley (1985) suggested four genetic categories, namely, (1) landraces, (2) those that have shown little change from their wild state, (3) advanced cultivars with a narrow genetic base, and (4) advanced diverse cultivars. The objectives of the Vegetable Gene Bank are to identify gaps in the collections, to collect relevant material, especially from areas that are not well represented, to store and document the material, and to encourage research into the taxonomy and phylogenetic relationships among crops and related wild or weed species. Collecting and preserving tree species is a major challenge.

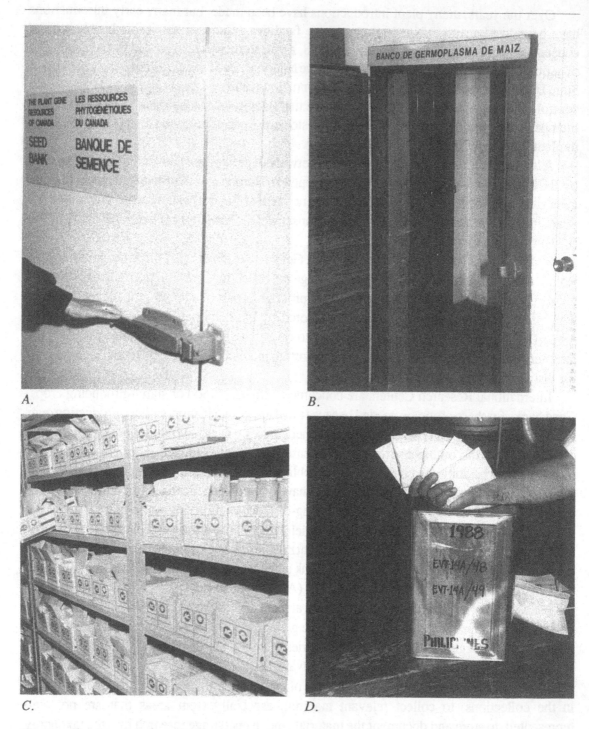

Figure 4.2 The interior of a working gene bank containing seed samples for distribution or revitalization. Genetic variation is a key to success in plant breeding. Gene banks are a means of conserving valuable genotypes in a controlled environment for present-day use and for future generations. Gene banks contain plant genotypes preserved in either seed form as living plants known as field gene banks or as dormant tissues in storage. The gene banks shown in these photos are examples

SOURCES OF GERMPLASM

Gene banks provide plant breeders with sources that may be of immeasurable value in plant breeding programs as reserves of genetic diversity. The unrestricted distribution of genotypes around the world is essential to continued success in plant breeding and food production. Seed and clonally propagated species from gene banks can be regarded as basic genetic stock to secure variation for hybridization and selection. The following are presented as **germplasm resources** for plant breeders:

1. Currently Grown Commercial Cultivars. This resource represents the most recent successes from breeding programs and is the most likely source of superior germplasm that is adapted to a specific environment. The commercial success of these genotypes is assurance that they are the best gene combinations available. Cultivars in this category normally would not be stored in gene banks because they usually can be obtained from commercial sources.

2. Obsolete Commercial Cultivars. Cultivars are rendered obsolete for commercial use with the release of superior genotypes, the outbreak of a virulent pest, or the discovery after release of an agronomic weakness. Such cultivars are an excellent genetic source when hybridized with a parent or parents chosen to overcome a recognized deficiency.

3. Breeding Lines and Stocks. Plant breeders know their material and recognize potentially useful genotypes that are either incompletely evaluated or offer a desirable genetic trait, but with other recognized deficiencies. Such materials can serve as parental stocks. Plant breeding is a long-term venture, and the best parental material is most desirable. This source of germplasm is changing constantly, and unless such lines exhibit a unique feature with lasting genetic appeal, they may not merit conservation in a gene bank. As a result, such plant material is often maintained in individual breeding germplasm collections. The long-term protection of these collections is questionable.

4. Landrace Cultivars. Crop domestication and settled agriculture meant that crops, through their seeds, became transportable commodities. Newly distributed crops were subject

---

Figure 4.2 continued

of many national units around the world, designed for either long-term storage or for filling requests of plant breeders seeking specific genotypes.

Photos A and B are doors to the refrigerated working gene bank units in Canada and Mexico, respectively. Seed may be stored in clearly labeled envelopes in boxes (Photo C) or in metal cans (Photo D). Metal cans help reduce the risk of insect damage.

Storage of plant germplasm is an almost overwhelming task in terms of collecting and of developing techniques to save space and reduce costs. Cryopreservation of seeds and vegetative materials stored at -196° C offers a possibility for extended storage.

The encroachment of civilization into centers of diversity threatens valuable genetic resources. Wild and weedy forms, often regarded as of immediate value, are preserved along with landraces as an irreplaceable reservoir of genetic material for genotypic improvement. Cultivars that are no longer grown commercially may be lost if efforts are not directed to preserving them.

The shift of agriculture to monoculture production in many developed areas of the world is associated with an ever-narrowing gene pool. Gene banks offer some insurance against an unanticipated widespread outbreak of a pest problem (Plucknett *et al.*, 1985).

Photo A courtesy Canada Agriculture, Ottawa.

Table 4.1 Network of International Agricultural Research Centers

	Name/Location	*Germplasm Collected*
CIAT	Centro Internacional de Agricultura Tropical - Cali, Colombia	Rice, forage grasses and legumes, grain legumes, cassava
CIMMYT	Centro Internacional de Mejoramiento de Mais y Trigo (International Maize and Wheat Improvement Center) - El Batan, Mexico	Wheat (bread and durum), maize
CIP	Centro Internactional de la Papa (International Potato Center) - Lima, Peru	Potato, sweet potato
IBPGR	International Board for Plant Genetic Resources - Rome, Italy	
ICARDA	International Centre for Agricultural Research in the Dry Areas - Aleppo, Syria	Wheat (bread and durum), barley, faba bean, lentil, chickpea, pigeonpea, forage grasses and legumes
ICRAF	International Council for Research in Agroforestry - Nairobi, Kenya	
ICRISAT	International Crops Research Institute for the Semi-Arid Tropics - Hyderabad, India	Sorghum, millet, chickpea, pigeonpea, groundnut
IFPRI	International Food Policy Research Institute - Washington, D.C.	
IIMI	International Irrigation Management Institute - Colombo, Sri Lanka	
IITA	International Institute of Tropical Agriculture - Ibadan, Nigeria	Rice, potato, sweet potato, groundnut, cowpea, soybean, maize, cassava
ILCA	International Livestock Center for Africa - Addis Ababa, Ethiopia	
ILRAD	International Laboratory for Research on Animal Diseases - Nairobi, Kenya	
INIBAP	International Network for the Improvement of Banana and Plantain - Montferrier-sur-Lez, France	Banana, plantain
IRRI	International Rice Research Institute - Los Banos, Philippines	Rice
ISNAR	International Service for National Agricultural Research - The Hague, Netherlands	
WARDA	West Africa Rice Development Association - Bouake, Cote d'Ivoire	

Note: A network of sixteen agricultural research centers around the world is under the direction of the Consultative Group on International Agricultural Research (CGIAR). CGIAR was formed as a nongovernmental, noncommercial network funded by many national governments. One center, the International Board for Plant Genetic Resources (IBPGR), functions to promote and coordinate a network of genetic resource centers or gene banks to collect, conserve, document, evaluate, and disseminate plant germplasm. Gene collections in this network are supplemented by national seed banks. The location and the germplasm collected for those with seed banks are listed. The efforts of these centers coupled with those of private industry have made commercial plant breeding truly international in scope.

A.

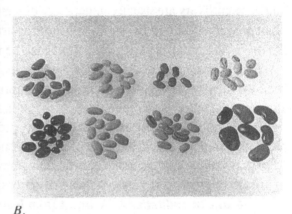

B.

Figure 4.3 Plant genetic resources are not dispersed uniformly around the globe. Areas such as the United States and Canada can claim no major food crops as native. Instead, most crops, including some featured in Photo A at a market in Mexico, evolved in the world's less-developed countries. The samples in Photo B were found in the lot shown in Photo A and represent the enormous variation found in crops where pure-line selection has not been practiced. As much as 96% of the genetic base of the food crops of the world have been contributed from less-developed regions of the world.

The seeds in Photo B are those of beans (*Phaseolus* spp), which represent an enormously diverse family, as indicated by seed size, morphology, and color markings ranging from white to black to red to variegated. Interfertilization and gene reassortment may produce valuable new genetic combinations.

If we manipulate both genotype through plant breeding and environment through agronomic practice, we can improve and stabilize crop yields by maximizing genetic potential and by minimizing constraints to the expression of that potential. In practice, these approaches are complementary and are pursued interactively and simultaneously. The possibility of increasing the production potential is enhanced by the discovery, even in well-researched crops such as grain legumes, of hitherto unrecognized adaptive responses that can substantially improve productivity.

Photo B courtesy Ontario Ministry of Agriculture and Food.

to local selection pressures in their new environments to produce landraces. Landraces are genetically diverse populations selected under low-input agriculture for dependability or yield stability rather than productivity or yield ability. Landraces may not have been subjected to conscious selection pressure and have been very useful in plant breeding programs for direct selection or as a broad gene base for wide adaptation. Under intense cropping systems or monoculture production, landraces may not perform as well as their modern counterparts and may be discarded.

Although many landraces have been utilized extensively by plant breeders, they continue to offer valuable germplasm for modifications in agricultural technology. Landraces remain a useful source of genetic variation when hybridized with a parent having more specific adaptation.

For purposes of introducing a new crop into a region, landraces are of great value. Single-head progenies derived from barley landraces collected in Syria and Jordan showed great diversity, useful for breeding purposes (Ceccarelli, Grando, and Van Leur, 1987). Since landrace cultivars are usually genetically diverse populations, selection might identify single plant progenies with larger yields and more desirable expressions of agronomic characters than the original landraces. Such selections may be narrowly adapted, however.

5. *Undomesticated Forms of Crop Plants*. These genotypes are distinct from landraces because they were never of commercial importance, may shatter their seed, and have other agronomically undesirable features. Such forms are a source of entire genomes or specific genes that may require **introgression** into an adapted cultivar. Undomesticated forms may be preserved in gene banks because of unique genetic features or for purely academic reasons. Changing technology on farms or in breeding procedures may turn specific unrecognized traits in primitive plant forms into valuable resources.

6. *Plant Introductions*. A wheat from Russia called Turkey, which was introduced into the central plains of North America, established hard red winter wheat production and became the cornerstone of an industry (Quisenberry and Reitz, 1974). Turkey wheat contributed an ancestral stamp to modern wheat cultivars because it entered into the lineage of most red winter wheats in North America. A monument to recognize the introduction of this wheat was erected by Mennonite farmers in 1942 (Figure 4.4). The contribution of introduced germplasm to the development of wheat cultivars in the United States was reviewed by Cox (1991).

There is little doubt that past introductions have played a major role in plant breeding programs around the world and that they will continue to be important. The desire for diversification in agriculture has given plant introductions a new meaning. Newly introduced crop species may be used directly as new crops, may be used as new cultivars of established crops, or may serve as new germplasm sources for hybridizing with established crops (Figure 4.5).

Plant introductions are likely to be successful if similarities exist in soil, climate, and photoperiodic conditions between the region of origin and the new area of production. An introduction with a broad gene base and wide adaptation is most likely to be successful. The soybean is an example of a successful introduction from northern China to North America (Figure 4.6).

Not all introductions are successful. Kudzu (*Pueraria lobata*) was introduced into the southern United States in the 1930s because of its ability to grow rapidly. Without natural

Figure 4.4   The monument depicting a Mennonite farmer shown in this photo commemorates the introduction of Turkey wheat to Newton, Kansas, about 1874. Mennonite settlers from Russia brought this introduction to central Kansas and named it Turkey wheat because it was obtained originally from a valley in Turkey.

Initially, plant introductions were of a haphazard nature because immigrants, uncertain of which crops existed in the New World, instinctively brought with them a small sample of seed. Sometimes these introductions were unadapted and failed; sometimes, as in the case of Turkey wheat, introductions proved to be of great importance.

The Turkey wheat genotype introduced into Kansas can be regarded as an unselected landrace form rather than a cultivar with narrowly defined characteristics because it had not been subjected to selection and purification. After its introduction, natural or deliberate selection within the variable genotype of Turkey wheat produced locally adapted strains that came to be known by 28 names in North America, including Kharkof, Crimean, Malakof, Red Winter, Turkey Red, Red Russian, and Tauranian. These selections are phenotypically similar, but no doubt genetic differences exist. Several of these selections may have been reintroduced from different geographical areas at different times. Kharkof, for example, was introduced in 1900 from a site north of where the original introduction occured and may be more winter hardy than the Turkey wheat initially introduced.

In 1971, the cultivar Centurk was released in the United States to commemorate a century of Turkey wheat in North America.

Photo courtesy Mennonite Library and Archives, Bethel College, North Newton, Kansas.

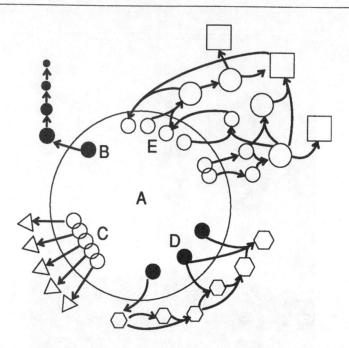

Figure 4.5 Germplasm introduction has played and will continue to play a vital role in the continuing improvement of agricultural plant species through plant breeding. This figure illustrates four germplasm flow models. The large circle, A, symbolizes the genetic diversity found in the world of a single crop species, and the smaller shapes represent the comparative genetic diversity in a limited number of introductions. For successful plant breeding research, a broad genetic base is essential, particularly as breeding objectives become more complex and more demanding.

Introductions are an important source of genetic diversity and may be the sole source of new materials in areas where native stocks do not exist. Such introductions may be deliberate or accidental, traded or freely given, legal or illegal. They may be intended for direct use as a cultivar or for use as a breeding parent.

Syme (1983) suggested four models to describe the use of introductions, namely, the hive-off model (B), the umbilical model (C), the booster model (D) and the symbiotic model (E).

The hive-off model occurs when a sampling of the world gene pool leads to separate evolution of local cultivars. It may occur when a single or a limited number of introductions are made that have undergone considerable selection and breeding work and that are adapted to the local environmental conditions. If subsequent cultivars are developed from these basic introductions in isolation from the world gene pool, elimination of unwanted genotypes leads to a reduction of genetic variation within the breeding population. As additional cultivars are hived-off from this limited genetic base, all cultivars grown in an entire region come to have a narrow, common gene base and are vulnerable to the outbreak of a virulent pest or climatic fluctuation.

In contrast, a new species in agriculture, triticale, can be used as an example of the umbilical model in which the world's germplasm may be sampled repeatedly to provide cultivars directly for use in agriculture in a country where basic genetic stocks of triticale have not been developed. Locally adapted triticale cultivars may be developed with various degrees of local selection from genetic material obtained directly from an international breeding program such as the International Maize and Wheat Improvement Center (CIMMYT) in Mexico.

For important crop species for which there are well-established breeding programs in various parts of the world, the booster model is common. Involved is a local breeding effort of intercrossing, selection, and release, with introduced germplasm providing a periodic boost through the infusion of genes

enemies, kudzu has adapted too successfully and in some regions of the southern United States has caused problems because it has become a major weed (Figure 4.7).

Most countries impose quarantine restrictions on plants and seeds to avoid the inadvertent spread of seed or plant-borne pests and to prevent undesirable species from becoming established.

7. *Related Species*. Since the development of agriculture, humans have used only about 3,000 plant species for food. Of these, about 150 plant species are commercially cultivated on an extensive or major scale at any one time. There are more than 300,000 species of plants in the world of which 20,000 or more may produce desirable edible products. An enormous potential exists for plant breeders to tap the botanical resources of the world for human needs. In addition to their use as food, wild plant materials may be used for livestock feed, medicine, spices, firewood, pulpwood, timber, shelter, pigment, and ornamentals.

N. I. Vavilov, a noted Russian plant explorer from 1916 to 1940, studied many plant species and, as noted earlier in this chapter, developed the concept that centers of diversity indicate centers of origin. Vavilov observed that many plant genera have common features, which he called **parallel variation**, and from these observations formulated the law of homologous variation (Vavilov, 1949-1950). Briefly, this law states that species and genera that are closely related have similar series of heritable variations. Thus, if variation for a trait occurs in a species, one would expect similar variation to occur in related species. For example, most species of *Gramineae* have annual and perennial forms. There is an annual and a perennial bluegrass; cereals have annual and winter habits. Maize, an annual plant grown extensively worldwide, must be planted every year at the expense of labor, erosion, and fossil fuel subsidies. A perennial form of maize offers advantages and disadvantages and needs to be considered carefully.

Teosinte was thought to have annual and perennial forms that might be used to produce a perennial maize cultivar. In 1910, a perennial (*Zea perennis*) tetraploid form was discovered, but differences in chromosome number prevented hybridization with cultivated maize. Because teosinte grows in limited areas of Mexico where it is regarded as a weed, it was feared that perennial diploid forms, if they ever existed, had been eradicated. The subsequent discovery of a perennial tetraploid form did, however, confirm that the law of homologous variation holds true in maize.

In 1977, a 20-chromosome perennial form of *Zea diploperennis* was found in a remote

---

Figure 4.5 continued

influencing desirable attributes. Unlike the hive-off model, which reduces genetic variation, the booster model increases genetic variability.

In addition to the processes of the booster model, the symbiotic model involves the distribution of elite germplasm to world collections and interested researchers. Plant breeders using this model are truly working on a worldwide basis for plant improvement in a manner that is beneficial to all concerned.

It is clear that a much wider range of germplasm is needed for modern breeding programs than for those of the past. Wild species and primitive cultivars that may still exist in undeveloped or remote areas of the world where ancient crop plants were first domesticated are valuable sources of germplasm. To ensure future generations of genetic diversity, collecting and conserving genetic stocks are necessary steps.

A.   B.

Figure 4.6   The plant twining around the grass plant in Photo A and the plant growing in a natural ground cover in Photo B are wild soybeans (*Glycine soja* Sieb and Zucc.). The soybean, *Glycine max* L., Merr., is believed to have originated in China, was domesticated about 1100 B.C., and was first introduced to North America in 1765 (Hymnowitz and Harlan, 1983). An expedition to north China from 1929 to 1931 resulted in the introduction of 4,451 soybean accessions, from which 41 cultivars emerged by direct selection (including landraces) (Hymnowitz, 1984). Subsequent development and distribution of high-yielding soybean cultivars displaced many landraces, which were lost because there was no system for collection and preservation. Fortunately, some of these genotypes were preserved in North America.

Introductions from China formed the genetic material on which the North American soybean industry is based. Since the original introductions, relatively few additional sources have contributed genetic material to present-day soybean cultivars. Genetic uniformity not only increases the hazard of the rapid and unchecked spread of a pest but also is a potential constraint to further genetic improvement. The introduction, preservation, and use of new germplasm now or in the future is a key to continued success in plant breeding.

Introductions of wild genotypes can be important in crop improvement despite the fact that such plants may lack many agronomic traits. Advances may be slow because considerable selection and backcrossing may be required.

Normal fertility of the hybrid between *G. max* and *G. soja* allows for the use of wild soybean resources. The 2n number of both species is 40 chromosomes.

A.   *B.*

Figure 4.7   The vine climbing over entire trees and shrubs in these photos is kudzu (*Pueraria lobata*), photographed in Japan where kudzu originated. Vines 20 m (60 ft.) in length are common, and the plant is so aggressive that in some areas it is regarded as a weed.

The introduction of kudzu to other parts of the world has met with mixed success. In the United States, the plant has passed the point of being considered beneficial (Tanner *et al.*, 1979). The plant is difficult to kill with conventional herbicides, and an abundant reservoir of starch on a thick taproot makes it difficult to kill by burning. On the other hand, a high biomass production makes kudzu a potentially important crop for carbohydrate production.

Introductions of new crop genotypes must be done with caution to avoid situations where they are no longer considered beneficial. The need for flexibility in breeding goals is exemplified by this long-lived, coarse, vigorous plant.

mountainous region of Mexico. The lesson to be learned, based on the law of homologous variation, is that specific forms of variation found in one species can be expected to occur in a parallel species. This holds true even though variation may not be apparent immediately at a particular location or in a specific cultivar.

In birdsfoot trefoil (*Lotus corniculatus*), seed pod shattering is a limitation to successful seed production. A range of different pod characteristics is found in alfalfa (*Medicago sativa*) and in other closely related species. In birdsfoot trefoil, seed pod shattering is a trait shared by closely related species, and therefore the expectation of finding a gene for nonshattering seed pods is more limited in *Lotus* than in *Medicago*. Initially, Vavilov's law established a parallel variability of homologous characters in taxonomically close species (Kupzow, 1975). Experience has shown that the law can be extended to include parallel variability of similar homologous characters in taxonomically remote species. The discovery that *Lotus conimbricensis* (2n = 2x = 12), although sexually incompatible with cultivated trefoil (*Lotus corniculatus*) (2n = 2x = 24), is a source of seed pod indehiscence adds further credence to the law of homologous characters.

Corresponding forms of variation require a persistent and thorough evaluation and may be obtained by the screening of existing genotypes, by plant exploration, or by hybridization and selection.

If a desirable trait is found in a wild species, or one that is distantly related, then it must be crossed to adapted crop plants. Such crosses are frequently sterile, and many crosses have been undertaken in the past solely to determine which species could be hybridized. More recent studies examine the cytogenetic and ultrastructural aspects of hybridization to determine why two particular strains cannot be hybridized. Some hybrids have led to commercial success. Biotechnological methods of circumventing barriers to hybridization may make weed and wild species of even greater potential as genetic resources than in the past.

8. Mutations. A pure line, defined as the progeny of a single homozygous individual, is in the short term genetically stable, but over a long time period, genetic variation may arise by mutation. Mutations are changes in genetic material and are a basic natural phenomenon of life. Naturally occurring variation assembled in a population is the basis of species improvement and as such is a valuable form of genetic variation for plant breeding.

Mutations can be induced by chemical agents or radiations of several types. The only apparent difference between induced and spontaneous mutations is the frequency of occurrence. Mutations may occur as point mutations involving a single gene or as changes in number and structure of chromosomes involving many genes. Mutations can either have very specific effects on a trait or have several interrelated effects on plant characteristics. A single allele may have several possibilities for mutation and may go through a series of mutation events that give rise to an allelic series. Traits of agricultural significance frequently involve a multiple allelic series.

Mutations do not occur entirely at random because certain loci are much more subject to change than others. Mutations may be deleterious or desirable, of practical or of no practical value, of major or minor significance. Their detection, isolation, and testing constitute a major task. Although the majority of mutations are unfavorable to the species and generally have not improved that plant for human use, over the years mutations have provided a source of variation that has been identified, selected, and used to advantage.

9. Interspecific and Intergeneric Hybridization. As the known natural variation within a species is exhausted, plant breeders turn to widely diverse materials to transfer specific desirable genes into a crop or to develop entirely new species (Figure 4.8). Relatively few viable wide interspecific and intergeneric hybrids have been developed because of gametic incompatibility and hybrid breakdown resulting in sterility. After years of selection, many sterility problems have been overcome in the crop triticale, the product of combining genomes from wheat (*Triticum*) and rye (*Secale*). Although triticale has been developed into an agronomically acceptable crop, its commercial role is yet to be defined. Tritordeum, a crop produced by

Figure 4.8   The sunflower plant (*Helianthus* spp.) in this photo offers a prime example of the potential that hybridization among different species of plants offers plant breeders. The genus *Helianthus* contains some 49 species, all native to the Western Hemisphere, adapted to a wide range of environments and exhibiting great diversity of morphological and physiological characteristics. *Helianthus annuus* is the common sunflower cultivated for its **achenes** (seeds), which are high in protein and edible oil.

Interspecific hybrids of *Helianthus* have provided the following advances (Laferriere, 1986):

- Starting in the 1930s, Soviet plant breeders turned to interspecific crosses to obtain rust resistance when sources of resistance in commercial cultivars proved fruitless. Other pest resistance associated with interspecific crosses included protection against downy mildew, stalk rot, mosaic virus, sunflower moth, and other minor pests.
- Interspecific hybridization was necessary in the formation of male-sterile hybrids, which revolutionized the sunflower industry by making the production of high-quality hybrids possible. Cytoplasmic male sterility was obtained by crossing with *Helianthus petiolaris*. Fertility-restorer genes are rare among cultivated *H. annuus* but common among wild populations. By 1976, as much as 80% of the North American crop was produced from hybrids that yielded 20% more than open-pollinated cultivars.
- Desirable agronomic traits have been obtained from wild *Helianthus* species, including genes for cold tolerance, for the elimination of far-red-induced seed dormancy, and for drought tolerance.
- Wild species have been used to increase protein level, oil content, and the balance of essential amino acids.

Photo courtesy Ontario Ministry of Agriculture and Food.

combining wheat (*Triticum turgidum*) with barley (*Hordeum chilense*), is being investigated, but when distant genera are combined, considerable evolutionary development is needed. As barriers to sexual union are overcome, gene banks will play an ever-increasing role in preserving and disseminating sources of genetic variation.

VARIATION

Genetic variations generally can be divided into two categories: those that can be classified into a few distinct classes; and those in which variation is continuous, ranging from one extreme to the other. Variation in the first category is qualitative and variation in the second category is quantitative.

Both qualitative and quantitative genetics follow the same laws of inheritance. In qualitative genetics in which clear dominance and recessive relationships are expressed, it is possible to describe easily distinguished phenotypic classes having a 3:1, 15:1, 9:7, or other discrete ratio in the F_2 generation. When many loci with small individual effects are involved in quantitative inheritance, variation is studied through statistics appropriate for continuous variables, such as the mean value of a population and the spread of the values on either side of the mean, which is called the variance or standard deviation. Variances are calculated as the squared deviations from the mean.

Most of the economically important characters in plants, such as grain yield, root size, quality, protein or oil content, maturity, height, winter hardiness, seedling vigor, and cold tolerance, exhibit a continuous range of expression and are quantitatively inherited. Continuous variation is the result of the involvement of many genes at various loci, which may have a major impact when combined with the environment. Both quantitatively and qualitatively inherited traits are expressed over a wide range of environmental conditions. If the impact of the environment is small, successful selection for multigene traits is possible, but a challenging task in plant breeding is to incorporate both desirable qualitative single gene traits with a desirable array of quantitatively inherited traits.

QUALITATIVELY INHERITED TRAITS

The phenotypic effects of qualitative genes may be observed and followed through the segregation process. Qualitative traits are conditioned by a few genes, possibly one to three, and most of these traits express clear dominance and recessive relationships. Because the environment plays a relatively minor role in most of these traits, the genotype of individuals is expressed faithfully by a phenotypic response, unless there is dominance.

An excellent example of a qualitatively inherited trait is height in sorghum. Sorghum cultivars range in height from 0.6 m to 4.8 m (2 to 15 ft.), with short cultivars desired in North America for direct combine harvesting for grain. Sorghum height is controlled by recessive genes at four loci (Table 4.2). Another qualitatively inherited feature is the classical example of wheat seed color. Simply inherited qualitative traits that have been incorporated easily into

Table 4.2   Gene Combinations for Sorghum Height

Gene Combination	Genotype	Average Height (cm	in.)
Recessive for one gene	Dw_1, Dw_2, Dw_3, dw_4	151	(59)
Recessive for two genes	$Dw_1, Dw_2, dw_3, dw_4;$		
	dw_1, Dw_2, Dw_3, dw_4	100	(39)
Recessive for three genes	$dw_1, Dw_2, dw_3, dw_4;$		
	dw_1, dw_2, Dw_3, dw_4	57	(22)
Recessive for four genes	dw_1, dw_2, dw_3, dw_4	No commercial cultivars	

Note: Four independently inherited genes and a modifying complex were reported by Quinby and Karper (1954) as influencing elongation of the internodes and height in sorghum. Shortening of the internodes is the only visible effect of these recessive genes because anthesis and leaf size are not affected. Tallness is partially dominant. At least one of the four genes is unstable in some strains and reverts to tall frequently enough to be a source of trouble to producers of pure seed. Dwarf cultivars are desired for grain production and direct combine harvesting in North America, where stover is not desired. A knowledge of height factors carried by cultivars is important when the cultivars are used as parents for hybrid sorghum production. Hybridization and selection is necessary to obtain parental lines that will produce dwarf hybrids suitable for combine harvesting.

wheat are dwarf and day-length insensitive genotypes. The inheritance of alkaloid level in tobacco consists of two major recessive genes that control nicotine level.

Other examples of traits that are controlled by qualitative genes include the following six maize endosperm types:

- Floury endosperm containing soft starch used as an easily ground maize by American Indians and given the gene designation *fl fl*.
- Sugary endosperm designated as *su su* and resulting in kernels high in sugar content, such as in sweet maize.
- Yellow and white endosperm designated as *Y-* and *yy*, respectively.
- Waxy endosperm, *wx wx*, which is high in amylopectin and used to make adhesives.
- Amylose endosperm, *al al*, used in the manufacture of sweeteners.
- High-lysine endosperm, *ly ly*, with an increased level of lysine, an essential amino acid.

Sweet clover (*Melilotus alba*) plants may contain coumarin, an organic compound with a bitter, stinging taste and a vanilla-like odor. The presence of coumarin is undesirable because it renders sweet clover relatively unpalatable to herbivorous animals. In addition, when certain conditions prevail in haying or ensiling the plant, an anticoagulant, dicoumarol, is produced. Genetic studies to develop coumarin-deficient cultivars have revealed that the level of coumarin is conditioned by a single dominant qualitative gene *Cu-*.

SELECTION FOR QUALITATIVE TRAITS

Qualitative traits usually can be identified readily in plants. In the situation where individuals with a recessive genotype are desired, selection of the recessives and their subsequent use as parents for the next generation will maximize the frequency of the desirable genotype. The initial population should be large enough to ensure that recessive individuals will be present. For example, when two parents homozygous for recessive and dominant alleles are crossed, 25% of the F_2 plants are expected to be homozygous recessive. If four loci are segregating, however, only one of 256 individuals is expected to be recessive at all loci. Once the desirable individual has been identified and if selfing occurs, the next generation of plants will all carry the desired trait. Selection for a recessive gene requires one generation of selection to be expressed in the phenotype in the homozygous state.

Selection for a dominant gene is more complicated. Assume that flower color is a simply inherited trait in which R conditions red color and the recessive rr conditions white flower color. In the self-fertilized petunia, a plant breeder would prefer to select for red flowers in the F_2 generation, but one-third of the plants selected would be homozygous for red flowers (RR) and two thirds would be heterozygous (Rr). The self-fertilized offspring of each individual plant are grown separately in the next generation. If the offspring of an individual plant, referred to as a family, are all red, then the parent must have been homozygous. Similarly, if a family exhibits both red and white flowers, then the parent was obviously heterozygous and would be discarded. Two generations are required to eliminate the recessive allele from the population: one generation of selection and one generation for a progeny test.

In a cross-fertilized species in which it is difficult to obtain seed by self-fertilization, progress in selection for the dominant gene is often much slower. Such species as red clover, rye, birdsfoot trefoil, and reed canary grass set little or no seed after self-pollination. If heterozygous parents for flower color (Rr) are selfed, selection in the F_2 generation would produce a genotypic frequency in the population of 0.25 RR, 0.50 Rr and 0.25 rr. If every individual showing the R- phenotype were selected, then the selected population would consist of 0.33 RR and 0.67 Rr. If every individual that exhibits the R- phenotype is mated to a heterogeneous population, then the frequency of homozygous RR individuals in the next generation will be increased, as shown in Table 4.3. Note that even after six generations of selection and mating selected individuals, only 77% of the population is homozygous for the dominant gene.

QUANTITATIVE CONSIDERATIONS

Generally, quantitative traits depend on the collective interaction of numerous genes or polygenic systems. Each gene separately has a small effect on phenotype, so that variation can be observed only for groups or populations, not for individuals. The expression of quantitative genes is probably also influenced by the environment. The plant breeder aims to quantify the impact of genetics and environment.

Phenotypic variation (V_p) can be partitioned into three components: genetic (V_g),

Table 4.3 Selection for Qualitative Traits in a Cross-fertilized Crop

Generation	Frequency		
	RR	Rr	rr
0	0.25	0.50	0.25
1	0.44	0.44	0.12
2	0.56	0.38	0.06
3	0.64	0.32	0.04
4	0.69	0.28	0.03
5	0.73	0.25	0.02
6	0.77	0.22	0.01

environmental (V_e), and the interaction of genotype and environment (V_{gxe}). It can be expressed as

$$V_p = V_g + V_e + V_{gxe}$$

Variation found among homozygous and heterozygous clones vegetatively propagated from one individual plant is due entirely to environmental factors (V_e). Among heterozygous plants the portions of variation due to environment and to genotype are more difficult to distinguish. To help plant breeders distinguish between genotype and environmental influences, a **heritability value** can be determined using the ratio of genotypic to phenotypic variation. Expressed algebraically, heritability is V_g/V_p x 100 or

$$\text{Heritability (broad sense)} = \frac{V_g}{V_g + V_e + V_{gxe}} \times 100$$

This equation may be simplified to

$$\text{Heritability (broad sense)} = \frac{\textit{Total variation due to genotype}}{\textit{Total variation due to phenotype}} \times 100$$

Heritability values that give the proportion of the total phenotypic variation that is due to all genetic factors are known as **broad sense heritability**.

Depending on the type of behavior the gene expresses, genotypic effects (V_g) can be subdivided into components: additive (V_a), dominance (V_d), and epistatic (V_z). The additive component describes the difference between homozygotes at any gene locus, dominance describes the interaction of alleles at the same gene locus (intragenic interaction), and espistatic action describes the interaction of nonalleles (intergenic interaction).

Additive gene action implies allelic performance that is invariable regardless of other alleles at the same locus. Gene *a*, for example, has the same relative value in homozygote *aa* as in heterozygote *Aa*.

Dominance variation is composed of two portions—that due to homozygous alleles (i.e., additive) and that due to heterozygous genotypic values. Allele *a* has one value as homozygous *aa*, but in combination with *A* in the heterozygous condition *Aa*, the allele *a* has the same value as the dominant allele. The effect of additive and dominance gene action at a single genetic locus for plant height is illustrated in Table 4.4.

Genes that act in an additive manner are preferred for selection because the superior

Table 4.4 Effect of Additive and Dominance Gene Action at a Single Genetic Locus for Plant Height (cm)

Genotype	Additive [a]	Dominance [b]
AA	90	90
Aa	60	90
aa	30	30
Mean of progeny *AA* x *aa*	60	90
Mean of progeny *Aa* x *Aa*	60	75
Mean of progeny *Aa* x *aa*	45	60

[a] Each *A* contributes 45 units, each *a* contributes 5 units.

[b] Homozygotes *AA* and *aa* have the same genotypic value as the additive model; genotypic value of *AA* = *Aa*.

Table 4.5 A Two-gene Model Showing Additive and Dominant Gene Action

	Genotypic Value	
Genotype	Additive	Dominant
	M^a+A + B	A+ B + M
AABB	10 + 6 + 4 = 20	20 = 6 + 4 + 10
AABb	10 + 6 + 2 = 18	20 = 6 + 4 + 10
AaBB	10 + 3 + 4 = 17	20 = 6 + 4 + 10
AAbb	10 + 6 + 0 = 16	16 = 6 + 0 + 10
AaBb	10 + 3 + 2 = 15	20 = 6 + 4 + 10
Aabb	10 + 3 + 0 = 13	16 = 6 + 0 + 10
aaBb	10 + 0 + 2 = 12	14 = 0 + 4 + 10
aabb	10 + 0 + 0 = 10	10 = 0 + 0 + 10

[a] Base value of 10 units.

phenotype will breed true in the next generation, provided the environment allows for phenotype to be expressed. Dominance would result in selection of both homozygous genotypes, which will breed true, and heterozygous genotypes, which will not breed true in the next generation.

The example just cited is identical to qualitative inheritance for a single gene with incomplete dominance and with dominance, respectively. A two-gene model showing additive and dominant gene action is presented. In this example, alleles *A* and *a* are equal to 6 and 3 units, respectively, and alleles *B* and *b* are equal to 4 and 2 units, respectively. In the case of dominance, the genotypic value of *AA* is equal to *Aa* and *BB* is equal to *Bb*. The genotypic values are given in Table 4.5. Note that the two genetic loci have different relative values with respect to their genotypes. On average, gene *A* contributes more than gene *b* to the genotypic value of an individual. Furthermore, in this particular model, the effect of an allele of gene *A* acts independently of an allele of gene *b*.

In a list of quantitatively inherited genes, many genes, each consisting of a number of alleles, can be included. Each gene could confer a relative genotypic value and the allelic series would also confer a relative value on the genotype within the framework of that genetic locus. In both the single and two-gene models, additive genetic variation offers the most promise for obtaining an array of favorable alleles of the genetic loci exposed to selection. Additive gene action results in the more rapid fixation of favorable alleles.

These illustrations of additive and dominance genetic variation are similar to the situation involving qualitatively inherited traits. In quantitatively inherited traits, both the degree of dominance and the number of genes that actually contribute to a trait differ from the examples given. Figure 4.9 illustrates the degree of dominance and is readily expanded to include a number of genes that control a specific trait. Emphasis is placed on the fact that in quantitative inheritance, measurements are made on many individuals in a population. These measurements would produce a frequency curve.

Figure 4.9 can be extended to include multigenic traits if $A_1 A_1, A_2 A_2$ indicate parents with many different genes and the $A_1 A_2$ genotype represents the performance of a hybrid between these two parents. In this case, a multigenic trait consisting of many genetic loci can exhibit all of the combinations mentioned in Figure 4.9, i.e., no dominance, incomplete, complete, or overdominance. An apparent lack of dominance for a quantitatively inherited trait does not imply that all of the component genes that give rise to the trait also show no dominance.

The degree of dominance can be estimated from the variances expressed as the ratio of dominance variation to additive variation, or V_d/V_a. Dominance variation is a more conspicuous feature of cross-fertilized crops, but even in these crops the average degree of dominance has a usual value between 0.05 and 0.20. A trait exhibiting complete dominance would have an average degree of dominance of 1.0.

A striking example of a high ratio of dominance variation to additive variation occurs for grain yield in F_1 maize hybrids. Specific maize hybrids have exhibited the unusually high value of 1.64 for the degree of dominance. However, most estimates have values below 1.0.

In quantitative genetic terminology, heterosis, which has been reported for a wide range of crop species, is usually measured as the superiority of the F_1 progeny over the average of its parents. Heterosis can be expressed in both self- and cross-fertilized crops. In commercial terms, heterosis is more meaningful if expressed as the superiority over the highest-yielding parent since hybridization between a high- and low-yielding parent may result in considerable measurable heterosis, which, however, is commercially unacceptable. For some traits, the expression of heterosis, is influenced greatly by the magnitude of genetic differences. As outlined previously, there are three possible genetic causes of heterosis: partial to complete dominance, overdominance, and **epistasis**.

Epistatic variance (V_n) can be defined as arising from interloci interactions, and dominance variance from intralocus interactions. The aggregate genotypic value of a quantitative trait, conditioned by genes at many loci, may or may not be simply an additive combination of the genotypic values for individual loci. Epistasis occurs when genes interact to produce a genotypic value that is equal to the sum of additive plus dominance effects of individual loci. When the total variation over loci is larger than the summation of additive and dominance variances for individual loci, the differences will be variances caused by epistasis (Table 4.6).

Since additive genetic variation is the portion of genetic variation for which selection is most effective and can be fixed in plant breeding, the proportion of additive genetic variation in total phenotypic variation becomes very important. Therefore, the most important component

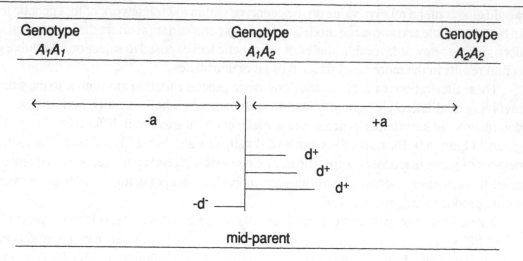

| Genotype A_1A_1 | Genotype A_1A_2 | Genotype A_2A_2 |

Figure 4.9 Dominance variance expressed as portion due to deviation from mid-parent value.

Note that -a and +a are equal to the additive deviation due to performance of homozygotes, and d is equal to the dominance deviation expressed as deviation from the mid-parent value.

The genotype value of homozygotes represented as a is the additive portion of the genetic deviation. As previously illustrated, homozygous alleles A_2A_2 have more value than A_1A_1, which is denoted as +a. If the heterozygote between two homozygous parents has a value that is the mid-parent or average between two homozygous parents, then the requirements for an additive genetic model are met.

Dominance of a genetic locus is detected on the basis of deviation from mid-parent values that would be expected in additive gene action. If the mid-parent value is greater than expected under the additive model, then d is positive. When the A_1A_2 genotype has the same genotypic value as the most favorable allele (A_2A_2), then dominance is said to be complete. Dominance that results in mid-parent values between A_1A_2 and A_2A_2 is referred to as incomplete dominance, and mid-parent values in excess of A_2A_2 are referred to as overdominance.

of the genotypic variation in terms of successful selection is V_a, or additive variation. The ratio of additive genetic variation to total phenotypic variation is referred to as **narrow sense heritability,** shown as

$$\text{Heritability (narrow sense)} = \frac{V_a}{V_p}$$

SELECTING PLANTS FOR QUANTITATIVELY INHERITED TRAITS

Hybridization and selection among segregating genotypes are essential features of any plant breeding program. Research in quantitative genetics has as one of its major objectives an understanding of the genetic consequences of plant breeding activities, and this was reviewed by Moll and Stuber (1974).

Plant selection is the most potent force available to the plant breeder for changing gene frequency. The aim of selection is to shift the population to a value that is more desirable than

Table 4.6 Plant Height Illustrating Epistatic Gene Action

Genotype	Plant Height (cm)	
AABB	100	——
AABb	80	——
AAbb	60	——
AaBB	60	...
AaBb	60	...
Aabb	30	...

Note: Additive gene action, denoted by the dashes following the 100, 80, and 60 cm height values, occurs at the *B* locus in the presence of *AA*. Dominance, denoted by the dots following the 60, 60, and 30 cm height values, occurs in the presence of *Aa*. The genotypic value of the *B* locus is affected by the alleles at the *A* locus and vice versa; and this is epistasis. Selection of plants where epistatic variance is occurring is difficult since particular allelic combinations at several loci condition performance. Performance in the next generation likely will deviate because of segregation and/or recombination. Where dominance variation or epistatic variance is important, its beneficial effects can be maximized but never fixed. Additive gene action, however, can be fixed. A common example in agriculture is single-cross maize hybrids in which the optimum performance achieved in the F_1 hybrids maximizes dominance and perhaps epistatic variance. If these F_1 hybrids are mated to produce double crosses, performance declines. Predicting progeny performance, if epistasis and dominance play a major role, is difficult and usually requires a progeny test.

the mean value of the parent population. For plant breeding to be effective, there must be phenotypic variation for the selected feature, and some of the variation must be genetic and not due entirely to the environment (Figure 4.10).

Plant breeders are interested in changing gene frequency, especially of desired traits, and in maintaining or increasing variation on which selection can be effective. Selection considerations should include selection response, selection intensity, the selection index, selection criteria, and the selection differential, all of which are described in the following paragraphs of this section.

The progress made by selection is measured by selection response. Selection response is the difference between the mean values of the progeny of one generation and the mean of the next generation. If the average height of 1,000 maize plants, for example, was 200 cm (79 in.) and if 100 of these were selected as parents with an average height of 240 cm (94 in.), the difference of 40 cm (16 in.) is called the selection differential. If these 100 maize selections are intercrossed and if their progeny in the subsequent generation average 225 cm (89 in.), the selection differential is 25 cm (10 in.), or 25 divided by 200 = 12.5%.

The progress on the response to selection is determined, in part, by the selection intensity, which is the number, or proportion, of individuals that are selected as parents for the next generation. When only a few individuals are selected, some good genotypes may be lost. Although this may result in good progress for a few generations, a population may eventually develop with little genetic variability.

Selection of many individuals representing a relatively high proportion of the population will mean that some genetically inferior or undesirable individuals will probably be selected.

Figure 4.10   The individual grass plants in this photo vary within the population. The challenge to the plant breeder is to determine if such differences are due to environment or genotype.

Genotypic variation can be divided into qualitative and quantitative categories with considerable overlap. Qualitatively inherited traits are those in which observed differences can be separated into discrete categories, or classes, that do not overlap. Pollen color in these grass plants, which is readily identified and which is not influenced dramatically by the environment, represents a qualitative character.

Quantitatively inherited traits such as hardiness, yield, longevity, or quality aspects cannot be divided into discrete classes. A gradation in such factors means considerable overlap from one class to another. Variation due to genotypes results from the interaction of many genes and may be expressed by additive, dominance, or epistatic gene action.

The proportion of total variation caused by the genotype is heritable and can range from a value of one, where all variation is genetic, to zero, where all variation results from the environment. Many important traits exhibit a low level of heritability, and progress through plant breeding is slow. As long as genetic variability exists, however, plant breeders will be attempting improvement.

Initially, this may result in slow progress, although intermating of selected individuals may result in new gene combinations that produce individuals superior to those observed previously. A second pitfall of making too many selections is that the program size is extended and becomes unmanageable; costs become prohibitive. The optimum selection intensity varies with the objectives of the program.

To select for several traits at the same time, a selection index can be calculated. Simultaneously improving more than one trait may prove to be difficult because some plants are superior for one trait but may be average or below average for another trait. The breeder may arbitrarily decide on a minimum or maximum level for each trait and discard plants that fail to meet the desired level. Another method is to select for one trait for a few generations and then for another trait for the next few generations. For each individual, a value is determined based upon the index used. Individuals with superior indices are selected as parents.

The selection index can be determined in a number of ways, the most common of which is to measure each trait and weight each measurement according to its relative importance or economic value. In a progeny test of birdsfoot trefoil, for example, traits measured included seedling vigor, seed yield, and forage yield. A selection index was formulated by summing the data as follows:

Seedling vigor x 100, or 4.52 x 100	=	452
Seed yield x 100, or 3.48 x 100	=	348
Forage yield in first year	=	2,050
Forage yield in second year	=	3,600
	Total	6,450

The value of 6450 is that of one progeny but could be compared to other progeny and the top individuals preserved as parents. Other techniques are available to derive selection indices, based possibly upon heritabilities and the relationships among the traits being considered.

Winter hardiness is a quantitatively inherited trait in which variance values can be used by the plant breeder (Figure 4.11).

Quantitative genetics is dependent upon many genes that influence individual plant characteristics, and an adequate population size must be used if specific criteria are to be selected successfully. Estimates of additive, dominance, and epistatic gene action are dependent on large populations (Comstock and Robinson, 1948). In cross-fertilized crops, an adequate amount of seed for each mating is produced easily. In self-fertilized crops, however, hybridization may be achieved by hand emasculation and pollination, which are costly and time consuming. The production of sufficient seed for replicated evaluation trials using conventional testing techniques precludes the use of many mating designs, such as **diallel crosses** in self-fertilized species. The limited number of parents normally utilized in studies with self-fertilized crops restricts the inferences that can be made.

Estimation of genetic variances requires some systematic mating scheme in which different types of relatives are developed. Results from numerous studies in maize led Moll and Robinson (1966) to suggest that 256 progenies from the reciprocal Jarvis x Indian Chief would be a minimum number for variance component estimations. An improved mating design for estimating genetic variances for self-fertilized crops was presented by Stuber (1970). Inbred relatives are generated by bulk selfing each of the F_1 (non-inbred) progenies in conventional mating designs. In this manner, production of F_1 seed should not be a serious handicap, as it is in many designs, but the number of crosses made should be large enough to sample the parental population adequately.

In conventional breeding programs, selection is generally conducted in a single environment (Figure 4.12). Attempts to assess progeny for yield and adaptation to various environmental conditions are not usually undertaken until later generations when sufficient seed is available to grow trials over several seasons at representative locations to achieve a selection differential.

Most breeders are anxious to produce cultivars adapted to a wide area that have the potential not only to yield well at the breeding station but also yield reliably each season. Cultivars that provide high-yield potential overall and that have low sensitivity to changes in the environment are desirable. To achieve these objectives, Hughes, Westcott, and Sharp (1987) jointly selected winter wheat families for high yield and low response to two important environmental variables.

Figure 4.11 Photo of a winter wheat plot severely damaged by winter injury in comparison with the adjacent plots. By using freezing tests under controlled conditions, Sutka (1984) determined that both the general combining ability and the specific combining ability were significant, indicating additive and nonadditive gene action in the inheritance of frost resistance. In the reciprocal crosses of a ten-parent diallel cross, a significant general reciprocal effect was found, indicating maternal cytoplasmic effects.

A graphical covariance-variance analysis indicated partial dominance of frost sensitivity. Frost-sensitive cultivars had the largest number of dominant genes, whereas frost-resistant cultivars had the highest number of recessive genes. The size of the additive component of variation was greater than that of the dominance component, and the overall measure of the degree of dominance was smaller than one, so average dominance was incomplete.

The values of narrow and broad heritability were high in this study of ten wheat cultivars. The results support the possibility of selecting for frost resistance and obtaining relatively rapid improvement in this character.

These families were grown in three contrasting environments. The procedure followed was that of Brumpton, Boughey, and Jinks (1977), whereby high yield and low sensitivity were selected simultaneously by choosing lines with a high-yield total combined with a small-yield difference from the trial mean yield under early- and late-seeded conditions. The assessment of advanced lines showed that three genotypes emerged, which were as high-yielding across the three environments as the highest-yielding check cultivar. Simultaneous yield and sensitivity selection exerts a much lower selection pressure than yield alone, a situation noted by Brumpton, Boughey, and Jinks (1977). It is possible that such a method of breeding is desirable to achieve yield stability.

A.

B.

Figure 4.12 The trees in Photo A may be found to vary in phenotype because of genotype and environmental modifications. If all the trees under test were measured for selected parameters, a range in phenotypes would be found. Each tree represents a single genotype, and distinct classes of plants (small, medium, large) might be established.

Because fluctuations in phenotypic expression are also influenced by the environment, class separations become less and less and when plotted are finally represented as a continuous set of values on the horizontal axis. Progress in plant breeding is possible if differences are genetic. Propagating the most desirable individuals selected from Photo A, based on their phenotype, runs the risk that the effect of environmental variation upon phenotypic expression may mask genotypic differences.

The cereal genotypes to be compared in Photo B are grown in close proximity and are affected by only a small portion of the variation in soil and microclimate over the entire field.

Quantitative genetics has the role of providing a genetic basis for the development of effective and efficient selection techniques to meet specified objectives.

Photo A courtesy Ontario Tree Improvement and Forest Biomass Institute, Ministry of Natural Resources, Maple, Ontario.

QUANTITATIVE GENETICS AND EVOLUTION

Plants respond to changes in the environment, especially to factors that affect survival, by phenotypic changes. Such changes represent a degree of plasticity within the existing genotype. Long-term evolutionary adjustments, however, involve changes in allelic frequency resulting from natural selection following hybridization or mutation.

The genetic constitution of a plant represents a highly integrated system. Any mutation or gene recombination must not only have a favorable effect on a particular aspect but must confer increased fitness in a number of genetic constitutions. For evolution to occur by natural selection, a sizable store of heritable variation must exist in any population. In general, natural plant populations have a great deal of heritable variation influenced by many quantitative genes.

Plant breeding is a form of evolution that is directed by humans rather than nature.

NUMBER OF GENES FOR QUANTITATIVE INHERITANCE

The extensive literature on plant genetics suggests that the number of genes controlling quantitative traits is very large. Genes are assigned to characters partly for methodological reasons. Even in the simplest case, a character may be regarded as the end result of the interactions of many biosynthetic pathways. For red flower color to be expressed, for example, a plant with roots, stems, leaves, conducting tissues, and various organs must be developed. There are numerous pathways involved for gene expression, and it is inconceivable that a single phenotypic character is determined exclusively by a single gene. Such a statement does not deny that in many instances a particular gene has a very major effect on a specific character.

Quantitative genetics may be viewed as a system whereby most if not all genes are integrated. Random inbred lines produced by doubled haploidy and single-seed descent were used by Powell, Caligari, and Jinks (1985) to estimate the number of genes or effective factors controlling quantitative characters in barley. Estimates of the number of genes involved in a specific character may be used to predict the number of plants expected to exceed the parental range or any given standard. The number of effective factors is generally underestimated and may be considered a unit only in a temporary sense. As such, estimates of the number of genes or effective factors may be of little value in plant breeding.

COMPARISON OF GENETICALLY UNIFORM POPULATIONS WITH SEGREGATING POPULATIONS

In a classic paper on genetics, East (1916) selected two true-breeding cultivars of a wild species of tobacco (*Nicotiana longiflora*) to study corolla tube length. The two parents differed in corolla tube length, a feature that is relatively unaffected by environment, and both parents were long-term inbreds, so that they were homozygous.

To determine heritability for corolla tube length, estimates of genetic variation (V_g) and phenotypic variation (V_p) may be obtained from the data in Table 4.7. Phenotypic variation (V_p) consists of both genetic variation (V_g) and environmental variation (V_e). An estimate of environmental variation (V_e) can be obtained by examining the variances of Parent 1, Parent 2, or the F_1 since individuals within each of these populations are identical in that they all share in some genotype. Any variation, therefore, must be environmental.

Individuals in the F_2 population will vary because of environmental variation as well as genetic variation associated with gene segregation, i.e., $(V_{F2}) = V_g + V_e$. The genetic variation (V_g) can be estimated by subtracting the environmental variation of a genetically fixed population (Parent 1, Parent 2, or F_1) from the genetically segregating population of the F_2 of the cross between the two parents, provided all of these populations are grown in the same environment. This relationship is shown by

$$V_g = \frac{V_{F2} - (V_{p1} + V_{p2} + V_{F1})}{3}$$

where V_{F2}, V_{p1}, V_{p2} and V_{F1} are the variances of the respective populations. Only the data from Parent 1 and Parent 2 obtained in 1912 can be used since only the F_2 generation was grown in 1912. Inclusion of other variances is not possible because they represent a different environment.

The estimate of genetic variance and heritability can be calculated from the data in Table 4.7 as follows:

$$\textit{Genetic variance } (V_g) = \frac{F_2 \textit{ variance} - (Parent\ 1 + Parent\ 2\ variance)}{2}$$

$$\textit{Genetic variance } (V_g) = \frac{34.93 - (4.00 + 4.97)}{2}$$

$$= 30.45$$

$$\textit{Heritability} = \frac{\textit{Genetic variance } (V_g)}{\textit{Phenotypic variance } (V_p)}$$

$$= \frac{30.45}{34.93} \times 100$$

$$= 87.2\%$$

Genetic variation accounts for 87% of the variation observed and is a more precise estimate of the proportion of the genetic variation present in the *Nicotiana* parents and their offspring than the birdsfoot trefoil example in Table 4.8. Selection for different corolla tube lengths in the F_2 population should be very successful, and, in fact, selection for widely divergent corolla tube lengths was later achieved from this population.

This system can be used to partition further genetic variance into additive (V_a) and dominance components (V_d). Partitioning genetic variance (V_g) is based on the relationship of the additive and dominance for a two-gene model and the expectations for mean squares for specific generations from a cross between homozygous parents. A number of assumptions are required, including specific mating plans that permit partitioning of the genotypic variance. One example of this technique is illustrated in Table 4.9, based on data presented by Allard

Table 4.7   Analysis of Corolla Tube Length in Two Parental Inbred Cultivars of *Nicotiana longiflora* and the F_1 and F_2 Generations from the Cross

Generation	Year	Number of Measurements	Mean Corolla Length (mm)	Variance
Parent 1	1911	125	40.5	3.06
Parent 1	1912	49	40.6	4.00
Parent 2	1911	88	93.2	5.24
Parent 2	1912	57	93.4	4.97
F_1	1911	173	63.5	8.52
F_2	1912	44	68.8	34.93

Source: Adapted from East, 1916, p. 166.

Table 4.8   Mean Seed Yields and Analysis of Variance for a Four-replicate Randomized Complete-block Design of Fourteen Birdsfoot Trefoil Cultivars

Cultivar	Mean Yield (kg/ha)
Leo	255
Cree	25
Recurrent selection 4	297
Recurrent selection 3	178
Selection RS3	1954
Carroll	308
KO$_4$	334
Mirabel	154
Empire	17
NC-83	229
Kentucky ecotype	146
Wallace	32
Maitland	366
Viking	143

Analysis of Variance

Source	Degree of Freedom	Mean Square
Replications	3	5,269
Cultivars	13	52,533 [a]
Error	39	6,683
Total	55	17,444

[a] Significant at the 0.01 level F test.

Table 4.9 Mean and Variance Values for Days to Heading in Wheat from an Arbitrary Date, and Calculations of Genotypic Variance and Heritability Estimates

	Parent		Generation		Backcross [a]	
Component	*1*	*2*	*F_1*	*F_2*	*1*	*2*
Mean	13.0	27.6	18.5	21.2	15.6	23.4
No. of plants	59	148	171	552	326	314
Variance	11.0	10.3	5.2	40.4	17.4	34.4

[a] B_1 = backcross of F_1 to P_2; B_2 = backcross of F_1 to P_2.

Calculations:

I. Expected components of variance in F_2 and backcross generations.

$$V_{F2} = V_a + V_d + V_e$$

$$V_{B1} + V_{B2} = V_a + 2V_d + 2V_e$$

$$\text{Environmental variance} = \frac{V_{p1} + 2V_{p2} + 2V_{F1}}{3}$$

$$= \frac{11.0 + 10.3 + 5.2}{3}$$

$$= 8.83$$

II. Additive variance obtained by simultaneous equations.

$$2V_{F2} = 2V_a + 2V_d + 2V_e$$

$$V_{B1} + V_{B2} = \frac{V_a + 2V_d + 2V_e}{V_a}$$

$$V_a = 2V_{F2} - (V_{B1} + V_{B2})$$

$$V_a = 2 \times 40.4 - (17.4 + 34.4)$$

$$= 80.8 - 51.8$$

$$= 29.0$$

III. Dominance variance (V_d) obtained by simultaneous equations.

$$2V_{F2} = V_a + 2V_d + 2V_e$$

$$V_{B1} + V_{B2} = V_a + V_d + V_e$$

$$= \frac{V_d + V_e - (V_{p1} + V_{p2} + F_1)}{3}$$

Table 4.9 continues overleaf

Table 4.9 continued

$$\text{Dominance variance} \quad = \quad \frac{(V_{B1} + V_{B2}) - V_{F2} - (V_{p1} + V_{p2} + F_1)}{3}$$

$$= \quad (17.4 + 34.4) - 40.4 - (11.0 + 10.3 + 5.2)$$

$$= \quad 51.8 - 40.4 - 8.83$$

$$= \quad 2.47$$

IV. Broad-sense heritability (h_B).

Broad-sense heritability $\quad = \quad \dfrac{\textit{Genotypic variance}}{\textit{Phenotypic variance}}$

$$H_B \quad = \quad \frac{\textit{Additive variance + dominance variance}}{\textit{Additive variance + genotypic variance + environmental variance}}$$

$$= \quad \frac{29.0 + 2.47}{29.0 + 2.47 + 8.83} \; \text{x } 100$$

$$= \quad 78.1\%$$

V. Narrow-sense heritability (h_n).

$$H_n \quad = \quad \frac{\textit{Additive variance}}{\textit{Phenotypic variance}}$$

$$= \quad \frac{V_a}{V_a + V_d + V_e}$$

$$= \quad \frac{29.0}{29.1 + 2.47 + 8.83}$$

$$= \quad 72.2\%$$

Source: Adapted from Allard, 1960, p. 106.

(1960, p. 106). In these data for heading time of wheat, genetic variation accounted for approximately 80% of the phenotypic variation; additive variance was of much greater magnitude than dominance variance. Broad-sense and narrow-sense estimates of heritability are similar.

The limitations to this type of genetic analysis are the assumptions that there is no linkage, which may not always be valid, that there is no epistasis, and that parents are homozygous. Furthermore, estimates of additive and dominance variances in such experiments are subject to rather large errors; hence, negative estimates of some components may be obtained. The mating designs involving F_1, F_2, and **backcrosses** 1 and 2 required to obtain the expected variances are specific and require a large expenditure of labor and space.

REFERENCES

Allard, R. W. 1960. *Principles of Plant Breeding.* John Wily and Sons, New York. 485 pp.

Astley, D. 1985. Conserving Genetic Resources in Vegetables. *Span* 28:75.

———. 1987. Genetic Resource Conservation. *Experimental Agriculture* 23:245-257.

Brumpton, R. J., H. Boughey, and J. L. Jinks. 1977. Joint Selection for Both Extremes of Mean Performance and of Sensitivity to a Macro-Environmental Variable. I: Family Selection. *Heredity* 38:219-226.

Ceccarelli, S., S. Grando, and J.A.G. Van Leur. 1987. Genetic Diversity in Barley Landraces from Syria and Jordan. *Euphytica* 36:389-405.

Comstock, R. E., and H. F. Robinson. 1948. The Components of Genetic Variance in Populations of Biparental Progenies and Their Use in Estimating the Average Degree of Dominance. *Biometrics* 4:254-266.

Cox, T. S. 1991. The Contribution of Introduced Germplasm to the Development of U.S. Wheat Cultivars. In *Use of Plant Introductions in Cultivar Development, Part I,* edited by H. L. Shands and L. E. Wiesner. CSSA Special Publication No. 17. Crop Science Society of America, Madison, Wisconsin, pp. 25-47.

Crisp, P., and D. Astley. 1985. Genetic Resources in Vegetables. In *Progress in Plant Breeding I,* edited by G. F. Russell. Butterworths, London, pp. 281-310.

East, E. M. 1916. Studies on Size Inheritance in *Nicotiana. Genetics* 1:164-176.

Harlan, J. R. 1971. Agricultural Origins: Centers and Noncenters. *Science* 174:468-474.

———. 1976. The Plants and Animals that Nourish Man. *Scientific American* 235:89-97.

Hawkes, J. G. 1985. *Plant Genetic Resources: The Impact of the International Agricultural Research Centers.* Study Paper No. 3. Consultative Groups on International Agricultural Research (CGIAR). World Bank, Washington, D.C. 115 pp.

Hawkes, J. G., and H. Lambert. 1977. Eucarpia's Fifteen Years of Activities in Genetic Resources. *Euphytica* 26:1-3.

Hughes, W. G., B. Westcott, and P.L. Sharp. 1987. Joint Selection for High Yield and Low Sensitivity in Winter Wheat (*Triticum aestivum* L.). *Plant Breeding* 99:107-112.

Hymnowitz, T. 1984. Dorsett-Morse Soybean Collection Trip to East Asia: 50 Year Retrospective. *Economic Botany* 38:378-388.

Hymnowitz, T., and J. R. Harlan. 1983. Introduction of Soybean to North America by Samuel Bowen in 1765. *Economic Botany* 37:371-379.

Kloppenburg, J., Jr., and D. L. Kleinman. 1987. The Plant Germplasm Controversy. *BioScience* 37:190-198.

Kupzow, A. J. 1975. Vavilov's Law of Homologous Series at the Fiftieth Anniversary of its Formulation. *Economic Botany* 29:372-379.

Laferriere, J. E. 1986. Interspecific Hybridization in Sunflowers: An Illustration of the Importance of Wild Genetic Resources in Plant Breeding. *Outlook on Agriculture* 15:104-109.

Moll, R. H., and H. F. Robinson. 1966. Observed and Expected Response in Four Selection Experiments in Maize. *Crop Science* 6:319-324.

Moll, R. H., and C. W. Stuber. 1974. Quantitative Genetics: Empirical Results Relevant to Plant Breeding. *Advances in Agronomy* 26:277-313.

Plucknett, D. L., N.J.H. Smith, J. T. Williams, and N. Murthi Anishetty. 1985. *Gene Banks and the World's Food Supply*. Princeton University Press, New Jersey. 247 pp.

Powell, W., P.D.S. Caligari, and J. L. Jinks. 1985. The Usefulness and Limitations of Estimating the Number of Genes in a Barley Breeding Programme. *Journal of Agriculture Science*, Cambridge 105:285-290.

Quinby, J. R., and R. E. Karper. 1954. Inheritance of Height in Sorghum. *Agronomy Journal* 46:211-216.

Quisenberry, K. S., and L. P. Reitz. 1974. Turkey Wheat: The Cornerstone of an Empire. *Agricultural History* 48:98-112.

Shulman, S. 1986. Seeds of Controversy. *BioScience* 16:647-651.

Stuber, C. W. 1970. Estimation of Genetic Variances Using Inbred Relatives. *Crop Science* 10:129-135.

Sutka, J. 1984. A Ten-Parental Diallel Analysis of Frost Resistance in Winter Wheat. *Zeitschrift fur Pflanzenzuchtung* 93:147-157.

Syme, J. R. 1983. Working with the World of Plant Improvement. *Journal of the Australian Institute of Agricultural Science* 50:181-190.

Tanner, R. D., S. S. Hussain, L. A. Hamilton, and F. T. Wolf. 1979. Kudzu (*Pueraria lobata*): Potential Agricultural and Industrial Resource. *Economic Botany* 33:400-412.

Vavilov, N. I. 1949-1950. *The Origin, Variation, Immunity and Breeding of Cultivated Plants*. Translated from the Russian by K. S. Chester. Chronica Botanica 13. Chronica Botanica Company, Waltham, Massachusetts. 364 pp.

CHAPTER 5

Statistical Applications
and Field Plot Technique
in Plant Breeding

RESEARCH METHODS

Plant breeding is a form of research that should be undertaken as methodically and as scientifically as possible to maximize progress. Science refers to systematized knowledge, and scientific research refers to systematized procedures to acquire knowledge. Statistical analysis offers plant breeders a valuable tool for the collection, classification, and analysis of data. It offers a means of determining if variation among measurements is due to random variation associated with biological data, to error in measurement, to environmental variables, or to meaningful variation resulting from genetic differences. In order to conduct meaningful scientific research, experiments must be designed carefully.

Many books have been written about statistical methods and procedures to be followed to account critically for random errors present in biological measurements. The statistical method is not a research method but a tool to aid in analyzing data. Caution must be exercised to ensure that research methods follow procedures that allow for analysis but do not make the researcher a slave to statistical procedures.

To understand the steps associated with designing valid research studies, the empirical and scientific methods are compared. The scientific method is based on inductive and deductive reasoning. Deductive reasoning, which begins with a generalization and infers a factual conclusion, provides a link between theory and observation. In contrast, inductive reasoning begins with a specific observation and moves to a generalization.

In deductive reasoning, a broad general principle or set of principles is used to determine what would happen in a specific situation. For example, given the generalization that all legume plants are dicotyledons, it can be deduced that alfalfa has two cotyledons. Deductive reasoning advances from a general situation to a specific one.

Inductive reasoning is the opposite; from a number of specific cases, a general principle is produced. For example, given that red clover beans and peas have two cotyledons and are legumes, it can be concluded that all legumes are dicots.

Both deductive and inductive reasoning can lead to hypotheses, a prerequisite for effective research and a part of the scientific method. The scientific method involves five steps:

Step 1: Collection of facts by observations, experimentation, or literature review to define or raise a question that needs to be resolved.

Step 2:   Formulation of a hypothesis or theory to serve as a tentative explanation of the problem, usually in terms of causes and effects.

Step 3:   Deduction of the consequences of the hypothesis if it proves to be valid.

Step 4:   Collection and analysis of data through observation, experimentation, and testing; often referred to as the experimental method.

Step 5:   Verification or rejection of the hypothesis on the basis of whether or not evidence was produced to support the suggested relationship between variables or whether new observations or experiments are needed.

The empirical method consists of conducting many tests and measurements without a theory or hypothesis. These are referred to as cut and weigh experiments. The empirical method may be used when little is known about a subject or when a topic is researched for the first time. Sometimes an investigation begins with empirical observations and leads to a hypothesis based on induction and deduction. If problems arise, empirical observations and experiments may again be undertaken.

Prior to the application of statistical analysis, researchers conducted large-scale tests over many years. By averaging the results, reasonably reliable judgments were made, but only by expending much money and time. In the main, researchers depended on good judgment based on careful observation, intuition, or common sense.

The use of improved experimental techniques, combined with statistical analyses, has enabled plant breeders and other researchers to reduce the number of experiments and hence the cost of testing a hypothesis and to distinguish between treatments with small differences. Good experimental techniques result in a low experimental error, and the use of the analysis of variance enables the researcher to distinguish between experimental error and treatment effects. Other statistical techniques such as correlation and regression are useful in plant breeding.

The application of statistical methods does not lead to proof of the hypothesis, because the real effect is unknown. The researcher formulates a hypothesis and then must prove (or disprove) that hypothesis, which may be difficult. For example, gliadin (protein) content in various wheat genotypes may vary, and such variation may affect patients suffering from the celiac sprue condition, but other factors may be important also (Hartsook, 1984; Powers Rickerd, 1982). The approach to this problem is based on the formulation of the **null hypothesis**, that is, the hypothesis that differences in gliadin content have no effect and that any variation among the patients is due to chance. If results indicate that differences in gliadin content have an effect, within certain probability limits, the null hypothesis is rejected. An initial step in formulating the null hypothesis is a thorough knowledge of the literature on the subject.

The application of statistical procedures is not a summary of the experiment. Statistics allow for statements about the results to be made with more authority. Careful examination and interpretation of the results must be described, but not only in statistical jargon.

This chapter, including a section on field plot techniques, concentrates on the plant breeding aspects of statistics and experimental design. Statistics, although a study unto itself, is included here because it is a valuable tool in plant breeding. A basic understanding of statistics is assumed; this chapter will serve as a guide and refresher.

MEASURES OF ASSOCIATION

The association between two and among more than two variables is of interest in plant breeding. Two of the most commonly used measures of association are **correlation** and **regression**. Correlation measures the association between two independent variables. Regression considers the change in one variable brought about by another variable or variables. One variable (dependent) is considered to be a function of another variable (independent). In regression, knowledge of the independent variable allows the prediction of the dependent variable. Correlation can tell the investigator whether or not two variables are related but is generally not used to predict a variable.

An example of the use of correlation is based on selection for high seed yield in a forage species. The relationship, if any, between forage yield and seed yield would be of value. A positive correlation indicates that as forage yield increases, seed yield increases; a negative correlation indicates that as forage yield increases, seed yield declines. A low or zero correlation coefficient would suggest no relationship. Although a correlation coefficient gives an estimate of the relationship between the two independent variables of seed and forage yield, it would not be useful in predicting seed yield on the basis of forage yield or forage yield on the basis of seed yield.

Regression, on the other hand, is useful as a predictive tool. Consider the association of seed size and seedling vigor in birdsfoot trefoil. From measurements of seed size and seedling vigor, it is possible to determine the association of seed size and seedling vigor by regression analysis. Once a regression coefficient is determined, seed size can be used to predict seedling vigor. Since seed size is easily measured, compared with the time-consuming measurements of seedling vigor, regression coefficients can be a valuable aid in plant breeding.

Correlation Coefficient

Correlation relates two independent variables, and the term used to describe the degree or amount of correlation is the *coefficient of correlation.*

The coefficient of correlation r between two variables x and y is given by

$$r_{xy} = \frac{C_{xy}}{\sqrt{V_x V_y}}$$

where C_{xy} is the covariance between x and y. Covariance is a measure of the variation of the product of two means. Covariance is calculated in the same manner as the variance of x and the variance of y, except in this case the product deviations from the mean of the two variables are used as follows:

$$C_{xy} = \frac{\Sigma(x - \bar{x})(y - \bar{y})}{N - 1}$$

If two characteristics are independent, the covariance will average to zero because two observations on a pair of variables will deviate from their averages in opposite directions just as often as in the same direction. Unlike variances, covariance can have a negative or positive value. The formula used to machine calculate the correlation coefficient is as follows:

$$r_{xy} = \frac{\Sigma(x - \bar{x})(y - \bar{y})}{\Sigma(x - \bar{x})^2 \Sigma(y - \bar{y})^2}$$

$$r_{xy} = \frac{\Sigma xy - (\Sigma x)(\Sigma y) / n}{\sqrt{[\Sigma x^2 - (\Sigma x)^2 / n] [\Sigma y^2 - (\Sigma y)^2 / n]}}$$

The t-test measures the statistical significance of the coefficient of correlation:

$$t = \sqrt{\frac{r(n - 2)}{(1 - r)^2}}$$

The value of t can be obtained from tables in standard statistical textbooks. The value r is the correlation coefficient, and n is the number of observations. The value obtained for t is compared to statistical t tables for the appropriate degrees of freedom (n-2) and probability levels to determine significance.

The use of correlations in seedling selection was considered by Gilbert (1961). He measured the existence of a genetic correlation between some seedling character x and a desirable adult character y. In the unselected population, little notice is given to a correlation of less than 0.50 since the fraction of the variation in x that can be attributed to y (or vice versa) is only r^2. When selecting more intensely, however, even small correlations can have important effects. As selection proceeds, the genetic part of a correlation may become less important, so that seedling selection will become less effective, but so will direct selection on the adult phenotype. Gilbert concluded that seedling selection is a remarkably powerful method when a correlation exists.

Regression

The regression coefficient gives the rate of change in one variable (dependent variable y) per unit rate of change in another variable (independent variable x):

$$b_{yx} = \frac{sum\ of\ products\ xy}{sum\ of\ squares\ of\ x}$$

$$b_{yx} = \frac{C_{xy}}{V_x}$$

where b_{yx} is the regression of y on x. Machine calculation of b_{yx} is given by the formula:

$$b_{xy} = \frac{\Sigma xy - (\Sigma x \Sigma y) / n}{\Sigma x^2 - (\Sigma x)^2 / n}$$

The regression coefficient is used to predict the value of y when the value of x is known. The prediction formula is given by

$$\widehat{Y} = \bar{y} + b(x + \bar{x})$$

where $\widehat{Y}$ is the desired prediction, $\bar{y}$ is the mean of y, b is the regression coefficient, x is the independent variable, and $\bar{x}$ is the mean of x. This form of regression, known as linear regression, attempts to explain the dependent variable (y) on the basis of the regression coefficient (b) and the observed independent variable (x). How well the value b accounts for the observed value of y can be evaluated statistically. Deviations of the observed value of y from the expected value of y (as calculated by the formula) determine if the regression coefficient is statistically valid. If the regression value b is zero, then there is no relationship between the two variables x and y. The regression can be positive or negative. If the regression coefficient explains all or a large proportion of the variation of y as x changes, then the regression is significant. However, if the calculated regression coefficient b does not explain the change in y as x changes, then it is nonsignificant. With a series of observations, a y value can be calculated for each x, and when this value is compared to the observed y value, the deviation from the expected can be summed over observations and displayed in an analysis-of-variance table (Table 5.1).

The F test measures the ability of the regression equation to explain the variation observed. The amount of the total variation explained by the regression is called r^2 and is calculated by the formula

$$r^2 = \frac{regression\ mean\ square}{total\ mean\ square}$$

STATISTICAL TECHNIQUES TO DETECT GENETIC VARIATION

The amount of genetic variation in breeding lines or populations, particularly the additive genetic variation, is a critical determinant of success or failure of selection. A number of techniques have evolved to estimate genetic variation, the magnitude of components of genotypic variation (additive, dominance, and epistatic), broad- and narrow-sense heritabilities, and, last, expected response to selection. Each of the estimates is subject to error, which means that usually the estimates are most accurate in the populations sampled. The estimates, however, serve a useful purpose in allocating resources in a breeding program and confirm the accumulated experience from many breeding programs.

Analyses of Variance

Different sources of variation can be isolated, identified, and measured through an analysis of variance. The analysis of 14 birdsfoot trefoil cultivars for seed yield was presented in Table 4.8. The objective of analysis is to determine if the variation observed in seed yield

Table 5.1 Analysis of Variance

Source	Degrees of Freedom	Sum of Squares	F Value
Regression	1	$\dfrac{(\Sigma xy - \Sigma x \Sigma y/n)^2}{\Sigma x^2 - (\Sigma x)^2/n}$	$\dfrac{MS\ regress}{MS\ dev.reg.}$
Deviation from regression	$n-2$	$\Sigma y^2 - \dfrac{(\Sigma y)^2}{n} - \dfrac{(\Sigma xy)^2}{(\Sigma x^2)}$	
Total	$n-1$	$\dfrac{\Sigma y^2 - (\Sigma y^2)}{n}$	

occurred by chance or if variation is genetic. The explanation is aimed at helping to understand the analysis of variance from a plant breeding, rather than from a statistical, viewpoint.

Variation among cultivars is highly significant, and the assumption can be made that a major portion of the variation is genetic. However, the proportion of this variation that is genetic and the proportion due to other causes are not known. The seed yield data combine differences in genetic potential for seed yield as well as differences not related to genetic seed yield. Cultivars Wallace, Empire, and Cree exhibit low seed yields, perhaps because of a lack of vigor, maturity differences, growth habit differences, or some other factor that may or may not be genetic. A more precise experimental design is required to obtain a better estimate of the genetic versus nongenetic variation.

Parent-Offspring Regression

Another method of estimating components of genotypic variation and heritability is the parent-offspring regression. Progeny from a genetically diverse set of parents would be expected to resemble their parents. For example, a parent with large seed size would be expected to produce progeny with a relatively large seed size if the variation in the parental population is genetic and, more specifically, influenced by additive genetic variance. Similarly, a parent in this population with low seed weight would be expected to have progeny with a relatively low seed weight. The relationship between a group of parents and their offspring reflects heritability of a specific trait and is illustrated in Figure 5.1.

The genetic relationship of the parent-offspring regression is given by

$$\beta_{op} = \frac{C_{ov}(P,O)}{V\,parent}$$

$$\beta_{op} = \frac{1/2\,V_a}{V\,parent}$$

where β_{op} is the regression of offspring on parent,
   $C_{ov}(P,O)$ is the covariance of parents and offspring,
   V is the variance of the parent, and
   V_a is the additive genetic variance.
Similarly, narrow-sense heritability is estimated by

$$\text{Heritability (narrow sense)} = 2(\beta_{op})$$

$$= \frac{V_a}{V_p}$$

The covariance of parents and offspring $C_{ov}(P,O)$ is 1/2 of the additive variance of diploids and is based on the genetic relationship of a parent (male or female) to its progeny. A more detailed explanation of these relationships can be found in most statistical genetics books. Simply stated, the covariance of parents and offspring is the deviation of parents from their mean times the deviation of progeny from their mean. Where performance of parents and offspring are distributed randomly about their respective means, the product of their deviations will be zero (Figure 5.1C). Where performance of parent and offspring are related, the covariance

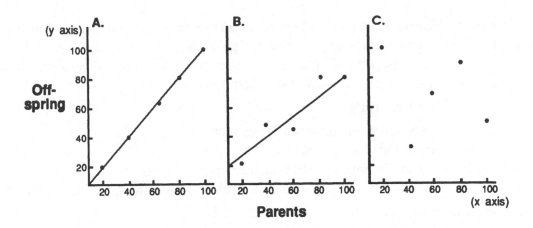

Figure 5.1   Parent-offspring regression as a measure of heritability. Seed yields for the parents are presented on the x axis and for offspring on the y axis. Graph A illustrates perfect heritability and indicates that genetic variation represented all the phenotypic variation of the parents. The slope of the regression line, *b*, has a value of 1. A high heritability indicates that progeny have performance similar to their parents. In this case, parents with high or low seed weight produce progeny with high or low seed weight, respectively.

A more realistic situation is illustrated in Graph B, which has a high heritability. The regression line represents the best fit of the data points. Such a situation means that a good-performing parent would produce offspring with relatively good performance.

Graph C illustrates a case where genetic variation accounts for none of the phenotypic variability present. A high-performing parent could have offspring with high, medium, or low performance.

---

would be positive (Figure 5.1B). As a statistical function, covariance can be positive, zero, or negative.

The parent-offspring regression is one of the more accurate estimates of additive genetic variance and narrow-sense heritability. It can be used with minor modification to determine narrow-sense heritability for autotetraploid species, as follows:

$$\beta_{op} = \frac{1/2\ V_a + 1/6\ V_d}{V\ parents}$$

Estimates of heritability may seriously underestimate total genotypic variation when the dominance variance (V_d) and variance (V_n) are large. Where a very high proportion of additive variance (V_a) exists, an estimate of narrow-sense heritability based on the parent-offspring regression would result in little change from broad-sense heritabilities. However, for normally cross-fertilizing species, such as maize, alfalfa, and birdsfoot trefoil, the narrow-sense heritability estimates obtained from the parent-offspring regression may be much less than the broad-sense heritability estimates obtained through estimates of components of variance.

PREDICTING GENETIC GAINS THROUGH SELECTION

Once heritability estimates are available for a trait in a specific population, predictions can be made of the amount of genetic gain expected for a given selection intensity. Either the variance or the mean of the population can be used to calculate genetic gain. The expected mean of a population following a generation of selection is given by

$$\bar{x}_e = \bar{x}(\bar{x}_s - \bar{x})h^2{}_\beta$$

where $\bar{x}_e$ is the expected mean in the next generation,

$\bar{x}$ is the mean of the base population,

$\bar{x}_s$ is the mean of plants selected from the population, and

$h^2{}_\beta$ is the heritability estimate for the trait under selection (narrow-sense heritability).

The relationship of $\bar{x}$, $\bar{x}_s$, and $\bar{x}_e$ is shown in Figure 5.2.

As an example, consider a base population consisting of 1,000 plants with a mean forage yield of 2,050 g/plant, the mean of 100 selected plants as 2,650 g/plant, and a heritability of 21% or 0.21.

$$
\begin{aligned}
\bar{x}_e &= \bar{x} + [(\bar{x}_s - \bar{x})\, h^2{}_\beta] \\
&= 2{,}050 + [(2{,}650 - 2{,}050) \times 0.21] \\
&= 2{,}050 + 126 \\
&= 2{,}176
\end{aligned}
$$

The latter part of this formula $[(\bar{x}_s - \bar{x})\, h^2{}_\beta]$ is the gain expected from selection, which in this example is 126 g. Selection gain depends on both the intensity of selection and the heritability of the trait.

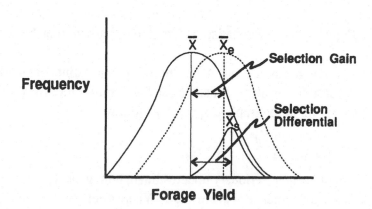

Figure 5.2 Expected genetic gain with selection from a base population for forage yield. In this figure, the curve $\bar{X}$ is the mean of the base population, and the curve $\bar{X}_e$ is the expected mean in the next generation. The small curve, $\bar{X}_s$, is the mean of the plants selected from the population, and if this has been successful, should be to the right of the base population. Selection gain and selection differential are shown.

Where predicted selection gain has been calculated, often the results are compared with the selection differential to determine how much of the selection differential is observed in the next generation. The comparison between the expected and observed gain is called the realized heritability.

$$\text{Realized heritability} \quad = \quad \frac{selection\ gain}{selection\ differential}$$

$$= \quad \frac{\bar{x}_{OBS} - \bar{x}}{\bar{x}_s - \bar{x}} \times 100$$

where $\bar{x}_{OBS}$ is the observed mean of the next generation,
$\quad \bar{x} \quad$ is the mean of the base population, and
$\quad \bar{x}_s \quad$ is the mean of selected plants.

Using the values in the above example, and if the observed mean of the population in the next generation is 2,100 g/plant, then the realized heritability would be:

$$\text{Realized heritability} \quad = \quad \frac{2{,}165 - 2{,}050}{2{,}650 - 2{,}050} \times 100$$

$$= \quad \frac{115}{600} \times 100$$

$$= \quad 19.2\%$$

Realized heritability is a useful tool in gauging the response to a selection scheme.

Obtaining estimates of the components of genotypic variance, heritability, predictions of selection gain, and realized heritability require well-planned experiments and often well-characterized breeding populations. Even so, the estimates often are more descriptive than predictive, or stated in another way, the information is more accurate in describing what changes have been made in a population as a result of selection than in predicting what progress will be made in future breeding efforts. Nevertheless, studies of well-characterized populations, such as the maize open-pollinated cultivar Iowa stiff stalk, have had broad and far-reaching implications for breeding efforts outside the population studied. Similar examples can be found in other crops as well.

The use of digital computers in statistical analysis has removed much of the drudgery associated with analysis and has emphasized the need for well-planned and well-executed experiments. In addition, computers have allowed for examination of an entire system, not just the individual parts (Porter, 1987).

The concept of a systems approach in plant breeding has opened the door to a comprehensive approach involving the entire system. Using linear regression models, it is possible to describe a series of sequences through the use of a **model** or modeling. In a scientific sense, models represent explicitly stated hypotheses. As indicated at the beginning of this chapter, clearly stated hypotheses are the essence of good research. The development of models based on genetic parameters may have application in plant breeding. Up to now, however, computerized models have had limited application in plant breeding but have been used for sophisticated management systems, for crops, especially cereal grains.

THE NEED FOR TESTING PROGRAMS

Plant breeding involves five basic steps: (1) identification or creation of variability; (2) selection of plants; (3) evaluation or testing of plants; (4) selection or combination of superior selections into cultivars; (5) testing of cultivars. Whatever the crop, these basic steps are an integral part of the process.

In self-pollinated crops, the wise handling of early-generation segregating material, early-generation performance trials, and eventual regional and possibly international testing schemes are major components of plant breeding systems, and it is these aspects that will be covered in this chapter.

Once a new cultivar is released, plant breeding activities do not cease on the premise that the objective(s) has been met and the needs of commercial agriculture have been fulfilled. Instead, plant breeding is an ongoing activity because newly released cultivars may exhibit shortcomings, if not in absolute then in relative terms, conditions are changing constantly, and new and exciting sources of germplasm may offer promise of advances. Hybridization is performed at every opportunity, and selections are made each season to generate hundreds of lines that must be evaluated for various criteria, including aesthetic and economic ones.

Yield is a dominant objective of many plant breeding programs, especially for field crops, and, accordingly, extensive yield trials are conducted to discriminate among selections and determine their potentials. Yield trials are assumed to predict performance and hence must be conducted so that reliable and consistent results are obtained.

Field plot testing would be simplified greatly if a uniform and consistent environment prevailed over a specific region year after year. Instead, testing every year at selected locations is required. But confounding testing to the greatest extent is the fact that genotypes are not consistent in their response to the environment and commonly exhibit a genotype x environment interaction, making identification of superior selections difficult and extrapolation to commercial agriculture tenuous.

Regional and international trials are an outgrowth of genotype x environment interactions. If such interactions did not exist, it might be possible to conduct a minimum of one year of testing to provide reliable and wide-scale recommendations.

The steps involved in plant breeding and the cyclical nature of plant breeding are illustrated in Figure 5.3.

EXAMPLES OF SELECTION PROCEDURES
IN SELF-FERTILIZED CROPS

Tens of thousands, even millions, of entries may be screened for each truly successful cultivar produced. Whole-plant screening methods are costly, require considerable field space, are time consuming, and may not be fully reliable because of environmental variables. Laboratory techniques associated with biotechnology may aid selection. Ideally, plant biotechnology may allow for millions of cells, each representing a potential plant, to be screened in a 9 cm flask in the laboratory by adding toxic or growth-inhibitory compounds to the medium. Only a few resistant cells may survive, and the regeneration of plants from these isolates may result in plants exhibiting tolerance to specific compounds or diseases (Chaleff, 1983).

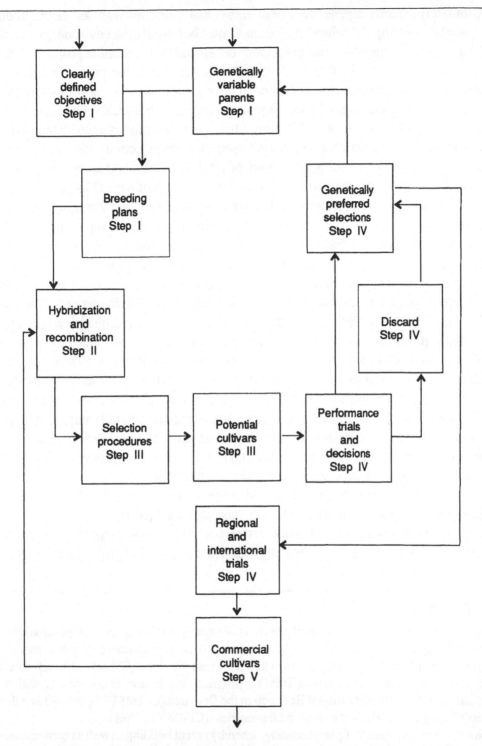

Figure 5.3 The five basic steps and the cyclical nature of plant breeding are illustrated in this diagram. Steps III and IV, involving selection and testing, are major aspects of plant breeding in terms of available resources, including labor, facilities, and expense. It is incumbent upon the plant breeder that these steps be conducted carefully and in the best manner possible.

Plant breeding programs may be funded on a regional basis and are concerned primarily with

Figure 5.3 continues overleaf

Ultimately, plants regenerated from individual cells, as well as those produced by conventional breeding and selection programs, must be tested in the environment in which they will be grown commercially. Except for crops developed for greenhouse production, most crop cultivars will be grown in fields where minimal control of the environment is possible. Measurement of a genotype x environment interaction can also be made in the field.

Selection of segregating, self-fertilized material is based on visual (phenotypic) differences or on measurements of several to many characters, sometimes considered sequentially, sometimes jointly. Selection normally is based on plant selection rather than reproductive development, but plant breeding costs time, effort, and money, and all breeders must find an optimal allocation of resources to the components of their programs (Figure 5.4). If selection is based on entire plants, more labor is required for harvesting and threshing, and in the next generation, plant selection will require more land and special seeding equipment. If single reproductive organs rather than plants are selected, a larger number of organs may be selected per population and a larger number of populations may be sampled.

Another selection procedure is to eliminate plants that do not meet specific criteria at various segregation stages or at various stages of growth and development. The concept of elimination by pulling undesirable plants in a multistage selection sequence was considered by Godshalk, Timothy, and Burns (1988). Breeders of field crops have tended to use single-stage selection indices, often based on the mature plant, to promote cultivar improvement. Multistage selection, based on establishing multiple criteria and eliminating unselected individuals from further testing, may be useful as a selection tool to reduce research costs since it reduces the number of entries under test. Multistage selection may be particularly important in perennial plant and tree breeding programs. Godshalk, Timothy, and Burns (1988) compared one-stage index genetic gains with two- and three-stage gains in the improvement of switchgrass (*Panicum virgatum* L.) forage quality and yield. Predicted genetic gains were generally higher in two- and three-stage indices than in one-stage indices. Consideration may be given to multistage selection as an alternative to one-stage index selection.

The F_1 nursery often is regarded as a seed multiplication stage preparatory to the important F_2 generation when segregation is maximized. The F_1 nursery is grown under widely spaced

---

Figure 5.3 continued

serving the needs of agriculture in specific regions. Programs funded under the Consultative Group on International Agricultural Research (refer to Table 4.1) may conduct tests on a global basis. In 1986, for example, collaborating plant breeders in 97 countries were sent 2,628 sets of bread wheat, durum wheat, and triticale nurseries from the CIMMYT program. In addition, in cooperation with ICARDA (International Centre for Agricultural Research in the Dry Areas), CIMMYT prepared and distributed 255 sets of barley nurseries for testing in 66 countries (CIMMYT, 1988).

Note the cyclical nature of plant breeding, whereby parent building as well as commercial cultivar release are inherent aspects. The cyclical nature does not mean that it is a closed system because new genetic material, some requiring considerable introgression before it is agronomically acceptable, should constantly be entering the system. Sources of genetic diversity are continually required to meet the changing conditions of agriculture associated with changing pest situations, mechanization, agronomic practices, or quality demands.

Source: Adapted from Simmonds, 1983.

A. *B.*

Figure 5.4 Early-generation selection techniques may vary from breeder to breeder as determined by resources available and the plant breeding methods involved. Selection methods include bulk, pedigree, honeycomb, mass, and other systems. Whatever system is used, it is essential that desired genotypes be identified. The plant breeder in Photo A is standing behind a semidwarf F_2 segregate of triticale, a highly desirable plant because excessive height has been a problem in triticale production. The single-stemmed barley plant in Photo B is exhibiting the uniculm characteristic. Such a plant might have been lost if careful early-generation selection had not occurred.

Selection of early-generation segregates may be based on several to many plant characteristics. Following the example of animal breeders, selection for multiple traits might be enhanced by selection indices. This would allow a decision to be made on the spot rather than by an extended evaluation of many traits individually. Perhaps plant breeders use an intuitive form of selection based on several criteria because formal indices may be slow and awkward. Achieving the objectives associated with a complex ideotype might require a more comprehensive approach. Possibly a portable microcomputer could be useful in selection involving indices (Jackson and Stone, 1987).

conditions to maximize seed increase and may be viewed as a routine stage. The F_1 generation, however, should be recognized as one of the most important generations, for it is during meiosis in F_1 plants that crossing over and recombination occur between the homologous chromosomes from the parents, and the only time that the chromosomes from each parent will pair is in the F_1 generation. To enhance the possibility of rare combinations occurring, a reasonably large F_1 population should be produced by a more extensive parental hybridization program, or the F_1 population should be as productive as possible to obtain desirable genes contributed by each parent in the F_2 generation (Akerman and MacKey, 1948) (Figure 5.5).

Figure 5.5   The spaced plants grown in this field nursery are F_1 barley plants. A reasonable number of F_1 seeds should be produced from the crossing program, and such seeds should be spaced to maximize seed production so as to allow crossing over and gene recombination to occur. It is in the F_1 generation that all the chromosomes of one parent pair with those of the other parent in an unaltered condition. Rare but desirable recombinations can occur in the F_1 generation. F_1 plants are phenotypically uniform and generally heterozygous. No opportunity normally exists for F_1 plant selection.

Population size of F_1 plants may be limited by the tedious and time-consuming effort associated with hand emasculation and pollination. In addition, seeds in subsequent populations can be produced beyond the capacity of the system to accommodate them.

In order to shorten the intergeneration time, seed may be harvested from the parent plants that were hybridized indoors and planted under controlled conditions as soon as possible. Such seeds may exhibit dormancy, but this can be overcome.

FIELD TRIALS

Purpose of Field Testing

Yield is the dominant objective of many plant breeding programs, and so considerable emphasis in this section is directed to field trials that determine yield. This should not imply that other factors are unimportant; see Figure 5.6 for additional valuable characteristics. Because of a genotype x environment interaction, new cultivars of any crop must be tested under the conditions in which they will be grown.

Field trials are not intended to maximize yield or to measure yield as such but rather to obtain a valid comparative yield relative to the best commercial cultivars. Likewise, other characteristics, such as lodging and pest resistance, winter hardiness, quality aspects, maturity, and plant height, are measured relative to a standard check cultivar (Figure 5.7).

Field trials are conducted to discriminate among potential new cultivars, and these tests are assumed to predict performance. Predictions of performance would be simplified greatly if no

A.   *B.*

Figure 5.6   A plant breeder wishing to evaluate the garden lilies (*Lilium* spp) shown in these photographs does not have to resort to elaborate yield trials in the field. Evaluation may be done visually, provided a clearly defined objective was determined for the cross.

The lily plants in Photo A are tall, subject to breaking over under high winds, and for the lily grower are disappointing in that the flowers are displayed poorly. Note how the majority of flowers face the soil. To appreciate the beauty of the flower, one must raise it by hand. In contrast, the lily plants in Photo B are borne on shorter, sturdy stalks, and the open flowers are displayed in an upright position.

Although the lilies in Photo B have considerable promise for lily producers, the plant breeder may wish to determine the presence of a genotype x environment interaction by growing the lilies in Photo B at several climatically diverse locations. Perhaps under some environments, these lilies are tall or the flowers change to a pendulous type. Before any statements can be made regarding the growth characteristics of these plants, regional testing is required.

---

genotype x environment interactions existed. The following example of forage dry matter yields at two test locations illustrates a genotype x environment interaction:

	Location One	Location Two
Cultivar A	10,000	11,000
Cultivar B	9,000	14,000

Cultivar A outyields B by 1,000 units at Location One. At Location Two, however, B outyields A by 3,000 units. Cultivar selection would depend on the location, as neither cultivar A nor cultivar B is superior at both locations.

Differences among cultivars under test may be small; potential may be based on yield performance equal to the check cultivar but with superiority in other agronomic aspects. Measurable differences in quantitatively inherited features are small, and sources of error must be eliminated by planting arrangements that allow for statistical analysis.

Replications

A statistical design with replications is a prerequisite to good field-plot technique. Replication, which represents one of the greatest advances in plot technique, achieves tw⁓

Figure 5.7 Photo of a winter cereal screening nursery following a severe winter in which many selections were destroyed. A check cultivar was seeded every ten rows, and its survival can be seen in the background, along with a winter-hardy selection in the foreground.

Hundreds of plant progenies consisting of a single row in the field may be evaluated visually or measured, counted, or scaled. Evaluation of numerous entries means that they are spread over a large area with the chance that soil, disease, snow cover, or fertility may vary. Under such circumstances, inclusion of a check cultivar at frequent intervals to provide a comparison is extremely important.

In replicated field-performance trials, one or more check cultivars should be included for comparison purposes. These check cultivars, which may include an adapted, currently produced commercial cultivar or a cultivar of acceptable quality, are included each year so that they are grown under the same conditions as the breeding material. Check cultivars serve as reference points to help the breeder make decisions on the status of his selection.

Plant yields may be higher than under commercial conditions because all heads are completely harvested. Although performance data are reported as they are measured, a well-known check cultivar provides a basis for comparison and recommendation to commercial producers.

objectives: reduction in the effect of random variation by averaging yields over several replications, and the ability to measure random variation. A disadvantage of replication is the increase in size of the test area, with an increased possibility of greater soil heterogeneity. To counter this problem, a reduction in plot size is desirable.

Replication number must be chosen with discretion. If large plots are used, three or four replications are suggested. If smaller plots are used, entries may be replicated five or six times. Replication number is also influenced by the degree of precision required in a test. Preliminary yield trials may be replicated twice, not only because of limitations on the supply of seed but also because a large number of entries are included and only major differences are sought.

After a preliminary screening when only the top 10% may be advanced, secondary screening tests may be conducted with four replications. In the final evaluation for licensing and recommendation purposes, six replications may be grown on off-station sites where less opportunity exists to select for soil uniformity or where no control on the previous cropping program is possible. Advanced evaluations are based on small differences, and reliable data are required to evaluate genotype x environmental interactions. Such advanced tests require the inputs of several collaborators and are commonly known as cooperative tests.

Cooperative tests conducted at several sites over several years generally reveal substantial genotype x environment interactions. Often a genotype x year x location interaction is found that is greater than a genotype x year interaction and that is also greater than the genotype x location interaction. Because of this, Simmonds (1983) reached what he termed an unwelcome conclusion, that trials should be replicated over years rather than sites. Dorst (1958) concluded also that a plant breeder should attach more value to environmental variation than to replications, especially in the first stages of experiments.

In every aspect of field testing, a consistent method of measuring a particular characteristic should be applied to each entry (Figure 5.8). Check or control cultivars are normally included for comparison purposes. If an entire test cannot be harvested on the same day, it is advisable to complete the operation on one or more entire replications on the same day. If subsequent rain and wind cause delays or lodging, the differences caused by the environment may be removed by replications. Associated differences do not affect comparisons among cultivars or treatments within replications.

Randomization

Competition between any two entries in a field trial must be eliminated or its effects controlled by randomization or by grouping cultivars of similar nature. A tall cultivar may compete strongly with an adjacent dwarf cultivar. Randomization of the experimental variables is essential for a valid test of experimental error and the reduction of bias in the results. The estimate of experimental error depends upon differences in plots treated alike.

If genotypes in a field trial are in striking and readily apparent contrast to others in a test, a restricted randomization may be appropriate whereby tall entries are randomized with tall entries and separated from dwarf entries. Larger plots to minimize interplot competition may be an alternative solution.

In cooperative tests that have identical entries grown at several locations and that will be summarized and analyzed collectively, a separate randomization for every test is imperative. Separate randomizations eliminate any bias associated with a particular randomization.

Plot Size and Shape

No single plot size or shape is best, and each breeder must consider the resources available, the statistical appropriateness, the amount of seed available, and special circumstances that may prevail. In the past 75 years, plots have been altered substantially.

The plots in Figure 5.9A are approximately 1/40 of an acre, a size that was considered a minimum under the imperial system of acres, rods, yards, and feet (Taylor, 1910). In 1906, the rod-row method of yield testing small grains was proposed by Norton (1907). It helped establish a trend toward smaller plot sizes in most crops. A major contribution of the rod-row method is that it allowed for replication without requiring an inordinate land base. For decades, the rod-row plot with four or five rows from which the center two or three were harvested

Figure 5.8   The technicians in this photo are measuring the height of barley genotypes in a replicated performance trial. Although entries may be generally homozygous, biological variability makes an apparently simple measurement such as height a challenging task.

Consistency of measurement among all entries in the test is important. Each plot in this photo consists of four rows. To avoid measuring the impact associated with an adjacent tall or short cultivar, measurements should be based on the center two rows. In addition, measurements should be taken a step into the plot to avoid differential border effect at the end of each plot. Height should be measured to the tip of the spike rather than to the tip of the awn because some cultivars may be awnless or display awns horizontally rather than vertically. In crops such as barley that exhibit a tillering characteristic, height per plot should be based on an intuitive mean rather than on the tallest or shortest spike. Late-formed **tillers** may be shorter than the primary shoot.

In cooperative test programs involving people at several sites, efforts should be made to standardize measurements so that an unaccountably large genotype x environment interaction does not arise. Field-plot error may result from faulty and careless field plot-technique, personal bias in interpreting data or measurements, incorrect or illegible recordings, or chance. Replication is a prerequisite that can increase the precision of an experiment. The mean yield of several plots of each entry provides a better estimate of the true yielding ability of a cultivar than the yield of a single plot.

Photo courtesy Ontario Ministry of Agriculture and Food.

A. *B.*

Figure 5.9 The cereal crop plots in Photo A are new introductions seeded in ten rows, approximately 5 m (16.5 ft. or one rod) in length. Large plots, along with many introductions, have consumed the entire area. With no replication, an analysis of variance is not possible. Such large plots may be justified, however, if seed increase and an opportunity for general adaptation is the objective.

In contrast to the situation in Photo A, hundreds of plots have been seeded in Photo B. Each plot consists of four rows, 3 m (10 ft.) in length, seeded side by side under replicated conditions. With four replications, the total area seeded per entry is identical.

The plots in Photo B allow for statistical analysis, and considerably more information might be obtained from them than from the plots in Photo A. Both require comparable amounts of time and effort. Note that in Photo B each plot in each replication is marked with a stake at the right to clearly identify the entry. All plots are weed free, well conducted, and seeded on what appears to be a uniform land area.

became the standard of excellence (Hayes, 1932; Klages, 1933; Hayes and Arny, 1917) (Figures 5.10 and 5.11).

Plot size, however, may vary with the crop under test. A three-meter-long (10 ft.) plot for small grains does not increase variability substantially if good plot techniques are practiced. Likewise, the custom of trimming the ends of plots to reduce differential responses to alleys and walkways among replications has been abandoned, with an enormous saving of time and effort.

In general, a shift to smaller and smaller plots has occurred over the years. In 1947, the hill plot was introduced as an innovative step toward microplots (Figure 5.12). Initially, hill plots were intended for testing head selections of small grains (Bonnett and Bever, 1947) but were extended to replicated yield trials by Ross and Miller (1955). Hill plots may be seeded with 25 seeds per 30-cm row and spaced 30 cm apart (12 in.). In time, hill plots came to be regarded as a supplement to conventional plots for screening a large number of lines when seed supplies are scarce and land is limited.

For soybeans, Green, Shibles, and Moraghan (1974) evaluated the performance of nine lines in several types of hill plots and short rows in comparison with large-yield plots with row spacings of 18, 51, and 102 cm (7.0, 20, and 40 in.). Relative rankings for height and maturity were consistent among hill plots, short rows, and large plots. Lodging and seed size predictions were inconsistent. Soybean seed yields were more accurately predicted in hill plots and short rows than in 51- and 18-cm-wide (20 and 7 in.) rows in large plots. The conclusion reached was that unbordered hill plots, short rows, and the 102-cm-wide (40 in.) row method may be suitable for predicting wide-row management performance. For selection of breeding material, hill plots appear satisfactory.

A.

B.

C.

D.

Figure 5.10   The four-row, rod-long cereal plots in Photo A are representative of a plot system conducted with sickle harvesting of the two center rows. Just before harvest, the ends of each plot were trimmed to a uniform length. Interplot competition was reduced by the fact that the two center rows were protected by a guard row on either side. The introduction of dwarf (plot on left), semidwarf (plot immediately to the right of the dwarf) and tall entries (lodged plot on the right) in one test introduced a need for the redesign of plots. The need to adopt machine harvesting when the sickle was abandoned gave further incentive to changes in plot design.

One solution to reducing interplot competition and facilitating machine harvesting is to include an unvernalized winter cereal, achieved by spring seeding, among the plots, as shown in Photo B. The

Early generation testing, the need for a plot suitable for testing methods, and plot size gave further impetus to hill plots. Hill plots allowed an increase in replications without extending the test site area. Jellum, Brown, and Seif (1963) compared hill plots replicated six times with single 8-ft.-row plots replicated three times. The hill-plot method was found satisfactory for evaluating yield, height, and maturity but not for lodging resistance.

The hill-plot technique was given a further boost by Frey (1965), who concluded the method could be used efficiently for early-generation testing of small grains, but that final evaluation should be done with conventional yield trials. Further, Frey suggested that much of the experimentation conducted with conventional plots could be done with hill plots.

The advent of a system for reliably producing doubled haploids in barley, whereby a large number of truly homozygous lines of barley following hybridization could be produced in just one generation, gave a strong push to hill-plot testing.

A major advantage of hill-plot testing is that it requires a relatively small amount of seed. A conventional row-plot test with four replications requires 85 g of seed. A hill-plot test with ten replications, considered appropriate, requires only 10 g (Walsh, Park, and Reinbergs, 1976). These workers observed a high selection efficiency, which indicated the hill-plot technique is satisfactory.

Experimental Design of Field Plots

An integral part of field-plot experimentation is the use of an appropriate experimental design that facilitates an analysis of variance and allows for the impact of various treatments and their interactions to be objectively identified. It is an accepted fact that the use of proper experimental design and appropriate statistical analysis is an integral part of field-plot technique because it allows all relevant information to be derived. Nevertheless, experimental design must be regarded as another breeding tool, and plant breeders must avoid becoming slaves to a particular design.

There are many experimental designs involving various arrangements of plots in terms of physical layout, replications, and randomizations (Cochran and Cox, 1957). The choice of a design may be determined by the annual or perennial habit of the plants being tested, the cultural factors imposed, or the size of the plants. In forest or horticultural tree-breeding studies, cumulative responses are typical. Studies with horticultural tree crops involving grafts to

---

Figure 5.10 continued

unvernalized winter cereal remains in the vegetative state and does not elongate. Machine harvesting is facilitated and weed growth between the plots is reduced.

In Photo C, the plants with the showy flowers are spring-seeded rapeseed or canola separated by a common spring-seeded cereal crop. The cereal crops reduce lodging, facilitate machine harvesting, and eliminate interplot competition. Competition is offered by the spring cereal but is uniform for all rapeseed entries.

Five-row cereal plots in Photo D, with a blank row between plots, facilitate machine harvesting of the entire plot. All five rows are harvested without regard to border effect.

Figure 5.11   The maize plots shown in Photo A are being evaluated for yield at a fixed-row width and similar plant population. With the large number of entries submitted by private corn-breeding companies for uniform yield testing, only one row per entry in each replication is used. To achieve the desired population, critical in maize testing programs, each plot is overseeded and thinned to a common density after complete emergence. In a test to compare 15 corn hybrids to tillage treatments, Carter and Barnett (1987) used a four-row plot.

Major differences in height of sunflowers (*Helianthus annuus*) means that large plots may be required for yield evaluations (Photo B). In forages (Photo C), plot size varies with the size of the plants but is commonly 1.5 m in width and 3 m in length. Small plot sizes are recommended for perennial forages, which may be tested for three to six years and are subject to the accumulated effects of soil heterogeneity and microclimate. With rapeseed and canola (*Brassica* spp), plot size varies from four rows to six rows, with a blank row between plots (Photo D). The number of rows and total plot length and width are decisions that must be made for the varying circumstances of each situation.

Figure 5.12 The hill plots in this photo are those of spring barley. Hill plots normally consist of single rows, 30 cm (1 ft.) in length, 30 cm apart, and seeded with 25 seeds per hill. Fourteen doubled haploids and six commercial cultivars of barley grown at two locations were seeded at 15, 25, and 35 seeds per hill using a jab planter, stove pipe, and hand-hoe seeder. Based on a combined analysis over locations and two years of testing, neither genotype x planting method nor genotype x seeding rate interactions were significant. Ten replications were found to be appropriate for hill-plot testing of barley in Ontario (Walsh, Park, and Reinbergs, 1976).

The routine production of doubled haploids in barley has made the hill-plot technique viable because a large number of entries can be evaluated with a minimum of seed. When haploid development is routinely practiced in other crops, changes in plot design may be investigated.

rootstocks introduce an element of variability. Because of extensive space requirements, the number of trees per plot is necessarily limited. Differences among trees under test are frequently a source of large experimental error. For long-term studies involving trees, a simple experimental design should be chosen.

The simplest possible design is the completely randomized design, although it is seldom used in field experiments because variation due to replications cannot be identified. The randomized complete block design (Figure 5.13) allows for greater efficiency because differences in soil heterogeneity that occur among replications can be removed. With this design, however, only a moderate number of treatments, possibly about 15 to 20, can be tested because a loss of efficiency is likely to occur as replications increase in size and greater soil heterogeneity is encountered.

Where soil heterogeneity causes large experimental errors, other experimental designs are available, such as the Latin square (Figure 5.14) and lattice designs. These designs are considered in most texts of experimental design, for example, Cochran and Cox (1957).

The most severe limitation of field tests is that results are applicable to only the particular area represented by the soil and climatic conditions of each particular test. Tests must be conducted in regions with different soil and climatic conditions and should be repeated for a number of years. Generally, screening tests are conducted on-station for a minimum of two

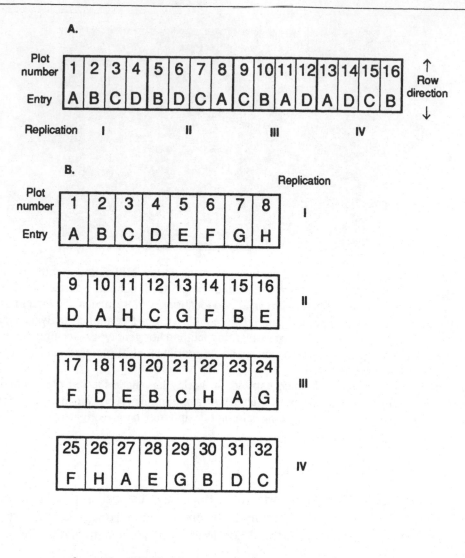

Analysis of Variance (8 cultivars)

Source of Treatment	Degrees of Freedom
Replications	3
Treatments	7
Error	21
Total	31

Figure 5.13   Illustration of the randomized complete block design shown with two layouts, each with four replications, and with four and eight entries or treatments in A and B, respectively. The analysis of variance is shown. The objective of the randomized complete block design is to keep each block as compact as possible so that variability caused by soil heterogeneity among plots is as small as possible. The layout in A is excellent for four or five entries but becomes cumbersome with more entries. Layout B can be used when more entries are involved, in this case, eight. Each entry occurs once in each replication, allowing for the data to be classified in two ways, by replications and by treatments.

In this design, soil heterogeneity can be removed from consideration. If a known gradient exists

years, and cooperative tests may be conducted at a number of off-station test sites for a minimum of three years before a decision on the status of a particular entry or a firm recommendation can be made.

In tests in which responses to different environments at several sites and over several years are involved, or in tests in any one year in which a number of factors are being imposed simultaneously, factorial experiments may be selected (Figure 5.15). In addition to years and sites, the experimenter may investigate different levels of fertility, growth regulators, pest control treatments, or different cultural treatments, such as conventional versus minimum tillage. Interactions among the treatments are important, and the experimental design should identify them.

Usually the interaction between two factors or among three factors may be as valuable as the main treatments themselves. In a factorial experiment, the treatment effects being compared are the result of all possible combinations of two or more factors, each of which is applied at two or more levels.

In experiments involving multiple factors, it is sometimes not practical to distribute all factor combinations at random over the replication. In an experiment with 15 maize hybrids tested under **conventional tillage** and **conservation tillage**, it was not possible to randomize the two tillage treatments (Carter and Barnett, 1987). In such a case, a split-plot design might be chosen (Figure 5.16) in which treatments are not arranged at random within each whole plot but are restricted so as to lie in strips across the entire experiment.

Off-season Nurseries

At high latitudes where one crop can be produced outdoors annually, off-season nurseries can serve a useful role in advancing seed one generation and in increasing seed. For selection to be effective for crops taken out of their area of adaptation, clear objectives must exist and a knowledge of plant reaction understood.

Maize produced at high latitudes may be triggered into reproductive development by short day lengths. At low latitudes, otherwise tall genotypes may be extremely short. Long-day-sensitive cereal plants may not reach reproductive development at locations near the equator. To avoid these problems, crops may be grown in the off-season at similar latitudes in opposite hemispheres.

Figure 5.13 continued

in a test area, it is advisable to place the replications across the gradient in order to obtain the major differences among replications. Plots running across the gradient tend to make soil conditions within each plot more uniform.

To assure a valid estimate of error, it is necessary to have several replications, but the number will be determined by the precision desired, by the labor, physical facilities, and resources needed to handle the plots, and by the seed supply. If the number of entries exceeds 15 to 20, a loss of efficiency may occur because each replication increases in size, and soil heterogeneity may introduce variability that cannot be removed.

A. A 9 x 9 Latin square

Column

	1	2	3	4	5	6	7	8	9	Row
Plot number →	A	B	C	D	E	F	G	H	I	I
	10	11	12	13	14	15	16	17	18	II
	B	C	D	E	F	G	H	I	A	
	19	20	21	22	23	24	25	26	27	III
	C	D	E	F	G	H	I	A	B	
Entry →	28	29	30	31	32	33	34	35	36	IV
	D	E	F	G	H	I	A	B	C	
	37	38	39	40	41	42	43	44	45	V
	E	F	G	H	I	A	B	C	D	
	46	47	48	49	50	51	52	53	54	VI
	F	G	H	I	A	B	C	D	E	
	55	56	57	58	59	60	61	62	63	VII
	G	H	I	A	B	C	D	E	F	
	64	65	66	67	68	69	70	71	72	VIII
	H	I	A	B	C	D	E	F	G	
	73	74	75	76	77	78	79	80	81	IX
	I	A	B	C	D	E	F	G	H	

B. A 6 x 6 Latin square (shown to facilitate randomization)

A	B	C	D	E	F
B	C	D	E	F	A
C	D	E	F	A	B
D	E	F	A	B	C
E	F	A	B	C	D
F	A	B	C	D	A

Analysis of Variance (for a 9 x 9 Latin square)

Source of Treatment	Degrees of Freedom
Rows	8
Columns	8
Treatments	8
Error	56
Total	80

Figure 5.14 In the Latin square designs shown for nine and six treatments, grouping of treatments is in columns as well as rows. Treatments occur only once in each row and once in each column. A principal requirement for the use of the Latin square is that the number of replications must equal the

The use of off-season nurseries with different day lengths has led to the development of day-length insensitivity by natural selection. Day length does not exert an exclusive control over flowering, however, and high temperatures have the effect of hastening flower initiation and development.

Techniques with Winter Annuals

Annual crops that have a winter form, called winter annuals, such as winter wheat, barley, rye, and triticale, require a period of cool temperatures and short days to initiate reproductive development. This treatment is known as **vernalization**. When seeded in the field in the fall, winter annuals are vernalized by natural low temperature, short days, and low light intensity over the winter period. In controlled environmental units, vernalization of winter annual seedlings can be achieved by temperatures maintained at 5 to 8° C (35 to 45° F) (Trione and Metzger, 1970) with eight to ten hours of low light intensity per day for a period of six weeks. Crops such as rye that are strongly winter hardy have a more demanding vernalization requirement than less-winter-hardy crops such as winter oat or barley.

Figure 5.14 continued

number of treatments, so that the total number of plots is the square of the treatment number. A Latin square allows for the measurement and removal of variability across the experimental area in two directions.

To avoid confusion in randomizing larger tests such as the 9 x 9 Latin square illustrated in A, use a systematic design such as that shown in B for a 6 x 6 Latin square. Randomizations can be obtained from books on experimental design, for example, that of Cochran and Cox (1957).

In the analysis of a Latin square, error variance is removed for both rows and columns and can be a major advantage if gradients exist on a greenhouse bench or field test. A limitation of the Latin square is that the number of replications equals the number of treatments. If 13 or more treatments are to be tested, the number of replications becomes impractical.

A. Factorial combinations with 4 cultivars and 4 nitrogen treatments

Nitrogen Treatment				
N_0	AN_0	BN_0	CN_0	DN_0
N_1	AN_1	BN_1	CN_1	DN_1
N_2	AN_2	BN_2	CN_2	DN_2
N_3	AN_3	BN_3	CN_3	DN_3

B. As a randomized block design

Rep I of Four

D N_1	B N_3	A N_3	C N_2	D N_3	C N_0	B N_0	B N_1	A N_1	C N_3	A N_3	B N_2	C N_1	D N_0	A N_2	D N_2

C. As a Latin square

A N_0	B N_1	C N_2	D N_3
D N_1	C N_2	B N_3	A N_0
C N_2	D N_3	A N_0	B N_1
B N_3	A N_0	D N_1	C N_2

Analysis of Variance (for C)

Source of Variation	Degrees of Freedom	
Rows		3
Columns		3
Treatments		15
Nitrogen	3	
Cultivars	3	
N x C	9	
Error		45
Total		63

Analysis of Variance (for B)

Source of Variation	Degrees of Freedom	
Replications		3
Treatments		15
Nitrogen	3	
Cultivars	3	
N x C	9	
Error		45
Total		63

Figure 5.15 A factorial experiment has two or more different treatments at two or more rates of application tested in all possible combinations. The examples shown in this illustration involve four cultivars under test with four levels of nitrogen application. The resulting 4 x 4 factorial and the 16

Figure 5.15 continued

treatments are shown in A and consist of all combinations that can be formed from the different factors. If five levels of nitrogen were applied, the experiment would be described as a 5 x 4 factorial. If another treatment, such as 4 levels of potassium, was added to the existing 4 x 4 factorial, a 4 x 4 x 4 factorial would result.

Factorial designs are highly efficient because a comparison of treatments can be made in a more precise manner than by separate experiments for each factor. A factorial experiment allows for the measurement of a differential effect or interaction. If cultivar A showed a specific response to treatments at one location but a contrasting response at another location, a cultivar x location interaction would be measured. If one cultivar showed a specific response to treatments and another showed a different response, a cultivar x treatment interaction would be measured.

Factorial experiments may be chosen when exploratory work is undertaken, when there is a possibility of interaction, or when recommendations over a large area are required.

A. Main plot

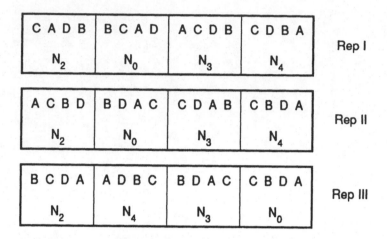

B. Analysis of Varlance (three reps) 4 cultivars and 4 nitrogern fertility levels

Source of Variation	Degrees of Freedom
Main Plots	11
Replications	2
Nitrogen (N)	3
Error (a)	6
Sub Plots	6
Cultivars (c)	3
Cultivars x N	9
Error (b)	24
Total	47 (48-1)

Figure 5.16   A split plot is a possible design for factorial experiments. In Figure 5.15, large plots were required to avoid uptake of N_0-treated plants adjacent to N_3-treated plants and so on. A restricted randomization may be required to prevent lateral roots from entering adjacent plots. In the factorial example in Figure 5.15, similar nitrogen treatments may be grouped to facilitate application and to minimize the effect on adjacent plots. Compromising the factorial design by a restricted randomization results in some loss of precision. This may not be serious because cultivars may show a major response to nitrogen levels that can be detected easily. When using a split-plot design, it is important that the precision of the treatment of greatest interest is not jeopardized.

A split-plot design may be useful for irrigation studies, soil tillage treatments, date of seeding or cutting studies, and studies involving chemical applications. Treatments of one factor are assigned to main plots, cultivars in this example, the ones in which some precision can be sacrificed. Treatments of the second factor are assigned at random to subplots within each main plot. Main plots may be arranged in randomized complete blocks or in a Latin square.

REFERENCES

Akerman, A., and J. MacKey. 1948. The Breeding of Self-Fertilized Plants by Crossing. *Svalof, 1886-1946,* edited by A. Akerman, C. Bloms, and A. Boktrycheri. Lund, Sweden, pp. 46-71.

Bonnett, O. T., and W. M. Bever. 1947. Head-Hill Method of Planting Head Selections of Small Grains. *Journal of the American Society of Agronomy* 39:442-445.

Carter, P. R., and K. H. Barnett. 1987. Corn-Hybrid Performance Under Conventional and No-Tillage Systems After Thinning. *Agronomy Journal* 79:919-926.

Chaleff, R.S. 1983. Isolation of Agronomically Useful Mutants from Plant Cell Cultures. *Science* 219:676-682.

CIMMYT. 1988. *1986 Annual Report.* International Maize and Wheat Improvement Center, Mexico, D.F. 83 pp.

Cochran, W. G., and G. M. Cox. 1957. *Experimental Designs. 2d ed.* John Wiley & Sons, New York. 611 pp.

Dorst, J. C. 1958. Use of Variation or Replication in Breeding Work. *Euphytica* 7:111-118.

Frey, K. J. 1965. The Utility of Hill Plots in Oat Research. *Euphytica* 14:196-208.

Gilbert, N. 1961. Correlations in Plant Breeding. *Euphytica* 10:205-208.

Godshalk, E. B., D. H. Timothy, and J. C. Burns. 1988. Application of Multistage Selection Indices to Crop Improvement. *Crop Science* 28:23-26.

Green, D. E., R. M. Shibles, and B. J. Moraghan. 1974. Use of Hill-Plots and Short Rows to Predict Soybean Performance Under Wide- and Narrow-Row Management. *Iowa State Journal of Research* 49:39-46.

Hartsook, E. I. 1984. Celiac Sprue: Sensitivity to Gliadin. *Cereal Foods World* 29(2):157-158.

Hayes, H. K. 1932. An Experimental Study of the Rod-Row Method with Spring Wheat. *Journal of the American Society of Agronomy* 24:950-960.

Hayes, H. K., and A. C. Arny. 1917. Experiments in Field Technique and Rod Rows. *Journal of Agricultural Research* 11:399-419.

Jackson, H. O., and J. A. Stone. 1987. Improved Data Collection Program for a Notebook-Size Microcomputer. *Agronomy Journal* 79:1087-1089.

Jellum, M. D., C. M. Brown, and R. D. Seif. 1963. Hill and Row Plot Comparison for Yield of Oats. *Crop Science* 3:194-196.

Klages, K. H. 1933. The Reliability of Nursery Tests as Shown by Correlated Yields from Nursery and Field Plots. *Journal of the American Society of Agronomy* 25:464-472.

Norton, J. B. 1907. Notes on Breeding Oats. *American Breeders Association Report* 3:280-285.

Porter, J. R. 1987. Modelling Crop Development in Wheat. *Span* 30(1):19-22.

Powers Rickerd, J. 1982. Catering to Special Diets. *Science Dimension* 14(1):1.

Ross, W. M., and J. D. Miller. 1955. A Comparison of Hill and Conventional Yield Tests Using Oats and Spring Barley. *Agronomy Journal* 47:253-255.

Simmonds, N. W. 1983. Plant Breeding: The State of the Art. In *Genetic Engineering of Plants: An Agricultural Perspective,* edited by T. Kosuge, C. P. Meredith, and A. Hollaender. Plenum Press, New York, pp. 5-25.

Taylor, F. W. 1910. The Size of Experiment Plots for Field Crops. *Proceedings of the American Society of Agronomy* 1:56-58.

Trione, E. J., and R. J. Metzger. 1970. Wheat and Barley Vernalization in a Precise Temperature Gradient. *Crop Science* 10:390-392.

Walsh, E. J., S. J. Park, and E. Reinbergs. 1976. Hill Plots for Preliminary Yield Evaluation of Doubled Haploids in a Barley Breeding Program. *Crop Science* 16:862-866.

CHAPTER 6

Hybridization and Selection in Self- and Cross-pollinated Crops

Hybridization is a common method of producing genetic variability for subsequent selection. Selection of superior plants is an integral part of a breeding program. As outlined in Chapter 4, variability may arise from qualitative or quantitative genes. Quantitative traits have a high value in breeding programs but were reported to be more difficult to select for than qualitatively inherited traits. Although qualitative traits exhibit additive and/or dominance variation, they can be expected to function in a predictable manner depending on choice of parents. The typical segregating sequence for corolla tube length in tobacco, a quantitatively inherited trait, is illustrated in Table 6.1 and was reported by East, (1916) using *Nicotiana longiflora*. The conclusions from East's experiments for corolla tube length in *Nicotiana* were as follows:

1. A cross between homozygous parents will result in an F_1 with the same degree of uniformity as the parents. Note that the mean corolla tube length of 63.5 mm in the F_1 is intermediate between the two parents and that variation in the F_1 was larger than the variance of the parent with the long corolla tube, the most variable parent. Selection in F_1 would not be productive because the F_1 generation is homogeneous.
2. Variation in the F_2 generation is much greater than in the F_1. The F_2 variance shown in

Table 6.1 Mean and Variance Values of Short and Long Corolla Tube Length of Parents, F_1, with Selection for Short or Long Corolla Tube Length in Segregating Generations Two to Five

Generation	Year	Number of Observations	Mean (mm)	Variance (mm)
Parent (short)	1911	125	40.5	3.06
Parent (long)	1911	88	93.2	5.24
F_1	1911	173	63.5	8.53
F_2	1912	444	68.8	34.93
Selection for short corolla tube				
F_3	1913	147	50.2	10.05
F_4	1914	189	46.2	3.50
F_5	1915	161	42.0	6.25
Selection for long corolla tube				
F_3	1913	162	80.2	22.66
F_4	1914	195	82.3	10.89
F_5	1915	125	87.9	30.47

Table 6.1 was approximately four times the variance found in the F_1 generation. The variation of 34.93 mm observed in the F_2 generation occurred because of gene segregation and recombination in the gametes of the F_1 generation, plus environmental effects.

3. If enough individuals are available in the F_2 generation, the parental types should be recovered. In this particular experiment, one extreme of the two parents was not observed, namely, short corolla length, although segregates with corolla tube length equal to the high parent were found.

4. Selection in the F_2 generation for specific characteristics should give rise to F_3 and later populations that have markedly different means. Progeny that were very different in their corolla tube length were found in the F_2 generation. The data in Table 6.1 indicate that selection in the genetically variable F_2 generation will result in individuals in the F_3 generation whose progeny will be very different. This principle emphasizes the possibility of selecting superior individuals with quantitatively inherited traits, such as yield. The possibility of positive **transgressive segregation** of quantitatively inherited traits is of interest to the plant breeder who works mainly with complex characteristics.

GENE SEGREGATION IN EARLY GENERATIONS

When parents differing in a number of genes are hybridized, the maximum amount of variability occurs in the F_2 segregating generation. With self-fertilization, the loss of heterozygosity and an increase in homozygosity occurs at a rate of 50% each generation. Therefore, the F_2 generation will be 50% homozygous, and the percentages of homozygosity for subsequent generations will be as follows:

F_3	75
F_4	87.5
F_5	93.75
F_6	96.88
F_7	98.44
F_8	99.32
F_9	99.66

The amount of variability in the F_2 generation is determined by the number of segregating genes (the number of genes that are different between the two parents), the number of alleles per gene that are segregating, and gene linkage. As the number of different genes distinguishing the parents increases, the number of possible genotypes expected in the F_2 generation increases markedly (Table 6.2). In maize, for example, which has ten chromosome pairs, if only one gene was different on each chromosome for a cross of two inbred lines, there would be a possibility of 59,049 different genotypes in the F_2 generation. In the case of hexaploid wheat with 21 chromosome pairs, if only one gene differed per chromosome pair between two parents, the F_2 generation would consist of 10,460,353,203 different genotypes. Even when the number of genes that distinguish parents is relatively small, the land area necessary to observe all possible genotypes becomes prohibitive. For many crosses, the F_2 population size is a compromise between having sufficient opportunity to expose the variation present in the F_2 versus the land

Table 6.2 Numerical Characteristics of F_2 Progeny Produced Between Parents Differing in n Allelic Pairs

Number of Heterozygous Loci	Number of Homozygous Genotypes in F_2[a]	Number of Different Genotypes in F_2	Minimum Population for Chance Presence of Each Genotype
n	2^n	3^n	4^n
1	2	3	4
2	4	9	16
6	64	729	4,096
10	1,024	59,049	1,048,576
15	32,768	14,348,907	1,073,741,824
20	1,048,576	3,486,784,401	$1,099 \times 10^9$
21	2,097,152	10,460,353,203	$4,398 \times 10^9$
30	$1,074 \times 10^6$	$2,059 \times 10^{11}$	$1,153 \times 10^{15}$
50	$1,126 \times 10^{12}$	$7,178 \times 10^{20}$	$1,268 \times 10^{27}$

[a] The 2^n value is also used to calculate (1) the number of different phenotypes in F_2 with complete dominance, (2) the number of different gametes in the F_1 generation, and (3) the number of different genotypes in backcross F_1.

Note: Plant breeders usually grow 2,000 to 5,000 plants from a particular cross in an F_2 nursery. Many of the homozygous genotypes will not occur or be identified if they do occur.

Source: After Briggs and Knowles, 1967, p. 124; Shebeski, 1967, p. 269.

area that is necessary to plant all possible genotypes. In principle, when a particular cross is between proven parents, the F_2 population should be larger than when parents of unknown performance are used if a desirable genotype is to be produced.

The number of alleles segregating for each gene will also determine the amount of variability in the F_2 generation. For a diploid inbred parent, the maximum number of alleles in the F_1 and F_2 generations is two. However, for noninbred parents or heterozygous parents, the theoretical maximum in the F_2 generation is four alleles. Three loci, for example, with two alleles segregating at each would result in 27 different genotypes in the F_2 generation. If three loci were segregating with four alleles, then the F_2 generation would have a theoretical possibility of 1,000 different genotypes for those three genetic loci.

Gene frequency in the F_2 and succeeding generations will remain constant if there is no selection, **directional mutation**, or **genetic drift**. Genotype frequency, on the other hand, will change with the progression through the segregating generations. As selfing occurs, there will be a 50% increase in homozygosity in each generation. Although there will be no change in gene frequency, the genotypes that compose each generation will consist of more and more homozygous genotypes as opposed to heterozygous genotypes as the segregating generations advance from F_2 to F_3 to F_4 and onward.

Linkage has no effect on gene frequency or the approach to homozygosity with generations of selfing. Linkage will, however, have an effect on the occurrence of specific gene combinations. Genes that are more closely linked in the parental combination will be found in succeeding generations at much higher frequencies than would be expected with random segregation. The effect of linkage on the occurrence of parental versus recombinant types is shown in Table 6.3.

Linkage may assist the aims of the plant breeder where two desirable traits occur in one parent and are closely linked. If two sources of disease resistance, for example, are ten map units apart, 20% of the individuals in the F_2 would have both resistant genes homozygous, whereas only 6.25% would have both genes homozygous resistant if the two genes were independent. In the case where a desirable gene is closely linked to an undesirable gene, linkage acts as a serious disadvantage to the aims of the breeding effort. Many of the stem-rust-resistance genes from *Triticum timopheevi* have been shown to be closely linked to other genes that condition late maturity, an undesirable combination. In order to break these linkages, very large populations must be produced so that the desirable disease resistance can be separated from the undesirable maturity difference.

For qualitatively inherited traits, such as vertical disease resistance, maturity, and some dwarfing genes, the plant breeder may wish to have an estimate of the population size necessary to find a gene or combination of genes in the segregating generation. The genetic composition in any inbred generation can be obtained by expanding the binomial

$$[1 + (2^m - 1)]^n$$

where n is the number of gene pairs involved and m is the number of generations of selfing. This binomial expansion, commonly used in genetics, gives the proportion of heterozygous and homozygous loci. The first exponent in each term gives the number of heterozygous loci and the second the number of homozygous loci.

With a population originally differing at four loci ($n = 4$), the fifth segregating generation (F_6, $m = 5$) would be expanded as follows:

$$[1 + (2^5 - 1)]^4 \quad = \quad (1 + 31)^4$$

$$= \quad 1^4 + 4(1)31 + 6(1)^2(31)^2 + 4(1)(31)^3 + (31)^4$$

The composition of this F_6 population is shown in Table 6.4.

Table 6.3 Linkage Effects upon the Proportion of *AB/AB* Genotypes Expected in the F_2 Generation from the Double *AB/ab* (Coupling) or *Ab/aB* (Repulsion)

Crossover Value (percent)	Percent Homozygous Individuals (AB/AB or ab/ab) in the F_2 if the F_1 is:	
	AB/ab[a]	Ab/aB[b]
0.60	4.00	9.00
0.50	6.25	6.25
0.30	12.25	2.25
0.15	18.06	0.56
0.05	22.56	0.06
0.01	24.50	0.0025

[a] $1/4 (1-p)^2 \times 100$.
[b] $1/4 p^2 \times 100$.
Source: Adapted from Briggs and Knowles, 1967, p. 125.

Table 6.4   The Composition of the F_6 Population When the F_1 Plant Was Heterozygous at Four Loci

Number of Loci		Calculation	Homozygous Plants	
Homozygous	*Heterozygous*	*Calculation*	*Number*	*Percent*
0	4	1^4	1	0.0001
1	3	$4(1)31$	124	0.012
2	2	$6(1)^2(31)^2$	5,766	0.60
3	1	$4(1)(31)^3$	119,164	11.40
4	0	$(31)^4$	923,521	88.10

Source: Briggs and Knowles, 1967, p. 129. Reprinted by permission.

When a plant breeder deals with a series of qualitatively inherited traits whose inheritance is known, then knowing the specific proportion of individuals that will have the desired combination of genes is of greater value than knowing the total array of possible gene combinations. There are several simple formulas that relate particular combinations of genes expected after a given number of generations of selfing. In each of the following formulas, m is the number of segregating generations and n is the number of segregating loci. The proportion of individuals in a segregating generation that have the desired alleles in heterozygous or homozygous form is given by

$$\frac{2^m + 1^n}{2^{m+1}}$$

The proportion of homozygous individuals from n segregating loci after m generations of segregation is given by

$$\frac{2^m - 1^n}{2^m}$$

The proportion of individuals that are either homozygous recessive or dominant for a given number of segregating loci (n) in a particular segregating generation (m) is shown by

$$\frac{2^m - 1^n}{2^{m+1}}$$

CHOOSING PARENTS FOR HYBRIDIZATION

The most important decision a plant breeder makes is the choice of parents to serve as sources of potential genetic variability. Parental selection is important because it determines the range and nature of the variability in the F_2 generation and sets the potential limit for successful selection in the segregating generations. In choosing parents for hybridization purposes, the objective of the cross must be clear. The objective may be direct cultivar development or the introgression of exotic genes for purposes of population building without the expectation of a cultivar release. Most examples of parental selection consider two parents, although many breeding programs rely on three or more parents, especially when associated with quantitatively inherited factors, broad adaptation, or yield stability.

The various types of crosses and the concomitant selection of parents include

- *Single crosses.* Single crosses involve hybridization of one cultivar or line with another cultivar or line. The female parent may be arbitrarily assigned, except where a plant has cytoplasmically inherited factors, in which case it automatically becomes the female parent.
- *Backcrosses.* A backcross is the cross of an F_1 to one of its **recurrent parents**.
- *Triple cross.* A triple, top cross, or three-way cross, is the cross of an F_1 to a cultivar or line.
- *Double cross.* A double cross is the crossing of two F_1 hybrids.
- *Complex or successive crossing.* Crossing single, triple, or double crosses among themselves or with additional parental cultivars at any stage from F_1 to F_5 or later. Such a procedure blurs the initial objective of a cross and requires accurate record keeping.
- *Multiple or composite crosses.* Among self-fertilized crops, elaborate pyramiding of germplasm from many cultivars into one cross may be employed. Multiple or composite crosses may be desirable under a system of evolutionary breeding. The procedure is to select cultivars with one or more desired characters and proceed to make single crosses, double crosses, quadruple crosses, and so on. The multiple-crossing method is not recommended because of the uncertainty of getting all the desired genes into the final cross and the complexity of conducting tests to reveal the presence of the many associated genes. Under systems of natural selection, the best competitors may be the poorest yielding under commercial conditions.

A number of techniques have evolved for parental selection. A comparison of various methods of selecting parents in common bread wheat was presented by Bhatt (1973). Twelve techniques are presented as follows:

1. Conventional Method. A commercial cultivar adapted to the region in question and selected for its proven performance in an area frequently is the choice for one parent. The other parent may be chosen because it complements a recognized and specific weakness in the first. Conventional selection of parents may be based on the experience of the individual breeder. Lines or cultivars that have previously resulted in successful combinations should probably be reused because proven parents may lead to success.

2. Random Selection. This method may be used when there is a pressing need to initiate a crossing program and where no information on the parental types exists. At best, this is a hit-and-miss method and may not be desirable.

3. Ecogeographic Diversity. Geographic isolation could be important in shifts in the genetic structure of plant populations, and therefore ecogeographic diversity is considered to be a reasonable index of genetic divergence. Electrophoretic variation was found within strains of the wheat cultivar Kharkof maintained at 11 locations (Cox and Worrall, 1987). Kharkof is a heterogeneous landrace, and ecogeographic diversity such as this may be useful in parental selection.

4. Early-Generation Testing. Another method of choosing parents is based on the performance of the F_1 and F_2 generations. F_1 generation progeny, for example, whose performance is superior to either parent for a particular trait or traits, would be a likely source for successful selection in the F_2 and subsequent segregating generations. The observation of heterosis (hybrid vigor) in a particular cross might not indicate a successful combination for selection purposes in subsequent generations if the superior performance was due to heterozygous loci in the F_1 generation. Predicting selection success based on the F_2 generation is probably a better guide.

Two types of information can be useful in predicting potential selection gain: the mean and the variance of the F_2. An F_2 generation that has a high mean performance compared to another cross should be a more productive population in which selection can be conducted. A high variance or spread about the mean in the F_2 generation indicates the potential of selecting individual lines that excel for a single trait or multiple traits. In this situation, a breeder might choose to keep a population whose mean performance was only average if the variance in the F_2 was large enough to warrant further selection. A cross with selection potential is one with a good mean F_2 performance and with a range of variability sufficiently great to indicate superior lines.

An attempt to predict the hybridization potential of parental materials through hybridization and subsequent selection was made by Thurling and Ratinam (1987) (Figure 6.1).

5. Vector Analysis. Vector analysis was proposed by Grafius (1965) and Grafius, Thomas, and Barnard (1976) as a means for selecting desirable parents. In the vector-analysis approach, individual parents are evaluated for variations in yield components, including such factors in cereals as tillers per plant, kernels per head, and kernel size. The objective is parental selection that will lead to increased yields through component complementation. Each of the yield components is regarded as a vector.

By calculating the average values of yield components of potential parents, breeders can select individuals on the basis of the mid-parent values for the traits of interest. The combination of parents that gives rise to the best combination of yield components and hence yield would be chosen. This type of analysis can be extended to select parents for hybridization purposes based on the complementation of components for yield for any specific trait.

If tillering (Figure 6.2) and other yield components are traits that are important in determining yield, vector analysis may be useful. The agreement of the predicted response with actual results will depend on the knowledge of the component traits under selection, and these can vary with the environment. The use of the parental component in determining parental selection in barley crosses was reported by Grafius, Thomas, and Barnard (1976). All parents had a satisfactory yield but differed widely in component values. In every case, the mean yield of the progeny exceeded that of the parents.

6. Diallel Cross. A **diallel cross** consists of all possible crosses among a group of cultivars. The evaluation of these crosses has led to a number of different ways of analyzing the data obtained (Hayman, 1954a, 1954b; Jinks, 1954; Griffing, 1956). Diallel crosses can be used to select parents (Whitehouse *et al.*, 1958) and have been used to identify general and specific combining abilities of maize inbreds (Sprague and Tatum, 1942). Diallel crosses have not been widely used in plant breeding because (1) many crosses must be made (for example, to evaluate ten parents, 90 F_1 hybrids are required, or if the reciprocal crosses are combined, 45 hybrids are needed); and (2) there have been of criticisms of the methods of analysis (Baker, 1978). Diallel analysis may be useful for determining general combining ability in the development of open-pollinated or **synthetic cultivars**.

A.

B.

Figure 6.1 The plants in these photos are cowpeas (*Vigna unguiculata* L. Walp.) exhibiting a bush habit in contrast to a vine habit. The crop is grown for its seed, which averages 23% protein, or for forage. For purposes of parental selection, Thurling and Ratinam (1987) studied three prediction tests based on data collected from experiments with ten cowpea lines. The F_1 and F_2 generations were evaluated based on all possible crosses among the ten lines. The three prediction tests were based on (1) yield of prospective parents, (2) general combining ability estimated from F_1 and parental data, and (3) frequencies of superior plants in F_2 populations.

It was concluded that parental yields provide a sound basis for an initial screening of prospective parents. Parental screening should identify a high-yielding potential parent that can be crossed with a number of contrasting lines to produce F_2 populations for the final screening phase. The complexity of yield, from both a physiological and a genetic viewpoint, means that parental performance data alone are inadequate as a basis for predicting cross potential. Thurling and Ratinam (1987) suggested that a relatively small group of F_2 populations be evaluated on the basis of parental yields in order to maximize the chances of identifying the cross with the highest yield potential.

Photos courtesy International Institute of Tropical Agriculture, Ibadan, Nigeria.

Figure 6.2   The basal part of the many-branched wheat plant (*Triticum aestivum*) shown in this photo was produced from a single seed and illustrates the phenomenon known as tillering. Tillering may be regarded as a component of yield, along with kernels per head and kernel size. Considerable attention has been directed to selection for increases in these components in an effort to increase yield. It may be argued that the proper combination of these components can result in yield increases through complementation.

Yield components have been observed to vary with competition, and as one is increased, the other components compensate, thereby giving cereal plants a degree of plasticity so that yield may be maintained with variations in the environment, including seeding rate (Adams, 1967). This fact can be explained on the basis of the sink-source relationship, whereby the amount of available photosynthate is the true determinant of yield, and yield components merely describe yield.

The tillering habit illustrated in this photo may be regarded as a remnant of perennial plant habit. Most cereal plants are grown as annuals, but a winter annual habit also exists. In perennial forage grasses, the perennial habit is common, and tillering provides a mechanism to develop not only a dense sward but also regrowth after a dormant or winter period.

In choosing cereal parents for hybridization purposes, plant breeders must assess the value of tillers. In annual crops where regrowth is not required and at seeding rates where tillering cannot be expressed, plant breeders may consider a uniculm as a parent. In short, parental selection is not a cut-and-dried decision but is contingent upon well-defined objectives.

Photo courtesy Ontario Ministry of Agriculture and Food.

7. Incomplete Diallel Cross. The practical application of diallel crosses is limited by the very large number of crosses, many of little potential value, that must be handled if a reasonably wide range of parents is to be considered. In order to avoid this difficulty, Lupton (1965) proposed a series of incomplete diallel sets of crosses in which potential parental cultivars were hybridized with five "tester" cultivars. These tester cultivars were chosen to represent the genotypes of winter wheats generally cultivated in the country. Trials conducted on F_2 and F_3 serve to identify superior crosses and parents.

8. Diallel Selective Mating System. To overcome defects of basic breeding procedures, often involving two-parent crosses and a relatively small gene pool, Jensen (1970) proposed a diallel selective mating system. The diallel selective mating system is designed to shift the attention of plant breeders from the concept of employing germplasm sources discretely and sequentially in the formation of numerous crosses. The system forces the simultaneous insertion of multiple genotypes into a few central populations.

Four stages are involved in the implementation of the diallel selective mating system:

- Choosing a basic parent series of crosses. Progeny from these crosses are used to set up the F_1 diallel series of crosses.
- Compositing the F_1 diallel crosses to form a parental block P_2, P_3, and so on.
- Practicing selective mating on the parental block based on mass and recurrent selection principles.
- Standard line selecting from the various F_5 composite populations.

The soundness and efficacy of selective mating procedures in the early-segregating generations, which involve mass and recurrent selection to channel progress in desired directions, were demonstrated by Redden and Jensen (1974). The selective mating system was shown to be efficient in germplasm use and effort. New cultivars can be developed in time periods comparable to other common breeding methods (Jensen, 1978). With this method, environmental variables such as competition may exert a helpful force.

The deficiencies Jensen saw in the use of basic breeding procedures can be summarized as follows:

- Multiple-parent participation in any one population is limited, despite large and growing germplasm collections. The sequential rather than simultaneous entrance of parents into progeny has the effect of blurring the initially sharp edge of a desired breeding objective.
- An initially restricted gene pool and the restriction of genetic recombination because of linkage inhibit genetic variability and recombination potential.
- Selfing and the low level of intermating intensify linkage and limit genetic recombination outside of narrow family lines of descent.

9. Haploid Method. The haploid breeding method (Kasha and Kao, 1970) provides a means of measuring the performance of different parental combinations. The system depends on the ability to generate haploids from F_1 plants of specific parents. In barley and possibly wheat (Barclay, 1975) this is achieved by use of *Hordeum bulbosum*. Anther culture, a method readily used in tobacco (*Nicotiana tabacum*) may have application in other crops (Collins, Legg, and Kasperbauer, 1974; Liang, Xu, and Hoang-Tang, 1987).

The haploid method involves a specific cross from which haploids are derived from F_1 plants with subsequent chromosome doubling to produce homozygous plants. Haploid gametes are representative of the potential gametic output of that cross. Homozygous progeny provide a sample that would be produced by any given cross.

Approximately 20 homozygous lines are enough to indicate the mean and range of the homozygous lines that would be selected from a cross (Reinbergs, Park, and Song, 1976; Simpson and Snape, 1979). This estimate of the mean and range of lines should be sufficient to indicate those crosses with superior characteristics. Once a desirable combination has been identified, then the parental cross would be made and a much larger population grown for subsequent selection.

Doubled haploids offer a precise method of cross prediction because they provide a means of estimating the population mean and additive genetic variance and hence the distribution of inbred lines that would be expected from a cross (Caligari, Powell, and Jinks, 1985). Doubled haploids can be multiplied readily to enable trials to be carried out over different seasons and sites so that genotype x environment interactions can be taken into account.

10. The F_2 Triple Test Cross. To overcome the problem of labor inputs associated with diallel analysis, Kearsey and Jinks (1968) suggested the triple test cross, whereby a random sample of n individuals from the population to be investigated are crossed as male parents to the same three testers. Two of these testers are homozygous, and the third tester is the F_1 produced from them. Hybridization results in 3n families, each family being replicated.

The triple test cross provides an efficient method for estimates of dominance also an unambiguous test for epistasis. The system provides satisfactory estimates of the additive genetic variance necessary for cross prediction and has been used for wheat (Snape, 1982) and barley (Thomas and Tapsell, 1983; Tapsell and Thomas, 1983). The F_2 triple test cross, however, remains too labor intensive, and therefore its practical use in plant breeding is somewhat limited.

11. F_3 System. To avoid the extensive crossing programs of diallel crosses and the F_2 triple test cross, Jinks and Pooni (1980) advocated growing samples of random F_3 lines and using twice the components of variance of F_3 family means as a measure of the additive genetic variance. Caligari, Powell, and Jinks (1985) showed that satisfactory estimates of the genetical parameters for cross prediction with spring barley can be obtained from F_3 lines. If such predictions, however, are to be of value in plant breeding, genotype x environment interactions need to be taken into account.

12. Multivariate Cross Prediction Method. Parental selection based on univariate methods are useful, and although it is important to examine the results for each character independently, breeding programs are rarely based on single characters. Bhatt (1970, 1976) suggested the use of multivariate analysis for yield and wheat quality. Bivariate, trivariate, and quadrivariate predictions have been made using genetic parameters estimated for F_3 and doubled haploid populations (Powell *et al.*, 1985).

The proportion of inbred lines satisfying two or more criteria simultaneously cannot be calculated simply as the product of two or more probabilities obtained independently, unless there is no genetic correlation between them. Pooni and Jinks (1978), however, described procedures whereby joint predictions for two or more characters can be obtained. Sample sizes must be larger than for univariate predictions. Whitehouse (1971) illustrated the use of multivariate or canonical analysis as an aid to selection at many stages of plant breeding, including prediction of parental combinations.

SELECTION IN SEGREGATING POPULATIONS

Nature has practiced plant selection since living organisms emerged. The criterion for selection was individuals possessing survival characteristics. When humans began selection, the objective became plants that best met human needs. Selection by nature involved millions of individuals and occurred slowly and methodically over hundreds, thousands, and perhaps millions of years. There is ample evidence of organisms that are adapted to extreme and unique environments. The lesson to be learned is that large populations are required if the most desirable individuals are to be found, but even then the task of identifying the most desirable segregates is a difficult one (Figure 6.3).

In discussing the need for large populations, Shebeski (1967) suggested that genes for yield are present on every chromosome in bread wheat. This statement is based on the fact that yield is an end product of the physiological development and fruition of a plant in its environment. Any gene that is involved in the development of the plant must be considered a yield gene. If, however, only one independently inherited gene was found per chromosome, that would mean 42 different loci. Crosses between two relatively unrelated wheat cultivars could involve more than 50 alleles affecting yield. Even for 50 allelic gene differences, the size of the population that would have to be grown in the F_2 for the chance presence of each possible genotype is staggering. Obviously, a plant breeder must work within limitations of available land and labor. To maximize population size with the resources available, plant breeders have turned to mechanization.

The question of when to select within a segregating population is of concern to plant breeders. If the parents of a cross differ by 25 important, independent yield genes, only 0.075% of the F_2, or one plant per 1,330 may be expected to contain all 25 of the desired traits (Shebeski, 1967). The remaining plants in the population would exhibit a lower number of the better alleles. If selection is delayed until F_4, only one plant in approximately 1.8 million would contain all 25 of the more desirable alleles. Early-generation selection therefore seems imperative.

If selection in F_2 is practiced, subsequent selection in F_3 is critical also. With 25 important, independent genes, one F_3 line out of 1,330 would contain all the desired alleles, as in the F_2 generation. Of the 25 alleles in question, the most probable genotype of the F_2 progenitor would be homozygous for eight and heterozygous for 17 genes. For the chance presence of one F_3 plant with all the desired alleles 133 plants per F_3 selection should be retained.

It is generally acknowledged that a plant breeder cannot reliably select for yield on a single-plant basis in the F_2 generation. Likewise, reliable selection is not possible in the F_3 or the F_4. In the F_4, Shebeski (1967) suggested that at least 100 plants should be selected to retain the critical genotypes. Failure to select a sufficiently large population means that the best genotypes are irretrievably lost.

The numbers just considered present the plant breeder with a dilemma. The logical solution would be to increase the number of plants grown from superior lines in each of the segregating generations. To do so would mean that fewer crosses could be handled. The ability to choose parents wisely becomes increasingly apparent. The limiting factor is the ability to recognize superior segregates, and this in turn emphasizes the need for a clear objective (Figure 6.4), aided perhaps by a well-planned and -designed ideotype to guide early-generation selection.

Figure 6.3 The clumps to the left in this photo are strawberry plants (*Fragaria* spp) that remain from a large population of plants following a cross. A total of five plants that have been selected out of thousands in the field can be seen in the photo. The discarded plants adjacent to these individual plants have been clipped to destroy them and to prevent asexual runners from getting mixed with the desired selections.

Strawberries are perennial, and the F_1 can be maintained by asexual propagation. Following hybridization, individual plants can be screened over a number of seasons for specific objectives, and only those that meet these objectives are saved.

To be grown economically, a cultivated strawberry must have favorable attributes such as desirable flavor, yield, color, size, and shipping quality, as well as disease resistance, ease of management, and, increasingly, adaptability to machine harvesting. The success of the plant breeder in achieving genetic gains through hybridization and selection rests on the ability to recognize and choose desirable genotypes from a large heterogeneous population.

In the strawberry, species exist that have two (diploid), four (tetraploid), six (hexaploid), and eight (octoploid) sets of chromosomes. The cultivated strawberry is an octoploid. By crossing a diploid strawberry with a hexaploid, a tetraploid can be produced. By treating this tetraploid with colchicine, an octoploid can be produced that can be crossed with a cultivated strawberry.

In most early-generation selection nurseries, no provisions are made for measuring, minimizing, or eliminating environmental effects such as plant competition, soil heterogeneity, and human subjectivity. As a result, it is impossible to identify and separate variances associated with genotype and environment, and consequently it is impossible to relate yields from one generation to the next.

A lesson may be learned from a study involving the normally asexually propagated potato (*Solanum tuberosum*) (Brown, 1987). Traditional potato breeding programs begin by growing potato seedlings from true seeds (Figure 6.5). On the basis of single- versus five-plant plots, Brown (1987) concluded that single-plant selection was generally ineffective. Error variances were high compared to the five-plant plots for total tuber weight, mean tuber weight, and

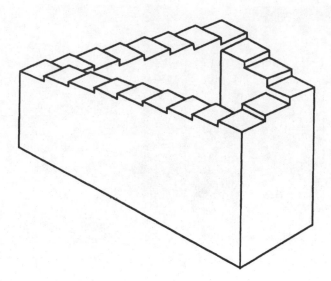

Figure 6.4   The challenge of locating the top or bottom step in this figure is similar to the challenge facing a plant breeder of selecting the highest or lowest segregate in a population. Without a well-defined objective, the plant breeder may find that one plant appears as good or as bad as the next. There is no apparent winner. Selection in plant breeding is not based on identifying a few plants from a small population. Usually thousands of segregates are involved, thereby emphasizing the need for a clear objective.

Plant selection for readily identifiable, qualitatively inherited traits may be simple and obvious. Selection for quantitatively inherited traits, however, is more complex, and guidelines developed from an objective, a hypothesis, or an ideotype are essential. Adding to the complexity of selection is variation resulting from the environment. A segregate may appear excellent because of a weak or missing plant adjacent to it.

Selection may be facilitated by growing plants under controlled environmental conditions, but commercial cultivars of most crops are grown in the field, and eventually selections must be field tested. Recognition of this fact means that selection is usually conducted under field conditions.

number of tubers per plant, but subjective, visual preference was not related. Correlation coefficients between single- and five-plant plots for breeder preference were inversely related, in magnitude, to the mean preference score for each group. Many clones would have been discarded on the basis of a single-plant observation but would have been selected based on the data recorded from two five-plant plots.

In self-fertilized crops, single-plant selection presents the plant breeder with an enormous challenge. Work by McGinnis and Shebeski (1968) corroborates the findings of many other investigators that there is no significant correlation between single-plant F_2 selections and F_3 line yield. These workers reported on a study involving 293 single-plant F_2 selections tested for yield in F_3 against controls of unselected plants, the controls representing the mean of the cross. Although no correlation was found between F_2 plants and F_3 line yield, some yield gain was achieved since in all cases the mean yield of the F_3 plots from selected F_2 plants was significantly higher than the control. The suggestion is that careful selection of F_2 plants will

Figure 6.5 The tomato-like seedballs on this potato plant in Photos A and B bear true seeds (Photo C) that are heterozygous. Steps toward cultivar development in the potato involve controlled pollination of an emasculated flower, shown in the upper left of Photo A, to form seed-bearing fruit containing up

Figure 6.5 continues overleaf

increase the general yielding capacity of the F_3 lines and will consequently be advantageous to a breeding program. Junior plant breeders should recognize the intense struggle associated with small yield increases.

For some agronomic features, such as winter survival or acid tolerance, natural selection can be useful in aiding human selection (Figure 6.6). The automatic elimination of undesirable plants can provide a screening technique, even when the desired plant is not phenotypically distinct. Introduction of biological agents such as insects or disease epidemics may aid natural selection. Competition, winter kill, or susceptibility to disease could screen a composite cross for tolerance or resistance to these factors. No evidence exists, however, that natural selection will produce a high-yielding genotype.

Although the plant breeder may have problems selecting for visual phenotypic traits above ground, even greater problems exist in attempts to select for below-ground features such as root growth, root morphology, or root depth. Selection may be enhanced by correlations between visual, above-ground features and specific root characteristics. Such correlations may be difficult to obtain or may be low in value. Plant breeders often have to resort to ingenious techniques to achieve a stated objective.

Robertson, Hall, and Foster (1985) reported on an example of a specialized technique in which an herbicide was banded into the root zone at seeding time at specific depths and lateral distances from the seed rows. As the roots of segregating populations of cowpeas (*Vigna unguiculata* L. Walp.) grew to contact the herbicide bands, the seedlings developed visual leaf symptoms, which allowed for a measure of root progress. Plants were effectively screened under field conditions.

Figure 6.5 continued

to 50 seeds. Each seed can produce a plant, but most researchers agree that no selection of individual genotypes for agronomic traits should be conducted at this seedling stage (Anderson and Howard, 1981; Tai and Young, 1984; Brown *et al.*, 1984).

The extreme heterozygosity of the autotetraploid potato requires the screening of large populations to secure the desired genotypes. To get one seedling of sufficient merit to warrant development as a new cultivar usually requires testing of about 100,000 seedlings.

Tubers harvested from seedlings produced from true potato seeds normally are planted in widely spaced single hills. These hills are given the first and most drastic culling, leaving only 2 to 5% of the original population. Screening usually is based on visual evaluation for disease resistance, plant and tuber type, maturity, and tuber yield. Selection of superior genotypes at this stage is critical.

In the following season, selections may be planted in four-hill units, followed by two to three years of growing, testing, and screening from ten-hill plots. Potatoes from ten-hill plots provide sufficient tubers for preliminary evaluations on quality. Tubers kept from the ten-hill plots go to 100-hill plots for additional evaluation over two to three years at various sites.

Photos courtesy Ontario Ministry of Agriculture and Food.

A.   *B.*

Figure 6.6   The perennial forage plot in Photo A has been severely winter killed in comparison with adjacent plots. Wheat plots in Photo B are grown under extreme levels of aluminum toxicity. In both cases natural selection has been an effective aid to human selection, provided hardiness and aluminum tolerance were the objectives.

In crops exhibiting winter habit, genetic differences exist among species and among cultivars within each species. Winter hardiness is a general term and is influenced by a number of factors, which makes selection, evaluation, and improvement through plant breeding difficult. The ultimate test for winter hardiness is field survival, but only rarely are environmental conditions suitable to obtaining the clear ranking of different genotypes evident in Photo A. Laboratory tests for cold tolerance are possible, but caution must be taken against screening for a single trait.

It is apparent from Photo B that genotypes exist for aluminum tolerance, but plant breeders should not attempt to solve all crop production problems. Perhaps a better solution to the problem exemplified in Photo B would be soil amendments that remedy the problem. Another alternative might be to grow another aluminum-tolerant crop or abandon the region for crop production purposes.   If an aluminum-tolerant wheat crop is produced, the question arises of its yield performance under the stress conditions associated with this situation.

Too frequently a plant breeder is ready to undertake finding a solution to a problem associated with producing a crop out of its range of adaptation. Plant breeding objectives must be carefully designed.

SCREENING METHODS FOR PLANT BREEDING PROGRAMS

The terms selection and screening often are used interchangeably, but for plant breeding purposes subtle differences exist. A dictionary defines screening as ascertaining the suitability of individuals, whereas selection implies a choice.   Frequently, plant breeders choose early-generation segregates that are later screened for special criteria, such as yield or quality aspects, as the last part of a breeding program.   If screening techniques were considered immediately after the objectives were established, a saving of time might result.

Screening procedures offer a means of helping the plant breeder select desirable individuals. As an aid to helping plant breeders develop systematic screening procedures, Britten (1960) classified screening techniques into five categories.

1. Direct identification. This may be accomplished by the unaided eye, may involve qualitative features, and may be used where the desired characteristic is of low frequency in the population.
2. Procedures involving physical comparisons. Specific plant parts may be measured individually by a technician to screen segregates for a specific objective.
3. Procedures involving chemical determinations. The ability to spray segregating plants with a specific chemical and cause a reaction useful for screening is a plant breeder's dream. Pigmented onions contain protocatechuic acid, which is toxic to the fungus causing onion smudge (Rieman, 1931). Screening for tolerance to soil salinity (Figure 6.7) may be achieved under saline conditions. In general, screening methods using chemicals progress from less complex to more complex tests.
4. Indirect procedures involving biological testing. The production of beer from malted barley or of bread from wheat are examples of indirect biological testing. Obviously, relatively few cultivars can be screened by indirect procedures because large volumes of seed are needed for effective screening. The importance of an initial screening by other methods of the large number of segregates to eliminate the undesirable genotypes early in the program is evident.
5. Combination methods. The production of beer and bread are examples of the combination method involving chemicals, yeast organisms, and physical conditions. The plant breeder should be constantly on the alert for new techniques that may be used to screen large numbers of individuals.

Figure 6.7   The cereal plants in this photo have been destroyed by toxic levels of soil salinity. Adequate screening methods are needed for isolating salt-tolerant genotypes of crops. One approach is to test the productivity of selected genotypes at a low salinity level; an alternative approach is to isolate genotypes that survive at very high salinity levels. Using the latter method, Kingsbury and Epstein (1984) eliminated 93% of the 5,000 accessions under test. This had the effect of reducing the number of salt-tolerant lines to a more manageable level. Productivity of the remaining 312 genotypes was then tested over a range of salinities and compared with that of other lines, ranging from salt-tolerant to salt-sensitive. The technique proved effective in identifying salt-tolerant genotypes.

Plant breeders using chemicals as a screening technique must decide whether to test at a juvenile stage, the adult stage, or over the entire life cycle. If genotypes are screened under the uncontrolled conditions shown in the photo, salinity may vary with slope and drainage from meter to meter.

Photo courtesy M. Dudas, University of Alberta, Edmonton.

REFERENCES

Adams, M. W. 1967. Basis of Yield Component Compensation in Crop Plants with Special Reference to the Field Bean (*Phaseolus vulgaris*). *Crop Science* 7:505-510.

Anderson, J.A.D., and H. W. Howard. 1981. Effectiveness of Selection in the Early Stages of Potato Breeding Programmes. *Potato Research* 24:289-299.

Baker, R. J. 1978. Issues in Diallel Analysis. *Crop Science* 18:533-536.

Barclay, I. R. 1975. High Frequencies of Haploid Production in Wheat (*Triticum aestivum*) by Chromosome Elimination. *Nature* 256:410-411.

Bhatt, G. M. 1970. Multivariate Analysis Approach to Selection of Parents for Hybridization Aimed at Yield Improvement in Self-Pollinated Crops. *Australian Journal of Agriculture Research* 21:1-7.

———. 1973. Comparison of Various Methods of Selecting Parents for Hybridization in Common Bread Wheat (*Triticum aestivum* L.) *Australian Journal of Agriculture Research* 24:457-464.

———. 1976. An Application of Multivariate Analysis to Selection for Quality Characters in Wheat. *Australian Journal of Agriculture Research* 27:11-18.

Briggs, F. N., and P. F. Knowles. 1967. *Introduction to Plant Breeding*. Reinhold Publishing Corporation. Davis, California. 426 pp.

Britten, E. J. 1960. A Proposed Classification of Screening Methods for Plant Breeding Programs. *Euphytica* 9:293-303.

Brown, J. 1987. A Comparison Between Single Plant Plots and Five Plant Plots for the Initial Selection Stage of a Potato Breeding Programme. *Euphytica* 36:711-718.

Brown, J., P.D.S. Caligari, G. R. Mackay, and G.E.L. Swan. 1984. The Efficiency of Seedling Selection by Visual Preference in a Potato Breeding Programme. *Journal of Agriculture Science* 103:339-346.

Caligari, P.D.S., W. Powell, and J. L. Jinks. 1985. The Use of Doubled Haploids in Barley Breeding. 2: An Assessment of Univariate Cross Prediction Methods. *Heredity* 54:353-358.

Collins, G. B., P. D. Legg, and M. J. Kasperbauer. 1974. Use of Anther Derived Haploids in *Nicotiana*. I: Isolation of Breeding Lines Differing in Total Alkaloid Content. *Crop Science* 14:77-80.

Cox, T. S., and W. D. Worrall. 1987. Electrophoretic Variation Among and Within Strains of "Kharkof" Wheat Maintained at 11 Locations. *Euphytica* 36:815-822.

East, E. M. 1916. Studies on Size Inheritance in *Nicotiana*. *Genetics* 1:164-176.

Grafius, J. E. 1965. *A Geometry of Plant Breeding*. Michigan State University Research Bulletin. East Lansing. 59 pp.

Grafius, J. E., R. L. Thomas, and J. Barnard. 1976. Effect of Parental Component Complementation on Yield and Components of Yield in Barley. *Crop Science* 16:673-677.

Griffing, B. 1956. Concept of General and Specific Combining Ability in Relation to Diallel Crossing Systems. *Australian Journal of Biological Science* 9:463-493.

Hayman, B. I. 1954a. The Analysis of Variance of Diallel Crosses. *Biometrics* 10:235-244.

———. 1954b. The Theory and Analysis of Diallel Crosses. *Genetics* 39:789-809.

Jensen, N. F. 1970. A Diallel Selective Mating System for Cereal Breeding. *Crop Science* 10:629-635.

———. 1978. Composite Breeding Methods and the DSM System in Cereals. *Crop Science* 18:622-626.

Jinks, J.L. 1954. The Analysis of Continuous Variation in a Diallel Cross of *Nicotiana rustica* varieties. *Genetics* 39:767-788.

Jinks, J. L., and H. S. Pooni. 1980. Comparing Predictions of Mean Performance and Environmental Sensitivity of Recombinant Inbred Lines Based upon F_3 and Triple Test Cross Families. *Heredity* 45:305-312.

Kasha, K. J., and K. N. Kao. 1970. High Frequency Haploid Production in Barley (*Horderum vulgare* L.). *Nature* 225:874-876.

Kearsey, M. J., and J. L. Jinks. 1968. A General Method of Detecting Additive Dominance and Epistatic Variation for Metrical Traits. *Heredity* 23:403-409.

Kingsbury, R. W., and E. Epstein. 1984. Selection for Salt-Resistant Spring Wheat. *Crop Science* 24:310-315.

Liang, G. H., A. Xu, and Hoang-Tang. 1987. Direct Generation of Wheat Haploids via Anther Culture. *Crop Science* 27:336-339.

Lupton, F.G.H. 1965. Studies in the Breeding of Self-Pollinating Cereals. 5: Use of the Incomplete Diallel in Wheat Breeding. *Euphytica* 14:331-352.

McGinnis, R. C., and L. H. Shebeski. 1968. The Reliability of Single Plant Selections for Yield in F_2. In *Proceedings of the Third International Wheat Genetics Symposium*, edited by K. W. Finlay and K. W. Shepherd. Australian Academy of Science, Canberra, pp. 410-415.

Pooni, H. S., and J. L. Jinks. 1978. Predicting the Properties of Recombinant Inbred Lines Derived by Single Seed Descent for Two or More Characters Simultaneously. *Heredity* 40:349-361.

Powell, W., P.D.S. Caligari, J. W. McNicol, and J. L. Jinks. 1985. The Use of Doubled Haploids in Barley Breeding. 3: An Assessment of Multivariate Cross Prediction Methods. *Heredity* 55:249-254.

Redden, R. J., and N. F. Jensen. 1974. Mass Selection and Mating Systems in Cereals. *Crop Science* 14:345-350.

Reinbergs, E., S. J. Park, and L.S.P. Song. 1976. Early Identification of Superior Barley Crosses by the Doubled Haploid Technique. *Zeitschrift für Pflanzenzuchtung* 76:215-224.

Rieman, G. H. 1931. Genetic Factors for Pigmentation in the Onion and Their Relation to Disease Resistance. *Journal of Agriculture Research* 42:251-278.

Robertson, B. M., A. E. Hall, and K. W. Foster. 1985. A Field Technique for Screening for Genotypic Differences in Root Growth. *Crop Science* 25:1084-1090.

Shebeski, L. H. 1967. Wheat and Breeding. *Canadian Centennial Wheat Symposium*, edited by K. F. Nielsen. Modern Press, Saskatoon, pp. 249-272.

Simpson, E., and J. W. Snape. 1979. Cross Prediction for Yield Using Doubled Haploid Lines. *Barley Genetics Newsletter* 9:95-97.

Snape, J. W. 1982. Predicting the Frequencies of Transgressive Segregants for Yield and Yield Components in Wheat. *Theoretical and Applied Genetics* 62:127-134.

Sprague, G. F., and L. A. Tatum. 1942. General Versus Specific Combining Ability in Single Crosses of Corn. *Journal American Society of Agronomy* 34:923-932.

Tai, G.C.C., and D. A. Young. 1984. Early Generation Selection for Important Agronomic Characters in a Potato Breeding Population. *American Potato Journal* 61:419-434.

Tapsell, C. R., and W.T.B. Thomas. 1983. Cross Prediction Studies on Spring Barley. 2: Estimation of Genetical and Environmental Control of Yield and its Component Characters. *Theoretical and Applied Genetics* 64:353-358.

Thomas, W.T.B., and C. R. Tapsell. 1983. Cross Prediction Studies on Spring Barley. 1: Estimation of Genetical and Environmental Control of Morphological and Maturity Characters. *Theoretical and Applied Genetics* 64:345-352.

Thurling, N., and M. Ratinam. 1987. Evaluation of Parent Selection Methods for Yield Improvement of Cowpea (*Vigna unguiculata* L. Walp.). *Euphytica* 36:913-926.

Whitehouse, R.N.H. 1971. Canonical Analysis as an Aid in Plant Breeding. In *Barley Genetics* II, Proceedings of the Second International Barley Genetics Symposium, edited by R. A. Nilan. Washington State University Press, Pullman, pp. 269-282.

Whitehouse, R.N.H., J. B. Thompson, and M.A.M. Do Valle Ribeiro. 1958. Studies on the Breeding of Self-Pollinating Cereals. 2: The Use of a Diallel Cross Analysis in Yield Prediction. *Euphytica* 7:147-169.

CHAPTER 7

Mass and Pure-line Selection

SELECTING FROM VARIABLE POPULATIONS

Self-fertilized crops may be **heterogeneous**, as exemplified by **landrace cultivars**, which may have originated from single plant selections but which have been grown over a long period of time. Landrace cultivars are a mixture of genotypes, perhaps because of breeders' lack of appreciation of selection techniques to produce a pure line, spontaneous mutations that subsequently were dispersed through the population, chance hybridization, or a combination of factors. Natural selection in landraces has resulted in a **cultivar** that is well adapted to the local environment.

Depending on the amount of human selection that has been conducted, differences among genotypes in a landrace cultivar may be small and inconspicuous, with minor differences in plant height because landrace cultivars tend to be tall, with similar dates of **anthesis** and maturity, and with similar seed size. On the other hand, a lack of human intervention may mean that differences among plants in a landrace cultivar will be conspicuous. Awned or awnless types may exist, or differences in kernel color may be found.

After years of self-pollination, individual plants are homozygous, and selection of such individuals would result in a faithful reproduction of their phenotype within the limitations imposed by the environment. Selection of different plants in the heterogeneous population, however, could result in contrastingly different plants.

Selection for a specific trait from a biologically variable population can be successful only if differences are genetic. The work of the Danish biologist W. Johannsen (1903), translated by J. A. Peters in 1959, serves to illustrate this point. His studies formed the basis for a plant breeding procedure known as **mass selection**. Johannsen noted that a self-fertilized cultivar of a common bean cultivar named Princess had seeds of different sizes. Experiments were conducted to determine whether the variability was due to inherited or environmental effects. To do this Johannsen selected 19 individual plants that exhibited a range in seed weight. Although the progeny of each plant produced seed of similar weight to that of the parent, differences were genetic and could be passed to the next generation.

MASS SELECTION

The oldest breeding method for crop improvement is mass selection, which is based on **phenotypic selection**. In ancient times, mass selection was conducted by separating the heaviest or plumpest seed by wind, hand, or sieves, harvesting plants that retained their seed, or cutting plants that matured early. Years of continuous selection eventually produced a crop that met human needs. Phenotypic mass selection may be based on visible characteristics such as bald, semi-awned, or awned type, flower color, or maturity. Although selection is based on phenotype, the plant breeder really selects for **genotype** because heritability is a measure of the degree of correspondence between genotype and phenotype.

The fact that crops could be improved by continuous selection led to the belief that selection could improve populations almost without limit, an idea that was stimulated by Charles Darwin's publication of *Origin of Species* in 1859. Subsequent understanding of genetics made it clear that improvement is limited to the genetic variability initially found in the original population. The work of Johannsen defined the limits and established mass selection as an effective first step in the improvement of self-fertilized crops.

Mass selection has been used effectively in the improvement of landrace cultivars because the lack of uniformity makes them unacceptable to modern industrial societies. Selection of homozygous and homogeneous cultivars from landraces helps preserve the wide adaptability, yield stability, and durability against disease, while eliminating obvious shortcomings and improving uniformity.

Mass selection may be achieved by identifying and marking desirable plants as they develop and harvesting only the phenotypically desirable plants at maturity. The same effect may be achieved by roguing undesirable plants in the field at various times during crop development. The remaining plants can be harvested and bulked. Either way, the best plants as determined by their ability to suit human needs are identified and bulked.

When the mass selection method is used, the number of individual plants or heads selected must be large enough to preserve the identity of the original variability of the landrace cultivar. The majority of plants selected are retained in mass selection, and if this is a large enough number, only minor modifications may occur from selection. As a result, a minimum of testing is required to ensure that performance has not been jeopardized.

In commercial agriculture systems in developed countries, landraces have been replaced by pure lines. The few landraces remaining in isolated areas of the world will inevitably be replaced as improved cultivars are produced.

Mass Selection as a Modern Breeding System

Mass selection historically has been a productive breeding method in self-fertilized crops. In modern breeding programs, mass selection remains a viable breeding system where the objective is population improvement associated with modifying the frequencies of desirable genes. Mass selection is not an outdated breeding system but one that is growing in importance because it permits the inexpensive propagation of both a large number of crosses and a large number of plants within crosses. Some examples of mass selection will help illustrate its application and principles of use.

   • Mass selection may serve as the initial means of improvement when a new crop is being investigated. A commercial cultivar may be developed rapidly to meet human needs by eliminating obvious off-types or poorly adapted plants (Figure 7.1).
   • Mass selection may be used to preserve the identity of an established cultivar or one about

A.   B.

Figure 7.1   Photos of vegetable-type pigeon peas (*Cajanus cajan*) of Kenyan origin being grown at ICRISAT (International Crops Research Institute for the Semi-Arid Tropics) in India, where about 90% of the pigeon pea crop of the world is grown. Pigeon pea is largely a self-fertilized crop with some natural hybridization. Traditional late cultivars mature in 250 days, but selections that mature in less than 100 days have been developed and are favored for double cropping.

Mass-selection techniques from existing landraces introduced into other parts of the world offer enormous scope for cultivar development. Included in mass-selection opportunities is the development of pigeon pea as a perennial crop achieved by **ratooning**. Perennial forms need resistance or tolerance to disease organisms to be suited to ratoon cropping.

Introductions of pigeon pea into potential production areas in the semiarid tropics could be mass selected for specific characteristics, such as date of maturity, response to ratooning, general adaptation, and agronomic traits, without hybridization.

Photos courtesy International Crops Research Institute for the Semi-Arid Tropics (ICRISAT), India.

to be released following any of the breeding schemes involved. The plant breeder selects perhaps 250 to 300 typical plants or heads from the most advanced plot and grows them in individual rows for comparisons among all the selections. Those rows that do not represent the population because of a noticeable difference in phenotype are eliminated, preferably before anthesis to remove any possibility of cross-fertilization. The remaining rows are harvested in bulk as **breeder seed** (Figure 7.2). Before harvesting, however, representative heads or plants are taken from each row to preserve the identity of all remaining rows so that breeder seed may be produced with reliability in the future.

• Mass selection may be applied to segregating populations following hybridization where the objective is one of increasing the gene frequencies of certain agronomic characters in a hybrid population. Derera and Bhatt (1972), for example, applied mechanical mass

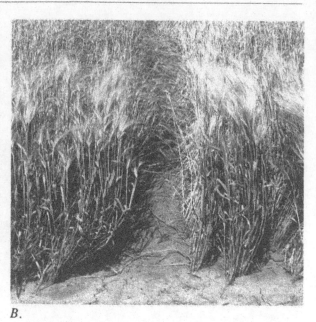

A.   B.

Figure 7.2   The development of breeder seed of self-fertilized crops may involve mass selection. The rows in these photographs of barley were each produced from seed of selected plants of an advanced selection destined to become a commercial cultivar. Each row is examined repeatedly for phenotypic differences in an effort to produce breeder seed that is uniform and homozygous, as is evident in Photo A. Rows that appear visually distinct are rogued, as illustrated in Photo B by the row that has been eliminated.

At maturity a representative plant is harvested from each row to reestablish long rows for production of a new lot of breeder seed as required. Rows are harvested in bulk and to avoid mechanical mixing or cross-fertilization with plants of the same species, a wide alleyway is maintained, as evident in Photo A. In normally self-fertilized crops, such as barley, the possibility exists that up to 4% cross-fertilization may occur and hence the wide alleys. Risk of cross-fertilization is reduced by the fact that the block in the background is triticale and that on the right is oats.

Instead of attempting to preserve the broad characteristics of a landrace cultivar by mass selection, the objective in breeder-seed production is to preserve the identity, but with a high degree of homozygosity, in a commercial cultivar.

selection to genetically heterogeneous and homogeneous populations of wheat to mass select for seed size. The results suggest that in genetically heterogeneous populations mechanical mass selection operates largely on the genetic variability for seed size, whereas in genetically homogeneous populations mass selection operates mainly on the nongenetic variation in seed size. Selection for large seed size resulted in a greater percentage of individuals possessing high kernel weight and high grain weight per ear, which in turn was reflected in higher grain yield. In addition to direct effects, Derera and Bhatt (1972) observed indirect effects. Mass selection for seed size eliminated from the segregating populations the genotypes that were not tolerant to disease such as rust, and those that resulted in shrunken seed and low yield. These workers suggested that an applied breeding program be so designed that the F_2 segregating populations are subjected to mechanical mass selection pressures for one or two generations. Desirable segregates are then selected from the resultant population. The elimination of undesirable genotypes producing an inherently low yield is expected to increase the probability of selecting promising individuals from the remaining population.

- Mass selection may be used to develop **horizontal resistance**, whereby general resistance rather than specific resistance exists (Figure 7.3). Mass selection provides a technique for assessing resistance quantitatively. With mass selection, the majority of the plants selected are retained.

The mass selection technique was applied to the oat crop (*Avena sativa* L.) by Frey (1967) to increase seed width by passing seed over a screen in each of the F_3 to F_7 generations. The mean yield of the final population was 9% higher than that of a comparable unselected population but was accompanied by a later heading date and increased plant height. Frey suggested that additional genetic advances through pure-line selection or additional mass selection should be possible because genotypic variances remaining within the population in the F_7 of the mass-selected line of descent were highly significant for all three traits—seed weight, heading date, and plant height.

In a subsequent program (Tiyawalee and Frey, 1970), mass selection was used to increase the gene frequency of crown rust (*Puccinia coronata*) resistance alleles in a genetically heterogeneous population of oat plants. The population was subjected to crown rust and the seeds subsequently screened for high density. Most of the increase in the frequency of the resistance alleles occurred in the first three cycles of selection. Additional screening did not increase further the gene frequency for crown rust resistance, possibly because major rust epidemics concentrated on **qualitative** rather than **quantitative** resistance genes.

Responses to two cycles of mass selection under two mating systems were compared in the F_2 and F_3 generations of one cross each of spring wheat and spring barley (Redden and Jensen, 1974). The number of green tillers prior to head emergence was used as the critical character, a feature with a low heritability of 0.10 to 0.25%. The purpose of this study was to examine the usefulness of mass selection for a character of low heritability and in particular to determine whether selection response was enhanced by **outbreeding** in comparison with inbreeding.

The results showed that mass selection with concurrent random mating could be a useful breeding strategy. Redden and Jensen suggested that efficient exploitation of a bulk gene pool, both genetically and in terms of time, may require early-generation selection in a **panmictic population**, achieved by means of genetically or chemically induced male sterility in self-fertilized crops. Through panmixis, individual selections taken from the populations for inbreeding and evaluation could contain input from several parents. These additional recombination cycles increase the chances for unfavorable linkages to be broken by recombination.

Figure 7.3 Photo of a chickpea plant (*Cicer arietinum*), a highly self-fertilized crop. Mass selection was applied to segregating populations of chickpea to achieve resistance to ascochyta blight (*Ascochyta rabisi*) of a quantitative nature that would be more durable and race nonspecific (Pieters and Tahiri, 1986).

To develop durable resistance, four methods were used to screen segregating populations—a seedling test, a germination test, a score of the percentage of infected pods, and a hair-density score. These screening methods were designed to detect major gene resistance but also the less clearly identified resistances of possibly minor genes. Instead of high inoculum densities of the disease organism, whereby disease development is strongly stimulated and disease assessment is predominantly qualitative, low inoculum densities were used to stimulate moderate disease development, which was scaled in a quantitative manner so that minor gene resistance could be detected.

The objective of mass selection in this program was to develop plants that could tolerate the disease with minor levels of disease buildup but that produce yield stability and durability. The chickpea plant in this photo shows abundant pods and is considered a promising line identified in breeding efforts.

Photo courtesy International Crops Research Institute for the Semi-Arid Tropics (ICRISAT), India.

Observations on Mass Selection

1. The purpose of mass selection is population improvement, as contrasted to pure-line or pedigree selection in which cultivars may be developed from a single plant selection.

2. Effective mass selection in heterogeneous bulk populations of self-fertilized crops offers a rapid and inexpensive method of achieving an increase in gene frequencies of certain characters. Plant breeders work with limited resources, and therefore decisions must be made regarding the most efficient allocation of these resources. Mass selection is both genetically and operationally efficient in exploiting valuable germplasm in a bulk gene pool.

3. Mass selection offers either mechanical (where some device determines what portion of the population will be retained) or visual (where retention of whole plants, inflorescences, or seeds is based on the judgment of the selector) opportunities for selection. Selection may be direct or indirect. Indirect selection occurs when a plant breeder improves one trait through selection pressure on a second. The success of indirect selection may be based on **pleiotrophy** or result from the fact that the selected trait is a component of the trait to be improved. Indirect selection may be more effective and less costly than direct selection for the primary trait.

4. When mass selection is practiced directly upon the trait for which improvement is sought, the effectiveness of the technique depends upon the heritability of the trait. With indirect mass selection, whereby selection for one trait is achieved through the expression of another, both the heritability of the selected trait and the genetic correlations between the two traits determine the effectiveness of the technique. Mass selection has limited success for traits with low heritabilities, such as yield. Caution must be exercised to guard against mass selection of a feature that is negatively correlated with yield or other desired features.

5. Mass selection can be applied to either self- or cross-fertilized species (Romero and Frey, 1962). The theory of mass selection is relatively more complete for outbreeding than for inbreeding systems of mating. Outbreeding systems will be covered in a subsequent section. Mass selection is most effective for **additive** types of gene action (Gardner, 1961) and should be especially suited for applications to advanced generations of bulk hybrid populations of self-fertilizing species because continuous selfing reduces the importance of nonadditive gene action very rapidly. Populations of cross-fertilizing species tend to contain a high degree of heterozygosity and thus afford great opportunity for the expression of nonadditive gene action upon which mass selection would be relatively ineffective.

6. The simplicity and the possibility of handling large populations and one generation per cycle are distinct advantages of mass selection. Mass selection may be imposed once or repeatedly. Repeated selection is known as recurrent mass selection. Mass selection with concurrent random mating could be a useful breeding strategy.

7. Mass selection predictably is effective with traits that have little interaction with the environment.

8. A disadvantage of mass selection is that plants are selected based on their phenotypic values. In order to achieve optimal selection for the genotype desired, some precautions should be taken. It is especially important in mass selection to diminish the influence of the environment on the phenotypic variation. This can be achieved through soil homogeneity,

uniform seedbed preparation and cultural practices, uniformity of stand, and a reasonable population size.

9. The environment must be appropriate to permit different genotypes to develop differential phenotypic expressions. When selecting for disease resistance, for example, effective selection is dependent upon the prevalence of the appropriate pathogen and its interactions with the host plant. Conditions must be appropriate to distinguish vertical or horizontal disease resistance.

10. In addition to the historical significance of mass selection, the method remains viable for modern-day plant breeders.

PURE-LINE SELECTION

In contrast to mass selection, in which the progeny of many pure lines are bulked to form a new cultivar, in pure-line breeding the new cultivar arises from the progeny of a single pure-line, or homozygous, plant. The procedure for pure-line selection includes individual plant selection from a genetically variable population, subsequent testing in progeny rows from each individual selection, and, finally, testing separate entries in replicated trials to identify superior genotypes.

Although John Le Couteur and Patrick Shirreff initiated the practice of selecting single heads or plants early in the nineteenth century, pure-line selection is an outgrowth of Johannsen's work with the Princess bean cultivar and inheritance of seed size. To demonstrate that each of the 19 distinct plant selections that Johannsen made had a genetically influenced seed weight, seeds of different weights were selected from each of the 19 lines. The seed weight produced on each of these mother plants was determined, and it was found that seeds had variable but similar bean weights. This variability was studied further by selecting the smallest and the largest beans from each of the 19 original selections for six years. Johannsen determined that continuous selection had not shifted the mean seed weight of the pure line. Existing variability was due entirely to environmental factors.

A pure line was defined by Johannsen as the progeny of a single, self-fertilized, homozygous individual. Johannsen appreciated the fact that the original Princess bean cultivar consisted of a mixture of pure lines. Johannsen's pure-line theory, plus accumulated experience, turned the attention of self-fertilized cereal crop breeders to the selection of pure lines. Many of the improved cultivars of the first three decades of the twentieth century were developed by this method. Individual homozygous plants were selected from landrace cultivars.

Johannsen recognized, contrary to Darwin's evolutionary theory of almost unlimited opportunity to improve populations through selection, that definite limits existed to the improvement of self-fertilized plants by selection alone. To induce greater variation, hybridization was needed to increase selection opportunities. Hybridization was practiced by a number of plant breeders before 1900, but parents were often chosen at random because inheritance of traits was not understood. Charles Saunders must have selected spring wheat parents because he recognized the need for early maturity when he crossed Hard Red Calcutta with Red Fife to produce the outstanding cultivar Marquis by pure-line selection.

Pure-line breeding procedures are based on individual plant selections from a genetically variable population, such as a landrace cultivar, with the objective of developing a cultivar from the best single homozygous plant, or pure line. The first step is to make a large number of either head or plant selections from the original heterogeneous population. If plants can be identified and selected, families can be established from each plant to verify homozygosity. The suggested number of individual selections is not a fixed amount but varies according to the resources available. Selections may number in the hundreds or even thousands, depending on the objective. In deciding on population size, plant breeders should recognize that the maximum genetic variation occurs in the original cultivar. If the number of selections is inadequate to include superior gene combinations, subsequent selection from within the progeny will not compensate for their absence. Selection of plants is based on phenotypic traits such as plant height, maturity, disease reaction, seed color, straw strength, or head type (length, degree, or type of awns, six- or two-row, lax or compact). Such selections will be homozygous unless chance hybridization has occurred recently.

Step two in the pure-line method is to plant the seed harvested from selections made from the mother plant in progeny rows for observation. Parameters that were originally selected are often better observed in the progeny rows, where the degree of uniformity can be determined. Those progeny with uniform phenotype and the desired traits are saved for further testing. Undesirable progeny rows are discarded.

Within a desirable progeny row, individual representative traits are saved for advancing seed to the next step. Several years of progeny row evaluation may be conducted to eliminate undesirable selections and to reduce large numbers of selections to manageable levels. Only those selections that have the potential of producing commercially acceptable cultivars are maintained. A common saying among plant breeders is, "If in doubt, throw it out." Replicated progeny field tests are conducted on remaining selections, and eventually tests may be conducted at more than one location to identify adaptation. Replicated yield trials may be conducted for three to four years along with appropriate quality determinations. A total of eight to ten years may be required with pure-line selection to produce a commercial cultivar.

Numerous examples of pure-line selections, usually from landrace cultivars, can be given (Anonymous, 1936). Prior to the application of genetics to plant breeding, landrace cultivar introductions were the most common method of crop improvement. Many introductions became commercially important cultivars. With the development of the pure-line breeding method, landrace cultivars became important sources of genetic variation. (It appears that all the recommended flax [*Linum usitatissimum* L.] cultivars in the United States as late as 1936 were developed by selection rather than hybridization [Culbertson, 1954].)

Pure-line selection has played a particularly important role in North America because many crops, with the exception of sunflower and maize, were introduced. In horticultural crops, plant breeders have adhered rather closely to pure-line selection, either from initially heterogeneous stocks or from early generations of a cross.

Ivers and Fehr (1978) investigated the pure-line-family method for yield improvement in soybean (*Glycine max* L., Merr.). Individual F_2 plants are harvested from a population, and bulk progeny from each plant (family) are selfed until an acceptable level of homozygosity is achieved, generally the F_5 generation. An individual homozygous plant (pure line) is harvested from each family, and the yield potential of each pure line is evaluated. If a pure line has superior yield, additional plants from its F_2 family are selected and evaluated as potential cultivars. If the performance of one of the original pure-line selections used to predict family performance is satisfactory, it could be released as a cultivar.

The pure-line method has been adopted as the basis for other breeding systems, such as the pedigree system or the bulk system followed by pure-line selection. The pure-line theory offers a valuable principle for selection in self-fertilized crops.

REFERENCES

Anonymous. 1936. *Yearbook of Agriculture*. United States Department of Agriculture, Washington, D.C. 1189 pp.

Culbertson, J. O. 1954. Seed-Flax Improvement. *Advances in Agronomy* 6:144-182.

Derera, N. F., and G. M. Bhatt. 1972. Effectiveness of Mechanical Mass Selection in Wheat (*Triticum aestivum* L.). *Australian Journal of Agricultural Research* 23:761-768.

Frey, K. J. 1967. Mass Selection for Seed Width in Oat Populations. *Euphytica* 16:341-349.

Gardner, C. O. 1961. An Evaluation of Effects of Mass Selection and Seed Irradiation with Thermal Neutrons on Yield of Corn. *Crop Science* 1:244-245.

Ivers, D. R., and W. R. Fehr. 1978. Evaluation of the Pure-line Family Method for Cultivar Development. *Crop Science* 18:541-544.

Johannsen, W. 1903. Heredity in Populations and Pure Lines. In *Classic Papers in Genetics,* edited by J. A. Peters. Prentice-Hall, Englewood Cliffs, New Jersey, 1959, pp. 20-26.

Pieters, R., and A. Tahiri. 1986. Methods of Screening Resistance in Chickpea to Ascochyta Blight. *Euphytica* 35:1007-1010.

Redden, R. J., and N. F. Jensen. 1974. Mass Selection and Mating Systems in Cereals. *Crop Science* 14:345-350.

Romero, G. E., and K. J. Frey. 1962. Mass Selection for Plant Heights in Oat Populations. *Crop Science* 6:283-287.

Tiyawalee, D., and K. J. Frey. 1970. Mass Selection for Crown Rust Resistance in an Oat Population. *Iowa State Journal of Science* 45:217-231.

CHAPTER 8

The Pedigree System

STEPS IN THE PEDIGREE SYSTEM

The **pedigree method** of plant breeding developed from the pure-line system and has become widely used for self-fertilized crops. It is used in handling the segregating generations following hybridization. The procedures are similar to pure-line selection, except that hybridization transforms selection into an active process because hybridization produces genetic variation. In contrast, selection in the mass or pure-line breeding systems is primarily a passive process that eliminates but does not produce variation. The problem of handling actively segregating populations was the primary force behind the development of the pedigree method, first described by Love (1927).

With the pedigree system, the F_2 generation represents the first opportunity for selection, but selection may not be practiced until the F_3 or F_4 generation. It is in the F_2 generation, however, that plant segregation is greatest following hybridization, and plant breeders generally want to take the opportunity for selection in the F_2 generation.

Following the initial selection, plants are reselected in each subsequent generation until a reasonable level of genetic homozygosity is reached and plants appear phenotypically homogeneous. The pedigree of each selection is maintained by a numbering system so that parent-progeny relationships can be traced back to an individual F_2 plant in any subsequent generation. Selection number 311 from cross 27 might be designated 27-311. If additional selections are made from the 27-311 family, a designation of 27-311-1 or 27-311-2 may be given.

General procedures for the pedigree system are outlined in Figure 8.1. The explicit steps are as follows:

First Year. The cross is made, and the F_1 plants are grown to produce as much seed as possible. These two steps may require two years, depending on location or indoor growth room facilities.

Second Year. Some 2,000 to 10,000 F_2 plants, as determined by seed, space, and time limits, are grown in space-planted rows under conditions that allow the plant breeder to select on an individual plant basis determined by the objective(s) and related to specific **autogamous** species.

Third Year. Individual plants selected in the F_2 generation are grown as families in the F_3 generation. Each family should be represented by enough plants to give an indication of the general features of the family: minimum of 10, but preferably 30 or more, individuals per family is suggested. For selection purposes, emphasis is placed on the appearance of the entire family

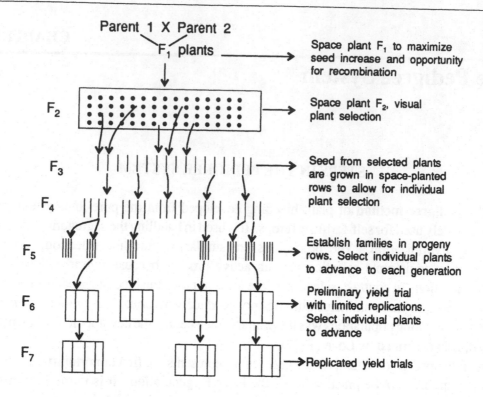

Figure 8.1 The pedigree method of plant breeding, outlined in this figure, is a widely used procedure for improvement and cultivar development of self-fertilized crops. The method consists of three fundamental steps: (1) making a cross using parents that together possess the features to be combined in a new cultivar, (2) growing the progenies of selected plants under space-planted conditions so that individual plants may be observed, and (3) keeping accurate records so that the individual plant may be traced from one generation to the next.

The system as outlined is not fixed and may be modified in a number of ways and still qualify as the pedigree method. Like plants, plant breeding systems are subject to evolution, with changes in objectives, physical resources, economic constraints, and human perceptions. Andrus (1963) reviewed several plant breeding systems and concluded that "there exists in the majority a colorful assortment of names for basically the same system of breeding" (p. 226). The student of plant breeding should attempt to understand the principles behind the development of various breeding systems.

For the practical plant breeder, the challenge is to develop systems for each crop that are manageable and effective in maintaining and exploiting genetic variation, yet are within the limits of time and space.

(Figure 8.2), and the most desirable plant is selected from each family. Some families may be discarded entirely; in other families, more than one plant selection may be made. Unless a particular cross appears to have high potential, the total number of selected plants should rarely exceed the number of families grown in the F_3 generation.

Single plants within a family continue to be selected in the F_4 generation but with a strong emphasis on the family unit, which might appear phenotypically homozygous but not

Figure 8.2   Photo of a pedigree family of spring barley consisting of four rows separated from adjacent plots by a spring-seeded winter crop (wheat or barley), which remains vegetative throughout the season. The family in this photo appears uniform in height, tillering, and general morphology. Based on phenotypic characteristics, the plant breeder may make a subjective decision to perpetuate this entry. Under such circumstances, a single plant is harvested as seed for the next generation, and the remaining plants are bulked to provide seed for testing a subsequent generation or for quality determinations.

If variation is present within the family, a number of preferred plants may be selected to establish a number of separate families in the next generation. Another choice is to discard the entire family. Discarding is a difficult task for a plant breeder because he grows to cherish the "babies" produced from hybridization programs. The fear exists that a genotypically superior plant is being abandoned because of a subjective decision based on phenotype. The ability to discard is essential in the pedigree breeding method to reduce populations to a manageable level.

sufficiently genetically homozygous to be mass propagated. If no apparent features distinguish two or more selections made from one family in the F_3, one selection usually will be sufficient to perpetuate the lineage.

Fourth Year.   By the F_5 generation, plants will be sufficiently homozygous so that F_4 plant selections may be planted in the F_5 at a commercial seeding rate. If plant density inhibits single-plant selection, there should be no problem if two or more plants are inadvertently used to perpetuate a family. Remaining F_5 rows may be harvested in bulk to obtain sufficient seed for quality tests and for F_6 yield trials.

In the F_5 generation, the breeder shifts attention from individual or qualitatively inherited traits to the overall features of a population. To do so, small blocks rather than individual rows are planted. Blocks provide a better opportunity to select for certain features, such as lodging resistance and uniformity. Some breeders use F_5 blocks for preliminary yield performance trials.

Fifth and Subsequent Years.   Segregation at any given **locus** will be virtually complete by the F_5 or F_6 generation, and yield trials can be conducted. To this point, selection for yield has been conducted on a subjective basis, but replicated F_6 and F_7 yield trials provide an objective criterion to advance superior selections for continued testing. Final evaluation involving precise on-site and regional field tests are conducted according to procedures dictated by experience or for collection of data for licensing and recommendation purposes.

Plant hybridization followed by pedigree selection offers plant breeders an opportunity to exercise selection skills, and for many years this system was the unchallenged method for breeding autogamous plants. The pedigree method was an effective means to improve specific or vertical disease resistance, to concentrate on improved straw strength, and to reduce plant height. Generally, these improvements added up to increased yield. The result was that the pedigree method of breeding remained unchallenged until the 1940s and dominated the scene into the 1960s, Powell *et al.* (1985) suggested the method predominated in the 1980s.

The pedigree system favors qualitative rather than quantitative disease resistance and therefore is not conducive to the development of horizontal or broad-based disease resistance. New sources of resistant genes generally cannot be secured rapidly enough and efforts to find new sources threatened to exhaust the readily available gene sources. The need for durable disease resistance and associated yield stability has meant a shift away from the classic pedigree system of plant breeding.

A second problem with the pedigree system is the pressure placed on resources associated with space plantings of selections, record keeping, individual plant selection, and harvesting. The pedigree system has survived in part because of the development of highly efficient seeding equipment (Figure 8.3).

A third problem with the pedigree system concerns the validity of early-generation yield selection. A major consideration in the pedigree system is the opportunity for selection of traits, including quantitatively inherited characteristics, as early as the F_2 generation. The slow and laborious effort required of pedigree system breeding programs cannot be justified if visual early-generation selection is not effective. For qualitatively inherited traits, such a system can be effective, but for quantitatively inherited traits, a knowledge of inheritance has caused most plant breeders to agree that F_2 yield selection may not be effective (Figure 8.4).

EARLY-GENERATION SELECTION

Since the highest frequency of genotypes possessing a desirable allele at all segregating loci occurs in the F_2 generation, it is logical that efforts should be directed to identifying high-yielding genotypes in the earliest possible segregating generation, possibly as early as the F_2 generation (McKenzie and Lambert, 1961; Shebeski, 1967; Sneep, 1977; De Pauw and Shebeski, 1973). In subsequent segregating generations, the frequency of genotypes possessing desirable alleles on all segregating loci rapidly declines. Within a fixed population, delaying selection for yield results in an irretrievable loss of valuable genotypes. Despite the fact that heritability estimates for single plant yield was 5.3% (Valentine, 1979), the conclusion was that if yield is to be maximized, no opportunity for selection in early generations should be lost. On the other hand, Knott (1972) noted that despite the risk of losing valuable early-generation

A.   B.

Figure 8.3 The unit in Photo A is a multirow seeder designed to plant seeds individually under spaced conditions. Distance between seeds can be selected readily (Photo B) through a choice of gears. Row width can also be adjusted, and the unit is self-cleaning. Once seeds have been prepared, many hundreds of seeds can be seeded in a day with two operators.

Units such as this greatly facilitate plant breeding efforts because all seeds are treated uniformly in terms of planting depth and distance between seeds and because of the ability to plant without the concern for delays associated with other less-rapid methods. The need for such units is not unique to the pedigree system of plant breeding, but perhaps no other plant breeding method places as great a demand on resources as the pedigree system.

Any plant breeding system is based on the ability to handle large numbers of individuals. Plant breeders confronted with high labor costs, the need for speed and accuracy, and the desire to handle ever-larger populations at a fixed cost are facing constant modification to a particular system or a change to a less-demanding system.

Photo courtesy Ontario Ministry of Agriculture and Food.

segregates by missing a selection opportunity, the effort was too great to be of value in a wheat breeding program. The value of F_2 selection remains a controversial point.

In a test with 440 single F_2 plant selections in which F_3 yield was tested against the mean yield of a cross with unselected controls, no advantage of the selected plants was observed (Shebeski, 1967). When three plant breeders each selected over 100 F_2 plants from a single spring wheat cross represented by about 8,000 plants, no advantage in selecting F_2 plants for yield was measured (McGinnis and Shebeski, 1968). Entries were tested in an F_3 yield trial along with a bulk composite consisting of a single head from each entry representing the total genotypic pool. Despite a lack of yield superiority of F_2 selections, however, the mean F_3 yield of selected F_2 plants was higher than the control. The conclusion reached was that selecting well-tillered, vigorous F_2 plants will increase the general yielding capacity of the F_3 lines and would be advantageous in a breeding program. Likewise, Hanson, Jenkins, and Westcott

Figure 8.4   The F_2 peanut (*Arachis hypogaea* L.) plants in this photo have been spaced in the field to provide the plant breeder with an opportunity to select on an individual plant basis. The peanut crop is a self-fertilized, full-season crop, and a pedigree breeding system provides the opportunity for the plant breeder to observe individual plant development over the entire season.

If the objective is increased peanut production, the challenge to the plant breeder is to identify, from the hundreds of segregating early-generation plants, individuals with the potential to produce high yields under field conditions at a commercial seeding rate. Other quantitatively inherited features are equally as challenging to identify.

For qualitatively inherited features, early-generation selection may be done with greater ease and success. Whether quantitative or qualitative features are considered, a clear objective is imperative so that the basis of selection is sound. Selection for quantitatively inherited features may be based on the experience of the breeder, intuition, or powers of perception.

Genetics has transformed plant breeding from the pure art form practiced prior to Mendel into a science. Many breeders would agree, however, that selection of superior plants among the population in the photo remains more of an art than a science.

(1979) found that visual selection for yield of single F_2 barley plants based on the individual conception of four plant breeders was not effective.

In an effort to improve the effectiveness of F_2 selection, Ismail and Valentine (1984) found that visual single plant yield assessment was moderately effective when based on tiller number. Tillering is known to be affected strongly by plant population, however.

Frey (1962) visually classified F_2 and F_5 oat plant progeny as good, random, or poor and compared yielding ability. In one cross, the mean yields for the three categories were approximately equal. In one out of two years, the poor category yielded lower than the other two categories. Lines derived from good F_5 progeny rows outyielded F_2 selections. Frey submitted that the phenotypic expression of single plants was so confounded by the environment that the criteria for visually selecting F_2 plants was ineffective.

Some encouragement for F_2 selection was reported by Frey (1954), who found that the performance of F_2-derived lines gives a good indication of the performance of lines that may be selected from them. The problem of identifying the best F_2 plants remains, however.

Using two barley crosses, McKenzie and Lambert (1961) obtained F_3-yield and other data for each F_2-selected family. Divergent visual selection for good and poor yield was practiced within families in the F_4, F_5, and F_6. Two years of divergent visual selection failed to improve yield. A visual yield score was not a satisfactory method for determining yielding ability, which confirmed the difficulty associated with selecting quantitative traits. Although a correlation for yield performance was found between the F_3 and F_6 generations, correlations were low, indicating that early-generation selections may deviate from their performance in later generations. This deviation may be due to segregation and the fact that selection occurred in a genetically heterogeneous population but subsequent yield tests were in monoculture. In addition, competitive forces under space planting and selection may produce a different expression than at commercial seeding rates. To survive pedigree selection, genotypes must perform well in genetically heterogeneous mixtures, in pure stands, and under different competitive conditions. The obvious question raised by early-generation selection is whether desirable genotypes are eliminated because of an inability to grow and compete well in a mixed F_2 population.

Space plantings for early-generation pedigree selection are preferred to facilitate single plant selection. Such conditions may offer a severe disadvantage in that they do not resemble commercial planting rates under which selections will eventually be produced. Powell *et al.* (1985) determined the presence of significant levels of interplant competition in spring barley, which reduces the effectiveness of early-generation selection. To maintain genetically heterogeneous competition under commercial agriculture conditions, these workers suggested cultivar mixtures, or **multilines**.

Good correlations between F_3 and F_6 generations were measured by McKenzie and Lambert (1961) for plant height, heading date, and seed weight, but such results are indicative of the more qualitative nature of these traits and the higher heritability values than for quantitative traits.

ADDITIONAL CONSIDERATIONS

A satisfactory theory of selection in segregating autogamous populations has not been developed. Breeding methods, including early-generation plant selection, are based on tradition, practical experience, and intuition.

When using the pedigree selection system, three problems arise: (1) how to choose the best crosses from among the many produced, (2) how to choose the size of early-generation segregating populations, and (3) how to choose the best lines within each population of each cross.

The potential of a cross may be apparent in early generations, and efforts may be directed effectively to such crosses. Shebeski (1967) and McGinnis and Shebeski (1968) emphasized the need for large F_2 and subsequent populations for the chance occurrence of a plant containing all the desirable genes. Schwarzbach (1981) calculated that selection efficiency, as measured by the response to the amount of selection, is inversely related to population size. The optimum proportion of selected individuals per generation was between 12 and 25% when selection was practiced from F_2 to F_7. At a selection pressure of 20%, an initial 8,000 F_2 plants would mean 1,600 in the F_3, 320 in the F_4, 64 and 13 in the F_5 and F_6 yield trials, respectively. To the beginner, this appears reasonable. If 55 crosses were begun simultaneously, however, there would be 88,000 entries in the F_3, 17,600 in the F_4, 3,520 and 715 in the F_5 and F_6 yield trials, respectively. Although this would not be an unreasonable program, it might not be manageable (Figure 8.5).

Plant breeders must adopt a ruthless attitude when selecting plants or be overwhelmed with materials. Limits of selection must be imposed. The limits suggested by Schwarzbach (1981) may be a surprise to plant breeders who have concluded that by selecting more segregates the probability of finding a desirable genotype is increased.

Figure 8.5   Photo of a breeding nursery using the pedigree system of plant breeding. The nursery is large, and there are many thousands of individual plants or families to be examined. Large F_2 populations are recognized as essential to securing plants with the chance occurrence of many of the desired genes. The risk in using the pedigree breeding method is the proliferation of selections out of manageable proportions. This prompted Riggs, Hanson, and Start (1981) to state, "In planning a plant breeding programme it is necessary to strike a balance between what is desirable for maximum effectiveness and what is manageable in terms of available resources" (p. 141).

It is not just the land area that must be considered. As the pedigree name implies, a record of the lines of descent are maintained on all individuals in each generation. Notebook entries allow for recording of observations on each individual. These notes are of value for multistage selection and for making decisions on final selections, but are time consuming.

Using the pedigree method of plant breeding, cultivar development takes 10 to 12 years, followed by 1 or 2 years of seed multiplication before a new cultivar becomes commercially available.

OBSERVATIONS ON PEDIGREE SELECTION

1. The pedigree method provides an accurate record of relationships that is essential in determining the inheritance of specific characteristics, but such extensive and detailed records are very demanding. The amount of material one breeder with technical help can handle is therefore limited. Inevitably, a conflict may develop between the necessity for exploring the potentialities of large numbers of crosses and the need to handle large selection numbers within each cross.

2. Effective selection is dependent upon the ability to distinguish desirable and undesirable plants on the basis of a single plant phenotype in segregating populations. The system, therefore, is very dependent on the skill and experience of the breeder, and even experienced breeders cannot reliably select high-yielding plants grown under space-planted conditions when the selection ultimately will be grown at a commercial seeding rate.

3. Family relationships established from single-plant F_2 selections provide a better indication of performance than a single plant or line, but a large area is required to maintain family relationships.

4. Based on family relationships, repeated selection is possible until families are homozygous, but if one generation is evaluated per growing season each year, the method requires several years before replicated trials for yield evaluation can be initiated. An advantage of the pedigree system is that a high degree of genetic purity is produced, which is useful where licensing agencies demand a high level of uniformity and genetic stability. Where a short time scale for cultivar production is desired, the system is slow, especially if yield trials are postponed until F_5 or F_6.

5. The low efficiency of selection among individual plants in the F_2 generation compared to selection among lines sown in rows in the F_3 and subsequent generations is a serious disadvantage. Greater precision in the pedigree system should be possible if the judgment of the breeder is supplemented by quantitative data on the material being handled. Objective data on yield and quality aspects would be of great value.

6. The inability of the breeder to select effectively for quantitative characteristics often has pushed the pedigree system into one of defect elimination. Individual plants within a family may be rogued on the basis of disease susceptibility, winter survival, or any one of a number of agronomically undesirable features. The remaining material is subjected to yield trials to check the performance and further eliminate the poorest individuals. An active rather than a passive approach is preferred, or else breeding could be placed in the hands of a technician.

7. Just as Darwin overemphasized the individual and underemphasized the population and the gene pool in his theory of evolution, so too the plant breeder using the pedigree system runs the risk of placing too much emphasis on individual parents, the individual plant, an individual character, and the individual gene. An agronomically desirable plant combines many factors and responds to a system of production. A holistic approach must not be overlooked.

8. Pedigree systems emphasize major gene resistance to pests, described as **vertical**, or narrow-base, **resistance** with limited durability. The pedigree breeding method is among the breeding systems least likely to accumulate a series of minor genes to provide horizontal resistance with greater durability.

9. A narrowing of the variability for yield may occur under the pedigree system if large numbers of selections are discarded in early generations on the basis of phenotypic observations for other traits, such as height, maturity, leaf morphology, or head morphology, among others.

VARIATIONS OF THE PEDIGREE SYSTEM

No doubt every plant breeder has adapted a particular breeding system to meet specific needs and resources. The pedigree method has numerous variations that have developed to meet specific circumstances, such as that of selecting uniculm cereal plants shown in Figure 8.6.

Harrington (1937) observed that the pedigree system is expensive and limits the amount of material carried. He also considered the mass-selection system to be too slow. Harrington suggested the mass pedigree method and may have been the first to recommend a modification to the pedigree system. He proposed selection whenever feasible during the segregating years and the conducting of plant progeny tests. Families in the F_4 generation are bulked, and the F_5 is produced in mass with opportunity for mass selection if desired. An alternative is to select individual F_6-generation plants for F_7 progeny tests.

Following F_2 selection of individual plants, Lupton and Whitehouse (1957) suggested F_3 families be grown to produce enough seed for F_4, F_5, and F_6 yield trials. Based on performance, entries showing the greatest promise are reselected. The system is known as the F_2 progeny method.

A second system proposed by Lupton and Whitehouse (1957) is known as the pedigree trial method of selection. In this modification, normal pedigree selection is carried out in F_2 and F_3. In the F_4 generation, single plants are selected from the better families for continuation of the pedigree process, and within each of these families the remaining plants are bulked to provide enough seed for a yield trial. This process is repeated in F_5 and F_6, the seed for performance testing being in each case obtained from the progeny rows and not from the preceding yield trial. Grain from the performance tests can be used for quality evaluations. A major advantage of the pedigree trial method of selection is the opportunity for elimination of segregating or atypical progeny where a high degree of genetic purity is desired. A disadvantage is that yield is delayed until F_5, and visual evaluation in F_4 of unreplicated progeny rows of spaced plants may not be effective in retaining potentially high-yielding families.

The ability to use bulk F_1 yield tests as predictive criteria for the potential yield levels of lines from the corresponding cross would be useful. Such a procedure would allow the elimination of crosses with a low likelihood of producing desirable lines and would permit a concentration of effort on crosses most likely to produce superior pure lines.

It is recognized that early-generation yield trials are subject to distortion by segregation and heterosis. Segregation can result in high standard errors and coefficients of variation. The results are not predictive of a homozygous, homogeneous population.

Positive assessments of early-generation bulk yield to preselect crosses have been reported for wheat (Busch, Janke, and Frohberg, 1974; Harrington, 1940), barley (Harlan, Martini, and Stevens, 1940; Smith and Lambert, 1968), and soybean (Leffel and Hanson, 1961).

Negative results for early-generation selection have been reported for wheat (Fowler and

Figure 8.6   The barley plants in the center of this photo exhibit a uniculm characteristic, even under widely spaced conditions. In the pedigree method of plant breeding, individual plants are harvested to establish families that provide an opportunity to examine a small lot of plants that can be examined collectively. If the uniculm characteristic is the breeding objective, a single head from each plant will not produce enough seed for subsequent steps, and modifications to the breeding system will be required.

Development of the uniculm characteristic represents a dramatic change in plant morphology. Associated with this change is a thick sturdy stem and a large head and seeds, but root lodging is a problem. Tillers normally establish secondary root systems. A plant breeder could discard outright the uniculm concept or modify an original objective to select for uniculm plants with large roots. Perhaps root size and the uniculm feature are linked, and large populations may be required to screen a massive number of segregates for root size, which generally can be detected by applying pressure on the upper part of the plant. The pedigree method offers an ideal breeding tool to screen segregates for clearly defined objectives.

The plants in this lot exhibit uniformity in height, leaf characteristics, the uniculm characteristic, and degree of awning. Perhaps the entire lot may be bulked to form a family unit in the next generation.

Heyne, 1955), oat, (Atkins and Murphy, 1949), and soybean (Weiss, Weber, and Kalton, 1947). Perhaps in these studies with negative results, there was insufficient sampling or a genotype x environment interaction occurred. Abandoning early-generation bulk hybrid tests because of large genotype x environment interactions seems unwarranted because a number of researchers (Allard, 1961; Busch, Hammond, and Frohberg, 1976; Finlay, 1964; Frey and Maldonado, 1967) have shown yields of bulks or mixtures of pure lines to be at least as stable across environments as pure lines.

Cregan and Busch (1977) argued that studies evaluating early-generation bulk hybrid testing often used parents chosen to represent a wide range of yield levels. A breeding program designed specifically for yield improvement may rely heavily on parents with high-yield ability. Yield comparisons were measured among parental lines, F_1 progeny from 29 diallel crosses, F_2 through F_5 bulks, and 519 F_5 selections. The results indicated that bulk hybrid yield tests can identify superior crosses when parents are selected for high-yield ability.

The effectiveness of early-generation yield selection of heterogeneous soybean lines was compared with the pedigree selection method in choosing superior segregates in four different crosses (Boerma and Cooper, 1975). The results indicated that the two systems were equally effective in identifying superior lines, but the major difference was the amount of time required to produce the same end product.

Pedigree breeding methods are common in horticultural crops, especially where readily identifiable qualitative traits are involved. Vegetable crops in which pedigree systems are used include tomato (Figure 8.7) (Tigchelaar, 1986), cucumber (*Cucurbitaceae* spp), pepper (*Capsicum* spp), pea (*Pisum sativum* L.), lettuce (*Lactuca* spp), and snap and field beans (*Phaseolus* spp), all self-fertilized crops. In general, selection is conducted on F_2 plants. Yield is important in many horticulture plants, but quality is equally or even more important. Quality may be associated with identifiable traits, making the pedigree system ideal for self-fertilized vegetable crops.

A.

B.

C.

Figure 8.7 The tomato (*Lycopersicon esculentum* Mill.) is largely a self-fertilized crop (Photos A and B show typical flowers) and exhibits considerable genetic variation (Photo C). Pedigree selection has been used widely in this crop. The pedigree method is preferred over mass selection because selection emphasis can be shifted gradually toward family performance rather than individual plant performance within segregating populations.

For a sprawling plant species such as tomato (Photo C), the number of segregating plants must be considerably restricted below the number desired from a genetic standpoint. Effective pedigree selection requires that the breeder identify superior individuals in early generations and yet maintain adequate genetic variability to permit later-generation selection based upon family performance for characteristics of low heritability.

Computer simulation was used by Casali and Tigchelaar (1975) to determine the effectiveness of various selection procedures. Pedigree selection was effective at high (75%) and moderate (50%) simulated heritabilities. Mass selection was found to be slightly inferior to pedigree selection at heritabilities of 75 and 50% but was significantly superior at 25%.

REFERENCES

Allard, R. W. 1961. Relationship Between Genetic Diversity and Consistency of Performance in Different Environments. *Crop Science* 1:127-133.

Andrus, C. F. 1963. Plant Breeding Systems. *Euphytica* 12:205-228.

Atkins, R. E, and H. C. Murphy. 1949. Evaluation of Yield Potentialities of Oat Crosses from Bulk Hybrid Tests. *Agronomy Journal* 41:41-45.

Boerma, H. R., and R. L. Cooper. 1975. Effectiveness of Early Generation Yield Selection of Heterogeneous Lines in Soybeans. *Crop Science* 15:313-315.

Busch, R. H., J. Hammond, and R. C. Frohberg. 1976. Stability and Performance of Hard Red Spring Wheat Bulks for Grain Yield. *Crop Science* 16:256-259.

Busch, R. H., J. C. Janke, and R. C. Frohberg. 1974. Evaluation of Crosses Among High and Low Yielding Parents of Spring Wheat (*Triticum aestivum* L.) and Bulk Prediction of Line Performances. *Crop Science* 14:47-50.

Casali, V.W.D., and E. C. Tigchelaar. 1975. Computer Simulation Studies Comparing Pedigree, Bulk and Single Seed Descent Selection in Self-Pollinated Populations. *Journal American Society of Horticulture Science* 100:364-367.

Cregan, P. B., and R. H. Busch. 1977. Early Generation Bulk Hybrid Yield Testing of Adapted Hard Red Spring Wheat Crosses. *Crop Science* 17:887-891.

De Pauw, R. M., and L. H. Shebeski. 1973. An Evaluation of Early Generation Yield Testing Procedure in *Triticum aestivum* L. *Canadian Journal of Plant Science* 53:465-470.

Finlay, K. W. 1964. Adaptation: Its Measurement and Significance in Barley Breeding. In *Proceedings of the First International Barley Genetics Symposium. Pudoc,* Wageningen, 1963, pp. 351-359.

Fowler, W. L., and E. G. Heyne. 1955. Evaluation of Bulk Hybrid Tests for Predicting Performance of Pure Line Selections in Hard Red Winter Wheat. *Agronomy Journal* 47:430-434.

Frey, K. J. 1954. The Use of F_2 Lines in Predicting the Performance of F_3 Selections in Two Barley Crosses. *Agronomy Journal* 46:541-544.

——. 1962. Effectiveness of Visual Selection Upon Yield in Oat Crosses. *Crop Science* 2:102-105.

Frey, K. J., and U. Maldonado. 1967. Relative Productivity of Homogeneous and Heterogeneous Oat Cultivars in Optimum and Suboptimum Environments. *Crop Science* 7:532-535.

Hanson, P. R., G. Jenkins, and B. Westcott. 1979. Early Generation Selection in a Cross of Spring Barley. *Zeitschrift für Pflanzenzuchtung* 83:64-80.

Harlan, H. V., M. L. Martini, and H. Stevens. 1940. *A Study of Methods of Early Breeding.* United States Department of Agriculture Technical Bulletin No. 720, Washington, D.C. 25 pp.

Harrington, J. B. 1937. The Mass Pedigree Method in the Hybridization Improvement of Cereals. *American Society of Agronomy Journal,* Washington, D.C., 29:379-384.

——. 1940. Yielding Capacity of Wheat Crosses as Indicated by Bulk Hybrid Tests. *Canadian Journal of Research* 18:578-584.

Ismail, A. B., and J. Valentine. 1984. Visual Assessment of Grain Yield and Its Components in Single Plants and Ears of Spring Barley. *Annals of Applied Biology* 104:367-373.

Knott, D. R. 1972. Effects of Selection for F_2 Plant Yield on Subsequent Generations of Wheat. *Canadian Journal of Plant Science* 52:721-726.

Leffel, R. C., and W. D. Hanson. 1961. Early Generation Testing of Diallel Crosses of Soybeans. *Crop Science* 1:169-174.

Love, H. H. 1927. A Program for Selecting and Testing Small Grains in Successive Generations Following Hybridization. *American Society of Agronomy Journal* 19:705-712.

Lupton, F.G.H., and R.N.H. Whitehouse. 1957. Studies on the Breeding of Self-Pollinating Cereals. I: Selection Methods in Breeding for Yield. *Euphytica* 6:169-184.

McGinnis, R. C., and L. H. Shebeski. 1968. The Reliability of Single Plant Selection for Yield in F_2. In *Proceedings of the Third International Wheat Genetics Symposium,* edited by K. W. Finlay and K. W. Shepherd. Australian Academy of Science, Canberra, pp. 410-415.

McKenzie, R.I.H., and J. W. Lambert. 1961. A Comparison of F_3 Lines and Their Related F_6 Plantings in Two Barley Crosses. *Crop Science* 22:840-843.

Powell, W., P.D.S. Caligari, P. H. Goudappel, and W.T.B. Thomas. 1985. Competitive Effects in Monocultures and Mixtures of Spring Barley (*Horderum vulgare*). *Theoretical and Applied Genetics* 71:443-450.

Riggs, T. J., P. R. Hanson, and N. D. Start. 1981. Modification of the Pedigree Method in Spring Barley Breeding by Incorporating Bulk Selection in F_4. In *Barley Genetics IV*, Proceedings of the Fourth Barley Genetics Symposium, edited by M.J.C. Asher, R. P. Ellis, A. M. Hayter, and R.N.H. Whitehouse. Edinburgh University Press, pp. 138-141.

Schwarzbach, E. 1981. The Limits of Selection in Segregating Autogamous Populations Under Ideal and Under Realistic Assumptions. In *Barley Genetics IV*, Proceedings of the Fourth Barley Genetics Symposium, edited by M.J.C. Asher, R. P. Ellis, A. M. Hayter, and R.N.H. Whitehouse. Edinburgh University Press, pp. 154-158.

Shebeski, L. H. 1967. Wheat and Breeding. In *Canadian Centennial Wheat Symposium,* edited by K. F. Nielsen. Modern Press, Saskatoon, pp. 253-272.

Smith, E. L., and J. W. Lambert. 1968. Evaluation of Early Generation Testing in Spring Barley. *Crop Science* 8:490-493.

Sneep, J. 1977. Selecting for Yield in Early Generations of Self-Fertilizing Crops. *Euphytica* 26:27-30.

Tigchelaar, E. C. 1986. Tomato Breeding. In *Breeding Vegetable Crops,* edited by M. J. Bassett. Avi Publishing Company, pp. 135-171.

Valentine, J. 1979. The Effect of Competition and Method of Sowing on Efficiency of Single Plant Selection for Grain Yield, Yield Components and Other Characters in Spring Barley. *Zeitschrift für Pflanzenzuchtung* 83:193-204.

Weiss, M. G., C. R. Weber, and R. R. Kalton. 1947. Early Generation Testing in Soybeans. *American Society of Agronomy Journal* 39:791-811.

CHAPTER **9**

The Bulk Population Method

DESCRIPTION AND APPLICATION

The classical approach to generation advance by the bulk population method, or **bulk breeding,** consists of parental hybridization, planting a number of seeds each generation, harvesting the plants in bulk, and planting a sample of seeds the following year. Bulk population breeding is an economic method of obtaining homozygous lines in self-fertilized crops.

Typically, the bulk breeding method involves seeding several hundred or even several thousand F_2 plants at a predetermined plant population (Figure 9.1). At maturity, the plot is harvested in bulk, and a sample of seed is used to plant a similar block the following season. Normally, **natural selection** is allowed to proceed for at least three generations, the F_2, F_3, and F_4. If individual plant selection is planned in the F_5, the population may be space planted to facilitate single plant selection. Selected plants are grown as a family seed increase plot for an F_7 yield trial. Conventional testing procedures are employed in the F_7 and until the eventual release of a cultivar.

Bulk population breeding takes advantage of selection pressure and attempts to increase the frequency of superior genotypes by the elimination of those types that are markedly inferior in their agronomic worth (Figure 9.2). Populations are propagated until segregation virtually has ceased, at which time single plant selections are made for evaluation and cultivar release. Bulk populations may be seeded under field space planted conditions, at commercial seeding rates, or under alternating rates. They may be rotated among several locations to provide contrasting environmental conditions (Patel *et al.*, 1987). They may initially be grown under controlled environmental conditions to screen seed germination or seedling development for a specific variable, followed by transplanting in the field to produce seed for the next generation if population numbers are not too large. Characteristically, selection pressure has been viewed as natural selection under field conditions, but increasingly, controlled environmental facilities are being used to exert specific selection pressures.

The bulk population method is used for small grain cereals, field bean (Figure 9.3), and soybean. Bulk breeding may be useful for a number of quantitatively inherited factors. Examples of traits that may be screened include winter hardiness following spring x winter or winter x winter crosses in which winter kill may eliminate the nonhardy types. Populations can be screened for response to **photoperiod** since those that do not reach reproductive development are automatically eliminated. Other factors include tolerance to soil salinity, horizontal or broad-based disease resistance, acid soil tolerance, and ability to withstand leaf- or soil-applied chemicals. In spring-seeded spring x winter crosses, segregates requiring

201

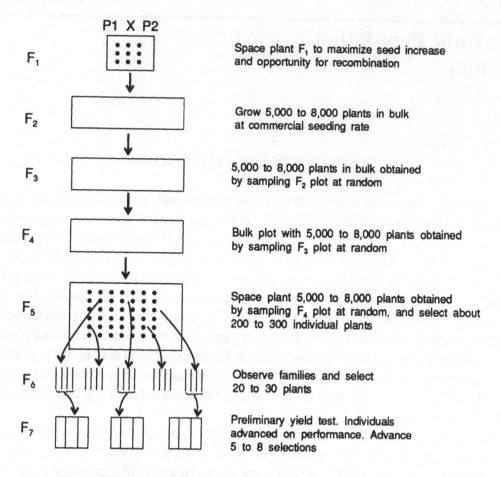

P1 X P2

F_1 Space plant F_1 to maximize seed increase and opportunity for recombination

F_2 Grow 5,000 to 8,000 plants in bulk at commercial seeding rate

F_3 5,000 to 8,000 plants in bulk obtained by sampling F_2 plot at random

F_4 Bulk plot with 5,000 to 8,000 plants obtained by sampling F_3 plot at random

F_5 Space plant 5,000 to 8,000 plants obtained by sampling F_4 plot at random, and select about 200 to 300 individual plants

F_6 Observe families and select 20 to 30 plants

F_7 Preliminary yield test. Individuals advanced on performance. Advance 5 to 8 selections

F_8 - F_{12} Cooperative tests at several locations, eliminate all but one or two entries, licensing, seed increase, cultivar release

Figure 9.1   The bulk method of plant breeding outlined in this diagram has been adapted to handle extremely large volumes of breeding materials in an inexpensive way and to take advantage of natural selection. Pedigree selection can be practiced among homozygous plants in the population to produce a cultivar. The Swedish plant breeder Nilsson-Ehle used the bulk system in 1908 to screen winter wheat progeny for high yield and winter hardiness (Newman, 1912).

Homozygous lines can be produced with minimum effort and expense, but caution must be exercised to sample adequately each generation to ensure that genetic variability is not reduced. Natural selection operating in the population may modify gene frequency in an undesirable direction.

The bulk system may be used to screen segregating progeny for winter hardiness following a spring x winter hybridization program, for general or horizontal disease resistance, and for grain yield.

The bulk method is not practical for developing high-yielding dwarf genotypes in crops in which dwarf and semi-dwarf stature are desired because such plants are noncompetitors in a population heterogeneous for plant height. The bulk system might be useful for screening dwarf plants in a population homogeneous for plant height.

Ikehashi and Fujimaki (1980) reported that in rice breeding, 21 out of 33 rice cultivars in Japan were developed using a rapid-generation-advance method of bulk breeding.

A.   B.

Figure 9.2 View of two segregating bulk plots of barley, both consisting of many individual segregating plants. Natural selection resulting from winter killing has eliminated many individual genotypes in the winter barley plots in Photo A. Harvesting the remaining plants and planting bulk breeding nurseries in subsequent generations should shift the population toward increased hardiness.

Photo B shows a bulk plot of spring barley in which the barley breeder hopes that natural selection will eliminate unadapted, genetically undesirable genotypes and low-yielding plants, in this case by competition. Normally, plants are grown in bulk breeding nurseries for four or five generations during the F_2 to F_6 in a manner in which natural selection will be expressed.

Barley plants have a high tillering capacity, and if seeded at a low density, profuse tillering may occur. At commercial seeding rates, plants may not produce many tillers, regardless of their genetic capacity to do so. Experience will determine an appropriate seeding rate and planting pattern for each crop.

Plants may be grown for additional years after the F_6 if deemed desirable and if time and resources permit. When plants have approached a reasonable level of homozygosity, pedigree selections are made, selections are tested in replicated yield trials, and a cultivar is released if the program has been successful.

vernalization will not produce seed. Bulk selection methods may be useful in selecting for seed yield if high-yielding segregates contribute more seed to the subsequent population.

The bulk population method generally is not applicable to screening fruit crops or breeding most vegetable crops because competitive ability is not desired, natural selection does not impact on quality aspects, or homozygosity is not an objective. Bulk breeding has been used to screen tomato seeds for low-temperature seed sprouting when hybridized with the cold-insensitive wild tomato (*Lycopersicon hirsutum*) (Ng and Tigchelaar, 1973). Cold tolerance is aimed at direct seeding, and screening could be conducted in an appropriately chosen field by early seeding under frost-free conditions or under controlled environmental conditions. The bulk breeding system was used by Patterson, Pauli, and Smillie (1978) to screen

Figure 9.3   A portion of a bulk breeding nursery of field beans (*Phaseolus vulgaris* L.). Bulk breeding methods that rely on natural selection have been common for self-fertilized crops for over 80 years, but bulk breeding in this crop has not been studied to any extent.

Hamblin (1977) studied the effect of bulk breeding on yield and components of yield for five generations (F_2 to F_6) in two high- and two low-yielding bean crosses. A steady increase in the mean values of the low-yielding crosses with advancing generation was observed, but bulk breeding did not alter the mean values in the high-yielding crosses. The five characteristics studied on a per plant basis were seed yield, seed numbers, pod numbers, seeds per pod, and seed weight. Bulk breeding shifted the mean value of all components of low-yielding crosses upward. Mean grain yield values were unaffected by bulk breeding methods in the high-yielding crosses.

Shifts in seed size when they occurred were sufficiently slow, which suggested that artificial selection for a preferred size might be conducted.

Since pedigree selection for yield in early generations generally is not effective in beans (Walker, 1969) and because bulk breeding does not lead to the elimination of high-yielding genotypes, the bulk breeding system is favored for this crop. When selecting for yield, resources may be used effectively by bulk breeding until homozygosity is approached and then by making pedigree selections for yield trials.

---

tomato seedlings for tolerance to low temperatures and could be applicable to adult plant cold tolerance.

Natural screening can be aided by artificial selection at any time during the production of bulk populations. Plants that have suffered winter damage but have survived may be rogued so that they are prevented from reproducing. Plants with major gene resistance may be discarded if general resistance is desired. Some factors, such as type of awns (smooth or rough), may be adaptively neutral with no net change expected from natural selection alone. If a preference exists, artificial selection may be imposed to emphasize those types that meet human needs.

If natural selection is to be aided by artificial means in the field, plants may be space planted;

if no roguing is intended, a commercial seeding rate may be used. The impact of aiding natural selection has not been studied fully. Careful manipulation of the population to aid natural selection, such as the removal of unwanted tall, competitive genotypes, may avert shifts of gene frequency in an undesirable direction.

The rate at which unadapted segregates are eliminated from a bulk population usually depends on the degree of selection pressure applied, the consistency of that pressure, and the heritability of the trait or traits under selection. It may not be worthwhile to extend bulk handling beyond homozygosity in the F_4 or F_6 generation or to grow a bulk population for up to 18 years (Degago and Caviness, 1987). Under conditions of a clear objective, aiding natural selection may be beneficial.

CONSIDERATIONS IN NATURAL SELECTION

The relationships among such factors as plant survival, interplant competition, and agronomic value are important considerations in using the bulk method of selection. Whether human selection should be imposed or natural forces allowed to act alone must be decided.

Whenever two or more lines compete in a bulk plot, the survival of one relative to the other(s) will depend on the number of seeds produced by each line. The proportion of those seeds capable of reaching maturity and the sampling procedure used to ensure offspring of each or of selected individuals only must be considered.

Harlan and Martini (1938) studied a mixture of homozygous barley cultivars in an attempt to secure information on natural selection. They suggested that the number of plants of a given line present in the population in any year will depend on two factors: the number, not the weight, of seeds sown, and the percentage of survival of the plants in competition. For theoretical purposes, this was reduced to the number of seeds that will produce plants the following year. Population shifts among ten lines grown in a mixture and subjected to natural selection are shown in Figure 9.4.

The theoretical curves in Figure 9.4 were compared with a mixture of 11 barley cultivars tested at ten widely varying locations across the United States for a period of four to twelve years. The results indicated that in a mixture of cultivars, inferior and nonadapted cultivars in a particular area decreased and/or were eliminated from the population. Cultivars that were adapted and agronomically superior increased and became the dominant cultivar in the particular area. This work strongly indicates that in a mixture of cultivars, and presumably in a segregating population following hybridization, natural selection can be effective in discriminating among those that are superior competitively and those that are inferior.

Caution, however, must be exercised. The cultivars used in Harlan and Martini's study were of comparable height. In a mixture of dwarf, semidwarf, and tall cultivars, dwarfs and semidwarfs may be eliminated due to excessive seedling and adult plant competition from the tall cultivars. When an economically important trait such as dwarfness is controlled by several recessive genes, the ratio of desired plants to those bearing dominant gene-governed characters will be small in the F_2 population. Natural selection by homozygous and heterozygous dominant plants may eliminate homozygous recessive plants as they emerge. Efforts to find a superior semidwarf might prove effective if the mixture did not comprise extremes in height.

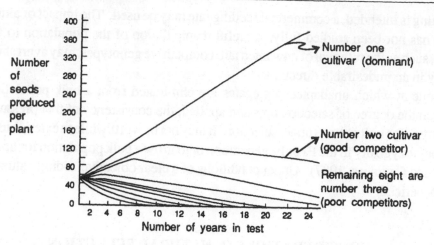

Figure 9.4 Theoretical curves of natural selection showing the number of seeds produced from poor to good competitors in a mixture of ten barley cultivars differing in seed productivity by five kernels each. The poorest competitor produced 45 seeds at the start of the test.

Three major types of competition (curves) are evident. The first is that of the dominant cultivar (number one), which consistently produces more seed per generation than the other cultivars in the mixture and increases progressively with each generation under natural selection.

The second type of curve is that of the good competitor, whose numbers increase in the early generations, until the tenth generation (number two). The good competitor shows competitive superiority over poorer competitors in early generations, with the decline after the tenth generation determined by competition with the dominant cultivar. Although the good competitor declines over time, it still remains a substantial portion of the mixture.

The remaining eight poor competitors give the third type of curve. These cultivars are unable to produce as many seeds as either the good competitor or the dominant cultivar in the mixture, and their numbers decrease rapidly in the population. However, many of these cultivars, even though they are unable to compete effectively, remain in the population for many generations at a low frequency.

Theoretically, the best cultivar at any location will eventually dominate the population. In time, some of the least competitive cultivars may be eliminated.

Source: Adapted from Harlan and Martini, 1938, p. 194.

Suneson (1949) studied a mixture of four homozygous barley cultivars with similar adaptation and found that after 16 years in a bulk plot, two of the component cultivars were practically extinct. One had a superior grain yield and leaf disease record over the others when grown in pure stands. The cultivar that ultimately dominated the mixture had the poorest leaf disease record and a mean yield below the median for the component cultivars. Although yield differences were small, this study supports the observation that the intangible character of competitive ability will not necessarily perpetuate either the highest-yielding or the most disease-resistant progenies.

Caution should again be exercised. When this study was reported, complete disease resistance (now termed vertical) was acclaimed over the presence of disease tolerance (now termed general or horizontal resistance). Application of a different set of standards with a different objective may have led to the conclusion that despite a slightly lower yield, yield stability emerged as the winner. Competitive ability may measure other very important plant characteristics.

NATURAL SELECTION IN SEGREGATING POPULATIONS

Natural selection among differing homozygous cultivars may not be similar to natural selection in a segregating population for the following reasons:

- Competitive forces in a segregating population become even more intangible because new types will occur each generation until selfing produces homozygosity.
- Consistency of competing genotypes will not occur until the F_5 to F_7 generation in self-pollinating crops, and then only with a highly variable population.
- A vastly larger number of competing genotypes will be acting in a population following hybridization at all stages of development.
- If artificial selection is applied by the complete removal of individual genotypes, a very unpredictable element is introduced.

The effect of natural selection on segregating genotypes can be observed by studying changes in gene frequency, by the appearance and performance of entire populations after natural selection has been allowed to act, and, finally, by the performance of pedigree selections from the bulk population.

The relative performance of different genes that occurred in segregating populations of barley were studied by Suneson and Stevens (1953). Among the comparisons made were those between rough and smooth awns, two- and six-rowed types, awned versus hooded condition, and the gene combination for lateral florets versus the gene for deficiens (absence of lateral florets). Each of these combinations was contrasting alleles of the same gene. The results indicated that adaptively neutral genes, such as rough and smooth awns, maintained the same relative proportions in the bulked population. There was no net change in the survival of one allelic form over another, and year-to-year variations were due to random changes.

In contrast to adaptively neutral genes, alleles of genes associated with poor adaptation were reduced in the segregating populations. Under the conditions of this experiment, six-rowed types performed better than two-rowed types, and in a relatively short number of generations, six-rowed types dominated the population. Perhaps this shift might have been reversed under space planted conditions because two-rowed selections normally exhibit a higher tillering capacity than six-rowed selections.

In the case of awned versus hooded types, the hooded allele was reduced in the population, indicating that this allelic combination was competitively inferior. In the case of the presence versus the absence of lateral florets, the deficiens gene (absence) was rapidly eliminated from the segregating populations.

These results of Suneson and Stevens (1953) support the work of Harlan and Martini (1938), which indicates that certain genes viewed by these workers as undesirable and inferior can be reduced or eliminated rapidly in a segregating population.

The bulk population breeding method should not be applied where natural selection might eliminate a certain genotype that may be useful under a different set of circumstances. Presumably, uniculm barley may not compete in a bulk population against tillering types, but in an era where weed competition can be reduced by herbicidal and cultural practices, a uniculm cultivar with suitable agronomic traits might perform satisfactorily in commercial seedlings.

The question whether natural selection in bulk populations is effective in selecting superior yielding types remains open to debate. The fact that a modified bulk breeding system has been used in Japan since 1947 and that 24 leading cultivars were grown on 44% of the total rice area of Japan in 1980 is indirect evidence that yield advances are being achieved (Ikehashi and Fujimaki, 1980).

Likewise, natural selection over time proved to be an effective means of selecting subtle differences in yield and maturity in barley (Suneson and Stevens, 1953). The study involved a composite bulk plot produced from diallel crosses of 28 cultivars to produce 378 crosses. The yielding ability of the segregating bulk was compared periodically with Atlas barley as a standard (Table 9.1). Again, perhaps general horizontal resistance was developed. It was noted by the authors that although certain types of diseases were prevalent in the area, disease resistance was not a conspicuous feature of the composite. They argued that disease resistance *per se* was not a particularly valuable asset in the population. This might be expected since a disease in any one year would not be so severe as to eliminate susceptible progeny. In years when disease is not severe, resistance genes would be adaptively neutral in the population. Indeed, horizontal resistance may have been a major asset in selection for yield stability. In the F_2 generation, random selections were made, and of 356 selections tested, all were of inferior yield to the cultivar Atlas. Among 50 selections made in the F_{20} generation, two were superior to Atlas in replicated trials. Natural selection appears to be an effective force for yield improvement over time.

The value of the bulk method of plant breeding as a means of improving gene frequency of seed yield in soybean (*Glycine max*) was demonstrated by Degago and Caviness (1987).

Table 9.1 Yield of Composite Cross II Compared with Atlas Barley in Replicated Field Trials

Test Year	Generation	Grain Yield in Percent of Atlas
1933-1934	F_3 - F_4	67.6
1937-1938	F_7 - F_8	85.0
1937-1940	F_{11} - F_{14}	88.8
1941-1946	F_{15} - F_{20}	106.0
1947-1949	F_{21} - F_{23}	101.3
1950	F_{24}	103.6

Note: In the early generations, the composite cross did not perform favorably in comparison with the check cultivar, Atlas. A number of the parental cultivars of the composite were not adapted to the environment in which they were tested with Atlas. However, after approximately 12 to 15 generations of natural selection, genotypes were produced with yields similar to Atlas. After several more generations of natural selection from the F_{15} to F_{20}, the composite exceeded the yield performance of Atlas. By the 1950 test year, in the 24th generation from the original composite, yield performance of the composite was superior, but increases had leveled off.

Source: Adapted from Suneson and Stevens, 1953, p. 11.

When four segregating bulk populations were grown at two diverse locations for 10 to 18 years, yield increased at the location where phytophthora root rot occurred over the same bulk population where the disease was not a problem.

The bulk method of breeding appears effective under a constant year-to-year selection pressure. Based on the results of Degago and Caviness (1987), the bulk system is effective when natural selection is exerted by stress conditions.

NUMBER OF GENERATIONS AS A BULK POPULATION

Bulk populations of soybeans were grown for 10 to 18 years by Degago and Caviness (1987) and for 24 years by Suneson and Stevens (1953). Very few plant breeders have the luxury of waiting so many generations before selecting pure lines. If the bulk population method is to be useful, three to five years might be regarded by many plant breeders as the maximum amount of time before pure-line selections can be made.

The desirability of any breeding method is determined by the number of superior genotypes for traits of high economic value that can be derived from a population and the expenditures of time and money necessary to handle the program. A major advantage of the bulk population method is that homozygosity can be achieved with a minimum of effort and expense. If achieving homozygosity is the primary goal, the number of generations that progeny are produced in bulk can be specified. In the F_6 generation a high level of homozygosity is achieved, and by the F_{10} generation, virtually 100% homozygosity should be achieved even with a large number of segregating genes.

In a segregating population, heterozygous plants exhibiting hybrid vigor will exert a powerful competitive force on homozygous plants. Unless widely contrasting parents are used, some alleles will be homozygous even in the F_1 generation. Subsequent segregation of individuals with a high degree of heterozygosity will produce new poor competitors until the inbreeding coefficient of the population becomes relatively high. Therefore, a population handled as a bulk for only a few generations changes very slowly in its population structure, unless there is a high degree of selection pressure eliminating the poor competitors. Many breeders have experienced this stagnation, especially for quantitatively inherited traits where significant shifts in the mean usually do not occur until about the F_{15} generation.

A possible solution is to make selections in the F_5 or F_6 generation and commence progeny testing, while at the same time maintaining the remaining population in the bulk for subsequent reselection if satisfactory results are not obtained from the first round of selection. If outstanding results are achieved, the breeder may wish to have the option of reselecting from the bulk population.

A second solution is to grow multiple generations per year. The advantage of the bulk population method is that natural selection can occur in the absence of the plant breeder, provided technical help is available for seeding and harvesting. The bulk method allows a winter nursery to be used effectively. Progenies from the bulk method might be planted under controlled environmental conditions, but genotypes might perform differently than under field conditions in its area of adaptation.

The single-seed-descent method was observed by Empig and Fehr (1971) to be the least influenced by natural selection of four methods of generation advance studied. The single-seed-descent method might be the most useful method in controlled environments or winter nursery environments. A combination of natural selection or bulk breeding followed by single seed descent may be useful.

The question of producing a bulk population between two alternate locations has received some study. Borlaug (1968) alternated wheat breeding material between two diverse locations in Mexico and was successful in developing lines with broad adaptation. Natural selection was important at each site. Alternating a barley pedigree selection program between two locations in the province of Quebec resulted in progeny that were more broadly adapted than those derived from continuous selection at either of the two locations (St. Pierre, Klinck, and Gauthier, 1967).

In contrast, Choo, Klinck, and St. Pierre (1980) observed that natural selection had little effect on yield, and alternating between locations had little effect on adaptability. Germplasm and locations, however, were suggested as insufficiently diverse to test the merit of alternating bulk populations. Natural selection in a doubled-haploid mixture of barley along with a barley composite cross was less effective when material was advanced outside of its intended region of adaptation rather than alternated between locations.

A third approach to the problem of reducing the generations required in a bulk population is to practice artificial selection on early-generation bulk progenies to shift the population toward agronomically desirable types. Artificial selection can aid natural selections for benign features such as seed color in wheat or fiber length in cotton. Likewise, plants with highly competitive features, such as indeterminate or vine types in soybeans, tall entries, or undesirable maturity, might be removed.

The Swedish pioneer breeder H. Nilsson-Ehle assisted with natural selection by discarding plants showing winter damage and thereby increased the rate of shift toward hardy types. The reasons for roguing plants lacking in winter hardiness and showing low survival ability may be less compelling than for roguing plants with an undesirable feature that is either neutral or has high survival value.

Jain (1961) suggested that artificial selection might be applied at suitable stages in the bulk breeding program to aid natural selection and to speed up the procedure. He cautioned against overly ambitious artificial selection that might result in slippage in the selective response of natural selection.

NATURAL SELECTION IN SEGREGATING BULK POPULATIONS

The question of natural selection with a segregating population of dwarf and tall genotypes was examined by Khalifa and Qualset (1975). They studied dwarf x tall wheat plants in which they assumed the dwarf condition was influenced by three genes. The results indicated that in four segregating generations, dwarf types would not be eliminated. Furthermore, if the assumption is made that among those dwarf types natural selection has acted to preserve those that are most suitable, then the most valuable genotypes would be preserved in the population. It is possible that continuation of the trends over an increased number of generations would eliminate dwarf genotypes. This conclusion is tenuous, however, because plant population can impact upon natural selection pressure.

In rice, Jennings and Aquino (1968) found that dwarf types were reduced drastically in frequency after five generations. Generally, the classical bulk breeding method is not considered satisfactory in breeding programs where short stature is desired and where the hybridization of tall and short stature is appropriate. Despite a conspicuous decline in dwarf genotypes in the presence of tall genotypes in a segregating bulk population, the mean yield of randomly evaluated lines showed a steady increase.

Trends in performance of segregating bulk populations in wheat (Khalifa and Qualset, 1975), rice (Jennings and Aquino, 1968), and barley (Allard and Jain, 1962) are similar. Perhaps such trends are explained by the hypothesis of Allard, Jain, and Workman (1968) that the evolution of such populations is characterized by natural selection, which eliminates unbalanced and inferior genotypes originating from segregation and recombination. There is little doubt that at an appropriate seeding rate, the progeny produced from morphologically contrasting parents set up several kinds of competitive interactions that may occur concurrently (Figure 9.5). When this happens, the final outcome will be determined by the balance among the competitive effects involved. The final outcome is not predictable. Dwarf plants in a variable-height population will be lost if the population is perpetuated for a long enough period, unless they are highly tolerant to stress associated with shading or other features. From a practical standpoint, dwarf plants might be lost as a result of sampling procedures when bulk populations are harvested. Machine harvesting at a fixed height may miss short plants; a technician sampling by hand may ignore dwarf plants.

In any event, the rate of improvement is slow, and modifications to the classical bulk breeding method may be useful.

MODIFICATIONS TO THE BULK POPULATION METHOD

Many plant breeders favor the bulk breeding method because of budgetary and record-keeping considerations, the ability to handle extremely large volumes of breeding material, and the opportunity to take advantage of natural selection. The classical bulk breeding method as first introduced by Nilsson-Ehle continues to be used by many breeders. Others have proposed modifications. Four of these modifications are described.

1. To prevent any drift in segregating populations associated with the classical bulk population method of breeding, the single-seed-descent method was applied (Empig and Fehr, 1971). Each generation a single seed is harvested from each plant in the bulk population and used to perpetuate the next generation under the bulk population method. The disadvantages of the single-seed-descent method include the time required for harvesting a seed from each plant in the population and the identification of single plants in a dense population. Use of single-seed-descent assumes space planted conditions.

Application of the single-seed-descent method may not prevent change in gene frequency if natural selection occurs in the seedling and juvenile stages or before flowering, as might occur with seedling mortality or with post-fertilization sterility.

To reduce harvest time with the single-seed-descent method and yet minimize the sampling problem of the classical bulk method, Empig and Fehr (1971) proposed the mechanical sampling of a small section of each plant. The technique was termed the restricted cross-bulk method.

Figure 9.5 The evening primrose (*Oenothera* spp) seedlings in this photo are planted in a configuration known as a wagon wheel, designed to provide preliminary information on plant population per unit area in which natural selection is expressed. The hub is in the center of the plot, with spokes radiating out. Selection pressure on plants increases toward the hub, and a suitable population can be determined for effective screening of a heterogeneous population.

Success in bulk breeding programs is dependent on several forces to identify superior genotypes and to eliminate less desirable types. Among the factors operating in bulk breeding plots are natural selection, winter kill, disease, insects, sterile plants, and other factors that screen plants, including artificial selection.

Natural selection is a form of **competition**, and plant breeders hope that natural selection will result in **complementary competition**. Bulk breeding programs must use a seeding rate at which stress will occur to aid in natural selection. Competition is not unique to bulk segregating populations; severe competition occurs within a homozygous cultivar seeded at a rate that produces the highest yield. Competition among similar plants in a homogeneous cultivar is termed **autogenotypic competition**, and in a heterogeneous population, **allogenotypic competition**. Natural selection through competition is highly variable and unpredictable and may require several generations to be expressed effectively, but it is a powerful tool in bulk breeding populations.

2. A second modification, termed rapid generation advance by Ikehashi and Fujimaki (1980), is not a breeding method but a means of growing bulk populations to shorten the growth duration of a given population. A major aim of bulk breeding is to attain homozygosity prior to pedigree selection, and therefore a rapid cycling through the generations would speed up cultivar development. The single-seed-descent method was suggested, making it possible to grow three to five generations per year in year-round outdoor field tests in the tropics. Harvesting of seed at or before **physiological maturity** is necessary for rapid cycling.

3. Success using the bulk breeding method might be enhanced by bulk generation yield trials to predict the best crosses. The evaluation of greater numbers of pure-line selections is necessary to find superior lines. Predictive tests would direct the greatest effort to the best combinations.

The question arises whether early-generation tests of bulk material are indicative of potential yield performance. Yield tests of early-generation bulk populations of wheat indicated that superior crosses can be identified even when parents are selected for their high-yield potential (Cregan and Busch, 1977). The results are similar to other tests (Busch, Janke, and Frohberg, 1974; Harrington, 1940; Leffel and Hanson, 1961; Smith and Lambert, 1968), except in the majority of those studies, the parents used had greater yield variations. Early-generation bulk testing appears to possess considerable merit.

The F_2 bulk generation was particularly effective in identifying those crosses from which the highest-yielding lines might be expected. As a basis for indicating superior crosses, the F_1 and midparent performances were of less value than a bulk generation performance (Cregan and Busch, 1977). Subsequent F_3 and later-generation bulk trials could confirm F_2 results as well as detecting possible genotype x environment interactions that would interfere with effective selection.

Bulk population tests, although useful in predicting crosses from which high-yielding selections might be made, may eliminate populations with low mean yield but with large yield variances. Such crosses may be discarded but have the potential of producing high-yielding lines following natural selection. Further, bulk populations that gave the highest yields in replicated tests in early-segregating generations were found by Atkins and Murphy (1949) not to produce the greatest proportion of high-yielding lines. Results were based on a one-year test, however. Ambiguous results regarding the value of unselected bulk progenies in evaluating the prepotency of a cross were obtained by Fowler and Heyne (1955) and by Grafius, Nelson, and Dirks (1952).

4. A bulked composite population consisting of the F_2 lines of several crosses was suggested by Suneson (1956). The essential features of this modified bulk breeding system are a broadly diversified germplasm and intensified natural selection. Suneson suggested that natural selection of such a composite might be prolonged to 15 generations in the area of contemplated use. A disadvantage of the composite bulk breeding method is a loss of parental identity.

The composite cross concept may be enlarged by the use of male-sterile plants. Male-sterile plants fail to set seed unless fertilized by foreign pollen. The resulting genetic recombinations increase genetic variability, which operates concurrently with natural selection. Although substantial variability was observed in composites with a high level of outcrossing, it was not accompanied by a corresponding increase in productivity.

These and other modifications to bulk breeding may provide ideas for additional modifications. Plant breeders may wish to integrate one or more of the bulk population method-related techniques to develop efficiency systems for any breeding objective or constraint (Figure 9.6).

BULK BREEDING SYSTEMS COMPARED

F_6 lines from bulk breeding and pedigree systems produced from four soybean crosses were studied for two years (Raeber and Weber, 1953). Mean yields indicated equal success, with minor differences in lodging, maturity, and height.

Likewise, Torrie (1958) compared bulk breeding and pedigree methods for F_6 lines from

Figure 9.6   The winter wheat plants in this photo have been broadcast-seeded onto the soil surface of an uncultivated field immediately following the removal of grain maize. The residue stover can be seen in the foreground, the edge of the field.

Plant breeding programs must attempt to keep abreast of changes in agriculture. The opportunity to seed wheat onto an uncultivated field has been made possible by complete weed control in maize, aided largely by herbicides. Weed seeds below the surface are contained by zero-tillage and erosion is reduced by zero-tillage seeding and the residue stover. Costs are reduced if tillage and/or residue removal is unnecessary.

Bulk breeding methods provide an opportunity to investigate the possibility of genotypes adapted to a specific cultural practice. Do genotypic differences exist for ability to germinate on the soil surface and to establish a root system under the conditions suggested in this photo? Perhaps maize stover residue produces allelopathic compounds as it decomposes. Do wheat genotypes respond differently to allelopaths? Surface residue may alter the microclimate, and the question arises whether differential responses occur in a segregating population. Bulk screening methods may provide an opportunity for selection of genotypes adapted to such conditions.

Bulk breeding methods involve a minimum of effort and expense because natural selection is at work for the plant breeder.

---

six soybean crosses. The mean seed yield was similar for both methods in the three maturity classes of all crosses, with two exceptions of superior lines from bulk breeding. Other agronomic features were similar.

Luedders, Duclos, and Matson (1973) found yield differences in F_6 and F_7 soybean selections advanced by modified pedigree and bulk breeding systems. Plant breeders were urged to discard more, perhaps 75%, of their yield-tested lines each year to permit larger populations to be evaluated and to retain only stable, high-yielding selections.

Based on 26 years of wheat breeding in New Zealand, Copp (1957) concluded that where disease is not a limiting factor the pedigree system is superior to the bulk method.

OBSERVATIONS ON THE BULK BREEDING METHOD

1. Attaining homozygosity with minimum effort, expense, and record keeping is a major attraction of bulk breeding.
2. Evidence exists that natural selection, a primary aspect of bulk breeding, can rapidly eliminate poorly adapted lines, but evidence also exists that gene frequency can be modified in an undesirable direction.
3. Large amounts of segregating material can be handled, so that more crosses can be attempted and evaluated. This is an important factor when time, space, and resources are limited.
4. Improper or inadequate sampling in each bulk generation can result in genetic drift.
5. For international collaborative programs, the bulk breeding method offers a means for selection on nearly fixed materials *in situ*, while ensuring that the maximum potential of the population is maintained during bulk growing. Generating diverse materials and selecting in a specific location can be integrated. Spring x winter crosses, conducted under field conditions, can be handled effectively at diverse sites by the bulk breeding method.
6. Unless there is a high degree of selection pressure eliminating the poor competitors, natural selection appears slow, especially for quantitatively inherited traits. Significant shifts may not occur until the F_{15} generation. This fact may make the system unpopular.
7. Plant breeders may effectively and efficiently handle the bewildering array of genotype variation involved in multiple crosses or composite populations.
8. Broad adaptation can be developed through natural selection when bulk populations are alternated at diverse locations.
9. Horizontal resistance to disease rather than vertical resistance can be developed with the bulk breeding method.
10. The method allows for the introduction of relatively more genetic variability through the evaluation of a large number of crosses or composites. Introgression of genes from agronomically unadapted strains may be handled by bulk breeding methods. Selected individuals exhibiting introduced traits may be backcrossed to agronomically adapted parents and subjected to further natural selection pressures.
11. Artificial selection can be applied effectively to segregating populations to assist natural selection and to speed up the elimination of agronomically undesirable genotypes or those not meeting the objective of a cross.

REFERENCES

Allard, R. W., and S. K. Jain. 1962. Population Studies in Predominantly Self-Pollinating Species. II: Analysis of Quantitative Genetic Changes in a Bulk-Hybrid Population of Barley. *Evolution* 16:90-101.

Allard, R. W., S. K. Jain, and P. L. Workman. 1968. The Genetics of Inbreeding Populations. *Advances in Genetics* 14:55-131.

Atkins, R. E., and H. C. Murphy. 1949. Evaluation of Yield Potentialities of Oat Crosses From Bulk Hybrid Tests. *Agronomy Journal* 41:41-45.

Borlaug, N. E. 1968. Wheat Breeding and its Impact on World Food Supply. *Proceedings of the Third International Wheat Genetics Symposium*. Australian Academy of Science, Canberra, pp. 1-36.

Busch, R. H., J. C. Janke, and R. C. Frohberg. 1974. Evaluation of Crosses Among High and Low Yielding Parents of Spring Wheat (*Triticum aestivum* L.) and Bulk Prediction of Line Performance. *Crop Science* 14:47-50.

Choo, T. M., H. R. Klinck, and C. A. St. Pierre. 1980. The Effect of Location on Natural Selection in Bulk Populations of Barley (*Horderum vulgare* L.). II: Quantitative Traits. *Canadian Journal of Plant Science* 60:41-47.

Copp, L.G.L. 1957. Bulk and Pedigree Methods of Wheat Breeding. *Wheat Information Service*, Biological Laboratories, Kyoto University, Japan, No. 5, p. 7.

Cregan, P. B., and R. H. Busch. 1977. Early Generation Bulk Hybrid Yield Testing of Adapted Hard Red Spring Wheat Crosses. *Crop Science* 17:887-890.

Degago, Y., and C. E. Caviness. 1987. Seed Yield of Soybean Bulk Populations Grown for 10 to 18 Years in Two Environments. *Crop Science* 27:207-210.

Empig, L. T., and W. R. Fehr. 1971. Evaluation of Methods by Generation Advance in Bulk Hybrid Soybean Populations. *Crop Science* 11:51-54.

Fowler, W. L., and E. G. Heyne. 1955. Evaluation of Bulk Hybrid Tests for Predicting Performance of Pure Line Selection in Hard Red Winter Wheat. *Agronomy Journal* 47:430-434.

Grafius, J. E., W. L. Nelson, and V. A. Dirks. 1952. The Heritability of Yield in Barley as Measured by Early Generation Bulked Progenies. *Agronomy Journal* 44:253-257.

Hamblin, J. 1977. Plant Breeding Interpretations of the Effects of Bulk Breeding on Four Populations of Beans (*Phaseolus vulgaris* L.). *Euphytica* 25:157-168.

Harlan, H. V., and M. L. Martini. 1938. The Effect of Natural Selection in a Mixture of Barley Varieties. *Journal of Agricultural Research* 57:189-199.

Harrington, J. B. 1940. Yielding Capacity of Wheat Crosses as Indicated by Bulk Hybrid Tests. *Canadian Journal of Research* 18:578-584.

Ikehashi, H., and H. Fujimaki. 1980. Modified Bulk Population Method for Rice Breeding. In *Innovative Approaches to Rice Breeding*. International Rice Research Institute, Los Banos, Philippines, pp. 163-182.

Jain, S. K. 1961. Studies on the Breeding of Self-Pollinating Cereals: The Composite Cross Bulk Population Method. *Euphytica* 10:315-324.

Jennings, P. R., and R. C. Aquino. 1968. Studies on Competition in Rice. III: The Mechanism Among Phenotypes. *Evolution* 22:529-542.

Khalifa, M. A., and C. O. Qualset. 1975. Intergenotypic Competition Between Tall and Dwarf Wheats. II: In Hybrid Bulks. *Crop Science* 15:640-644.

Leffel, R. C., and W. D. Hanson. 1961. Early Generation Testing of Diallel Crosses of Soybeans. *Crop Science* 1:169-174.

Luedders, V. D., L. A. Duclos, and A. L. Matson. 1973. Bulk Pedigree and Early Generation Testing Breeding Methods Compared in Soybeans. *Crop Science* 13:363-364.

Newman, L. H. 1912. Plant Breeding in Scandinavia. *Canadian Seed Growers' Association*, Ottawa.

Ng, T. J., and E. C. Tigchelaar. 1973. Inheritance of Low Temperature Seed Sprouting in Tomato. *Journal American Society of Horticulture Science* 98:314-316.

Patel, J. D., E. Reinbergs, D. E. Mather, T. M. Choo, and J.D.E. Sterling. 1987. Natural Selection in a Doubled Haploid Mixture and a Composite Cross of Barley. *Crop Science* 27:474-479.

Patterson, B. D., R. Pauli, and R. M. Smillie. 1978. Chilling Resistance in *Lycopersicum hirsutum* Humb. and Bonpl., a Wild Tomato with a Wide Altitudinal Distribution. *Australian Journal of Plant Physiology* 5:609-617.

Raeber, J. G., and C. R. Weber. 1953. Effectiveness of Selection for Yield in Soybean Crosses by Bulk and Pedigree Systems of Breeding. *Agronomy Journal* 45:362-366.

St. Pierre, C. A., H. R. Klinck, and F. M. Gauthier. 1967. Early Generation Selection Under Different Environments as it Influences Adaptation of Barley. *Canadian Journal of Plant Science* 47:505-517.

Smith, E. L., and J. W. Lambert. 1968. Evaluation of Early Generation Testing in Spring Barley. *Crop Science* 8:490-493.

Suneson, C. A. 1949. Survival of Four Barley Varieties in a Mixture. *Agronomy Journal* 41:459-461.

———. 1956. An Evolutionary Plant Breeding Method. *Agronomy Journal* 48:188-191.

Suneson, C. A., and H. Stevens. 1953. *Studies with Bulked Hybrid Populations of Barley.* United States Department of Agriculture Technical Bulletin, Washington, D.C. No. 1067. 14 pp.

Torrie, J. H. 1958. A Comparison of the Pedigree and Bulk Methods of Breeding Soybeans. *Agronomy Journal* 50:198-200.

Walker, J. T. 1969. Selection and Quantitative Characters in Field Crops. *Biological Review* 44:207-243.

CHAPTER 10

The Backcross Breeding Method

The **backcross breeding** method was suggested by Harlan and Pope (1922) and was studied extensively by Briggs (1930, 1935, 1938, 1941, 1958). This method can be used in both self- and cross-fertilized crops and is analogous to line breeding in livestock and to introgression of natural genes.

Following hybridization between a **donor** and a **recurrent parent**, the recurrent parent is hybridized again with the F_1 progeny of the cross and repeatedly with the F_1 backcross progeny designated BCF_1, BC_2F_1, BC_3F_1, and so on. The features for which improvement is sought are maintained by selection. After a sufficient number of backcrosses, which varies from two to ten, the progeny will be heterozygous for the alleles involved in the transfer, but homozygous for all others. Selfing the last backcross generation, coupled with selection, will produce some progeny homozygous for the genes being transferred and identical with the original superior or recurrent parent in all other respects.

The prerequisites for a successful backcross program include

- the ability to identify the characteristic being transferred in the successive backcrosses, even though it may be temporarily diminished in expression;
- a simply inherited source of the plant characteristic to be transferred from a donor parent; and
- a sufficient number of backcrosses to reconstitute the recurrent parent.

DESCRIPTION OF A TYPICAL BACKCROSS BREEDING METHOD

The recovery of the recurrent parent and the incorporation of the plant feature to be transferred are the two objectives of a backcross breeding program. Highly heritable, simple traits, which are commonly controlled by one or three genes and which are visually identifiable among the progeny, are most suitable for transfer. Regardless of whether the gene being transferred is dominant or recessive, the same principle applies for the recovery of the trait, but the procedure varies slightly. Transfer of a dominant gene is accomplished more readily than transfer of a recessive gene because the phenotype of the backcross progeny reveals which individuals are carrying the gene in either the homozygous or heterozygous condition (Figure 10.1). Following the final backcross, progeny are heterozygous for the gene in question, but other genes are homozygous. Selfing will stabilize the allele in the homozygous condition following the final backcross.

When the allele to be transferred is in the recessive state, selfing in each generation is essential to allow the recessive gene to occur in the homozygous condition so that phenotypic

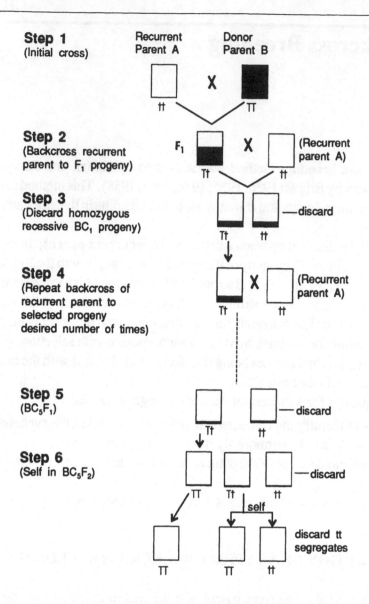

Figure 10.1 Scheme illustrating the transfer of a dominant gene from donor parent B to the recurrent parent A using the backcross system. When a dominant gene is being transferred, phenotypic selection for the trait can be achieved in each generation. Only those plants that exhibit the feature and hence carry the dominant gene are used for further backcrossing. Recessive genotypes are discarded. There is no selection for recurrent parent traits.

The average proportion of the donor parent in the hybrid population is reduced by one-half with each backcross. After the desired number of backcrosses have been completed, the plants carrying the desired trait will be heterozygous for that trait but homozygous for other genes.

After the last backcross, selfing will produce homozygosity for the specific trait and a cultivar will be produced with the same performance as the original cultivar but with the addition of the selected trait. Self-fertilization in the last step stabilizes the gene in the homozygous condition. Only those progeny that do not segregate for the trait are saved.

Each backcross generation to the recurrent parent results in an additional 50% of the recurrent parent being incorporated. After the third backcross, the progeny phenotypically resemble the recurrent parent.

selection can take place (Figure 10.2). Selfing does not change the basic premise of the backcross system that the recurrent parent is recovered. Although selfing does extend the program and interrupts the backcross series, it is necessary to ensure the presence of the recessive allele in the end product. Lengthening the backcrossing program could allow for critical selection to improve the recurrent parent if superior progeny can be identified, but this may not be possible or necessary.

When several genes are involved in a backcross program, larger plant populations are necessary. The backcross system is not effective for the transfer of quantitatively inherited characteristics involving numerous genes. Successfully incorporating quantitatively inherited traits is challenging with any breeding system.

The examples illustrated in Figures 10.1 and 10.2, are for self-fertilized crops, and although the backcross method commonly is viewed as having application to self-fertilized crops, it can also be used in cross-fertilized crops. Methods for self- and cross-fertilized crops do not differ materially, and both have the same objective. The major differences are those that take into account the heterozygous and heterogeneous nature of cross-fertilized species.

In cross-fertilized species, a series of crosses must be made in each and every generation to preserve vigor associated with heterozygosity and heterogeneity. Instead of hybridizing one selected donor and one recurrent parent as in self-fertilized species, at least 15 to 20 individual plants of each of the donor and recurrent parents should be used for making crosses. Another requirement is larger F_1 progenies.

Backcrossing was used to transfer the cytoplasmically inherited feature **triazine** resistance into oilseed rape (*Brassica napus*) and the summer turnip rape (*Brassica campestris*), largely self- and cross-fertilized species, respectively. The backcross screening procedure is outlined in Figure 10.3.

BACKCROSSING THEORY

Backcrossing in self-fertilized crops is based on the fact that a heterozygous F_1 population backcrossed to a homozygous parent will become homozygous for the genotype of the recurrent parent. The recovery of the recurrent parent is assumed, therefore, when the progeny of the initial cross is repeatedly backcrossed to the recurrent parent.

The proportion of genes from the donor parent is reduced by one-half following each generation of backcrossing, as illustrated in Figures 10.1 and 10.2. The precise content of the donor parent is given by the relationship $(1/2)^n$, where n is equal to the number of crosses and backcrosses to the recurrent parent. In the BC_2, for example, the donor parent contribution is $(1/2)^3 = 1/8 = 12.5\%$; in the BC_4, the value is $(1/2)^5 = 1/32 = 3.125\%$.

Figure 10.1 continued

After five or six backcrosses, most of the recurrent genes have been recovered. The shaded portions of the diagram illustrate the declining genetic influence of the donor parent.

The backcross method provides the plant breeder with a high degree of genetic control and is a repeatable process.

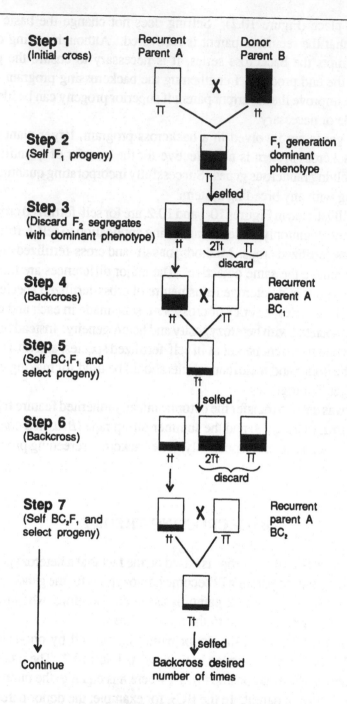

Figure 10.2 Scheme illustrating the transfer of a recessive gene from the donor parent B to the recurrent parent A. Progeny will automatically contain the recessive gene, but unlike the dominant gene situation in Figure 10.1, phenotypic identification is not possible. The key to identifying the homozygous recessive plants is self-fertilization. Backcrossing in series without selfing could result in the loss of the desired gene with no chance of success. Only those progeny known to contain the recessive gene are backcrossed to the recurrent parent.

Selfing, however, is time consuming and slows up the backcrossing program. The first backcross could be made onto the F₁ progeny without selfing, in which case 50% of the progeny would be homozygous dominant and discarded and 50% would be heterozygous and containing the desired

Figure 10.3 The transfer of triazine resistance into oilseed rape (*Brassica napus*) is an excellent example of the successful transfer of a single gene using the classical backcross breeding method (Beversdorf et al., 1980). Resistance to specific analogs of the triazine family, a widely used agricultural herbicide, was found in a weed plant, wild turnip (*Brassica campestris*). The resistant gene, however, was found to be carried in the chloroplast of the cell rather than in the nucleus. In sexual crosses, the chloroplast is transmitted by the maternal line alone, and the male contributes only nuclear DNA. To transfer this trait, the resistant weed species was used as the female parent and an agronomically suitable cultivar was selected as the recurrent male parent.

Following each cross, a triazine herbicide was applied to the progeny, and only those that showed resistance were fertilized. Four triazine-treated rows are shown in this photograph. Rows two and four were killed by the chemical; rows one and three are resistant.

After five to seven generations, a triazine-resistant oilseed rape was developed. The ability to apply triazine compounds to field-grown rapeseed allows for the removal of broad-leaved weeds as well as wild mustard, a *Brassica* species that contaminates the oil produced from rapeseed with a bitter compound.

Photo courtesy D. Hume, University of Guelph.

Figure 10.2 continued

recessive gene. A large enough population would ensure the presence of the gene, but to ensure further against its loss, a selfing operation must be conducted. After two or three backcrosses followed by selfing and selection, the odds are increased that the recessive allele will be present. With an adequate plant population, the selfing interruptions may be reduced.

The shaded portions of the diagram illustrate the declining genetic influence of the donor parent.

Furthermore, the proportion of homozygosity in a particular generation, m, is given by the relationship

$$\frac{2^m - 1}{2^m}$$

Since in each generation the progeny are backcrossed to the recurrent inbred parent, the homozygous loci are those of the recurrent parent. In the BC_5F_1 generation, for example, the proportion of homozygosity is given by

$$\frac{2^5 - 1}{2^5} = \frac{31}{32}$$

This means that 31/32, or 96.875%, of the loci are homozygous and that these homozygous loci are those from the recurrent parent.

The proportion of homozygous genotypes after m number of backcrosses and with n number of heterozygous loci is given by the relationship

$$\frac{2^m - 1^n}{2^m}$$

After the BC_5 generation, for example, with five heterozygous loci, the percentage of homozygous individuals for ten loci of the recurrent parent is $(31/32)^{10} = 72.8\%$. As the number of backcross generations increases, the proportion of homozygous loci that are identical to those of the recurrent parent increases. Because the recovery of the recurrent parent genotype is a function of the number of crosses to the recurrent parent, selection for the recurrent genotype usually is unnecessary.

A continuous series of backcrossings to transfer a monogenic dominant allele requires fewer plants than any alternative mating system. Briggs and Allard (1953) calculated that a theoretical minimum of 53 plants from backcrossed seeds, 96 F_2 plants, and 68 F_3 rows are required to complete the transfer through six backcrosses. Since cultivars of self-fertilized crops may consist of a mixture of closely related pure lines, unless they were developed through a doubled-haploid breeding system, too small a population is not recommended because the genetic base of the recovered cultivar may be too narrow.

The number of backcrosses may vary from 2 to 12. Six backcrosses were considered satisfactory by Briggs and Allard (1953) when coupled with rigid selection in the early generations. These workers suggested that selection for the type of the recurrent parent is equivalent to one or two additional backcrosses in a continuous series if a moderate-sized population exists. By the third backcross, further selection for the recurrent parent type is not possible because the population resembles the recurrent parent so closely. If no selection is practiced for the recurrent parent and if populations are small, ten backcrosses may be used. Six to ten backcrosses may be required if the donor parent has an agronomically poor genotype. Four to five backcrosses are used commonly when an agronomically acceptable donor parent is used.

Borlaug (1958) attempted to obtain from 100 to 200 backcrossed seeds on each selected parental F_1 plant. The use of such a large population allowed the number of backcrosses, originally planned at four to six, to assure the recovery of the phenotype if the recurrent parent was reduced to two or three backcrosses. If the original parents are similar in agronomic type, fewer backcrosses can be considered.

Atkins and Mangelsdorf (1942) suggested the use of **isogenic lines** to compare effects of alternative characters in plant performance. Isogenic lines are produced by backcrossing and lead to the emergence of two cultivars that contrast with each other for the feature in question (Figure 10.4). Ten isogenic lines produced by eight to ten backcross generations demonstrated that removing awns reduces wheat yields an average of 4.1% and over 8% under dry conditions (Atkins and Norris, 1955).

Isogenic lines may differ from their counterpart by one or a very small number of alleles. To produce a true isogenic line with a one-gene difference, 10 to 12 backcrosses may be required, but a near isogenic line may be formed with four or five backcrosses.

Peterson (1957) suggested that some breeders use only two or three backcrosses to retain some of the benefits of transgressive segregation for agronomic characters such as yield and adaptation without attempting to reconstitute the recurrent parent. Three to five backcrosses were suggested by Lawrence and Frey (1976) as necessary for the introgression of plant characteristics from a wide oat cross using *Avena sterilis* L. to produce progeny appropriate for a practical breeding program. Nine backcross-derived wheat cultivars tested by Suneson (1947) following six backcrosses were similar in yield performance to their progenitor cultivars under pest-free conditions.

Backcrossing can be used for incorporating plant characteristics that are more complex than simply inherited traits. The system is a combination of backcrossing and convergent crossing and is used where the characteristic to be improved can be followed through the generations (Figure 10.5).

BACKCROSSING CROSS-FERTILIZED CROPS

A rapid loss of vigor normally occurs when cross-fertilized and heterozygous crops are selfed, and therefore the use of a recurrent parent in a backcross program is equivalent to one generation of inbreeding. The proportion of homozygosity, however, would not increase beyond that reached by a single inbred generation with recessive backcrosses.

The loss of vigor following a backcross to the recurrent parent means that when the backcross system is used, large populations are important for two reasons: (1) to adequately sample and maintain the diversity of the cultivar, and (2) to insure against the harmful effects of inbreeding.

If the trait under transfer is conditioned by a single dominant gene, those individuals that exhibit the trait can be used for the next generation of backcrossing and those without the trait can be discarded. Where **polysomic inheritance** is involved, an extra generation of **intercrossing** may be necessary to identify individuals that carry a sufficient expression of the trait under transfer.

For the situation in which a recessive gene is being transferred to a recurrent parent, each backcross generation must be followed by a generation of intercrossing so that individuals with the recessive phenotype can be identified for the next backcross generation (refer to Figure 10.2).

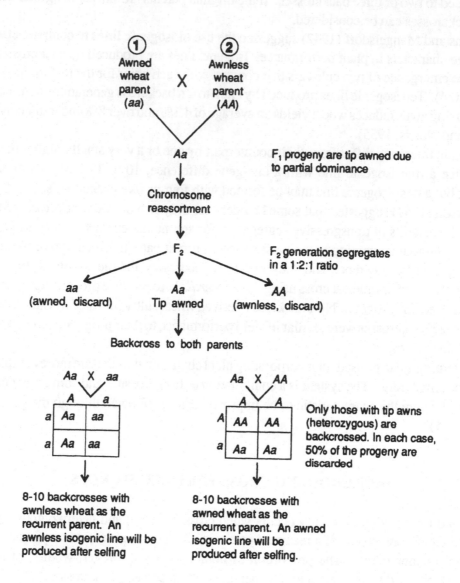

Figure 10.4 Scheme showing the development of isogenic lines in which two cultivars are produced that contrast for a particular feature, in this case awned and awnless wheat. Awned cultivars are homozygous recessive; awnless cultivars are homozygous dominant. In the heterozygous condition, partial dominance is expressed as tip awns, making it easy to select progeny for backcrossing. To develop two cultivars, one with awns and one without awns, both parents are used as the recurrent parent so that comparisons can be made between awned and awnless parent 1 and awned and awnless parent 2. Similar procedures could be used to produce isogenic lines for hooded versus awned, hooded versus awnless, and rough awns versus smooth awns.

Isogenic lines may be developed for other features, such as maturity, disease resistance, plant height, or leaf width. Backcrossing is applicable to any situation where the expression of a character is dependent primarily upon one pair of genes. When the heterozygote is easily identified and self-fertilization is the mode of reproduction, the system is especially useful.

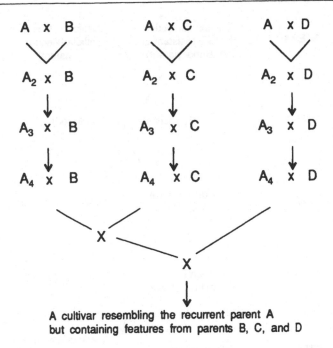

A cultivar resembling the recurrent parent A
but containing features from parents B, C, and D

Figure 10.5   Illustration of a backcross breeding system where parent A is the recurrent parent to be improved and parents B, C, and D are the donor parents.

In self-fertilized polyploid crops, such as wheat where genic buffering occurs, a straight A x B cross generally will produce progeny intermediate between the two parents. In advanced breeding programs where highly adapted and agronomically superior cultivars are common, such a result is seldom adequate. A more desirable objective would be to recover an agronomically desirable cultivar, the recurrent parent, with some transgressive improvements added. This may be achieved by a combination of backcrossing and convergent crossing.

Convergent improvement has been investigated more extensively in cross-fertilized maize for inbred line improvement than in self-fertilized crops. A slight, although consistent, improvement was reported in inbred maize lines, but not enough for commercial applications (Briggs and Allard, 1953).

To improve an inbred maize line, backcrossing offers a dependable method, and in this case, inbreds are essentially the same as self-fertilized crops and the same procedures prevail.

A plan for convergent maize improvement that is based on backcrossing coupled with selection was presented by Richey (1927). He concluded that this system offers the most definite and certain means of increasing the productiveness of inbred maize without interfering with its behavior in hybrid combinations. Richey's ultimate objective was to produce inbred lines capable of yielding as much as F_1 hybrid combinations to overcome the problem of single-cross seed costs that was limiting hybrid maize use at the time.

Stanford (1952) reported on the use of the backcross method to incorporate bacterial wilt resistance into the alfalfa cultivar California Common using a simply inherited source of resistance in a cultivar from Turkey. In addition to incorporating bacterial wilt resistance, leafspot and mildew resistance were simultaneously incorporated from a line of California Common alfalfa. The procedure, which was successful, is outlined in Figure 10.6.

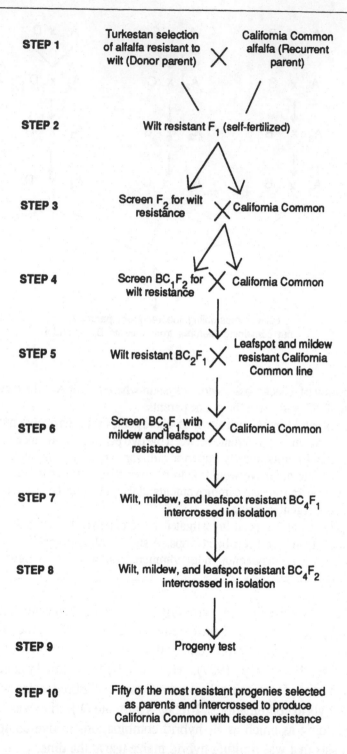

Figure 10.6   Outline of a backcross procedure presented by Stanford (1952) to incorporate bacterial wilt (*Corynebacterium insidiosum*) resistance from an unadapted cultivar from Turkey into California Common alfalfa, a cross-fertilized crop. An additional step of incorporating leafspot (*Pseudopeziza medicaginis*) and mildew (*Peronospora trifoliorum*) resistance was undertaken. This was achieved by substituting the recurrent parent in step 5 at the time of the third backcross.

Because alfalfa will not tolerate inbreeding, backcrosses were made to many, generally unrelated,

Details of cultivar production with cross-fertilized crops through backcrossing may differ with individual species. The essential features for success are the maintenance of a broad gene base by dealing with large segregating populations and the use of an adequate number of backcrosses to ensure success in regaining the recurrent parental genotype. In practice, however, the laborious crossing schedule associated with the large numbers needed for both the donor and recurrent parents means that a minimum number of backcrosses may be used.

Three backcross generations and 500 interpollinated clones were used to develop bacterial-wilt-resistant Iroquois alfalfa (Murphy and Lowe, 1968). Two backcrosses and 40 plants were used by Sherwood *et al.* (1967) to incorporate stem nematode (*Ditylenchus dipsaci*) resistance into alfalfa. To produce wilt-resistant Saranac alfalfa, three backcross generations were used (Murphy and Lowe, 1966).

After the last backcross generation, and at times during the backcross procedure, a number of generations of intercrossing and selection may be conducted for both the trait under transfer and the phenotype of the recurrent parent. These techniques make it possible to recover the desirable adaptation and performance of the recurrent parent while still maintaining the good performance associated with the gene under transfer.

A modified backcross method was used to incorporate a dominant gene for apple scab resistance into commercial apple cultivars (Hough, Shay, and Dayton, 1953). The modification involved a different but commercially acceptable apple cultivar as the recurrent parent in each backcross generation. Crossed fruit were harvested when seeds were mature, and initial scab resistance was determined within two weeks. Seedlings took five years to reach the reproductive stage before a backcross was possible. Large populations were used, which helped reduce the number of backcrosses required to obtain scab resistance in a commercially acceptable cultivar.

Figure 10.6 continued

plants of the parental population. An average of 125 plants were used in each generation as female parents and 275 plants as male parents. Cross-pollination was achieved by collecting flowers from a random selection of male parent plants and using them to pollinate about 150 florets on each female parent plant. The pollen used represented a good sample from the entire male population. Flowers were not emasculated prior to pollination, but tests showed that 95% of the progenies were hybrid. The small percentage of selfed plants were eliminated in the wilt test. Since the objective was to reproduce the California Common cultivar without any change, there was no rigid selection of parent plants.

Tests for disease resistance were conducted in the greenhouse. All plants in the BC_2F_1 and subsequent generations were screened for bacterial wilt resistance. Plants in the BC_3F_1 and subsequent generations were screened for mildew and leafspot resistance. In step 7, resistant plants were allowed to intercross. Intercrossing of plants heterozygous for the same disease resistance factors is just as effective as selfing to produce homozygosity. A total of seven years were required, including seed increase, to transform California Common into a cultivar with resistance to three specific problem diseases.

LINKAGE AND RECOVERY OF
DESIRABLE VERSUS UNDESIRABLE TRAITS

The statement that the recurrent parent automatically is assured by continued backcrossing does not hold when genes are linked on the same chromosome. **Linkage** occurs when a gene is 50 map units or less from an adjacent gene; genes in excess of 50 map units apart are not linked and segregate as if they were on different chromosomes. When an undesirable gene(s) is linked to the gene being transferred in a backcross breeding program, the undesirable gene(s) may be transferred also.

The backcross procedure can have an effect on the probability of losing an undesired gene and at the same time maintaining the desired gene. Two different situations may occur: (1) selection is practiced for the desirable allele of the gene under transfer, with no selection against an undesirable allele on the same chromosome; (2) selection is practiced for the desirable allele of the gene under transfer and against an undesirable gene linked to the gene under transfer.

The important consideration in the case of no selection is the proportion of the progeny from a given backcross that will result in the desirable gene being separated from the undesirable gene. The probability of eliminating the undesirable gene is given by the formula

$$1 - (1 - p)^{m+1}$$

where p is the recombination fraction and m is the number of backcrosses.

After five backcrosses where genes are separated by 50 map units, the probability that the undesirable gene would be eliminated if selection was practiced for the desirable gene is 98%. Under self-fertilized conditions, the probability of elimination is 50%. The probability declines for genes more closely linked. If, for example, an undesirable gene is ten map units from a gene under transfer, the probability of elimination after five backcrosses is about 47%. After five generations of selfing, the probability is only 10%. However, when selection for the desirable allele under transfer is accompanied by selection against the undesirable gene, self-fertilization is more likely to give the desired combination.

Under conditions of self-fertilization, effective recombination can occur in the F_1 gametes from both parents and in subsequent generations. In contrast, backcrossing results in effective recombination only in the gametes of the hybrid progeny and not in the gametes coming from the recurrent parent. The lack of effective recombination in the recurrent parent is due to the homozygous nature of that parent in a self-fertilized species.

Where no selection is practiced against the undesirable gene, each generation of backcrossing allows more opportunity for the rare recombination to occur by chance in the recurrent parent such that the undesirable gene is lost. Fifteen years were required before the tight linkage between resistance to the root knot nematode (*Meloidogyne incognita*) found in a wild tomato (*Lycopersicum peruvianum*) and undesirable fruiting characters was finally broken (Rick, 1967).

It is possible that two desirable genes may be linked, and this should aid the program. If the donor parent containing the two desired linked genes is an agronomically poor parent, the segment of the chromosome between the two loci should be eliminated. This can be accomplished only if a strategically located crossover could occur, an unlikely but possible situation.

CONSIDERATIONS OF A BACKCROSS BREEDING PROGRAM

Traits that normally are transferred from a donor parent to a recurrent parent in backcrossing are those that are highly heritable, that are controlled by one or a few genes, and that are identifiable in each generation. Linkages of desired genes with undesired genes can disrupt the ease of transfer.

When a number of genes influence a particular trait, a greater probability exists that some of these genes will be linked to other genes that might be undesirable. In addition, traits are less clearly expressed if conditioned by more than one gene, making selection less precise in any backcross generation.

The backcross breeding method can be conducted independently of the environment where the recurrent parent is adapted, provided the characteristic to be transferred can be identified. Since selection during the backcross generation usually is conducted only for the trait being transferred, the recovery of the genotypic and phenotypic characteristics of the recurrent parent is a simple matter of the number of backcross generations that have been completed.

Advanced progeny of a backcross program do not require as extensive testing as a new cultivar because the vast majority of the genes are those of the recurrent parent, except for the gene being transferred and perhaps other closely linked genes. A new cultivar produced through a backcross breeding program closely resembles the phenotype of the recurrent parent.

The backcross breeding method has not gained wide acceptance, perhaps because breeders seek positive transgressive segregation. In other words, backcrossing is regarded as too conservative because the method makes only a minor adjustment to a recurrent parent following a cross, and genetic variability between the two parents is removed quickly. Plant breeders generally hope to develop a new cultivar improved in several aspects and often seek improvement in quantitative characteristics. The backcross breeding method, however, is recognized as useful in a large number of self- and cross-fertilized crops, including tree species (Thomas, 1952).

BACKCROSSING TO TRANSFER MORE THAN ONE GENE

The backcross method has been used mostly to transfer monogenic agronomic features, but it can be used successfully for cultivar improvement for two or more characteristics. In other words, the accumulation of additional desirable genes governing other single plant characteristics can be accomplished with the backcross breeding method.

When two or more genes are to be transferred to a recurrent parent, a stepwise progression of backcross programs may be used in which one gene is transferred from one donor parent that excels for the trait, followed by a second backcrossing program where a second parent is chosen for its excellence for the trait under transfer. The procedure outlined in Figure 10.6 is a two-step approach but involves substitution of the recurrent parent. A two-step approach would likely be adopted even in the situation where a single donor parent carried two genes for desired features.

More than one gene may be incorporated by conducting two simultaneous backcross programs and merging the end products. This might involve crossing two isogenic lines and selecting plants containing both agronomic features. Two simultaneous backcross programs were used by Paris *et al.* (1985) to transfer the concentrated yield of the birdsnest melon to prostrate vine melons (Figure 10.7). The trait is inherited as a single recessive with modifier genes. To transfer all those genes, some of which might be cytoplasmically inherited, two reciprocal backcross programs were conducted, and these were merged after a limited number

A.   B.

Figure 10.7   Photo A shows a vine melon (*Cucumis melo* L.) (muskmelon, cantaloupe) that has a prostrate vine habit. The birdsnest melons in Photo B possess a compact, upright growth, resulting in a bushy appearance, and exhibit a uniform development and maturation of fruit. The birdsnest plant type is controlled by a single recessive gene plus modifiers (Paris, Nerson, and Karchi, 1984).

To incorporate the dramatically different birdsnest characteristics into adapted vine melons, Paris *et al.* (1985) used a backcross-pedigree system. The initial step consisted of crossing a birdsnest melon as the male donor parent and a vine melon as the female recurrent parent. The cultivar name of the birdsnest melon was Persia 202 and that of the vine melon was Galia. Following the initial cross, pedigree selection was practiced in the F_3, and three backcrosses were completed. Pedigree selection was practised in the F_3 of a similar cross but with Galia serving as the male parent and with four backcrosses. Prototypes from both crosses BC_3F_3 x BC_4F_3 were hybridized to produce pedigree progeny for evaluation.

The concentrated yield characteristic of the birdsnest melon was transferred successfully into horticulturally acceptable material, but further backcrossing may be needed to improve fruit quality. Total yield per unit land area from birdsnest cultivars was found to be comparable to vine plants, but optimum plant densities and irrigation and fertilizer regimes need to be determined.

The backcross-pedigree system is ideal for transfer of the birdsnest characteristics into vine plants. Pedigree selection is appropriate because of the wide differences in the two types, and backcrossing is appropriate because phenotypic selection is possible for the recessively inherited birdsnest feature. Further backcrossing to a melon with fruit quality as the recurrent parent may improve quality aspects.

of three and four backcrosses so that transgressive segregation would allow yield advances. Pedigree selection was conducted on the resulting progeny. The method, proposed by Allard (1960), is known as the backcross-pedigree method. The precise control of genetic forces following backcrossing is sacrificed, but the birdsnest character was incorporated into a compatible genotype.

Two or more genes may be transferred in a backcross program by simultaneous selection among the progeny. Such an approach requires a larger population than if two genes were transferred independently. The complexity involved with two- and three-gene transfers increases exponentially as does population size. Adding to the complexity is the need for screening conditions that allow two or three traits being transferred simultaneously to be expressed, and this may not always occur.

APPLICATIONS OF THE BACKCROSS BREEDING METHOD

Multiline Development

The precision and reliability with which a single deficient agronomic feature can be incorporated into an otherwise acceptable cultivar has made the backcross method a useful breeding tool. The backcross system has been used in many cases to incorporate a single gene for disease resistance into an otherwise acceptable cultivar only to have a **virulent** form of the disease arise. Lack of yield stability has discouraged use of the backcross program, which by definition incorporates specific resistance influenced by a single allele. A matching virulent form of the disease can arise to counter this specific allele. Under such conditions, backcross breeding for disease resistance could be a never-ending race unless greater stability can be incorporated. In addition, new sources of disease resistance are not readily available.

The need for a desirable form of resistance to wheat stem rust (*Puccinia graminis tritici*) was dramatically illustrated by the sudden and widespread appearance in 1950 of race 15B of stem rust. All of the commercial cultivars that prior to 1950 were resistant were attacked (Knott, 1958). Backcrossing to incorporate resistance provided useful tester stocks but was a temporary solution.

Multilines were suggested as an approach to develop a more lasting form of resistance, a method first proposed by Jensen (1952). A multiline is defined as a blend or mixture of cultivars that are genetically very similar but each has a different gene for resistance to the same pest organism. Each line composing a multiline is an isogenic line derived from a series of crosses incorporating different sources of genetic resistance, each of which are backcrossed to a common parent. A number of workers have endorsed the concept of a multiline as a means of greater yield stability (Borlaug, 1958; Suneson, 1960; Browning and Frey, 1969; Parlevliet, 1979).

Multilines consist of a number of phenotypically similar lines that differ genotypically for resistance to a specific disease. The lines included in a multiline can be determined by the relative prevalence of the different races. A multiline cultivar could be modified by simply removing lines that are susceptible and substituting other backcrossed lines with different genetic resistance. Components of a multiline would be grown separately and blended accordingly.

When plant breeders are limited by a restricted gene source, single genes can be rotated when developing new cultivars, thereby reducing the opportunities for pests to overcome resistance. Multiline cultivars are more expensive to develop than a single cultivar, but the expense may be offset by improved yield stability.

An alternative to multilines is to **pyramid** desirable genes to provide a wide array of horizontal resistance. Multiple gene resistance, however, may not provide a solid foundation for sustained agricultural production.

A multiline may consist of three or more lines. Borlaug (1958) suggested a minimum of eight to ten, thereby requiring eight to ten backcrossing programs simultaneously and screening for genes resistant to a specific race. Multiline development is laborious and time consuming, and progress may be hindered by linkage of undesirable genes. *In vitro* screening techniques (Figure 10.8) could greatly assist backcross breeding of multilines.

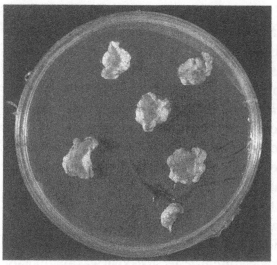 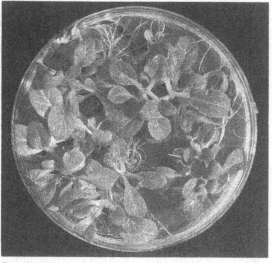

A.   B.

Figure 10.8 Genetic engineering may facilitate backcross breeding programs by identifying resistance to a specific race of a disease. Calli from hybridized plants may be screened for resistance to a specific disease or race by culturing them on a culture filtrate produced by the fungus. Plants lacking resistance will find the substrate toxic and perish, as demonstrated in Photo A. Resistant forms, as in Photo B, flourish in the same medium. The plantlets in Photo B are tobacco.

In vitro selection of barley and wheat for resistance against pathotoxins of *Helminthosporium sativum*, common root rot, was reported by Chawla and Wenzel (1987). In cereals, regeneration of plants as shown in Photo B is a limiting factor. The selection system of embryogenic calli induced from miniature embryos was used by Chawla and Wenzel.

Plant tissue culture could meet the requirements for screening a large number of individuals in a small space and could greatly assist backcross breeding programs and multiline development.

Source: Petri plates produced by Dr. R. Nazir; photos courtesy H. Rauscher.

Introgression of Genes from Wide Crosses

Backcross programs offer a means for the **introgression** of weedy, unadapted, exotic, or wild germplasm into a suitable background for subsequent crossing or further backcrossing programs. Economically important traits found in distantly related species, but in agronomically undesirable genotypes, make such **wide crosses** a potential source of useful traits for breeding purposes. Such crosses often result in negative **transgressive segregation**, and repeated backcrosses are needed. The backcross system remains a traditional system for the introgression of exotic germplasm (Hawkes, 1977).

Backcrossing in wide crosses for the introgression between the hexaploid (2n = 42) oat (*Avena sativa*) and two tetraploid (2n = 28) species, *Avena magna* and *Avena murphyi*, is illustrated by the work of Ladizinsky and Fainstein (1977). A high degree of self-sterility of the pentaploid F_1 hybrids was overcome by a very large number of backcrosses to the adapted recurrent parent. The objective was the transfer of genes without attempting to release a commercial cultivar. Backcrossing improved the cytological stability of the oat progenies.

Because wild plant species possess a whole range of undesirable characters, such as low yield, unacceptable quality aspects, and other unfavorable agronomic features, backcross programs involving wide crosses may be long and tedious. Crosses may be achieved with difficulty. Large populations are necessary to allow for intense selection following each backcross.

When one parent is markedly inferior to the other parent, the backcross-pedigree system is advocated. In an effort to find the best crossing method to incorporate unadapted germplasm, Eaton, Busch, and Youngs (1986) compared four crossing systems. A problem in the introgression of unadapted germplasm is that favorable genes or gene complexes may be masked by major genes for adaptation to different environments. Based on line performance, the backcross and the three-way cross were preferred methods of incorporating unadapted germplasm.

OBSERVATIONS ON THE BACKCROSS BREEDING METHOD

1. The method of backcross breeding is based on clear-cut genetic principles that lead with precision to predictable results in that characteristics of the derived cultivar are known in advance. No other plant breeding method can provide these qualifications.
2. The backcross method offers a means of improving an existing cultivar that is deficient in one or a few features without the risk of breaking up desirable existing gene combinations.
3. Fewer backcross plants are required to transfer a monogenic dominant allele than in any alternative mating system.
4. The backcross procedure is most useful where the trait under transfer has few genes and a high heritability. The procedure offers the possibility of accumulating a series of desired monogenic or **qualitative traits** but is valueless in dealing with a **quantitative trait**.
5. The location where the backcrossing procedure is conducted is independent of the area of adaptation as long as selection can be carried out for the trait being transferred.
6. Since the derived cultivar is like its progenitor in all respects except for the specific

improvement, a backcross cultivar requires less testing for commercial release than a cultivar developed through other breeding systems.

7. Backcrossing is a conservative method of plant breeding. There is no possibility of a desirable new recombination of genetic traits.

8. Depending on linkages of undesirable genes with those desired for a transfer from donor to recurrent parent, the backcross method may result in a cultivar that does not regain all of the performance of the original recurrent parent.

9. Other uses of backcross breeding include introgression of specific features from wide crosses, production of individual components of a multiline, and isogenic line development, which offers physiologists and other researchers an opportunity to measure the effect of particular alleles on plant performance or treatment response.

10. Backcrossing offers a method of developing vertical pest resistance, in which resistance is inherited as a single gene, but is valueless for developing horizontal resistance, in which a large number of genes gives generalized resistance.

11. Selection of the recurrent parent is extremely important since the derived cultivar cannot be any better than the original, except for the characteristic to be transferred.

12. Backcrossing procedures may be combined with other breeding methods, such as the backcross-pedigree system, to meet specific objectives.

REFERENCES

Allard, R. W. 1960. *Principles of Plant Breeding.* John Wiley and Sons, New York, p. 156.

Atkins, I. M., and P. C. Mangelsdorf. 1942. The Isolation of Isogenic Lines as a Means of Measuring the Effects of Awns and Other Characteristics on Small Grains. *Journal of American Society of Agronomy* 34:667-668.

Atkins, I. M., and M. J. Norris. 1955. The Influence of Awns on Yield and Certain Morphological Characters of Wheat. *Agronomy Journal* 47:218-220.

Beversdorf, W. D., J. Weiss, L. R. Erickson, and V. Souza Machado. 1980. Transfer of Cytoplasmically Inherited Triazine Resistance from Birds' Rape to Cultivated Rapeseed. *Canadian Journal of Genetics and Cytology* 22:167-172.

Borlaug, N. E. 1958. The Use of Multilineal or Composite Varieties to Control Air-Borne Epidemic Disease of Self-Pollinated Crop Plants. In *Proceedings First International Wheat Genetics Symposium.* University of Manitoba, Winnipeg, pp. 12-31.

Briggs, F. N. 1930. Breeding Wheats Resistant to Bunt by the Back-Cross Method. *Journal American Society of Agronomy* 22:239-244.

———. 1935. The Backcross Method in Plant Breeding. *Journal American Society of Agronomy* 27:971-973.

———. 1938. The Use of the Back-Cross in Crop Improvement. *American Naturalist* 72:285-292.

———. 1941. The Use of the Backcross in Plant Breeding. *Proceedings of the Seventh International Genetics Congress,* Edinburgh and London, Oliver and Boyd, pp. 81-82.

———. 1958. Backcrossing: Its Development and Present Application. In *Proceedings of the First International Wheat Genetics Symposium.* University of Manitoba, Winnipeg, pp. 8-11.

Briggs, F. N., and R. W. Allard. 1953. The Current Status of the Backcross Method of Plant Breeding. *Agronomy Journal* 45:131-138.

Browning, J. A., and K. J. Frey. 1969. Multiline Cultivars as a Means of Disease Control. *Annual Review of Phytopathology* 7:355-382.

Chawla, H. S., and G. Wenzel. 1987. In Vitro Selection of Barley and Wheat for Resistance Against *Helminthosporium sativum*. *Theoretical and Applied Genetics* 74:841-845.

Eaton, D. L., R. H. Busch., and V. L. Youngs. 1986. Introgression of Unadapted Germplasm into Adapted Spring Wheat. *Crop Science* 26:473-478.

Harlan, H. V., and M. N. Pope. 1922. The Use and Value of Back-Crosses in Small-Grain Breeding. *Journal of Heredity* 13:319-322.

Hawkes, J. G. 1977. The Importance of Wild Germplasm in Plant Breeding. *Euphytica* 26:615-621.

Hough, L. F., J. R. Shay, and D. F. Dayton. 1953. Apple Scab Resistance from *Malus floribunda* Sieb. *Proceedings of the American Society for Horticultural Science* 62:341-347.

Jensen, N. F. 1952. Intra-varietal Diversification in Oat Breeding. *Agronomy Journal* 44:30-34.

Knott, D. R. 1958. The Inheritance of Stem Rust Resistance in Wheat. In *Proceedings of the First International Wheat Genetics Symposium*. University of Manitoba, Winnipeg, pp. 32-38.

Ladizinsky, G., and R. Fainstein. 1977. Introgression Between the Cultivated Hexaploid Oat *A. sativa* and the Tetraploid Wild *A. magna* and *A. murphyi*. *Canadian Journal of Genetics and Cytology* 19:59-66.

Lawrence, P. K., and K. J. Frey. 1976. Inheritance of Grain Yield in Oat Species Crosses. *Egyptian Journal of Genetics and Cytology* 5:400-409.

Murphy, R. P., and C. C. Lowe. 1966. Registration of Saranac Alfalfa. *Crop Science* 6:611.

———. 1968. Registration of Iroquois Alfalfa. *Crop Science* 8:396.

Paris, H. S., T. G. McCollum, H. Nerson, D. J. Contliffe, and Z. Karchi. 1985. Breeding of Concentrated-Yield Muskmelons. *Journal of Horticultural Science* 60:335-339.

Paris, H. S., H. Nerson, and Z. Karchi. 1984. Genetics of Internode Length in Melons. *Journal of Heredity* 75:403-406.

Parlevliet, J. E. 1979. The Multiline Approach in Cereals to Rusts: Aspects, Problems and Possibilities. *Indian Journal of Genetics* 39:22-29.

Peterson, R. F. 1957. Accomplishments of the Pedigree and Backcross Methods in Wheat Breeding. *Report of the Third International Wheat Rust Conference*. University of Minnesota, St. Paul, pp. 5-9.

Richey, F. D. 1927. The Convergent Improvement of Selfed Lines of Corn. *American Naturalist* 61:430-449.

Rick, C. M. 1967. Exploiting Species Hybrids for Vegetable Improvement. *Proceedings of the XVII International Horticulture Congress*. Michigan State University, East Lansing, Vol. 3, pp. 217-229.

Sherwood, R. T., J. W. Dudley, T. H. Busbice, and C. H. Hanson. 1967. Breeding Alfalfa for Resistance to the Stem Nematode, *Ditylenchus dipsaci*. *Crop Science* 7:382-384.

Stanford, E. H. 1952. Transfer of Disease Resistance to Standard Varieties. *Proceedings of the Sixth International Grassland Congress*. Pennsylvania State College, College Park, pp. 1585-1590.

Suneson, C. A. 1947. An Evaluation of Nine Backcross Derived Wheats. *Hilgardia* 17:501-510.

Suneson, C. A. 1960. Genetic Diversity of Protection Against Plant Diseases and Insects. *Agronomy Journal* 52:319-321.

Thomas, M. 1952. *Backcrossing: The Theory and Practice of the Backcross Method of Breeding Some Non-Cereal Crops*. Commonwealth Bureau of Plant Breeding and Genetics, Cambridge, England, Technical Bulletin No. 16. 139 pp.

CHAPTER 11

Single-Seed-Descent and Recurrent Selection Breeding Methods

SINGLE SEED DESCENT

Two main functions in breeding self-fertilized crops are progress toward homozygosity by selfing following hybridization and the selection of agronomically superior lines from the segregating populations. In the pedigree system, these two functions are performed simultaneously. As outlined in Chapter 8, the effectiveness of early-generation selection for quantitatively inherited traits is questionable. Generally, the pedigree method restricts the plant breeder to a single generation per year because selection is preferred under environmental conditions representative of those under which a new cultivar will be grown commercially. Family size in the pedigree method must be large enough to evaluate family performance with confidence.

The single-seed-descent method separates the step toward homozygosity and the selection step. The initial step is to enhance progress toward homozygosity by advancing the generations as rapidly as possible and subsequently to select among the randomly derived inbred lines. To maintain manageable levels of segregating populations, one randomly chosen seed is planted from each previous-generation plant. After reaching homozygosity, conventional performance trials are conducted. The method is illustrated in Figure 11.1.

Although this plant breeding system was first outlined in the late 1930s by Goulden (1939), the method did not find wide acceptance until the 1960s when Brim (1966) presented the system as a modified pedigree method. Kaufman (1961, 1971) investigated single seed descent in oat breeding but called it the "random" method because a single random seed is harvested from each plant in each generation to propagate the next generation and because there is no selection in segregating populations. Kaufman concluded that single seed descent is superior to the pedigree method when breeding for increased yield in the oat crop because handling of segregating generations is simplified and larger numbers of selections can be evaluated more thoroughly.

Issues in Single Seed Descent

Plants from each segregating generation may be grown in a limited area, sufficient only to ensure that a few or even a single seed is produced (Figure 11.2). An exception is in the F_5 generation when the entire plant is harvested to plant a progeny row or family block for preliminary evaluation in the following generation.

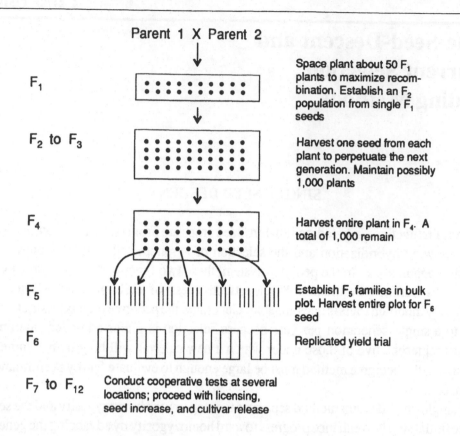

Figure 11.1 Schematic presentation of the single-seed-descent method initially presented by Goulden (1939) and reintroduced by Brim (1966) as a modified pedigree system. During the F_2 to F_4 or F_5 generations, one seed from every plant is used to plant the next generation. When the desired level of inbreeding is attained, each plant is used to establish a family to aid in selection and to produce sufficient seed for subsequent generation yield trials.

Like any plant breeding system, the classical single-seed-descent method can be modified to suit specific crops, circumstances, or objectives. The single-seed-descent method has been used extensively with soybeans and cereals and is starting to be used in tomato, lettuce, and safflower breeding programs. An attractive feature of the single-seed-descent method is the time and space saved in reaching homozygosity.

Experiments to reduce the generation time of wheat as much as possible were conducted by Mukade, Kamio, and Hosoda (1973). Accelerating seed development and maturation, and germinating immature seeds allowed for four to six consecutive generations to be produced within a year.

The size of the population to be handled with the single-seed-descent method needs consideration. To ensure adequate recombination among parental chromosomes, the F_1 population should be as large as time and facilities permit for hybridized seed to be produced. The size of the F_2 generation depends on the number of plants required to form a random sample of the genetic output of the F_1 generation. The number of segregates in subsequent generations depends on the number of homozygous lines desired for F_6 testing. Rather large populations

A.   B.

Figure 11.2   The barley and soybean plants in these photos represent a generation advance in the single-seed-descent method of plant breeding. The plants are growing in soil in flats under continuous light and temperature regimes of up to 25° C (77° F). Techniques for rapid development in a limited area were outlined by Riggs and Hayter (1975). Densities as high as 2,500 plants/m^2 with low nutrient supplies resulted in four to five generations being produced in one year.

At high density, low nutrient fertility and a long photoperiod lead to considerable stunting in plants. Cereal heads typically produce only four to five grains and soybeans only a few pods. Under such intense competition, plant losses of 20% may occur. Because such losses may not be distributed randomly, different plant culture techniques may be conducted similar to the flats in these photos where plants are reduced in height and size but are still capable of producing seed with a minimum of losses. A system of barley culture was described by Grafius in which 8,000 plants could produce seed in an area 1.1 x 8 m (3 ft. x 25 ft.) (Grafius, 1965). Seed was harvested two weeks after the average date of anthesis.

usually are carried through the segregating generations so that an adequate representation of the potential of a cross can be found among the homozygous lines tested in F$_6$ and later generations. Inadequate sampling of segregates may result in the irretrievable loss of desirable alleles due to the problem of **genetic drift** (Sneep, 1977).

Theoretical investigations of the expected length of parental linkage blocks by Hanson (1959) suggested that a considerable amount of genetic recombination can be expected in the F$_2$ and F$_3$ populations with self-fertilization. Based on that information, Ikehashi (1977) estimated the frequency of genotypes of linked loci through succeeding generations at varying levels of linkage and chromosome numbers (Table 11.1).

Single-Seed-Descent Method Compared

Yield testing is the costliest aspect of cultivar development. Unless a selection procedure isolates superior yielding germplasm, increased yield testing cannot be justified to compensate for failure to identify superior segregates. The effectiveness and efficiency of the single-seed-descent method were compared with the pedigree selection method and

Table 11.1 Frequency of Genotypes of Linked Loci Through Succeeding Operations of Various Population Sizes

Population Size	Recombination Value	Number of desirable recombinants[a]						
		F_2	F_3	F_4	F_5	F_6	F_7	F_8
200	0.3	0.3	1.5	2.4	2.5	2.0	1.7	1.7
	0.5	2.4	5.2	6.0	6.4	5.5	5.1	5.0
400	0.3	6.8	2.9	5.4	4.6	4.5	4.3	3.9
	0.5	5.4	9.31	2.5	12.2	10.5	10.3	9.8
800	0.3	1.8	10.1	10.1	9.5	8.5	8.0	7.9
	0.5	10.8	20.1	24.3	23.6	20.0	17.8	17.1
1,600	0.3	3.4	16.0	20.8	21.9	20.3	19.8	19.3
	0.5	20.7	40.7	48.3	45.1	40.4	37.0	36.3

[a] Mean of 10 trials.

Note: Population size and the number of desirable recombinants from F_2 through to F_8 shown here are based on the computer-generated number of recombined individuals. These data are supported by the evidence of Hanson (1959) that a considerable amount of genetic recombination is expected in F_2 and F_3 segregating generations of self-fertilized crops. On purely theoretical grounds, the results suggest that the proportion of good recombinants from undesirable linkages can be increased up to F_4 or F_5 through additional recombination. To maximize the number of desirable recombinants, selection should be delayed until later generations, such as the F_4 or F_5. Although these theoretical results have not been verified, Ikehashi and Fujimaki (1980) reviewed reports in which the variance of some agronomic features in rice increased from F_2 to F_5. The conclusion was that the segregating genotype in each generation had been accumulated in the bulk-harvested population in the course of its fixation. At a population size of 200 and with a recombination value of 0.5 and selecting in the F_5 generation, 6.4% of the plants would have desirable genetic recombinations. The challenge is to identify such quantitative changes and maintain them in the population. A large and unmanageable population of 1600 may be less efficient in identifying desirable recombinants than a population of 200 to 400.

Source: Adapted from Ikehashi, 1977, p. 372.

early-generation yield testing using four segregating soybean populations (Boerma and Cooper, 1975). Among the three methods, no consistent differences were measured among the means of all selected lines, the means of the five highest-yielding lines, or the means of the top line from each population. The number of plots required by each procedure for each type of selection were totaled over the four crosses and compared (Table 11.2).

The early-generation test material was yield tested in the F_3, F_4, F_5, and F_6 generations for a total of 6,136 plots. The pedigree selection method required one-tenth as many plots as early-generation testing. Lines were obtained in the F_4 generation. Lines for the single-seed-descent method were obtained in the F_5 generation, and this method required less than one-third of the lines of the early-generation test method. The single-seed-descent method emerged as the most efficient procedure because it requires less selection effort, allows for a means of rapid advance of early-generation segregating populations, and does not require expensive yield testing until later generations when yield testing is more efficient.

In four oat crosses (Kaufman, 1971), the F_6 yield trial represented a random sample of the gametes from the F_2. Individual lines yielded 8% above the highest parent. One line outyielded the highest parent by 3.5% over a two-year test period, a small but meaningful yield advance.

Table 11.2   Relative Amount of Selection Effort as Measured by the Number of Plots Evaluated for Three Selection Procedures, Totaled over Four Crosses of Soybeans

Plot Type	Early Generation	Pedigree Selection	Single Seed Descent
Yield tests			
Short rows	2,572	612	1,900
Long rows	3,564	none	none
Phenotypic observations			
Pedigree selections	none	2,579	none[a]

[a] In the F_6, 949 plots were observed for phenotype and 116 lines discarded.

Note: These data indicate that the single-seed-descent method requires an intermediate amount of short-row yield testing in the F_6 generation. The early-generation testing method requires far more observations for yield, particularly in the segregating generations. Although the number of yield tests in the early generations for the pedigree selection method is fairly small, the method requires a large number of observations of parent-progeny relationships and phenotypic notes to be taken on individual progeny rows. In the single-seed-descent method, phenotypic observations were made in the F_6 generation as indicated but did not require extensive note taking.

Source: Adapted from Boerma and Cooper, 1975, p. 226.

In a study of two breeding procedures, Knott (1972), and Knott and Kumar (1975) compared the early-generation testing method of selection with the single-seed-descent method for two different wheat crosses. A single seed was taken from each F_2 plant to produce the F_3, from each F_3 plant to produce the F_4, and the entire F_5 plant was harvested for F_6 yield levels. In the second procedure, F_3 lines were yield tested with pedigree selections in the F_3 and F_4. Each F_4 line was bulked to provide seed for an F_5 yield trial. Yield comparisons were made between the F_6 single-seed-descent line and F_5 pedigree selections.

The early-generation testing method resulted in higher yields than selections from the single-seed-descent method. Early-generation testing identified inferior lines, which were discarded following F_3 selection and evaluation. From the original 300 F_2 lines, about 30 were eliminated in the single-seed-descent method in the F_3 and F_4 generations because of late maturity, weak straw, or extreme dwarfness, but this mild selection had little, if any, effect on the range and variability of the lines tested in the F_5 and F_6 generations.

When the top 20% of the lines were compared, no differences were found between the two methods (Table 11.3). Single-seed-descent-derived lines were viewed as being equivalent to those from early-generation yield testing methods. Superior genotypes did not appear to be lost by random segregation away from agronomically desirable types. The single-seed-descent method required less human effort than early-generation testing procedures in advancing material from the F_2 to the F_6 generation. Accurate yield trials on homozygous material can be conducted in the F_6.

Fifty-two barley lines developed by the single-seed-descent method from each of two crosses were compared with a similar number of lines developed by the doubled-haploid technique and by pedigree selection methods (Park *et al.*, 1976). No differences were detected

Table 11.3   Yield Comparison of the Top 20% of Lines from Single-Seed-Descent and Early-Generation Yield Trials of Two Spring Wheat Crosses Tested at Saskatoon, Saskatchewan

Cross	Breeding Method	Number of Lines	Yield (% of Checks)	
			Mean	Range
A	Single seed descent	46	98.5	91.5 to 117.6
	Early generation	36	92.7	80.6 to 114.9
B	Single seed descent	38	109.5	103.0 to 126.0
	Early generation	44	108.3	75.8 to 131.2

Note: In the early-generation testing method, 402 F_3 lines and 80 F_5 lines were yield tested, compared to 442 F_6 single-seed-descent lines. The results favor the single-seed-descent method. Although early-generation testing identified a significantly higher number of high-yielding lines, the real question is whether the differences were worth the cost of conducting over 0.8 ha (nearly two acres) of F_3 yield trials.

Source: Knott and Kumar, 1975, p. 297. Reprinted by permission.

in the mean and range of yield of lines developed by the three methods when tested at two locations using ten replications in hill plots.

Comparisons of population size and chance of success in barley breeding programs developed from the single-seed-descent and doubled-haploid methods were made by England (1981). It has generally been assumed that as far as population size is concerned, the two breeding methods are equivalent. The single-seed-descent method, however, has a reduced chance of achieving homozygosity at any locus, which means population size must be increased in order for the plant breeder to have the same chance of success. England (1981) calculated the population size should be, on the average, 2.13 times larger than for the doubled-haploid method.

Spring barley lines produced by single seed descent, pedigree inbreeding, and doubled haploidy were compared for a range of agronomic characters (Powell, Caligari, and Thomas, 1986). Genotypes capable of outyielding the best parent in each cross may be produced by any of the three methods. The choice of method can be based upon nongenetical considerations such as cost and resources available.

Field and computer simulation studies were used by Tigchelaar and Casali (1976) to compare single seed descent and pedigree selection for selecting quantitative traits with low heritability in self-pollinated crops. At moderate and high heritability levels of 50 and 75%, respectively, pedigree selection was superior. At a 10% heritability level, single seed descent produced the best single F_6 line. Use of strict single seed descent requires that the plant breeder maintain large numbers of random lines to F_6 to compensate for breeding progress with pedigree or mass selection methods.

For plant characteristics with heritabilities below 50%, the number of lines required with single seed descent becomes prohibitive, especially for crops like tomatoes that require a lot of space. At low heritabilities, where genetic advances are relatively slow, the possibility of maintaining large numbers of lines with single seed descent more than compensates for limited progress using pedigree or mass selection.

Progress in tomato breeding for several selected characteristics using single seed descent was compared to progress using pedigree selection (Casali and Tigchelaar, 1975a, 1975b). As expected, single-seed-descent lines generally were inferior to F_5 pedigree lines for qualitatively inherited characteristics for which early-generation pedigree selection had been effective. Conversely, for more complex quantitatively inherited plant attributes, early-generation pedigree selection was ineffective and a large percentage of F_5 single-seed-descent lines were superior.

Favorable results with the single-seed-descent method were reported by Fernandez Martinez *et al.* (1986) for safflower (*Carthamus tinctorius* L.), a partially cross-fertilized crop for which breeding methods for self-fertilized crops have been used (Figure 11.3). Snape and Simpson (1984) considered the single-seed-descent method to be easier and cheaper to perform than the doubled-haploid system. With winter crops, where vernalization is required and

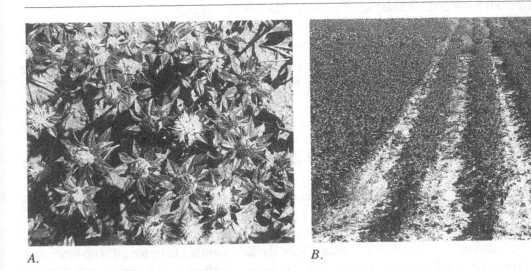

A.   B.

Figure 11.3 Photos showing safflower plants (*Carthamus tinctorius* L.) with spiny outer bracts on the flower (and with stiff spiny leaves, Photo A). The crop is grown for oil produced in the seed. Although safflower is a partially cross-fertilized species, traditional breeding methods for self-fertilized crops have been used. The single-seed-descent method was studied as a means of producing homozygous lines by advancing segregating generations under controlled environmental conditions (Fernandez Martinez *et al.*, 1986).

Progeny were grown 6 cm (2 1/4 in.) apart under controlled environmental conditions, allowing 600 plants to be raised per generation in a 3 m^2 space. Reproductive organs developed in 80 days under an 18 and 25° C night-day temperature regime and 14-hour daylength. Several seeds were sown from each plant, and after emergence seedlings were thinned to leave one plant at each site, thereby ensuring one plant in every generation from each plant in the previous generation. Plants were selfed for the F_2 and F_3 generations. In the F_3, individuals were spaced 10 cm (4 in.) apart in order to obtain larger plants with sufficient seed for a 5 m (20 ft.) row in the F_4 generation in the field. A 12 x 12 lattice square with three replications was used to evaluate F_5 lines.

Transgressive segregation for yield and oil content produced lines with considerably higher values than the better parent. These results and the great savings in time and effort led to the conclusion that single seed descent is a desirable tool for safflower breeders.

generation turnover is slow, doubled-haploid systems are to be preferred. In lettuce (*Lactuca sativa*), the efficiency of the single-seed-descent method was improved when F_2 selection was practiced first (Casali and Tigchelaar, 1975b).

Observations on the Single-Seed-Descent Method

1. Single seed descent allows for plant selection to be conducted on homozygous plants rather than segregating material. Without efficient early-generation selection, genetic drift may cause the loss of desirable alleles. However, practice has shown that positive transgressive segregants can be obtained. Comparative data on breeding methods indicate that the homozygous lines obtained in the F_6 and later generations are a representative sample of the gametes produced by F_1 plants (Maxon Smith, 1984). The possibility does exist that if the sample size is too small, superior genetic combinations may be lost because a single seed from each plant is used to increase plants from F_2 to F_5.
2. The single-seed-descent method is becoming widely accepted and is being used extensively by breeders of annual self-fertilized crops. A major attraction is that limited space is required for the segregating generations F_2 to F_5 since dense populations can be used rather than progeny rows. Progeny rows may be required in the last generation, before yield testing, to produce enough seed.
3. Controlled environmental conditions can speed up generation advances with only a limited number of seeds produced on each plant.
4. The method requires less note taking than the pedigree method since the only data required are the identity of the cross and the generation of selfing.
5. Time and effort in harvesting are considerably less because only one seed from each plant is required.
6. **Artificial selection** for qualitatively inherited traits can be practiced; undesirable plants are not included in the generation advance. Since the final number to be tested is no more than the number of plants sampled each generation, selection should not be too severe. Genetic drift may occur with intense selection pressure in early generations.
7. If early-generation pedigree selection is followed by single seed descent, the identity of superior F_2 plants is lost in succeeding generations. Individual notes are not taken on parent-progeny relationships, and therefore remnant seed of superior F_2 plants cannot be grown to search for additional superior lines that might be produced from superior plants. This problem could be eliminated by harvesting superior F_2 plants and handling them as pedigree lines.
8. F_5 and F_6 selection based on individual plant phenotype may not be as effective in heterogeneous populations as early-generation yield evaluation techniques.
9. For simply inherited plant characteristics that are discontinuous in expression, less effort is spent in obtaining homozygous plants. With factors involving recessive genes, the propagation of homozygous recessive plants increases with every generation of selfing.
10. Selection for some agronomic features, such as lodging resistance, is not as effective in space plantings as at commercial seeding rates.

Suggested Plant Breeding Systems Based on the Single-Seed-Descent Method

The classical single-seed-descent breeding method is unique in its simplicity, and to modify it is to devise an entirely new breeding method. Instead, single seed descent can be applied to other systems of plant breeding.

The need to select effectively for qualitative and quantitative traits prompted a combination of pedigree selection and single seed descent for tomato breeding. The requirement for large land areas for sizable tomato populations led to the development of a system in which pedigree selection is practiced in F_2 and F_3 to concentrate plants with superior characteristics, followed by single seed descent to facilitate attainment of homozygosity. The result is a scheme that has proved effective when several characteristics must be selected simultaneously (Figure 11.4).

A system proposed for wheat breeding, which also has merit for other autogamous plants, was outlined by Knott and Kumar (1975). The initial steps involve individual plant selection, which the authors recognize may not be effective but for which there is no good alternative, followed by single-seed-descent procedures to F_5 or F_6. The detailed steps are as follows:

- Yield test the F_1 and/or F_2 generations of crosses as bulks to give an indication of the potential of the crosses to help direct effort.
- Space plant large F_2 populations for individual plant selection. The size of the population needed will vary with selection intensity for simply inherited and highly heritable characters. The objective should be to save enough F_2 plants to produce the number approaching homozygosity that can be handled in the F_5 or F_6 generation.
- Grow F_3 families, if desired, to allow for further selection for highly heritable qualitative traits.

RECURRENT SELECTION

The success of breeding programs depends on accurate evaluation of genotypic potential and on effective recombination to form new gene combinations for further selection. Breeding procedures for crop improvement whereby commercially adapted cultivars are frequently used as parental material severely limit germplasm diversity. A means to improve adapted genotypes is needed. **Recurrent selection** offers a system of cyclical improvement.

Much emphasis has been placed on the use of recurrent selection in cross-fertilized crops using broad-based genetic populations (Sprague and Eberhart, 1977). This cyclical breeding method has been extensively and effectively applied to cross-fertilized crops such as maize. Recurrent selection involves the establishment of a genetically broad-based population, the identification of individuals from this population with superior genotypes, and subsequent intermating to produce a new population and to begin another cycle. The name grew out of the cyclic, recurring population improvement concept that tends to concentrate desirable alleles. Programmed intercrosses inject new vigor and the opportunity for genetic recombination. Cross-fertilized crops are considered more fully in Chapter 16.

The recurrent selection procedure has been outlined as an alternative to traditional selection procedures for a number of self-fertilized crops (Compton, 1968; Hanson, Probst, and Caldwell,

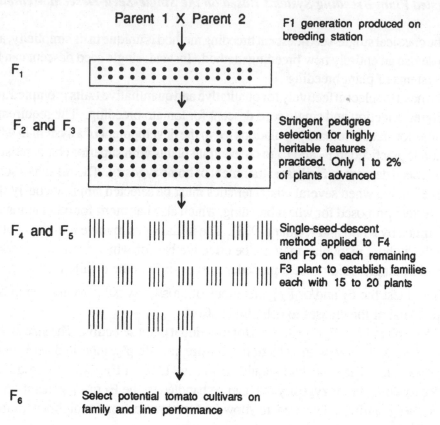

Figure 11.4 Illustration of a breeding procedure for tomato improvement combining stringent early-generation pedigree selection to F_4 followed by single seed descent to F_6, adapted by Tigchelaar and Casali (1976).

The potential of a cross is visually determined in the F_1 generation and emphasis directed toward the most promising. Individual plant selection is conducted on large F_2 populations for highly heritable traits, such as vine type, fruit characteristics, disease resistance, and specific genetic features. Generally, 1 to 2% of the plants in F_2 are retained for evaluation of F_3 populations in the intended area of production. Stringent pedigree selection on highly heritable traits is conducted on the F_3 population.

Fifteen to 20 seeds are taken randomly from each F_3 plant, and all are advanced by single seed descent to establish F_4 and F_5 families and lines. In the F_6, selection is based on family and line performance. Seven to ten plants are grown for each F_6 single-seed-descent line. Promising lines are harvested for yield and fruit quality to determine which lines are to be advanced as cultivars.

The advantages of this combined breeding method were outlined as follows:

- The time required to advance from F_1 to F_6 is reduced because single-seed-descent lines can be rapidly produced in controlled environments.
- Larger numbers of advanced lines can be developed to maintain a broad genetic base to F_6. About ten F_6 single-seed-descent lines can be developed for each F_6 pedigree line without altering field space requirements.
- Selection efficiency is increased by greater selection emphasis on F_6 family performance.
- Opportunity exists to impose special conditions to aid selection. High or low temperatures may be imposed to measure the germinating seeds or fruit set in adult plants.
- A minimal amount of record keeping is required.
- Multiple feature selection is enhanced. Opportunity exists to select at different times, especially when a desired characteristic is best expressed, possibly in the F_6 generation.

1967; Matzinger and Wernsman, 1968; Miller and Rawlings, 1967). A problem with recurrent selection is the large number of crosses required, which normally are done by hand. In some crops such as soybean, emasculation and pollination is a tedious process, and few seeds are obtained per cross. Despite these problems, recurrent selection procedures in self-fertilized crops are growing in use where an increase in the frequency of favorable alleles is desired.

More recently, recurrent selection has been proposed by several authors as a basic breeding approach in self-fertilized crops (Jensen, 1970; Hallauer, 1981, 1986; Ramage, 1981; Frey, 1984) to overcome the problems associated with pedigree or modified pedigree selection, such as the relatively small size of the gene pool, the lack of intermating after the initial cross, and the occurrence of linkage blocks that restrict recombination.

Delogu *et al.* (1988) reported positive results with a recurrent selection scheme for grain yield in winter barley (*Hordeum vulgare* L.). These workers indicated the potential usefulness of the system in developing parents or lines with superior grain yield and with minor changes in the expression for several other agronomic traits. They suggested that if the gene base of the population narrows too rapidly after selection cycles, variation can be increased by adding new germplasm to the population at the intercrossing phase. To overcome the problem of not producing enough seed in each cycle after intercrossing, described by Hallauer (1981), genetic or chemically induced male sterility can be used to facilitate intercrossing (Avey *et al.*, 1982; Ramage, 1981).

Positive results by other workers have increased interest in recurrent selection in self-fertilized crops (Burton, Wilson, and Brim, 1983; Busch and Kofoid, 1982; Gupton, 1981; Loffler, Busch, and Wiersma, 1983; Sumarno and Fehr, 1982).

Recurrent selection procedures do not follow classical steps. The number of steps before progeny are evaluated depends on objectives, resources, the extent of testing before parents are selected for recombination, the number of cycles, and the number of parents involved. Figure 11.5 illustrates a three-cycle recurrent selection program that uses single seed descent to facilitate steps in inbreeding and is fashioned after the outline of Payne *et al.* (1986).

Recurrent selection provides an opportunity to observe progressive changes associated with the primary objective, perhaps yield. Proper analysis can detect the dynamic processes in the causal mechanisms through which progress generated by selection takes place. This is demonstrated by the work of Payne *et al.* (1986).

Rather than concentrate on static parameters such as grain yield, grain yield components, and harvest index that are usually measured by pragmatic plant breeding programs, Payne *et al.* (1986) undertook to perform a physiological and constitutive analysis of grain yield changes in the oat crop. Concomitant physiological changes were measured in a 12% oat yield increase achieved by means of a three-cycle recurrent selection program. This increase was described by an 8% increase in kernel number and a 3% increase in kernel weight. But these parameters describe only an outcome and fail to indicate the all-important pathway by which that outcome was achieved. Grain filling rate was found to be one factor. Changes in 14 parameters are shown in Table 11.4.

A clear superiority of recurrent selection systems in self-fertilized crops over classical breeding programs of hybridization when linkage disequilibrium exists has been established (Silvela and Diez-Barra, 1985). One- and two-loci models were studied, and genotypic

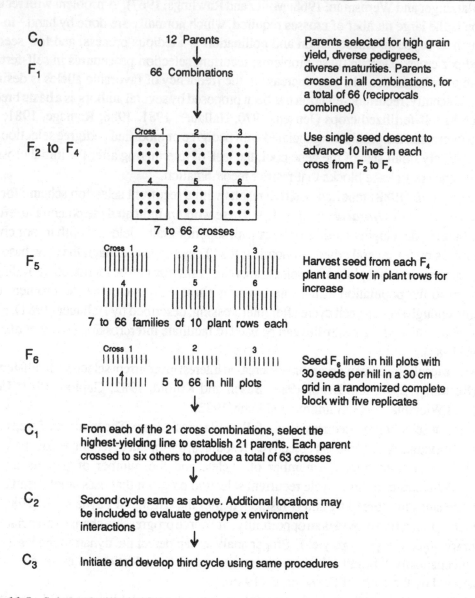

C_0	12 Parents	Parents selected for high grain yield, diverse pedigrees, diverse maturities. Parents crossed in all combinations, for a total of 66 (reciprocals combined)
F_1	66 Combinations	
F_2 to F_4	Cross 1 2 3 4 5 6 7 to 66 crosses	Use single seed descent to advance 10 lines in each cross from F_2 to F_4
F_5	Cross 1 2 3 4 5 6 7 to 66 families of 10 plant rows each	Harvest seed from each F_4 plant and sow in plant rows for increase
F_6	Cross 1 2 3 4 5 to 66 in hill plots	Seed F_6 lines in hill plots with 30 seeds per hill in a 30 cm grid in a randomized complete block with five replicates
C_1	From each of the 21 cross combinations, select the highest-yielding line to establish 21 parents. Each parent crossed to six others to produce a total of 63 crosses	
C_2	Second cycle same as above. Additional locations may be included to evaluate genotype x environment interactions	
C_3	Initiate and develop third cycle using same procedures	

Figure 11.5 Schematic illustration of a three-cycle recurrent selection program for grain yield improvement in oat. Each cycle of recurrent selection consists of three operational phases: (1) derivation of individuals or families for evaluation, (2) evaluation of individual plants or families and selection of parents, and (3) intermating of selected parents to form the population for the next cycle of selection. The objective of recurrent selection is to increase the frequency of favorable alleles in a population by gene recombination.

In this example, choice of the original 12 parents was based on high grain yield, genetic diversity, and range of maturity. The objective was to increase grain yield, and following three cycles of selection, grain yield was increased 12%.

Following the third cycle of recurrent selection, replicated tests were conducted for two years on selected progeny, along with the 12 original (C_0) and 21 third-cycle (C_3) parents. Alternate hill plots of wheat and oat plants provided a constant competitive buffer when two hills were harvested at each of five intervals for physiological determinations. Single seed descent was used to facilitate handling of segregating generations.

Table 11.4 Results of Three Cycles of Recurrent Selection in the Oat Crop

Characteristic	Units	Check	Cycle C_0	Cycle C_3	C_3 in Percent of C_0
Productivity					
Grain yield	g/hill	34.9	34.7	38.7[a]	111.5
Biomass	g/hill	88.6	86.1	98.7[a]	114.7
Growth rate					
Vegetative	g/culm/day	--	0.0676	0.0687	101.6
Grain	g/panicle/day	--	0.0548	0.0613[a]	111.9
Maturity					
Days to heading		61.2	60.3	62.6	103.8
Days to physiological maturity		82.3	82.2	84.1[b]	102.3
Grain-fill period		21.2	21.9	21.6	98.6
Yield components					
Panicles per hill	number	29.4	29.0	29.4	101.4
Kernels per panicle	number	40.2	41.0	44.1[b]	107.6
Kernel weight	g	2.98	2.95	3.03[a]	102.7
Morphology					
Culms per hill	number	28.6	29.2	29.8	102.1
Plant height	cm	97.7	100.6	104.3[a]	103.7
Yield indices					
Harvest index	%	39.4	40.3	39.2	97.3
Partitioning	%	--	82.6	90.4[a]	109.4

[a] Significant from C_0 at 5%.
[b] Significant from C_0 at 10%.
Note: Table shows means of C_0 and C_3 parents and check cultivars grown over a two-year period. A substantial 12% increase reflects the potential that recurrent selection has for improvement in self-fertilized crops. Equally exciting is the opportunity to gain an insight into the causes of this increase through a cyclical, structured recurrent selection scheme rather than using a purely pragmatic program. The physiological approach illustrated in this table, along with recurrent selection procedures, may be useful tools for **ideotype breeding** programs where conscious objectives and limits are established for selected parameters. To test the concept that knowledge of the existence and extent of correlated responses can be useful in ideotype breeding, research was performed by Bregitzer *et al.* (1987). The results shown in the table reveal that morphological changes may be accomplished by recurrent breeding programs but that construction of an ideotype based on certain levels of expression, such as an increase in leaf area and hence photosynthetic area or a change in individual morphological traits, is not likely to be effective in increasing grain yield. This view is based on the lack of consistent and strong relationships between grain yield and individual morphological traits. Grain yield is the result of many interdependent developmental processes that may be too complex to be manipulated easily with a simplified ideotype model.
Source: Payne *et al.*, 1986, p. 735. Reprinted by permission.

interactions from allelic and nonallelic loci were considered, as well as genes with a large effect. The main advantages of recurrent selection from a theoretical standpoint include better breeding opportunities

- if dominant or overdominant alleles are segregating at low frequencies in the original population, and
- when below-equilibrium variance or negative linkage disequilibrium is either present or generated.

In a purely self-fertilized system following hybridization, the main advantages are

- improved breeding limits for recessive alleles segregating at low initial frequencies, and
- a faster initial response, provided the population is not loaded with negative linkage disequilibrium.

When epistatic interaction enhances the selective advantage of new recombinants, the superiority of recurrent selection is even greater.

To overcome the problem of extensive crossing needed in recurrent selection programs, Brim and Stuber (1973) outlined a recurrent selection scheme for soybean that uses a genetic male-sterile factor. Male-sterile plants do not set seed if self-pollinated, and under most conditions, natural crossing will occur aided by wind or insects.

Genetic male sterility is inherited as a single recessive *(ms ms)* gene pair. The basic steps are as follows:

Step 1: Establish a basic population. This is normally done by crossing selected parents in diallel as two-way crosses, backcrosses, or other schemes. In this instance, the male-sterile plant is introduced. The contribution of the cytoplasm from the male-sterile source is minimized by using heterozygous fertiles from a maintainer line as the male parent. Hand pollination will be required to establish the basic population. Backcrossing helps minimize the genetic contribution of the male-sterile genotype.

Step II: Conduct a basic selection scheme as follows:

Generation 1: Allow natural cross-pollination to occur in plants segregating for male sterility.

Generation 2: Allow progenies of male-sterile plants to self-fertilize to provide seed for evaluation in the field.

Generation 3: Field test plants from generation 2. Bulk remnant seeds from parents of selected progenies or seeds produced by male-fertile plants in the selected progenies to serve as parents for the next cycle of intermating.

Cycle I

Generation 1: If seed is harvested from only male-sterile plants in this generation, the progeny will be male sterile and no reintroduction of male sterility will be required, even after repeated cycles.

Generation 2: Produce sufficient selfed seed on male-sterile plants for yield trials. Male-sterile plants will be cross-fertilized and produce both fertile and sterile plants.

Generation 3: Conduct replicated yield trials on progenies of male-fertile plants produced in generation 2.

Cycle II: Repeat procedure.

Male sterility in barley was used to facilitate crosses for a recurrent selection program aimed at improving shrunken seed and grain weight (Habgood and Uddin, 1984) Efforts were made to increase the quality of barley protein by incorporating increased lysine levels, a step that drastically modifies seed protein composition. In the presence of the lysine gene, grain weight exhibits 10 to 34% reductions. Selection followed by recombination was undertaken, and after two recurrent cycles, some improvement was made for both characters, but progress was slow. Repeated cycles may be required to increase both factors slowly.

Recurrent selection for yield in peanut (*Arachis hypogaea* L.) using a wild parent to establish a broad genetic base was studied by Guok, Wynne, and Stalker (1986). Recurrent selection was successful in increasing yield and improving disease resistance. A positive gain for grain yield was predicted by Patel, Reinbergs, and Fejer (1985) from recurrent selection in doubled-haploid populations.

Recurrent selection as a breeding method in self-fertilized crops appears to be well established and has produced positive results. The problems of crossing and producing adequate seeds have been reduced by the installation of controlled growth rooms that extend the crossing period and provide a pest-free environment under controlled conditions.

REFERENCES

Avey, D. P., H. W. Ohm, F. L. Patterson, and W. E. Nyquist. 1982. Three Cycles of Simple Recurrent Selection for Early Heading in Winter Wheat. *Crop Science* 15:908-912.

Boerma, H. R., and R. L. Cooper. 1975. Comparison of Three Selection Procedures for Yield in Soybeans. *Crop Science* 15:225-229.

Bregitzer, P. P., D. D. Stuthman, R. L. McGraw, and T. S. Payne. 1987. Morphological Changes Associated with Three Cycles of Recurrent Selection for Grain Yield Improvement in Oat. *Crop Science* 27:165-168.

Brim, C. A. 1966. A Modified Pedigree Method of Selection in Soybeans. *Crop Science* 6:220.

Brim, C. A., and C. W. Stuber. 1973. Application of Genetic Male Sterility to Recurrent Selection Schemes in Soybeans. *Crop Science* 13:528-530.

Burton, J. M., R. F. Wilson, and C. A. Brim. 1983. Recurrent Selection in Soybean. IV: Selection for Increased Oleic Acid Percentage in Seed Oil. *Crop Science* 23:744-747.

Busch, R. H., and K. Kofoid. 1982. Recurrent Selection for Kernel Weight in Spring Wheat. *Crop Science* 22:568-572.

Casali, V.W.D., and E. C. Tigchelaar. 1975a. Breeding Progress in Tomato with Pedigree Selection and Single Seed Descent. *Journal American Society of Horticulture Science* 100:362-364.

―――. 1975b. Computer Simulation Studies Comparing Pedigree, Bulk, and Single Seed Descent Selection in Self-pollinated Populations. *Journal of American Society of Horticulture Science* 100:364-367.

Compton, W. A. 1968. Recurrent Selection in Self-pollinated Crops Without Extensive Crossing. *Crop Science* 8:773.

Delogu, G., C. Lorenzoni, A. Marocco, P. Martiniello, M. Odoardi, and A. M. Stanca. 1988. A Recurrent Selection Programme for Grain Yield in Winter Barley. *Euphytica* 37:105-110.

England, F.J.W. 1981. Population Size and Chance of Success in a Barley Breeding Programme Based on the Use of Doubled Haploids (DH) or Single Seed Descent (SSD). In *Barley Genetics IV* Proceedings of the Fourth Barley Genetics Symposium, edited by M.J.C. Asher, R. P. Ellis, A. M. Hayter, and R.N.H. Whitehouse. Edinburgh University Press, pp. 176-178.

Fernandez Martinez, J., J. Dominguez Gimenez, A. Jimenez, and L. Hernandez. 1986. Use of the Single Plant Descent Method in Breeding Safflower (*Carthamus tinctorius* L.). *Plant Breeding* 97:364-367.

Frey, K. J. 1984. Breeding Approaches for Increasing Cereal Crop Yields. In *Cereal Production*, edited by E. J. Gallagher. Butterworths, London, pp. 47-68.

Goulden, C. H. 1939. Problems in Plant Selection. *Proceedings of the Seventh International Genetical Congress*, Edinburgh, Scotland. Cambridge University Press, Cambridge, pp. 132-133.

Grafius, J. E. 1965. Short Cuts in Plant Breeding. *Crop Science* 5:377.

Guok, H. P., J. C. Wynne, and H. T. Stalker. 1986. Recurrent Selection Within a Population from an Interspecific Peanut Cross. *Crop Science* 26:249-253.

Gupton, C. L. 1981. Phenotypic Recurrent Selection for Increased Leaf Weight and Decreased Alkaloid Content of Burley Tobacco. *Crop Science* 21:921-925.

Habgood, R. M., and M. R. Uddin. 1984. The Effect of Genetic Background on Grain Shrivelling in Lys3 Barley, and its Possible Modification by a Recurrent Selection Programme. *Zeitschrift für Pflanzenzuchtung* 93:65-77.

Hallauer, A. R. 1981. Selection and Breeding Methods. In *Plant Breeding II*, edited by K. J. Frey. Iowa State University Press, Ames, pp. 3-55.

———. 1986. Compendium of Recurrent Selection Methods and Their Application. *CRC Reviews in Plant Science* 3:1-33.

Hanson, W. D. 1959. The Breakup of Initial Linkage Blocks Under Selected Mating Systems. *Genetics* 44:857-868.

Hanson, W. I., A. H. Probst, and B. E. Caldwell. 1967. Evaluation of a Population of Soybean Genotypes with Implications for Improving Self-pollinated Crops. *Crop Science* 7:99-103.

Ikehashi, H. 1977. Simulation of Single Seed Descent in Self-pollinating Populations. I: Advance of Polygenic Recombination Through Generations. *Japanese Journal of Breeding* 27:367-377.

Ikehashi, H., and H. Fujimaki. 1980. Modified Bulk Population Method for Rice Breeding. In *Innovative Approaches to Rice Breeding*. International Rice Research Institute, Los Banos, Philippines, pp. 163-182.

Jensen, N. F. 1970. A Diallel Selective Mating System for Cereal Breeding. *Crop Science* 10:629:635.

Kaufman, M. L. 1961. A Proposed Method of Oat Breeding for Central Alberta. *Cereal News* 6(2):15-18.

———. 1971. The Random Method of Oat Breeding for Productivity. *Canadian Journal of Plant Science* 51:13-16.

Knott, D. R. 1972. Effects of Selection for F_2 Plant Yield in Subsequent Generations of Wheat. *Canadian Journal of Plant Science* 52:721-726.

Knott, D. R., and J. Kumar. 1975. Comparison of Early Generation Yield Testing and a Single Seed Descent Procedure in Wheat Breeding. *Crop Science* 15:295-299.

Loffler, C. M., R. M. Busch, and J. W. Wiersma. 1983. Recurrent Selection for Grain Protein Percentage in Hard Red Spring Wheat. *Crop Science* 23:1097-1101.

Matzinger, D. F., and E. A. Wernsman. 1968. Four Cycles of Mass Selection in a Synthetic Variety of an Autogamous Species *Nicotiana tabacum* L. *Crop Science* 8:239-243.

Maxon Smith, J. W. 1984. Is Early Generation Selection Necessary? In *Efficiency in Plant Breeding*, Proceedings of the 10th Congress of the European Association for Research on Plant Breeding. EUCARPIA. Pudoc, Wageningen, p. 328.

Miller, P. A., and J. O. Rawlings. 1967. Selection for Increased Lint Yield and Correlated Responses in Upland Cotton, *Gossypium hirsutum* L. *Crop Science* 7:637-640.

Mukade, K., M. Kamio, and K. Hosoda. 1973. The Acceleration of Generation Advancement in Breeding Rust-Resistant Wheat. *Proceedings of the Fourth International Wheat Genetics Symposium*. University of Missouri, Columbia, Missouri, pp. 439-444.

Park, S. J., E. J. Walsh, E. Reinbergs, L.S.P. Song, and K. J. Kasha. 1976. Field Performance of Doubled Haploid Barley Lines in Comparison with Lines Developed by the Pedigree and Single Seed Descent Methods. *Canadian Journal of Plant Science* 56:467-474.

Patel, J. D., E. Reinbergs, and S. O. Fejer. 1985. Recurrent Selection in Doubled-Haploid Populations in Barley (*Hordeum vulgare* L.). *Canadian Journal of Genetics and Cytology* 27:172-177.

Payne, T. S., D. D. Stuthman, R. L. McGraw, and P. P. Bregitzer. 1986. Physiological Changes Associated with Three Cycles of Recurrent Selection for Grain Yield Improvement in Oats. *Crop Science* 26:734-736.

Powell, W., P.D.S. Caligari, and W.T.B. Thomas. 1986. Comparison of Spring Barley Lines Produced by Single Seed Descent, Pedigree Inbreeding and Doubled Haploidy. *Plant Breeding* 97:138-146.

Ramage, R. T. 1981. Comments About the Use of Male Sterile Facilitated Recurrent Selection. *Barley Newsletter* 24:52-53.

Riggs, T. J., and A. M. Hayter. 1975. Practical Aspects of the Single Seed Descent Method in Barley Breeding. In *Barley Genetics III*, Proceedings of the Third International Barley Genetics Symposium, Garching, Verlag Karl Thiessig, Munchen, pp. 708-717.

Silvela, L., and R. Diez-Barra. 1985. Recurrent Selection in Autogamous Species Under Forced Random Mating. *Euphytica* 34:817-832.

Snape, J. W., and E. Simpson. 1984. Early Generation Selection and Rapid Generation Advancement Methods in Autogamous Crops. In *Efficiency in Plant Breeding*, Proceedings of the 10th Congress of the European Association for Research on Plant Breeding. EUCARPIA. Edited by W. Lange, A. C. Zeven, and N. G. Hogenboom. Pudoc, Wageningen, pp. 82-86.

Sneep, J. 1977. Selection for Yield in Early Generations of Self-fertilizing Crops. *Euphytica* 26:27-30.

Sprague, G. F., and S. A. Eberhart. 1977. Corn Breeding. In *Corn and Corn Improvement*, edited by G. F. Sprague. American Society of Agronomy, Madison, pp. 305-362.

Sumarno, and W. R. Fehr. 1982. Response to Recurrent Selection for Yield in Soybean. *Crop Science* 22:295-299.

Tigchelaar, E. C., and V.W.D. Casali. 1976. Single Seed Descent: Applications and Merits in Breeding Self-pollinated Crops. *Acta Horticulturae* 63:85-90.

CHAPTER 12

Heterosis and Inbreeding in Cross-fertilized Crops

CHARACTERISTICS OF CROSS-FERTILIZED CROPS

In cross-fertilized crops, individual plants are heterozygous and many of the cultivars are very heterogeneous. Some cultivars of cross-fertilized species are hybrids and hence very homogeneous. Individual plants in an alfalfa cultivar or an **open-pollinated maize**[1] cultivar exhibit a wide range of phenotypic variation for qualitative and quantitative traits. Individual plant yield, height, maturity, morphological features, and many other traits exhibit a range of variation much greater than that found in self-fertilized crops. Nevertheless, cultivars of cross-fertilized crops are recognized that exhibit distinct characteristics. Even though a large amount of plant-to-plant variation occurs within cultivars, this variation is less than the variation that is exhibited among cultivars of the same species.

Selection in cross-fertilized crops differs from that employed in most self-fertilized crops because emphasis in self-fertilized crops is placed on individuals whose progeny are believed to be superior. Selection within a genetically heterogeneous self-fertilized population is aimed at selecting superior individual genotypes that will reproduce their characteristics faithfully because of their homozygosity. For such desirable genotypes, the gene frequency of the desirable genes and gene combinations is fixed, with a gene frequency equal to one.

In cross-fertilized crops, the gene frequency of desirable alleles or a complex of desirable alleles is increased in the population but seldom reaches the point where quantitative traits are fixed. Cross-fertilized crops, therefore, consist of a genetically variable population in which genes are exchanged between individuals according to the cross-fertilizing model of reproduction. The plant breeder is not interested in the genotype of a heterozygous plant because it will not breed true but instead is interested in the genotypes of the gametes that the plant will contribute to the next generation.

Dobzhansky (1950, 1951) proposed the gene pool concept for the situation that exists in randomly mated populations. He considered the gene pool as being formed at the end of the sporophytic generation where gametes represent genes from all the individuals in the population. The gene pool concept specifies that all the gametes in the population are available, and because of random mating, can give rise to the genotypes of the next generation. The population can change from generation to generation, depending on the composition of the gametes that

---

1. The term open-pollinated maize has come to encompass an entire class of maize. Normally, open-pollinated means natural cross-pollination. A more appropriate term would be cross-fertilized. Open-pollinated maize stands in contrast to hybrid maize.

are available. The number of genotypes that can arise under such a mating system becomes infinite.

Genotypes that do relatively better than other genotypes in the population are able to contribute more genes into the gene pool because more gametes are produced by virtue of the fact that superior genotypes are in greater abundance. Therefore, the gene frequency in any generation is also dependent on genotypic performance. Gene and genotype frequencies are said to be coadapted. Stated otherwise, favorable or superior genotypes produce more gametes; thus a higher gene frequency in the following generation and conversely favorable genes confer favorable genotype performance (Figure 12.1).

Gene pool equilibrium results when both gene and genotype frequencies remain constant in the population. A change in either gene or genotype frequency will result in corresponding changes of other genes and genotypes in the population and in a new gene pool equilibrium. Breeding cross-fertilized crops dictates manipulation of gene frequency toward agronomically desirable types.

The survival of a gene in a population depends on the effect of the gene in specific genotypes. A gene in a population, therefore, can have several states in a genotype, as follows:

- The gene effect on the genotype when in the homozygous condition. This was referred to previously as the additive portion of genotypic variance.
- An allele of a specific gene can have any effect, depending on its interaction with other alleles at the same locus. This is referred to as intragenic interaction, or dominance genotypic variance, and is analogous to heterozygote performance for a specific allelic combination of a single gene.
- Intergenic interaction. This refers to the interaction of a specific allele of a gene with other genes at different genetic loci in the genome and was previously referred to as epistatic genotypic variance.

In addition to the effect of a specific allele of a gene on other alleles of the same gene and on other genetic loci, gene linkage may exist. Linkage restricts the free combination of genes, and a favorable gene may be reduced in frequency or eliminated in a population if it is closely linked to an unfavorable gene. Some breeding and genetic studies have indicated that the linkage of a series of genes together has an important effect on the survival of specific genes of a series

---

Figure 12.1   The individual alfalfa plants in these photos are the remnants of the entire plant population of a cross-fertilized alfalfa crop. The plants in Photos A and B exhibit white and dark purple flowers, respectively, a characteristic of a cross-fertilized crop.

If a plant breeder gathered seed from one of the plants shown in Photo C, which have been cross-fertilized by the individual plants, the genotype would not breed true. Instead the plant breeder is interested in collecting seed from all of the plants to sample the gametes that might exhibit longevity.

Seed from such cross-fertilized plants represents a gene pool of favorable genes because alfalfa plants lacking persistence have died out. The concept of a gene pool is useful when considering breeding programs in a cross-fertilized crop. Repeatedly sampling these plants for seed would continue to provide changing assemblies of genotypes as gametes combine at fertilization to form the next sporophytic generation. Removal of one of the plants in Photo C as a pollen source because of an undesirable feature will allow changes of other genes and genotypes in the population. The possibilities for genetic combinations are endless.

A.

B.   C.

Figure 12.1 continued

Linkage may restrict the free movement of genes in the gene pool. A plant breeder, therefore, is more interested in the kind of progeny an individual plant is capable of producing than in the individual phenotype.

in a population. Because of these considerations, the breeder of cross-fertilized crops is not so much interested in the individual phenotypes of a plant as in the kind of gametes the plant is capable of producing. In other words, what kind of progeny can an individual plant produce?

INBREEDING DEPRESSION

Inbreeding results when individuals are mated that are more closely related than individuals in a large, randomly mated population. Normally, cross-fertilized crops are regarded as outcrossed, that is, cross-fertilized to a plant of a different genotype. Inbreeding results in a loss of heterozygosity and is measured by the coefficient of inbreeding (F). The coefficient of inbreeding is based primarily on the number and closeness of ancestral connections between the two parents. The F value depends on the mating system and on the degree of inbreeding in the previous generation.

Self-fertilization is the most severe form of inbreeding and results in a gain of 50% homozygosity in each generation. The coefficient of inbreeding, F, for self-fertilization is given by the formula

$$F = 1/2 (1 + F^1)$$

where F^1 is the coefficient of inbreeding in the previous generation. The F value increases with each generation of self-fertilization. F has a value of zero in randomly mated populations and increases toward 1 as the proportion of heterozygosity goes toward zero.

There are a number of different mating systems that result in inbreeding. The following mating systems are listed in decreasing order of inbreeding severity:

self
brother-sister
double first cousins
quadruple second cousins
octuple third cousins

These mating systems would require many generations before the inbreeding coefficient approached 1, or complete homozygosity.

OBSERVATIONS OF INBREEDING AND HETEROSIS
IN CROSS-FERTILIZED CROPS

Selfing of naturally cross-fertilized crops results in a dramatic decrease in plant vigor as homozygosity is reached (Figure 12.2). Jones (1939) reported on the effects of inbreeding maize begun in 1904 and conducted for 30 generations. Variations of a degenerative nature were observed and caused the extinction of one of the original lines. This loss could not have been prevented by selection of normal plants as progenitors. Selfing resulted in 25% reduction in height; height reduction ceased after five generations. Yield was reduced 75% but ceased to decline after 20 generations. After 20 generations of self-fertilization, inbred lines appeared uniform and constant for all visible characters and homozygous for all loci that have any effect

A.

B.

Figure 12.2   Photo of technicians selfing inbred maize plants, which show short stature, reduced vigor, and low yield (Photo A). Inbreds are produced by selfing individual plants by hand. Pollen is placed on the silks of the same plant from which it came. Ear shoots are protected from stray pollen by covering them with glassine bags before silks emerge and after pollination. The best of the ears are planted, and good plants within each row are self-pollinated. With continued inbreeding, there is a marked increase in plant uniformity, and after five to six generations of self-pollination, individual plants generally breed true.

Inbreeding reveals grossly unfavorable characteristics, and these plants are discarded. Once homozygosity is achieved, pollination among plants of a strain is essentially the same as self-pollination. Crossing plants between unrelated rows, however, restores vigor. This is the essence of hybrid maize breeding programs.

Shull (1909) noted that the objective of the plant breeder is not to find the best pure line but to find and maintain the best hybrid combination. In this observation, Shull stated the main reason for

Figure 12.2 continues overleaf

upon hybrid vigor. No variations were reported throughout the 30 generations of selfing, which could be interpreted as being favorable to survival.

Shull (1908) reported that the individuals in an ordinary field of maize are genetically very complex and that the deterioration that takes place as a result of self-fertilization is due to the gradual change to a homozygous condition. He reasoned that the objective of the maize breeder should not be to find the best pure line but to find and maintain the best hybrid combination. It was this latter point that was addressed in a subsequent paper (Shull, 1909) in which the procedures for producing hybrid maize were spelled out.

Shull (1909) observed that heterosis, or performance above the high-yielding parent, occurred when certain of the individuals from self-pollinated homozygous families were crossed. Crosses between members within the same line did not offer any improvement over self-fertilization, but crosses between families restored vigor and productivity to that of the original line. Indeed some crosses had yields superior to the original parental populations. The observations of Shull are summarized in Table 12.1 and are the basis of maize breeding. Shull noted that the yield and quality of the crop were functions of the parental combinations and remained the same whenever the parental genotypes were crossed. This means that a particular hybrid can be produced at will from selected inbreds, which can be maintained indefinitely in the homozygous state.

Heterosis also can be obtained by hybridizing heterozygous, open-pollinated maize cultivars. Robinson *et al.* (1956) reported yield performance in progeny produced by crossing three open-pollinated cultivars to produce three crosses (Table 12.2). These data indicate that several combinations of crosses between open-pollinated maize cultivars were capable of producing a heterotic response, defined as an increase in yield above the high parent. A yield increase was observed more frequently when cultivars being crossed came from distinctly unrelated cultivars, such as flint x dent, flour x flint, or flour x dent endosperm types.

Richey (1922) surveyed experiments involving 244 crosses with open-pollinated parents and found that 82.4% of the crosses produced more and 17.6% produced less than the average of the parents. Further, 55.7% of these crosses produced more than either parent.

Heterosis and inbreeding depression also are conspicuous features of other cross-fertilized crops, such as alfalfa (Table 12.3) and birdsfoot trefoil. The results with alfalfa confirm that heterosis produced by crossing heterozygous parents is present as in maize. The effect of degree of unrelatedness of parents on hybrid vigor is not as clearly illustrated as in maize, although higher yields were produced in the Flemish x Vernal crosses.

An indication of inbreeding and heterosis in cross-fertilized birdsfoot trefoil is shown in Table 12.4. Both heterosis and inbreeding depression are readily apparent in birdsfoot trefoil

---

Figure 12.2 continued

practicing inbreeding. Jones (1945) described in detail six recessive variations that appeared in five inbred maize lines, including narrow leaf, dwarf plants, pale chlorophyll, crooked stalks, blotched leaf, and late flowering.

Note the greater height of the hybrid corn in the background relative to the inbreds in the foreground in Photo B.

Table 12.1 Conclusions of G. H. Shull on Selfing and Hybridization in Maize

1. The progeny of every self-fertilized maize plant are of inferior size, vigor, and productiveness as compared with the progeny of a normally cross-bred plant derived from the same source. This is true both when the chosen parent is above the average condition and when the parent is below it.
2. The decrease in size and vigor that accompanies self-fertilization is greatest in the first generation and becomes less and less in each succeeding generation until a condition is reached in which there is (presumably) no further loss of vigor.
3. Self-fertilized families from a common origin differ from one another in definite hereditary morphological characters.
4. Regression of fluctuating characters has been observed to take place away from the common mean or average of the several families instead of toward it.
5. A cross between sibs within a self-fertilized family shows little or no improvement over self-fertilization in the same family.
6. A cross between plants belonging to two self-fertilized families results in progeny of as great vigor, size, and productiveness as are possessed by families that had never been self-fertilized.
7. The reciprocal crosses between two distinct self-fertilized families are equal and possess the characters of the original maize with which the experiments were started.
8. The F_1 from a combination of plants belonging to certain self-fertilized families produces a yield superior to that of the original cross-bred stock.
9. The yield and the quality of the crop produced are functions of the particular combination of self-fertilized parental types, and these qualities remain the same whenever the cross is repeated.
10. The F_1 hybrids are no more variable than the pure strains that enter into them.
11. The F_2 shows much greater variation than the F_1.
12. The yield per acre of the F_2 is less than that of the F_1.

Note: These observations are the basis of the commercial hybrid maize industry.
Source: Adapted from Shull, 1909.

Table 12.2 Yield Performance of Maize Produced by Hybridizing Three Open-pollinated Cultivars

Parent or Cross Parents	Percentage of the High Yielding Parent
Jarvis (2,916 kg/ha; 2,430 lb./ac.)	---
Indian Chief (2,554 kg/ha; 2,128 lb./ac.)	---
Latham (2,056 kg/ha; 1,714 lb./ac.)	---
Jarvis x Indian Chief	121
Indian Chief x Latham	131
Jarvis x Latham	94

Note: These results are based on replicated trials conducted at three locations in North Carolina over a two-year period. Heterosis in crosses of open-pollinated cultivars suggests that cultivars are not genetically identical. As cross-fertilization occurs in open-pollinated cultivars, shifts in the gene pool might be expected, so that similar genotypes cannot be repeated.
Source: Adapted from Robinson *et al.*, 1956, p. 128.

and are similar to that in maize under space planted conditions. The same results would not have been expected under broadcast conditions because forage species are sown in broadcast **swards** in which somewhat less than 20% of the number of seeds actually planted contribute to yield, whereas the others are suppressed by competition. Those seeds that contribute to forage yield from a sward may not be representative of the performance of the entire population. The

Table 12.3   Measure of Heterosis for Forage Yield Produced in Crosses of Open-pollinated Cultivars of Alfalfa

Cross	Number of Crosses	Percentage Forage Yield of Check Cultivars
Flemish x Flemish	10	119
Vernal x Vernal	19	122
Flemish x Vernal	22	124

Note: The yield of the best hybrid combination was 137% of the check cultivars and 118% of the best check cultivar. These results are based on broadcast plot yields, which produce a competitive sward that limits expression of heterosis. Heterosis is not as adversely affected as inbred progeny, however. The data are from yield trials conducted at the University of Guelph.

Source: Adapted from Onokpise, 1980.

Table 12.4   Inbreeding and Heterosis for Seedling Vigor, Forage Yield, and Seed Yield Expressed as a Percentage of an Adapted Cultivar, Leo

Trait	Crossed Progeny (%)	Selfed Progeny (%)
Seedling vigor	111	82
Forage yield	129	75
Seed yield	116	36

Note: Cross- and self-fertilized progeny are compared in a birdsfoot trefoil population. The data are based on 21 polycross and selfed parents evaluated in space-planted conditions.

Source: Adapted from Onokpise, 1980.

recommended seeding rate for maize is such that each plant is capable of producing an ear and contributing to economic yield.

The yield of self-fertilized progeny will tend to be closer to that of the original population because segregants that retain relatively more favorable genes will survive in contrast to those that carry deleterious or subvital genes.

PERSPECTIVE ON THE GENETIC BASIS OF HETEROSIS IN A CROSS-FERTILIZED CROP

Classical explanations for hybrid vigor or heterosis are based on either the dominance or the overdominance concept. The dominance concept suggests that hybrid vigor is the result of the action and interaction of dominant, and often linked, growth factors. Heterozygosity *per se* is not essential as long as the parents of the hybrid maximize the number of dominant alleles that confer rapid growth. The dominance concept predicts that homozygous individuals that have similar performance to the F_1 should be observed in the F_2 generation and that the F_2 distribution should be skewed because of dominant alleles. Observations of metrical traits indicate that neither the F_2 generation yield nor a skewed F_2 hold. The F_2 distribution observed in crops clearly shows a symmetrical pattern that would be expected because of incompletely dominant genes, a large number of genes, and the relatively low heritability for yield.

Linkage would also have an effect on F_2 symmetry; linkage would reduce the expected symmetrical distribution because unfavorable recessive genes could be linked to favorable dominant genes. This would result in a reduced frequency of multiple dominant alleles on a given chromosome. The theory of the effect of linkage on heterosis and the F_2 distribution was proposed by Jones (1917).

The overdominance theory of heterosis states that heterozygosity *per se* is a necessity for the full expression of heterosis. A striking example of overdominance is that of grain yield in maize hybrids, which depends on specific, somewhat rare parental combinations.

There are examples of single genetic loci in several species where overdominance has been observed. A number of explanations for single-gene heterozygotes have been proposed. In one, the heterozygote is regarded as supplementary. Leaf rust in flax, for example, is conditioned by dominant genes. When the genotype M_1M_1, which is resistant to race one of flax rust, is crossed with an M_2M_2 genotype, which is resistant to race two, the M_1M_2 hybrid is resistant to both races. Other examples of the supplementary type of gene action are found in the leaf mark character in red and white clover (*Trifolium pratense* L.) and among human blood groupings.

Another explanation for single-gene heterozygotes is that of the alternative pathway. A number of temperature-sensitive mutants have been observed that require specific temperatures for the expression of a phenotype. Genotype R_1R_1, for example, produces a red pigment in leaves when the temperature is about 27° C (80° F) and a second homozygote, R_2R_2, produces a red pigment only when the temperature is approximately 10° C (50° F). The R_1R_2 hybrid produces the red pigment at both 10 and 27° C (50 and 80° F).

Another example of the alternative pathway concept can be found in the response of inbred lines and their hybrids to environmental stress. Several observations have been made in which the inbreds themselves are very sensitive to environmental changes, whereas their hybrids are stable over a series of environments.

The concept of optimal amount provides a third explanation for single-gene heterozygotes. One homozygote contains an inactive allele that is lethal or sub-vital, the other homozygote an allele unit in excess of gene product to give a wild-type phenotype. The heterozygote contains one active and one inactive allele to produce an "optimal amount" for excellent performance.

The explanation for heterosis may lie somewhere between the dominance and overdominance theories. Heterosis resulting from hybridizing two parents is due to a unique series of alleles in the F_1. An F_2 plant that contains all of the superior performance genes of an F_1 would be rare because of recombination and crossing over in the formation of gametes from the F_1 and a loss of unique gene combinations.

APPROACH TO HARDY WEINBERG EQUILIBRIUM

In a diploid species, a number of genotypes *Aa*, when mated at random, would reach equilibrium after one generation of mating. After a single generation of random mating, the genotypic frequencies would be 1/4 *AA*: 1/2 *Aa*: 1/4 *aa*. The normal assumptions of the **Hardy**

Weinberg equilibrium—random mating, no selection, no genetic drift, and no directional mutation—apply.

In an autotetraploid in which a series of individuals have genotype *AAaa* and the gene frequencies of *A* = 0.5 and *a* = 0.5, the equilibrium frequencies in the different genotypes would be as follow (F is the coefficient of inbreeding):

$$F(AAAA) = (0.5)^4 = 0.0625$$
$$F(AAAa) = 4(0.5)^3 \times (0.5) = 0.25$$
$$F(AAaa) = 6(0.5)^2 \times (0.5)^2 = 0.375$$
$$F(Aaaa) = 4(0.5) \times (0.5)^3 = 0.25$$
$$F(aaaa) = (0.5)^4 = 0.0625$$

The approach of an autotetraploid to equilibrium with generations of random mating is shown in Table 12.5. For quantitatively inherited traits, an autotetraploid may vary in performance from generation to generation as the population approaches equilibrium. Measuring parameters such as forage yield or plant height in an autotetraploid in a population that has not achieved equilibrium can cause inaccurate estimates of population performance.

INBREEDING DEPRESSION WHEN AUTOTETRAPLOIDS ARE SELFED

Inbreeding depression in alfalfa (Figure 12.3), birdsfoot trefoil, and other autotetraploids is similar to that observed in maize. A striking loss of vigor can be observed after self-fertilization. The loss of vigor associated with inbreeding maize and other diploid species follows the loss of heterozygosity expected when selfed. In a diploid species, for example, there is an increase in homozygosity of 50% in each selfed generation. A similar explanation, however, does not hold for autotetraploids since the approach to homozygosity is much slower (Table 12.6).

Table 12.5 Approach to Equilibrium Under Random Mating with Chromosome Segregation

Generation	Genotypic Frequencies				
	AAAA	*AAAa*	*AAaa*	*Aaaa*	*aaaa*
0			1.0		
1	0.028	0.222	0.500	0.222	0.028
2	0.049	0.247	0.408	0.247	0.049
3	0.058	0.250	0.384	0.250	0.058
4	0.061	0.250	0.378	0.250	0.061
5	0.062	0.250	0.376	0.250	0.062
6	0.063	0.250	0.375	0.250	0.063

Note: After five generations of random mating, the population has not reached equilibrium. However, values of the genotypes approach the equilibrium value after approximately four generations of random mating. This illustrates that autotetraploids reach equilibrium values at a much slower rate than diploids.

A.   B.

Figure 12.3 Photos of alfalfa racemes with good and poor seed set resulting from cross-fertilization (Photo A) and self-fertilization (Photo B), respectively. Cross-fertilization of alfalfa results in a higher percentage of flowers that form pods and a larger number of seeds per pod (Photo A) than in self-fertilization (Photo B). The alfalfa ovary is a pod coiled like a ram's horn with three or four spirals and containing several seeds. In alfalfa, 10 to 12 ovules develop per ovary.

Within the *Medicago sativa-falcata* (alfalfa) genetic complex, both diploid (2n = 16) and tetraploid (2n = 32) forms exist, which will cross to produce fertile hybrids when the diploid is raised to the tetraploid level. Individual alfalfa plants are almost entirely cross-fertilized, so that the level of heterozygosity in a population is high. Self-fertilization is less effective in accomplishing fertilization because of a partial self-incompatibility system and because self-fertilized ovules abort more frequently than cross-fertilized ovules. For additional information on self- and cross-compatibility in alfalfa, consult Whitehead and Davis (1954) and Wilsie (1951).

Photo courtesy Ontario Ministry of Agriculture and Food.

ALLELIC INTERACTIONS IN AUTOTETRAPLOIDS

Despite a slower progress toward homozygosity in autotetraploids than in diploids, inbreeding depression in alfalfa, birdsfoot trefoil, and potatoes is just as severe, if not more so, after a number of selfed generations. Inbreeding depression and heterosis observed in autotetraploids can be explained, in part, by the allelic interactions. Consider a diploid in which gene locus A has four different alleles, designated as A_1, A_2, A_3, A_4. Any single individual can carry only two alleles at any one time, and individual genotypes such as $A_1A_2, A_3A_3, A_1A_3, A_2A_4$,

Table 12.6   Genotypic Frequencies and Percent Homozygosity Under Self-fertilization of Duplex *AAaa* Selfed for Several Generations, Assuming Chromosome Segregation

Generation	Genotypic Frequencies					Percent Homozygosity
	AAAA	AAAa	AAaa	Aaaa	aaaa	
0			1.0			
1	0.028	0.222	0.500	0.222	0.028	6
2	0.097	0.222	0.361	0.222	0.097	20
3	0.163	0.191	0.292	0.191	0.163	33
4	0.218	0.160	0.241	0.160	0.218	44
5	0.265	0.134	0.201	0.134	0.265	53
6	0.500				0.500	100

Note: The proportion of homozygosity even after four generations of inbreeding in an autotetraploid is not as great as that observed in a diploid after a single generation of inbreeding. Inbreeding depression should not be as severe for the same generation of selfing because the approach to homozygosity is not as rapid. Therefore, the coefficient of inbreeding F_1 is not as high.

and so on may be found to account for a number of variations, all of which produce additive or dominant gene interactions.

In an autotetraploid, five different classes of genotypes are possible with four different alleles of gene *A* (A_1, A_2, A_3, A_4) in a population. The five possible classes of genotypes are

Monoallelic	$A_1A_1A_1A_1$, $A_2A_2A_2A_2$, . . .
Diallelic simplex	$A_1A_1A_1A_2$, $A_2A_3A_3A_3$, . . .
Diallelic duplex	$A_1A_1A_2A_2$, $A_2A_2A_3A_3$, . . .
Triallelic	$A_1A_2A_3A_3$, $A_2A_2A_3A_4$, . . .
Tetrallelic	$A_1A_2A_3A_4$, . . .

Thirty-five possible genotypes can be observed with four segregating alleles. Experimental evidence indicates that heterosis and superior performance are associated with tetrallelic combinations. Similarly, loss of tetrallelic and triallelic combinations results in a loss of performance. A single generation of selfing would result in relatively fewer tetrallelic and triallelic interactions and in a likely reduction in performance. The loss of tetrallelic and triallelic interactions occurs much faster than progression toward homozygosity or increase in the coefficient of inbreeding value, F. Breeding strategy in autoploids is often focused on maximizing the number of allelic interactions of the triallelic and tetrallelic form, especially in commercial seed increase.

RESPONSE TO SELECTION

Gene and genotype frequencies change more slowly in autotetraploids than in diploids. Response to selection, therefore, would be expected to be much slower in an autotetraploid than in a diploid. Selection for a recessive gene would, however, progress just as rapidly in an autotetraploid as in a diploid provided the recessive phenotype could be identified clearly.

Individuals selected for the recessive phenotype would be expected to maintain that phenotype. Fixation of a recessive gene is accomplished after a single generation of selection. Selection for the dominant phenotype (see Table 12.5 for example) reduces gene frequency of the recessive allele much less rapidly because a small proportion of the recessive gene occurs in a homozygous recessive individual.

If there is difficulty in recognizing individuals that have the recessive genotype and if the desired dominant gene is in low frequency, a slow increase in the dominant gene frequency will be observed. If the recessive genotype can be recognized by phenotype, however, progress toward changing the gene frequency of the dominant allele will be rapid, as long as the gene frequency of the dominant gene is below 0.5. At a frequency above 0.5 for the dominant gene, very slow progress will occur and will be a result of the elimination of recessive individuals, because the proportion of recessive individuals will be less than 5% in any generation. Further progress at this point will require some form of progeny testing to reduce the frequency of the recessive undesirable allele.

EVOLUTIONARY ASPECTS

The nature of gene segregation of autotetraploids creates a vast array of genes and genotypes. The autotetraploid condition allows for the accumulation of a large number of allelic forms of a particular gene, many of which are not particularly well adapted to one environment but may be adapted to other environments. Autopolyploids tend to be stable under widely varying conditions and as a rule have very wide adaptation. Although they have wide adaptation, autotetraploids are not capable of making rapid changes to specific environments because of the variety of allelic forms that can be maintained in a population and because of the complex segregation patterns. However, because of the large genetic reservoir, autotetraploids are capable of wide distribution and are often found in extreme areas.

REFERENCES

Dobzhansky, T. 1950. Mendelian Populations and their Evolution. *American Naturalist* 84:401-418.
――――. 1951. *Genetics and the Origin of Species*. Columbia University Press, New York. 446 pp.
Jones, D. F. 1917. Dominance of Linked Factors as a Means of Accounting for Heterosis. *Proceedings of the National Academy of Science* 3:310-312.
――――. 1939. Continued Inbreeding in Maize. *Genetics* 24:462-473.
――――. 1945. Heterosis Resulting from Degenerative Changes. *Genetics* 30:527-542.
Onokpise, O. U. 1980. Evaluation of Self (S₁) and Polycross Progeny Performance in Two Populations of Birdsfoot Trefoil (*Lotus corniculatus* L.), Cultivar Leo. MSc Thesis, University of Guelph. 62 pp.
Richey, F. D. 1922. The Experimental Basis for the Present Status of Corn Breeding. *Journal of the American Society of Agronomy* 14:1-17.
Robinson, H. F., R. E. Comstock, A. Khalil, and P. H. Harvey. 1956. Dominance Versus Over-dominance in Heterosis: Evidence from Crosses Between Open-pollinated Varieties of Maize. *American Naturalist* 90:127-131.
Shull, G. H. 1908. The Composition of a Field of Maize. *American Breeders' Association* 4:296-301.

————. 1909. A Pure-Line Method in Corn Breeding. *American Breeders' Association* 5:51-59.

Whitehead, W. L., and R. L. Davis. 1954. Self- and Cross-Compatibility in Alfalfa, *Medicago sativa*. *Agronomy Journal* 46:452-456.

Wilsie, C. P. 1951. Self-fertility and Forage Yields of Alfalfa Selections and Their Progenies. *Agronomy Journal* 43:555-560.

CHAPTER **13**

Selection Methods in Cross-fertilized Crops

The ultimate effectiveness of selection in both cross- and self-fertilized crops is demonstrated by the performance of cultivars resulting from the selection procedures. From an experimental viewpoint, the effectiveness of selection can be determined in three distinct ways. First, changes in gene or genotype frequencies indicate when selection has been successful. A selected population, for example, would be expected to have a higher proportion of desirable phenotypes, or in cases where genes can be identified, changes in gene frequency would confirm the effects of selections. A higher proportion of desirable genotypes or phenotypes would be expected also to cause a shift in the population mean. Successful selection for increased seedling vigor in a population of birdsfoot trefoil, for example, should result in a higher level of seedling vigor than in the original population.

A second way of determining the effectiveness of selection is based on the presence of new genotypes or phenotypes not present in the original population. Selection for high oil or high protein in maize kernels, for example, would likely result in individual plants that have oil or protein content higher than any observed in the previous generation, as shown in Figure 13.1.

A third measure of the effectiveness of selection is the relative change in genetic variability. A selection method, though effective, should preserve genetic variability. If selection has resulted in a population in which genetic variability is reduced, further selection likely will not be as successful. Preservation of sufficient genotypic variability for selection is an important determinant of the success of a selection method, particularly in the case of quantitative traits, because a single cycle or generation of selection often produces minor changes in population. Only after a series of cycles would major changes in the population be expected to occur. In this situation, a selection method that rapidly reduced genotypic variability would not be judged effective.

UNCONSCIOUS SELECTION

Unconscious selection is defined as nonintentional human selection and was considered in the evolution of domesticated plants in Chapter 1. A key aspect of the early domestication of crops was that unconscious selection did not operate until seeds were deliberately planted. Seed shattering would not be eliminated by mere collection of wild stands, but harvesting and planting the harvested seed would favor plants that retained their seed (Heiser, 1988).

Unconscious selection generally is associated with early domestication, but the fact that unconscious selection may occur in modern breeding programs should not be overlooked. Deliberate selection for one factor followed by planting these selections may be accompanied

271

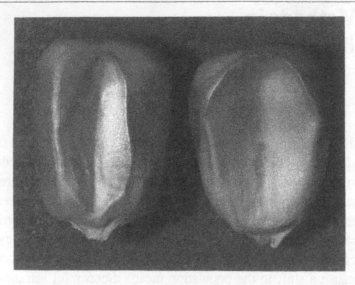

Figure 13.1 The impact of selection is evident in the morphology of these maize kernels. The kernel on the left has 4% oil and the one on the right 14%. An increase in oil content of 10 percentage points represents a major change and was achieved over many years of selection pressure. Note the morphological change and the large scutellum, a component of the embryo.

After ten generations of mass selection in an open-pollinated maize cultivar, Smith (1908) reported a change in oil content from 4.7 to 7.4%. Continuous selection since then has resulted in strains with 14% oil, but techniques switched from mass selection to backcross recoveries. Efforts were not directly fruitful in an economic sense (Alexander, 1971), primarily because the hybrids were either lower in yield or otherwise less satisfactory than ordinary hybrids. Later, recurrent selection was started in a broad genetic pool.

Past studies have shown that cultivars with high oil content yield less than traditional hybrids. Perhaps the synthesis of a fixed amount of oil requires twice the energy as the same amount of starch. With a fixed amount of energy available from the sporophyte, total grain production in high-oil strains would probably be lower.

Experiments to test for possible exclusiveness of high-oil content and high-grain yield suggest that calorie yield per plant is probably independent in most stocks but may be limited in some stocks by inefficient oil synthesizing systems (Alexander and Lambert, 1968; Alexander, 1986). Conversely, high-oil strains may be limited in total calorie storage by high-oil synthesis.

Techniques to improve selection using a nuclear magnetic resonance process have produced selections in which oil contents were raised from 4.6 to 20% in 20 generations compared with 80 generations to achieve the original increase.

Courtesy D. E. Alexander with photo by W. O. Scott.

by unconscious selection, also referred to as automatic selection. Selection for drought tolerance in maize, shown in Figure 13.2, may result in unconscious selection.

A factor as simple as seed color can result in unconscious selection. Black and red beans, for example, have higher levels of tannins than do white beans. Tannins react with protinase to reduce digestibility, and thus selection for light-colored or white beans may have unconsciously resulted in an improvement of their nutrient value (Elias, Fernandez, and Bressani, 1977; Powell, Oliveiro, and Matthews, 1986). Seeds of white beans, however, have less vigorous and lower seed emergence than do black and brown beans. A lack of awareness would result in unconscious selection for low seedling vigor. An awareness of the situation could result in screening white beans for seedling vigor to break any possible linkage effects.

Unconscious selection can occur in seed-propagated plants over several generations of selection, replanting, and reselection. A shift in an undesired direction may alert the plant breeder that unconscious selection is occurring. In vegetatively propagated plants, however, selection is absolute and its effects immediate. Asexual propagation to fix the genotype of an otherwise heterozygous and heterogeneous crop is preferred in cross-fertilized crops.

In field cultivation systems, repeated application of a single and specific pesticide may result in unconscious selection for a resistant form of the pest. Since the 1960s, agriculture has benefited from the development of selective herbicides and technically successful insecticides and fungicides, including products that work through systemic chemical action. Resistant forms of the target organisms have resulted in very serious problems (Mlot, 1985; Dover, and Croft, 1986; Gressel, 1978; Perrin, 1983; Wade, 1982).

The phenomenon of unconscious selection of resistant pest strains has been recognized since the 1940s when the first flies became resistant to DDT. In the mid-1950s, 25 species of pests or pathogens were known to have developed resistance to pesticides prescribed for their control. Concern rests with the rapidity with which resistant strains are developing and the evident associated unconscious selection pressure. By 1980, more than 430 resistant species had been identified (Anonymous, 1982).

Associated with unconscious selection is the risk of **cross-resistance**, in which change in one genetic factor results in resistance to different chemical pesticides (Wade, 1982). Cross-resistance usually occurs between structurally similar chemicals or ones with similar modes of action. Cross-resistance is known to occur with insects, pathogenic bacteria, and weeds and probably applies to all organisms. Cross-resistance explains how the effectiveness of a pesticide may be diminished or lost even before it has been applied.

Plant breeders may associate unconscious selection with crop domestication and overlook its importance in present-day breeding programs. Unconscious selection may be beneficial or detrimental, but the plant breeder must monitor all aspects of the system to be aware of and to understand changes that occur as a result of selection.

MASS SELECTION

Mass selection in cross-fertilized crops is essentially similar to mass selection in self-fertilized crops, except that the plants selected are randomly mated. In both self- and cross-fertilized crops, selection is based on individual plant performance. In cross-fertilized crops, the selected plants are allowed to pollinate each other, resulting in a new population.

Mass selection can be used to maintain existing cultivars and to develop new and improved cultivars. In cross-fertilized crops, as in self-fertilized crops, it has been one of the most effective breeding methods in developing new open-pollinated cultivars, as illustrated with black spruce in Figure 13.3.

This technique was used by the Indians in North America to transform maize into an adapted, high-yielding crop. Perhaps it was intuitively obvious for North American Indians and

Figure 13.2   The possible consequences of unconscious selection are illustrated by the maize plants being developed for their response to moisture stress. Note leaf rolling, a consequence of moisture stress, in Photo A. If these plots are evaluated and selected on the basis of grain yield alone, unconscious selection may take place for the dates of anther and seed development. In Photo B, a perfect nick occurred between male and female development, both at 74 days after planting, as shown on the tag. If there are 15 days or more between male and female development, grain yield may be greatly reduced. Selection for grain yield without an awareness of male and female development might result in unconscious selection for proper nicking.

Unconscious or automatic selection for a specific root morphology may stem from deliberate

subsequently for European settlers to select the most desirable plants or ears of maize to serve as seed for the next generation. The result is that mass selection may have been practiced since the early development of agriculture in the Americas. Richey (1950) suggested that using seed from unselected ears in a heterogeneous mixture of shelled corn is inconceivable. Producers who selected maize for ear size, ear shape, number of kernel rows on a cob, stalk strength, or other features became plant breeders. Often a specific selection was named after its originator, and many cultivars emerged with definite adaptations and characteristics.

Competition for seed sales and superiority of performance led to corn shows at county and state fairs. Many large ears that were exhibited reflected **environmental influences** achieved by wide spacings of large plants. Specific trait selection often produced nothing more than eye appeal, and experiments over a number of years demonstrated that these traits had little effect on yield. The most striking effects of mass selection in maize were the changes in visible plant features such as ear height, ear shape, husks covering the ear, and other genetically controlled morphological features, as illustrated in Figure 13.4.

Mass selection in forage crops was the most common method of producing new cultivars until the 1950s. Many of the commercial forage cultivars released between 1900 and 1950 were the result of local or farm selections, mass selected from heterogeneous cultivars. Some of these cultivars have been very productive over wide geographic areas, perhaps because of the heterozygous and heterogeneous nature of cross-fertilized crops.

Selection based on phenotypic performance of individual plants has been generally ineffective for quantitative traits such as yield, which are difficult to identify based solely on phenotype. Cross-fertilization in an uncontrolled population involving superior and inferior genotypes means that progeny often deviate from the parental type. Strict selection with few selected plants per generation results in loss of vigor for quantitative traits because of inbreeding.

A modified mass selection program in an open-pollinated maize cultivar called Hays Golden was used by Gardner (1961). With this selection method, he divided the population into blocks of 40 plants each and restricted selection of superior plants within each block to one or two plants. Environmental effects due to location in plot area were minimized, and after a four-year selection program, grain yield of Hays Golden had increased from 5,329 to 6,545 kg/ha (79.3 to 97.4 bu./ac.), an increase of 22.8%. Restrictions on phenotypic selection allowed better identification of superior genotypes based on phenotype. This method is known as stratified mass selection.

Figure 13.2 continued

selection for nicking and resulting grain yield. The cylinders in Photo C are designed to evaluate possible differences in root morphology. Other tests at this site employed core samplings to a depth of 120 cm (47 in.) near the crown of the root to determine root volume at depths of 0 to 30, 30 to 60, 60 to 90, and 90 to 120 cm.

Awareness of selection for one criterion that automatically emphasizes another factor has been discussed for self-fertilized crops. If the breeder is aware of correlations between two criteria, a more fitting term is indirect selection. Plant breeders must be cautious about unconscious selection that may have an adverse impact on overall plant performance.

Figure 13.3   An example of a modern mass selection program is provided by the thousands of black spruce (*Picea mariana* Mill. B.S.P.) seedlings in Photo A, which were asexually produced from trees selected for their superiority based on borings to determine age and relative size. Based on seedling growth rates under controlled environmental conditions, further selections were made and superior families propagated for outdoor planting (Photo B).

Preliminary results suggest that enormous genetic variation exists among native stands of black spruce and that mass selection can identify genotypes for rapid growth rates. Seedlings in Photo B are 19 weeks old, in Photo C, 12 months old, and in Photo D, 24 months old. Note the contrasting types

A.   B.

Figure 13.4   These photos show ears of maize suspended on the stalk, a desirable feature because it protects the ear from weathering, bird damage, and possible entry of unwanted organisms. Developing ears of maize are protected from weathering and pests by glumes and husks that adhere tightly during grain development. As the ear matures, the husks loosen and drying is facilitated. The ear is supported in the early stages of development in an upright position but relaxes as maturity progresses so that it shifts to a suspended position. These photos were taken in April, following a severe winter in the Great Lakes region of North America.

Suspended ears, sturdy stalks that support the ears and keep them from falling on the ground, and loose husks are features that were probably developed by mass selection. Since the ancestry of maize is obscure, it is not clear when the suspended ear position at maturity was developed.

Mass selection provides an effective means of selecting visible traits from a heterogeneous population, but during the mass selection era of maize breeding, yield was not improved. The reasons for this failure are the inability to distinguish visually between environmental and genetic influences, and the fact that pollination is completely uncontrolled, with both superior and inferior genotypes contributing pollen, so that the yield potential of a superior genotype was not perpetuated.

Figure 13.3 continued

in Photo D. The specimen on the left is characterized by an upright lead shoot and rigid branches. The specimen on the right has a droopy lead shoot and main branches.

Although these studies with mass selection of black spruce are in the early stages, extrapolating observed comparative growth rate suggests a harvestable tree may be produced at a rate well beyond the majority of trees now grown. Mass selection has revealed the advances that can be made in black spruce. A logical step is to hybridize superior and perhaps genetically diverse genotypes in Photo D, followed by further screening and selection for superior performance, and quality tests for paper products produced.

The selection of individual plants based on phenotypic performance is one of the simplest means of cultivar improvement, and after stratified mass selection was introduced by Gardner (1961), breeders showed renewed interest in mass selection for maize improvement. Odhiambo and Compton (1987) completed twenty cycles of mass selection for seed size in the open-pollinated maize cultivar Krug yellow dent in a single isolation block, measuring 21 x 42 m (23 x 46 yds.), composed of 3,500 plants. Selections were made in two sets of at least 100 ears each for those that exhibited uniformly smaller or larger seed. Two divergent populations, for small and large seed size, were thus formed.

Following the first year of selection, the two populations were grown in separate isolation blocks, and each year 100 ears from a total of 1,200 to 1,500 plants were selected for seed size. Equal numbers of seed from each selected ear were composited to use for planting the next generation. After several cycles of selection, about 100 representative plants were randomly sib mated, that is, small x small seed, large x large seed, small x large seed. Progeny from these tests were yield tested. After 20 cycles, pollen from the large-seeded population increased seed size by 0.12 grams per kernel when used to pollinate plants from the small-seeded group. Pollen from the small-seeded population reduced seed size 0.15 g per kernel when used on the large-seeded group. These very large effects were significant. Small seeds significantly reduced yield, but selection for large seeds had no effect on yield.

Computer simulation was used to study mass selection with parental control in populations of maize, sorghum, wheat, and other species. Although stratified mass selection is designed to increase the frequency of favorable genes, genetic drift and inbreeding will also occur. Genetic drift could result in a loss of genetic variance and consequently a loss in potential for grain from selection for quantitative traits. Inbreeding may result in a loss of vigor. The results of the study indicate that equating effective population size with the number of parents will lead to a substantial underestimation of the cumulative effects of genetic drift and inbreeding depression when mass selection is conducted for a moderately or highly heritable trait. Plant breeders are advised to consider heritability, selection intensity, and parental population size when estimating the variance and need an effective size of population to do so. Hallauer (1990) presented an excellent general reference on inbred development in maize.

EAR-TO-ROW SELECTION

Early observations that selection for quantitative traits was ineffective when based on individual plant phenotype led to breeding methods that capitalized on some form of progeny test. Testing systems for improving sugar content and seed yield in sugarbeets used by Louis de Vilmorin in 1856 were among the first progeny tests conducted (Hayes, Immer, and Smith, 1955).

The Vilmorin selection method was in use before Mendel discovered the genetic basis for inheritance. With the Vilmorin method, individual mother sugarbeets were selected on the basis of their vigor, shape, and sugar content. These selections were allowed to produce seed, and those with acceptable sugar content and seed yield were retained and increased. The pollen parent of these sugarbeets was not controlled, that is, pollen from selected and unselected plants contributed to the genotype of the next generation. The progeny from each of the sugarbeets were grown and evaluated, and the progeny that exhibited the highest sugar content and seed yield were increased for the next year. Using this type of progeny testing method, the average sugar content was increased from 5 to 6% to about 16 to 18%. This, incidentally, is the approximate sugar content of beets grown commercially today.

In 1897, C. G. Hopkins devised the ear-to-row selection method for high and low oil and high and low protein in maize (Hayes, Immer, and Smith, 1955). In the method used by Hopkins, selected plants were progeny tested, and the best progeny rows were used to indicate superior mother plants. Remnant seeds from the best mother plants were then mixed together, allowed to interpollinate, and in the next generation, plants were selected for progeny testing. The experiments of Hopkins resulted in consistent and dramatic changes in oil and protein levels in maize. Both are quantitatively inherited traits but with a relatively high heritability.

The success of Hopkins led to further experiments with the ear-to-row method by other investigators (Richey, 1950). In a series of experiments in Nebraska from 1911 to 1917, the ear-to-row method was used to select for high yield from the open-pollinated cultivar Hogue's yellow dent. The initial yield of this cultivar was 3,582 kg/ha (53.3 bu./ac.) and after six cycles of selection was 3,602 kg/ha (53.6 bu./ac.). The reason for this disappointing lack of success lies in the low heritability associated with grain yield. Superior maternal plants were chosen based on the results of a single progeny row, and from this single progeny row, it was impossible to separate genotypic and environmental responses. An additional explanation is that both genetically inferior and selected plants can serve as the pollen parent with the ear-to-row selection method.

The Hopkins experiments were successful, whereas the Nebraska experiments were ineffective. A third possible outcome is one in which an initial response is noted followed by a plateau in which no selection gains are made. Sometimes a series of generations of increase followed by a plateau will again be followed by an increase for the trait under selection. A genetic explanation for this observation is that new gene combinations or rare combinations of crossovers result in new genetic variability. Genetic variability can be of two types: free and potential. A single genetic locus that is homozygous exhibits free genotypic variability, and selection of a stable plant can be made. A heterozygous locus, on the other hand, exhibits potential genetic variability because it can give rise to two different genotypes. These two genotypes would then express a phenotype that could be selected (free variability).

When two or more genetic loci are linked, the genes that are linked are said to have potential variability. A crossover or recombination between different alleles of those genes can result in the expression of free variability because of that recombination. This is consistent with observations in natural populations in which crossovers or rare combinations place new allelic combinations together and result in the ability to select further extremes as a consequence of exposing genetic variability.

The essential steps in the ear-to-row breeding system in maize are as follows:

1. Separately shell 50 to 100 ears and use part of the seed to plant a row (ear-to-row). Store remnant seed.
2. Identify superior rows by observation and harvested yield performance.
3. Mix remnant seed from the top 10 to 20 rows and establish a plot. Select ears from this plot and repeat the process in succeeding years.

A modification of the ear-to-row method, in which progeny rows are grown at several locations and selections are made within the best progeny rows, was proposed by Lonnquist (1964). The male pollen source is a population sample, restricted to selected plants. By growing progeny rows at a number of locations, restricted pollen sampling is avoided.

The ear-to-row-to-ear method, also termed the within-half-sib family selection method, has proved effective in selecting for traits of low heritability in species such as maize, alfalfa, and birdsfoot trefoil.

DESCRIPTION OF SELECTION METHODS USING A PROGENY TEST

Half-Sib Progeny Test

Half-sib progeny tests include the **top cross** progeny test, open-pollinated progeny test, some forms of maternal line selection, and the polycross progeny test. These progeny testing methods share the principle that the identity of the maternal parent is known, but the specific identity of the male pollen source is not known. The methods differ in the source of the pollen parent.

In maize breeding, the ear-to-row progeny test was used initially to select superior mother plants. For forage crops, a similar type of test is used and is referred to as maternal line selection. With the interest in hybrids, maize breeders crossed inbred lines to produce hybrids. Not all F_1 hybrids, however, are of value, so maize breeders began to look for more efficient methods to screen inbred lines in large numbers.

Richey (1950) suggested the use of a top cross test. Using this test, each selected inbred line is crossed with an open-pollinated cultivar and the top cross progeny evaluated. Progeny performance was assumed to estimate the average performance of a line in a series of single crosses.

Forage crop breeders use the top cross test and also two other types of progeny test: the open-pollination progeny test and the polycross progeny test. The open-pollination progeny test is similar to the ear-to-row method, except superior mother plants are identified. In perennial forages, the superior mother plants are then interpollinated to create a new population. In the polycross test, plants are selected on the basis of phenotype, and then a large number of these are planted together in isolation and allowed to interpollinate. The seed is harvested from each plant separately and used to plant a polycross progeny test.

Full-Sib Progeny Test

In full-sib progeny tests, the identities of both the male and female parents are known. The most common form of full-sib progeny test is the specific mating of two individuals, that is, a controlled mating in which both parents are known. This method has been used in most cross-fertilized crops to choose combinations that result in high or exceptional vigor. Another type of test in which the identity of both parents is known is the S_1 or S_2 progeny test. In this case, the plant is selfed for one or two generations. The S_1 progeny test has been used in most cross-fertilized species, although it is used most commonly in maize. Top crossing, using an inbred line on specific females, is also a full-sib progeny test.

CONCEPT OF COMBINING ABILITY

The principle introduced in this chapter is of the necessity of using progeny testing methods to identify superior parents for traits of low heritability, such as grain yield, plant persistence in forages, and feeding quality. A combining ability test is employed to specify the ability of an individual parent, or in some cases two specific parents, to produce high-yielding progeny. A parent that has high combining ability is inferred to be a parent whose progeny have good performance.

Two types of combining ability are general and specific combining ability, which will be discussed more fully in subsequent chapters. General combining ability refers to the ability of an individual parent to produce progeny of a given performance when mated with a number of other parents. A parent that has high general combining ability produces high-performing progeny when mated with a series of other parents.

Specific combining ability refers to the performance of a specific parental combination and can deviate from the performance assigned by a general combining ability test. For example, two specific parents may have very high-yielding progeny, whereas one of the parents when crossed with a third parent results in lower-yielding progeny.

DEVELOPMENT AND MAINTENANCE
OF OPEN-POLLINATED MAIZE CULTIVARS

F_1 hybrids have dominated maize production in the developed world, but open-pollinated maize cultivars commonly are grown in the developing countries. Open-pollinated maize cultivars are well suited to the vast regions where traditional agricultural practices prevail for the following reasons:

- Maintenance and seed production is relatively simple.
- Superior new cultivars developed from an ongoing population improvement program can replace old cultivars when desired, either as new cultivars or as improved versions of existing cultivars.
- Open-pollinated cultivars have a distinct advantage where seed distribution is difficult and costly and where a distribution system does not exist. Seed from open-pollinated cultivars can be saved from year to year and can move among farmers, both of which have a multiplicative effect on area coverage.
- Seed production costs are low, and seed supplies build up rapidly. Commercial grain production is only two generations away from breeder seed.
- Exchange of germplasm among national programs is easier with open-pollinated cultivars than with some closed-pedigree maize materials that involve proprietary rights.

The term open-pollinated cultivar traditionally refers to a landrace that exhibits high genetic variability as a result of cross-fertilization. Genetically diverse populations may be crossed to produce composites, germplasm complexes, and advanced generations of cultivar crosses that may be mass selected to produce a new open-pollinated cultivar by releasing the bulk from the latest cycle of improvement. Such loosely defined cultivars are variable in agronomic attributes and lack phenotypic appeal.

An open-pollinated maize cultivar was redefined by CIMMYT (1984) as a superior fraction of an improved population that is stable, relatively uniform, and different because it possesses traits that distinguish it from other named cultivars. Variation for significant agronomic traits is reduced, and it is relatively stable in terms of the expression of these traits over time in its area of adaptation. A cultivar constituted by recombining eight to ten selected families from a family-structured population would meet the new definition of an open-pollinated cultivar. This would be assured by selecting families that are similar in plant and ear height, general morphology, and maturity.

Cultivars of this type can be developed in several ways (Figure 13.5), using half-sib, full-sib, S_1, or S_2. F_1 seed should be advanced to F_2 to provide adequate seed quantities for cultivar evaluation and to reduce any effects of heterosis in the cultivar. Any of the following procedures can be used to prepare the F_1 sample for advancing to F_2:

- If ten families were used to form the cultivar, there would be 45 possible crosses. A roughly equal number of ears can be selected and shelled from each possible cross. Each cross bulk is kept separate. To make the F_1 bulk for planting and advancing to F_2, a specified seed number or volume can be taken from each cross and mixed. Alternatively, each of the 45 possible cross bulks can be planted separately. Misfit crosses can be rejected.
- An approximately equal number of selected ears from each family that is crossed with all other families can be shelled and bulked. If ten families were involved, there would be ten family-cross bulks. Equal seed quantities can be taken from each family bulk to make a balanced F_1 bulk for planting to obtain the F_2 seed. Family bulks also can be planted individually, and if any family bulk is exhibiting undesirable features, it can be discarded. This does not preclude the participation of a poor family in the recombination process because it has already contributed as a pollen source to other bulk families.
- A simple procedure is to bulk all the selected ears from all family crosses, being careful to ensure that roughly equal numbers of ears are taken from each family. The resulting bulk is shelled and thoroughly mixed to produce the F_1 bulk. A sample of the bulk is planted to produce F_2 seed.

The quantity of F_1 seed should be such that enough F_2 seed is produced for cultivar evaluation and future multiplication of the cultivar selected for farm production. No matter which of the above alternatives is chosen, the F_1 seed should be advanced to F_2 through controlled pollination. For example: seed approximately 1,500 F_1 plants; rogue out all undesirable plants; select and tag desired plants; use bulk pollen only from the selected plants for pollination; shell selected ears in bulk to produce F_2 seed.

A cultivar should show little change in its phenotypic attributes when reproduced for maintenance and/or seed increase from one cycle to another. This stability is highly desirable because only a stable cultivar will perform in accordance with expectations from year to year. Although uniformity is important in any good open-pollinated cultivar, seldom will it be as uniform as a single-cross hybrid. A cultivar should exhibit some distinct genetic feature(s) to distinguish it from other cultivars. Upon release, a description of the cultivar should be given to serve as a guide for its maintenance and for seed certification. Standards for certification should not be too stringent.

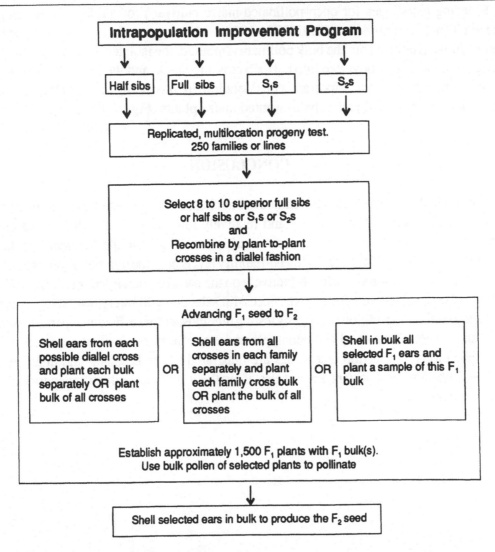

Figure 13.5 Illustration of different ways in which open-pollinated maize cultivars can be developed.

1. A cultivar can be developed using a half-sib scheme by recombining six to eight half-sib families. Replicated multilocation progeny trials are used to identify superior half-sib families. Remnant seed of these families is used in the recombination process, either by a diallel crossing system or by bulk sibbing. Uniformity in plant characteristics is important, and it is critical to rogue out off-type plants and even to eliminate complete families and their crosses if a mistake was made in their selection.

2. In full-sib family selection, about eight to ten superior families are identified on the basis of replicated, multilocation progeny trials. Remnant seed can be used to recombine these families, as described in item 1 above.

3. Plants are selfed once or twice to form an S_1 or S_2 selection scheme. Multilocation, replicated progeny trials are used to select S_1 or S_2 lines. Remnant seed is used to recombine eight to ten of the superior lines by making all possible crosses among them.

4. Recurrent selection for general combining ability and reciprocal recurrent selection can be used. With this method, remnant S_1 seed from the selected eight to ten lines is used in the recombination process, and the same procedure should be used for cultivar formation as outlined for S_1 and S_2 selection schemes in item 3.

Source: CIMMYT, 1984, p. 3.

Breeding techniques for open-pollinated maize cultivar production were developed by CIMMYT maize breeders (CIMMYT, 1984). Four procedures for maintaining an open-pollinated cultivar are the bulk pollination method, the isolated mass selection plot, the isolated bulk planting converted into half-sib crossing block, and the isolated half-sib ear-to-row crossing block. An important consideration in producing open-pollinated seed is isolation to prevent fertilization by unwanted maize plants (Figure 13.6).

CONCLUSION

A number of selection methods outlined in this chapter can be used in cross-fertilized crops. For those traits with high heritability and relatively few genes, a straightforward selection approach without progeny testing is usually sufficient to increase the frequency of desirable genes in the population. With declining heritability levels or as the number of genes controlling a trait increases, the ability to select effectively on the basis of phenotype is reduced. With low heritability or quantitative gene action, successful selection will likely result from some form of progeny testing. In addition to progeny testing methods, a more efficient experimental design to partition the environmental and genotypic variance effectively is often necessary.

Applications of selection techniques are illustrated by the breeding and maintenance of open-pollinated maize cultivars, known also as synthetic cultivars, which are considered in the following chapter.

Figure 13.6   The light-colored rectangular block in the center of this photo is an isolation block of corn surrounded by soybeans. The purpose of isolation is to eliminate or reduce cross-fertilization from adjacent fields or strips of maize. A distance of about 300 m (975 ft.) from other maize is suggested as safe.

Trained technicians rogue these increase blocks throughout the season and at harvest to eliminate individual plants that do not fit the cultivar description. Roguing prior to pollination will eliminate outcrossing from undesirable plants. In an open-pollinated maize cultivar, as many as 10 to 15% of the plants may be rogued out to maintain the cultivar's distinctive features.

The seed increase block in this photo is used to produce foundation seed maize. In open-pollinated maize, this is the first increase of breeder seed. A second generation of foundation seed can be produced from the first generation foundation seed when large quantities are required. Maintenance of genetic purity and cultivar uniformity is important, and such plots should be monitored closely by the breeder.

Certified seed of open-pollinated maize cultivars is produced from foundation seed and is grown in isolation. Certified seed is sold for commercial production. Isolation for certified seed is 200 m (650 ft.).

Isolation can also be achieved by time of sowing, allowing two fields to be planted side by side with an appropriate time interval between the sowings. Isolation by time of planting is suitable if stigma emergence and pollen shed in the first planted isolation are complete by the time tassels in the second isolation start emerging. A combination of dates of sowing and distance can also be used in planning isolation. Careful planning based on the number of days to pollen shedding may allow for several isolation blocks to be arranged in the same field. A time isolation can be inserted between two distance isolations, with an appropriate interval so that flowering times do not coincide.

Photo courtesy Pfister Hybrid Corn Co., El Paso, Illinois.

286   *Selection Methods in Cross-fertilized Crops*

REFERENCES

Alexander, D.E. 1971. Progress in Breeding Maize for Oil Content. In *Proceedings of the Fifth Meeting of the Maize and Sorghum Section of Eucarpia,* edited by I. Kovacs. European Association for Research and Plant Breeding. Akademia Kiado, Budapest, Hungary, pp. 74-78.

Alexander, D. E. 1986. High-Oil Corn. *Crops and Soil* 38(9):11-12.

Alexander, D. E., and R. J. Lambert. 1968. Relationship of Kernel Oil Content to Yield in Maize. *Crop Science* 8:273-274.

Anonymous. 1982. Pest Resistance Poses Challenge to Chemical Control. *Ceres* 15(4):4-5.

CIMMYT. 1984. *Development of Maintenance and Seed Multiplication of Open-pollinated Maize Varieties.* International Maize and Wheat Improvement Center, El Bator, Mexico. 11 pp.

Dover, M. J., and B. A. Croft. 1986. Pesticide Resistance and Public Policy. *BioScience* 36(2):78-85.

Elias, L. G., D. Fernandez, and R. Bressani. 1977. Possible Effects of Seed Coat Polyphenolics on the Nutritional Quality of Bean Protein. *Journal of Food Science* 44:524-527.

Gardner, C. O. 1961. An Evaluation of Mass Selection and Seed Irradiation with Thermal Neutrons on Yield in Corn. *Crop Science* 1:241-245.

Gressel, J. 1978. Factors Influencing the Selection of Herbicide-Resistant Biotypes of Weeds. *Outlook on Agriculture* 9(6):283-287.

Hallauer, A. R. 1990. Methods Used in Developing Maize Inbreds. *Maydica* 35:1-16.

Hayes, H. K., F. R. Immer, and D. C. Smith. 1955. *Methods of Plant Breeding.* McGraw-Hill Book Company, New York, p. 95.

Heiser, C. B. 1988. Aspects of Unconscious Selection and the Evolution of Domesticated Plants. *Euphytica* 37:77-81.

Lonnquist, J. H. 1964. Modifications of the Ear-to-Row Procedure for the Improvement of Maize Populations. *Crop Science* 4:227-228.

Mlot, C. 1985. Managing Pesticide Resistance. *BioScience* 35(4):216-218.

Odhiambo, M. O., and W. A. Compton. 1987. Twenty Cycles of Divergent Mass Selection for Seed Size in Corn. *Crop Science* 27:1113-1116.

Perrin, R. M. 1983. A World-Wide Problem: Pesticide Resistance in Aphids. *World Crops* 35:93-94.

Powell, A., M. Oliveiro, and S. Matthews. 1986. The Role of Inhibition Damage in Determining the Vigour of White and Coloured Seed Lots of Dwarf French Beans (*Phaseolus vulgaris*). *Journal of Experimental Botany* 37:716-722.

Richey, F. D. 1950. Corn Breeding. *Advances in Genetics* 3:159-192.

Smith, L. H. 1908. Ten Generations of Corn Breeding. *Illinois Agriculture Experiment Station Bulletin* 128:454-575.

Wade, M. 1982. Resistance to Fungicides. *Span* 25(1):8-10.

CHAPTER 14

Hybrid Development and Parental Selection

Stated simply, a hybrid is the progeny produced by a cross of two different parents. In actual practice, hybrid development has become considerably more complex than this simple statement would suggest. After eight decades of experience with hybrid maize, the various ramifications of hybrid development have been identified, solutions found, and practical techniques developed. A wealth of information exists on hybrid maize, and because of this, hybrid maize is the focus for hybrid development for cross-fertilized crops.

Following his work on heterosis and inbreeding depression in maize, Shull (1908) first proposed the use of hybrids for commercial maize production. In his work with inbred lines and their performance, he observed a number of principles to be used when selecting inbred lines as parents of hybrids. These principles form the basis for much of this chapter.

Not all inbred parents produce superior hybrids, and each should be chosen carefully on the basis of its performance in hybrid combinations. Certain inbred families were observed to give rise to progeny that were higher in performance than the original open-pollinated cultivars. Shull (1908) proposed using inbred lines as parents of maize hybrids for two practical reasons: (1) Inbred lines reproduce themselves faithfully from year to year, and (2) hybrids can be produced consistently year after year. Both are prerequisites for commercial use in an annual species.

Shull also pointed out that a major difficulty in selling hybrid seed to farmers was producing the seed economically. The problem arises because of the low vigor and low yield of inbreds produced after a series of selfing generations. Seed for commercial use would be produced on a female plant that bears a small ear with small kernels. This situation lacked appeal for farmers, who had been encouraged to select seed on the basis of large ears and large kernels. In the early years of hybrid maize, single-cross hybrids of corn did not appear to be commercially feasible and the commercial acceptance of hybrid maize was delayed.

A solution to the problems of poor quality seed and low yield was presented by Jones (1918). He proposed a double-cross system for producing commercial hybrid seed. The double-cross procedure combines four inbred lines, A, B, C, and D, first by developing two single crosses, A x B and C x D, and then crossing the F_1 of these single crosses. The double cross is designated as AB x DC. From his experiments, Jones (1918) observed that double-cross performance was very close to that of the single cross. Double-cross seed was produced on F_1 hybrid plants that produced large ears and kernels, which had producer appeal and was economical for commercial purposes.

Even before double-cross seed was developed and before hybrid corn was grown commercially, private enterprise recognized the potential of commercial hybrid seed production. Henry

A. Wallace, an innovative farmer, started studies of self-pollination and selection in 1913. This work led to the formation of the Pioneer Hi-Bred Corn Company of Iowa in 1925.

Control of parental materials used to develop exclusive F_1 hybrid seed stocks has allowed maize breeding companies to realize a profit, which in turn has been used for further research and the development of superior hybrids. Commercial maize breeding companies have made an enormous contribution to agriculture through the development and successful marketing of such hybrids. Hybrid seed production involves a system for developing inbreds, making crosses, and evaluating and licensing superior F_1 progeny. Without a complete system for handling the steps from breeding to farm sales, F_1 hybrid production could not be successful. The final step that completes the task of the plant breeder is for commercial producers to plant seeds of crops adapted to their region and to apply sound cultural practices that allow for their full genetic potential to be expressed.

Average grain yields of maize in North America did not increase appreciably until after the use of double-cross hybrids was initiated in the late 1930s. From 1940 to 1960, hybrids improved very little, and gradual increases in maize yields were a reflection of increased use of adapted hybrids and improved farm practices. Douglas *et al.* (1961) noted that the best inbred lines in the early 1960s were the ones developed in the early years of maize breeding. Reasons for the low rate of improvement in maize hybrids were suggested by Douglas *et al.* (1961), as follows: (1) the rapid fixation of genes in conventional breeding procedures, which reduces the efficiency of selection for combining ability; and (2) the extremely large number of plants that must be sampled to identify favorable genotypes. Visual phenotypic selection tends to favor lines from the model classes, since characters other than yield are not highly correlated with combining ability.

In order to overcome the problem of slow progress, other hybrid types were developed. A three-way cross [(A x B) x C] can be produced on an F_1 hybrid female parent (A x B). Modified crosses are produced by crossing two closely related inbred lines (A^1 x A) and using the resulting single cross as the female parent. A modified single cross [(A^1 x A) x B] or a modified three-way cross [(A x B) x (C^1 x C)] can be produced with this method.

With improvements in seed corn production techniques and better inbreds (Jones, 1958), the cost of single-cross seed was reduced (Eberhart, 1969) and it was accepted commercially. By the early 1980s about 70% of the maize seed sold in North America was single-cross seed, and during that decade single-cross seed continued to grow in importance.

The relative performance of three types of crosses with 108 entries tested at two locations for a two-year period showed that, on average, single crosses outyielded three-way and double crosses. The superiority of single crosses is still pronounced when the best crosses are compared.

Average yield, however, is not the only measure of performance. Consistency of performance across years and locations is important also. The hybrid x environment mean square for single crosses was more than twice that for doubles, and the mean square for three-way crosses was intermediate between that for double and single crosses. These results support the findings of other workers (Eberhart and Russell, 1969; Eberhart, Russell, and Penny, 1964; Rojas and Sprague, 1952; Sprague and Federer, 1951) that single crosses are more sensitive to environmental conditions than three-way or double crosses. As discussed by Allard and Bradshaw (1964), there are two ways of achieving stability. In the population buffering method, stability results when a population is composed of different genotypes, such as in a three-way or double cross. In a second method, known as individual buffering, stability is achieved when members of a population are all alike, as in a single cross, but each member is adapted to a wide range of environments. Weatherspoon (1970) noted that among 36 single crosses under test, high-performing combinations could be found among the high-yielding crosses that showed no significant interaction deviates. Moreover, a high-yielding single cross could be poor in a given environment relative to its average yield over environments and still be superior in that environment to the best double cross.

In the Great Lakes region of North America, yield performance was determined for all single-cross, three-way cross, and double-cross maize hybrids recommended by the Ontario Corn Committee between 1968 and 1972. Single-cross hybrids outyielded three-way and double-cross hybrids, and by an increasing margin as the potential of the environment increased. All entries in these tests were licensed for sale in Canada, a procedure that requires testing at a minimum of two locations for one year. Apparently, this screening is capable of identifying single crosses with individual buffering ability since single crosses outyielded three-way and double crosses in 22 out of 25 tests. No differences in yield stability among the three types of hybrids were measured as determined by a regression coefficient and the standard error of this regression (Lynch, Hunter, and Kannenberg, 1973).

From other reports in the literature, it can be concluded that single-cross hybrids generally are more productive than three-way and double-cross hybrids. The difference in commercial seed production potential of single-cross hybrids is a direct result of the improved performance of inbred lines. The inbred lines observed by early maize breeders such as Shull (1908), Jones (1918, 1939), and Richey (1922) had such low vigor and yield that economical seed production from single-cross hybrids was not possible. Improved inbred lines available since the early 1960s have allowed the production of single-cross hybrids on a commercial scale. The shift to improved inbreds has focused attention on selection of improved inbred lines.

SELECTION OF INBRED LINES

The most commonly used sources for selecting inbred lines for hybrid development are single crosses, progeny of backcrosses to existing inbreds, hybrids between adapted inbreds and new introductions, improved open-pollinated populations, and **synthetic cultivars**. The procedure used in developing (Figure 14.1) and testing inbred lines is similar to that used in a self-fertilizing species, with the exception that self-pollination must be performed on each individual plant and protection from foreign pollen must be accomplished.

Many of the breeding methods discussed for self-fertilized crops are applicable to an inbred selection scheme. For example, a pedigree-type system could be used along with mandatory self-fertilization in each generation. As a rule, an open-pollinated cultivar, a synthetic cultivar, or a hybrid may be selfed for five to seven years, during which time selection may be carried out for a number of agronomic traits (Figure 14.2). As in self-fertilized crops, selection in the early generations is primarily for qualitative traits such as plant height, ear height, ear type, kernel characteristics, stalk strength, or disease resistance. Qualitative traits generally are passed from inbred lines to F_1 hybrids and hence perform similarly in hybrids as in inbred lines.

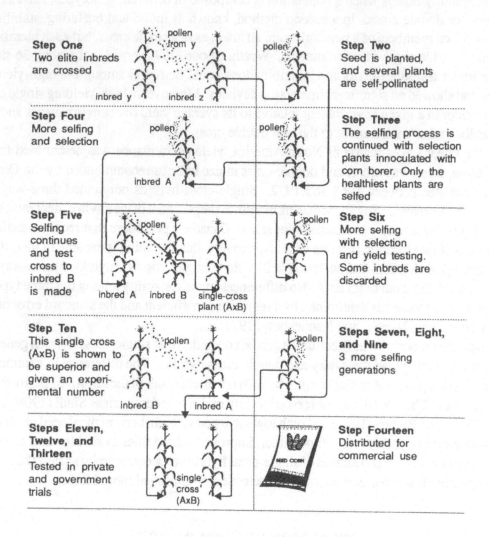

Figure 14.1 Plant breeding procedures for hybrid maize development are shown schematically in this figure. Three basic steps are involved, namely, (1) development of inbred lines by controlled self-pollination, (2) identification of the best inbred lines and test crosses to determine combining ability and productivity, and (3) increase of inbred lines to produce commercial seed.

Considerable maize breeding in North America is conducted by private commercial enterprises, which are highly competitive. Breeders tend to avoid looking back because a competitor may be gaining. Introgression of unadapted maize lines is analogous to looking back. As a result, maize breeders in North America have emphasized selection within adapted genotypes, which has resulted in a small sampling of the total available gene pool, probably less than 5%. Despite this fact, breeders have effectively used this limited gene base for genetic gains.

Concern exists, however, about whether genetic advances can continue without a broadening of the gene base. New gene sources may be required to meet long-term goals. A tremendous range of germplasm is available to maize breeders, but much of it is unadapted to specific locations or to meet the level of performance required in modern-day corn production systems.

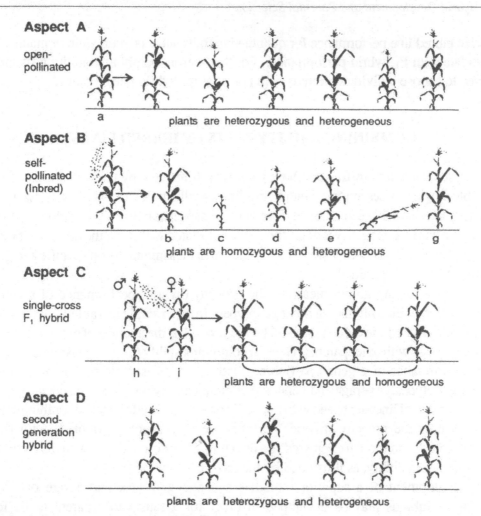

Aspect A

open-pollinated

a

plants are heterozygous and heterogeneous

Aspect B

self-pollinated (Inbred)

b c d e f g

plants are homozygous and heterogeneous

Aspect C

single-cross F_1 hybrid

h i

plants are heterozygous and homogeneous

Aspect D

second-generation hybrid

plants are heterozygous and heterogeneous

Figure 14.2 Four aspects of maize breeding. The heterogeneous aspect of open-pollinated maize cultivars is shown in A. Progeny produced from the open-pollinated plant a vary in stalk height, ear height, leaf morphology, size of ear, yield performance, leaf number, and response to the environment. Open-pollinated cultivars may have wide adaptation, but breeding efforts to improve performance are difficult because of the heterozygous nature of the plants.

To produce a pure line, selfing is forced upon the plant by fertilizing the ear with pollen from the same plant. The resulting progeny, illustrated in B, show reduced vigor as homozygosity is achieved. Some plants, such as b, are better than others. Some, such as c and d, fail to maintain themselves. Others, such as e, f, and g, show undesirable agronomic traits and may be discarded.

F_1 hybrids are produced by combining selected inbreds, h and i, in C. Plant h is designated as the male plant, and plant i, the female plant, is detasseled. Only seed on the female plant i is harvested for commercial use. The progeny of this single cross are uniform, have vigor equal to or greater than plant a, and are highly uniform. Plants are heterozygous but homogeneous. The heterozygous nature of F_1 plants means that segregation will occur, and if used as F_2 seed, plants in the population will be variable and have a lower yield. The heterozygous and heterogeneous nature of an F_2 population shown in D is sufficient encouragement to obtain F_1 seed for each commercial crop.

Mean F_2 grain and fodder yields of single crosses were reported by Kiesselbach (1951) to be 73 and 75% of the F_1, respectively; for double crosses, they were 85 and 87%.

Hybrid corn is defined as the utilization of F_1 hybrid vigor following a cross of two unrelated inbreds.

Maize inbred line performance for quantitative traits such as grain yield or maturity may not correlate with F_1 hybrid performance. For that reason, combining ability tests must be employed to choose individual inbreds with potential for hybrid performance.

COMBINING ABILITY TESTS OF INBRED LINES

The first approach to testing combining ability is illustrated by the diallel cross, a test of all possible single crosses among *n* number of lines, as illustrated in Table 14.1. Each parent is shown in this table to have a certain performance for a specific hybrid. In addition, each parent or inbred line has an average level of performance associated with all of the other combinations of lines with which it has been tested. The result is that measurements on specific and general combining ability are obtained.

Average combining ability measures the progeny (hybrid) performance of a genotype (inbred) with a specified number of other genotypes. In the example in Table 14.1, the average performance of Inbred 1 is 6,480 kg/ha (96 bu./ac.), which is the average of every combination involving Inbred 1 with other parents. General combining ability (GCA) measures the progeny (hybrid) performance of a genotype (inbred) with a random sample or a large number of individual genotypes. Average and general combining ability are similar, especially where a very large number of lines are tested with a specific parent or inbred. Specific combining ability (SCA) measures the progeny (hybrid) performance of a genotype (inbred) with another genotype (inbred) and is often expressed as a deviation of performance predicted by average or general combining ability, as illustrated by the cross 6 x 1 in Table 14.1.

The development of a superior inbred is a challenging task and a rare occurrence. Thousands of inbreds may be developed in order for a satisfactory parent to be found. Kiesselbach (1951) estimated that in the first 50 years of corn research in the United States, one hundred thousand different inbred lines had been developed.

A reliable method of selecting superior inbred parents is therefore critical to success in maize breeding. Although statistics are difficult to obtain in North America, about 70% of corn hybrids are based on no more than half a dozen inbreds. This observation raises the practical question of how best to screen huge numbers of inbred lines efficiently and to identify reliably the few that are to survive as parents. Initially inbred lines were tested as potential parents by crossing all inbreds. It quickly becomes evident that this is an impossible task. The number of hybrid combinations can be calculated from the following formulas:

single crosses	=	$n(n-1)/2$
three-way crosses	=	$n(n-1)(n-2)/2$
double crosses	=	$n(n-1)(n-2)(n-3)/8$

From 50 inbreds, it is possible to make 1,225 different single crosses, 58,800 three-way crosses, and 690,900 double crosses. These figures do not include reciprocal crosses.

A possible solution to the problem was presented by Davis (1929), who proposed crossing each inbred with the same selection cultivar. This is called a top cross. Data to support this proposal was provided by Jenkins and Brunson (1932). They reported on 37 inbred lines that were crossed with an open-pollinated mixture of 25 adapted corn cultivars. Based on these

Table 14.1 Example of a Diallel Cross with Six Inbred Lines of Maize (parental combination shown above the bar, and the yield in kg/ha [bu./ac.] shown below the bar)

Part A

Parent Number	1	2	3	4	5	6
1	1 x 1 / 4,500 (67)					
2	2 x 1 / 7,500 (112)	2 x 2 / 3,000 (45)				
3	3 x 1 / 6,500 (97)	3 x 2 / 6,700 (100)	3 x 3 / 5,000 (74)			
4	4 x 1 / 5,500 (82)	4 x 2 / 5,400 (80)	4 x 3 / 5,900 (88)	4 x 4 / 4,100 (61)		
5	5 x 1 / 4,800 (65)	5 x 2 / 7,200 (107)	5 x 3 / 6,500 (97)	5 x 4 / 6,100 (91)	5 x 5 / 4,000 (60)	
6	6 x 1 / 8,500 (126)	6 x 2 / 7,400 (110)	6 x 3 / 8,200 (122)	6 x 4 / 7,500 (112)	6 x 5 / 6,500 (97)	6 x 6 / 3,200 (48)

Part B

Average performance of each line in kg/ha (bu./ac.):

$$\text{Line 1} \quad \frac{7{,}500 + 6{,}500 + 5{,}500 + 4{,}400 + 8{,}500}{5} \quad = \quad 6{,}480 \ (96)$$

Line 2 = 6,840 (102)

Line 3 = 6,760 (101)

Line 4 = 6,080 (90)

Line 5 = 6,140 (91)

Line 6 = 7,620 (113)

Average yield performance = average combining ability.

Note: The average performance of inbred 1 is 6,480 kg/ha (96 bu./ac.), which is the mean of every combination involving inbred 1 with other parents. In the cross between parents 6 and 1, a yield of 8,500 kg/ha (126 bu./ac.) was achieved. This illustrates that specific combining ability may deviate from the performance predicted or from general or average combining ability. Yield performance of cross 1 x 6 is much higher than that predicted on the basis of general combining ability. The terms *general* and *specific combining* ability were offered by Sprague and Tatum (1942).

results, it was suggested that a commercial cultivar be used to cross with inbred lines for the rapid, preliminary testing of new inbreds. Such crosses can be made economically in an isolated crossing plot in which a selected cultivar is used as the pollen parent. In the following season, these line x cultivar crosses can be evaluated for field performance. The results provide a fairly reliable measure of the relative prepotency or general combining ability of the inbred lines. The high-combining lines can be further tested for specific combining ability.

A top cross gives a measure of general combining ability and is primarily a measure of additive gene action. The rationale for an open-pollinated cultivar to indicate additive gene action or GCA is that the sample of gametes of a widely based synthetic or open-pollinated cultivar would give the same result as crossing an inbred line with a large series of other inbred lines. The open-pollinated cultivar, or in some cases the double-cross hybrid, used as the pollen parent is termed the **tester parent**.

For measuring general combining ability, no general agreement exists as to the best type of tester to use. The choice of the tester parent determines how well the top-cross method measures GCA, and after a good tester is found, the method can be used to discard possibly 50% of inbred lines tested. This form of testing represents a fairly sizable saving of time and labor. One commercial company, however, may choose a particular synthetic cultivar, whereas another company may select a different one.

Prior to the 1970s broad-genetic-base testers were used almost exclusively to improve GCA (Hallauer and Miranda, 1981). It was accepted that a tester with a narrow genetic base, particularly an inbred line, would improve combining ability with the specific tester but would have little value for measuring GCA. A homozygous individual or an inbred line would be a relatively better indicator of SCA because every gamete (homogeneous) would indicate the ability of specific lines to produce high-performing progeny.

The procedure of using inbred lines as testers was modified by three studies (Horner *et al.*, 1973; Russell, Eberhart, and Vega, 1973; Walejko and Russell, 1977). Use of inbred lines as testers resulted in significant improvement not only of combining ability with the specific testers but also of GCA as measured by crosses with unrelated broad-base populations.

The fact that genetic variance among test-cross progenies using inbred testers was about twice as large as when broad-base testers were used (Darrah, Eberhart, and Penny, 1972; Horner *et al.*, 1973), prompted Russell and Eberhart (1975) to suggest the use of inbred lines extracted from the selection population as testers in reciprocal recurrent selection studies instead of individuals from the population itself.

Additional support for the usefulness of inbred lines as testers for the improvement of GCA in maize was provided by Zambezi, Horner, and Martin (1986). One hundred crosses were tested at two locations. Significant estimates of GCA effects were obtained at both locations for grain yield, ear height, husk rating, and percentage erect ears at harvest. SCA effect estimates were significant but much smaller than those for GCA. Correlation coefficients for the 18 inbred parents were similar to those for the two broad-base populations, which suggests that inbreds were in general as effective as broad-base populations for ranking parents for GCA. The conclusion reached was that inbred testers can be used successfully for improving GCA, as well as SCA, in maize.

Two practical reasons were given by Zambezi, Horner, and Martin (1986) for preferring inbred (narrow-base) testers to broad-base testers: (1) sampling errors associated with hetero-geneous testers exist, unless large samples of the tester are used, which requires a great deal of labor; and (2) use of an inbred-line tester may permit quicker utilization of new lines in commercial hybrids, especially if the tester is already in commercial use.

A top cross of a specific inbred-line tester (inbred line is used as a pollen source) can measure both additive and dominant gene action.

ESTIMATING THE PERFORMANCE OF DOUBLE CROSSES IN MAIZE

In the era when double-cross hybrids were the usual outlet for hybrid seed corn, not only was performance as a single cross important but so too was predicting the performance of the double cross. The final stage in picking superior inbred lines or parents for hybrids returned again to making all of the possible single crosses to select those parents that perform best in hybrid combinations. The problems involved in testing all of the possible double crosses are even more severe than in testing all possible single-cross combinations because of the additional number of double crosses that must be tested to determine potential performance. Thirty inbred lines, for example, can produce 435 different single crosses and 82,215 different double crosses.

In an experiment involving 53 of 55 possible single-cross combinations among 11 inbred lines, the inbred-cultivar crosses of these lines, and 42 double crosses among the 990 possible combinations, Jenkins (1934) used four different methods to predict double-cross hybrid performance tested over a 5 year period. The four methods were based on inbred performance from (1) the six possible single crosses among the four inbred parents of the double cross, (2) the four single crosses not used in making double crosses, (3) diallel crosses, and (4) top crosses. Correlations between the actual and predicted yields of 42 double crosses using the four methods produced correlation values of 0.75, 0.76, 0.73, and 0.61, respectively. Based on these correlations, it has become accepted practice to predict the yields of double crosses from the yields of the single crosses not used in the double cross. The decline of vigor in advanced generations is an indication of double-cross hybrid performance.

PRODUCING F₁ HYBRIDS

The responsibility of producing F_1 hybrid maize seed to plant each crop produced throughout the developed, and in parts of the developing, world is a massive undertaking. This task has largely been assumed by maize seed companies and involves a myriad of parent stock strains required to produce F_1 hybrids adapted to specific environmental conditions throughout the maize-producing regions.

Once superior inbreds have been identified, they become the base population or breeder seed stock. This base population represents the genetic constitution of the inbred, is perpetuated from bulked, self-fertilized seed, and is used as seed for inbred maintenance increases. Inbred maintenance seed usually is produced in isolated blocks, which, because they are homozygous, are fertilized by natural random sib mating. All parent stock increases are produced from maintenance seed stocks. Parent stocks are used to plant and produce commercial F_1 hybrid cultivars and may be used directly as inbreds to produce single crosses for direct F_1 consumption, to produce a related-line cross in a modified three-way cross, or by combining single crosses to produce a four-way cross.

Throughout the growing season, careful plant and ear selection is practiced to eliminate individual plants with phenotypes deviating from the accepted phenotype of the inbred. Ear harvest rather than shelled grain is practiced at harvest to permit a final examination of the ear before shelling takes place (Figure 14.3).

F_1 hybrid seed production is a competitive operation, and every effort is made to maximize

A.

B.

Figure 14.3   The maize ears in the lower portion of Photo A, taken in a hybrid seed maize processing plant, are being examined by trained technicians to remove undesirable off-type or diseased ears as indicated by discoloration, lack of uniformity in ear size and shape, kernel size and shape, and differences in maturity. The ears in Photo B are being harvested as intact ears, including the husks, to reduce shelling and to provide an opportunity to remove undesirable ears before further processing. At the seed processing plant, a husking bed removes the majority of the husks prior to the ears, passing over the sorting table in Photo A. Loose seed is discarded, and unhusked ears are passed back over the husking bed.

As much plant selection and roguing as possible should occur prior to pollination to eliminate outcrossing of genetically undesirable plants. If seed maize is harvested as shelled grain, no opportunity is provided for sorting off-type ears. Under such conditions, field roguing is critical to ensure that seed fields are free of volunteer and off-type maize plants.

Harvesting the seed crop is the responsibility of the contract grower and can begin as soon as the kernels have reached physiological maturity, usually about 30 to 38% moisture. If seed corn is harvested as shelled grain, moisture levels should be 20% or less to minimize mechanical damage. Harvest, however, usually is coordinated by fieldworkers to avoid congestion and mixing and to ensure a flow to the processing plant.

seed harvest of cross-fertilized seed produced on the female plant. Yields on female plants may be maximized by the following methods:

- Maximizing plant populations in seed fields. Populations of 50,000 plants/ha (20,000 plants/ac.) may be used for seed parents, levels that parallel commercial maize production.
- Increasing the number of female plants relative to male plants.
- Solid planting of the female seed parent in wide rows, with every fourth or fifth row space interplanted with the pollen parent.
- Eliminating competition from the male pollen parent during grain filling by removal of the male parent after pollination is complete. This facilitates harvesting because the risk of seed contamination is eliminated.

Systems of alternating male and female rows are illustrated in Figure 14.4.

Differences in maturity between the parents may require different planting dates to ensure that the male and female organs nick, that is, simultaneously reach development to complete fertilization. Dates of planting are determined by the number of **heat units** that have elapsed since the first parent was planted. Several systems of planting are illustrated in Figure 14.5.

To enforce cross-pollination between male and female plants of the intended cross, pollen control is essential. Hybrid maize is feasible because of its special kind of monoecism, whereby pollen can be controlled by pulling the immature tassels. As F_1 hybrid production expanded, the laborious task of detasseling was replaced by the conversion of inbred parents to cytoplasmic male sterility, which was isolated at the Texas Agricultural Experiment Station (Rogers and Edwardson, 1952) and is known as the Texas, or T, cytoplasm.

In double-cross production, only one inbred line that would serve as the female parent of the seed parent single cross would need to carry the cytoplasmic male-sterile character. This line could be maintained and increased by planting it in a crossing block, where its female counterpart would be used as the pollinator. Both sterile and fertile strains of a particular line could be maintained in this manner. Such a male-sterile line would then be crossed to a normal line, which does not restore fertility to produce a male-sterile single cross for use as the seed parent. Fertility is restored if one or two of the inbreds carry fertility-restoration genes and are used as the pollinator parent of the double cross. An alternative is to blend one part normal seed as a pollen source with two or three parts male-sterile, single-cross seed to ensure the user of normal seed set. In this way, detasseling is eliminated.

In certain genotypes, cytoplasmic male sterility can be used to delay pollen shed until the tassel has extended above the leaves, followed by cutting with minimal leaf removal. In such a case, cytoplasms that produce only partial male sterility may be used (Duvick, 1965).

To produce single-cross seed, fertility restoration can be incorporated into the male parent,

---

Figure 14.3 continued

Drying temperatures are controlled within a range of 38 to 46° C (100 to 115° F) to preserve seed quality.

Sizing of seed stocks after shelling offers a means of quality control because genetic impurities, especially those resulting from outcrossing, may be concentrated in specific seed sizes. Grow-outs of seed lots during the winter season provide an estimate of genetic purity prior to use.

Photo B courtesy Pfister Hybrid Corn Co., El Paso, Illinois.

C.

Figure 14.4 This sequence of photos shows systems of commercial hybrid maize seed production under field conditions. The parent that produces the seed is designated the female, and the pollen parent the male. The choice of male and female parents is arbitrary or based on morphological features, such as the ability to produce abundant pollen or large ears. Seed bags are clearly marked to avoid confusion (Photo A). Cross-fertilization is achieved by alternate planting of male and female parents. The ratio of male to female rows is a compromise between adequate cross-pollination and maximum seed yield on the female parent.

To produce double-cross seed, a 2:6 pattern is used. Photo B shows an aerial view of a hybrid seed production field with six detasseled female rows alternated with two pollen parent rows. In this case, provided a reliable genetic restorer mechanism exists for the particular male-sterile cytoplasm used. Although this method is attractive for the seed producer, restored hybrids do not always shed adequate pollen (Duvick, 1959), which may be risky for the farm producer. As a consequence, single-cross production is likely to be of a nonrestored genotype, in which case 25 to 50% of fertile hybrid seed produced by detasseling is blended with 50 to 75% of the identical hybrid produced by the cytoplasmic male-sterile method with the result that 25 to 50% of the plants will shed pollen normally in the grower's field.

Among the male-sterile cytoplasms available, the Texas source proved to be the most satisfactory because inbreds generally were completely sterile and fertility restoration was managed easily in this cytoplasm. Cytoplasmic male sterility offers the most satisfactory technique available to produce hybrids, and for two decades was used widely. A severe setback occurred, however, with the discovery in 1970 that plants with Texas cytoplasm are susceptible to race T of southern leaf blight (Tatum, 1971; Ullstrup, 1972). Other sources of male-sterile cytoplasm exist, but their impact on grain yield remains uncertain. Inability to rely on cytoplasm shifted attention to chemical male sterility, previously considered for cereals and other crops. Detasseling is once again an accepted means of pollen control, until alternative forms are found.

To control pollen effectively, all tassels must be removed from the female rows before any pollen is shed and as soon as they are free from the uppermost leaf sheath. Detasseling may be achieved by hand, as illustrated in Figure 14.6. The tassel on an inbred plant may be pulled during a period of six to eighteen hours, but because plants differ in time of tasseling, the period involved in detasseling a field may be spread over a five- to fourteen-day period to remove late tassels or tillers. Detasseling crews must patrol each row four to seven times during the season and go over the field every 24 to 48 hours.

Mechanical detasselers can be used to either cut or pull the tassels, as shown in Figure 14.7. Detasseling must be achieved with a minimum of plant damage so that seed parent plant yields are not reduced. Tassel removal without any leaf loss or damage causes a minimum of physiological damage to the plant and results in an increase in ear size and grain yield. There are several reasons for this, including improved light penetration to leaves when the tassel is removed and the fact that photosynthetic energy not required to produce 25 million pollen grains is directed to grain yield (Hunter *et al.*, 1969; Hunter, Mortimore, and Kannenberg, 1973;

Figure 14.4 continued

the seed parent is never more than three rows from the pollen parent, but at harvest the male rows, representing one-eighth of the field, are lost for seed production purposes.

With single cross-seed, pollen production on inbred plants may be less than on F_1 hybrids of a double cross, and a higher male to female ratio is used. A common planting pattern is one row of pollen parent to four rows of seed parent, so that the seed parent is never more than two rows from the pollen parent. A male to female planting pattern of 1:2:1:4 offers the advantage that two-thirds of the seed parent is adjacent to a pollen parent, compared to one-half in a 1:4 pattern.

Photo C shows an arrangement whereby every female row is adjacent to a male row. This is achieved by a solid planting of the female parent, with the male parent interplanted between every other female row.

Photo B courtesy Pfister Hybrid Corn Co., El Paso, Illinois.

A.

B.

C.

Kiesselbach, 1945). Removal of one or two leaves with the detasseling process can cause a yield reduction sufficient to warrant serious consideration in seed production fields. A common practice in hybrid seed fields is to eliminate the pollen parent after pollination is complete.

In maize breeding plots, pollination of individual plants is achieved by bagging male and female reproductive organs (Figure 14.8) and manipulating pollination as desired. Pollination is performed on an individual plant basis (Figure 14.9) by a crew of technicians who must work rapidly as a team to complete the daily tasks associated with bagging, pollinating, and recording events in a large maize breeding nursery.

EXTENSION OF MAIZE TECHNIQUES TO OTHER CROPS

Hybrids are commercially available in sorghum (*Sorghum bicolor* L. Moench), a normally self-fertilized crop, which can be induced to cross-fertilize. The panicle of sorghum is borne at the top of the plant and contains both male and female organs. To effect cross-fertilization, cytoplasmic male sterility is used to prevent viable pollen from being produced on female parents, or the A-line. Evidence of cytoplasmic male sterility was announced by Stephens, Kuykendall, and George (1952) and more clearly defined by Stephens and Holland (1954). The introduction of hybrid sorghum in the United States in 1956-1957 had an impact on agriculture comparable to that of hybrid corn.

Commercial hybrids are also produced in pearl millet (*Pennisetum americanum* L. Leeke.), a normally cross-fertilized crop that has both male and female reproductive organs contained in each flower. The inflorescence is a prominent structure on the top of each plant. Pearl millet exhibits protogyny that serves to reduce self-fertilization. For commercial hybrid seed production, sources of cytoplasmic male sterility and a source of fertility restoration were reported by Burton (1958).

Male sterility in onion was discovered by Jones and Emsweller (1936) in 1925 and led to the development of hybrid onions.

Figure 14.5 The maize seedlings shown in these photographs illustrate the practice of split-date planting of parents to ensure the synchronization of flowering. Photo A shows four female rows on either side of the photo, with two male rows in the center. Seeding of the male rows occurred after the female rows were seeded. Note that the male row on the right was seeded earlier than the male row on the left, constituting a three-seeding-date planting system. This system of double-delayed male rows assures adequate pollination but does not maximize commercial seed production because rows are a meter apart.

Photos B and C illustrate a planting pattern that is designed to maximize seed production and that involves a 96 cm (38 in.) spacing between all the female rows. The male parent is planted alternately between the female rows, resulting in a 48 cm (19 in.) spacing (note the meter stick). In Photo B, seeding was delayed in the female rows; in Photo C, seeding was delayed in the male rows. In some cases, male and female rows may be seeded on the same date.

As soon as pollination is complete, all male rows are removed, including those in Photo A, to reduce competition with the female rows and to eliminate the risk of mechanical seed mixing at harvest.

Problems encountered with delayed seeding include excessive delays in the late-planted rows because of wet soil conditions.

A.

B.

Figure 14.6   Photos of a crew of workers detasseling maize, that is, removing the tassel from the plant. The rig is a detasseling cart, which carries ten people on elevated platforms and is propelled at a steady pace through the field. The newly emerging tassel can be readily grasped from the elevated position and removed by a firm, smooth, upward thrust. The teenage work crew in this photo is typical of people employed for hand detasseling, an activity that may last for one to five weeks, depending upon the number of fields to be detasseled and the spread of seed parent maturities within the area.

Pollen control is dependent upon complete tassel removal from all female parent plants in a timely manner prior to pollen shedding. To ensure complete removal, it is necessary to go over the same field several times, and a seven-day work week is not uncommon. The time period from tassel emergence to pollen release is limited. If tassels are pulled too soon, one or two leaves may be removed with the tassel or the tassel may break and not be removed completely. Careless detasseling may result in seed pollinated by the female parent, with the result that hybrid vigor and uniformity are lost.

Rigid standards of pollen control for pedigreed status must be met. If on any one inspection more than 1% of the female seed parent plants possess tassels, or if the total for three inspections on different dates exceeds 2% when 5% or more of the female seed parents have receptive silks, the field will not be eligible for pedigreed status. Tassels are considered to be shedding pollen when 2.5 to 5 cm (1 to 2 in.) of tassels on tillers, portions of tassels, or tassels on the main plant have anthers extruded from their glumes.

Figure 14.7   The unit in Photo A, operating in a maize seed production field, is a mechanical detasseler designed to remove the tassels on female seed rows to reduce production seed costs associated with the more expensive hand detasseling. When the tassel is well extruded from the leaf whorl and before pollen shedding occurs, over 70% of the tassels can be removed mechanically with one or two passes over the field. Hand detasseling may be required to remove tassels on short plants, on late-formed tillers, on missed plants, or on plants with delayed development.

Two types of mechanical detasselers commonly are used in hybrid seed production. The unit in Photo B consists of a rotating cutter blade or knife (right side) that operates in conjunction with a rotating disc (left side). As the top of the maize plants pass between the plates, the tassels are severed. The unit can be adjusted for height; however, a problem arises when any remaining uncut portion produces pollen. On the other hand, if the unit is set lower, leaf removal is increased. By delaying mechanical detasseling for as long as possible prior to silk emergence, extrusion of tassels is maximized and leaf damage is minimized.

The unit in Photo C exerts a pulling action on the tassel, resulting from two counter-rotating rollers that grasp the tassel and upper leaves and pull them vertically in a manner approximating a hand detasseling operation. The tassel breaks under pressure at the soft meristematic region just above the upper node.

Of greater importance than the type of mechanical detasseler used are the many variables in the seed field. These include the machine operator's skill, the degree of tassel emergence at the time of cutting, climatic conditions that may speed up or delay pollen development in relation to tassel emergence, differences in parental genotypes and their morphological development, and the number of times the mechanical detasseler is operated over the field.

Photos courtesy Pfister Hybrid Corn Co., El Paso, Illinois.

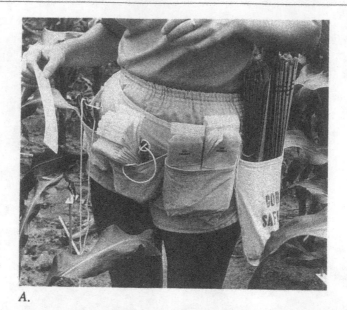

A.

B.   C.

Figure 14.8   The apron worn by the technician in Photo A provides a convenient tool for rapidly bagging and tagging maize plants involved in controlled breeding programs. Ear shoots are covered with a glassine or parchment bag (Photo A), such as those shown in the three central pockets of the apron and in the technician's right hand. This waterproof bag is placed on the ear shoot one or two days before the silks emerge. One day before pollination is performed, the silk shoots may be cut back about 2 or 3 cm (about 1 in.) below the tip of the husk with a sharp knife, the handle of which is attached to a string (seen in the center pocket of the apron). The following day, the silks will have grown about 2 to 4 cm (1 to 1-1/4 in.) and will have formed a brush, permitting pollen to be placed effectively and with considerable ease by simply powdering it on. The freshly pollinated silks are covered with a bag (Photo B) to exclude foreign pollen.

Figure 14.8 continued

Carried on the technician's left are kraft paper bags, which are used to cover the tassel at the time the silks are cut back. The following day, the pollen is collected in the tassel bag and poured over the brushlike silks of appropriate plants. Following pollination, the tassel bag or another waterproof bag is used to cover the ear and is stapled at the base and around the stalk to hold the bag in place securely (Photo C). The stapler and a waterproof marking pencil to identify the cross or to indicate other pertinent information are shown in the right-hand pocket of the technician's apron.

Figure 14.9 Pollen is collected by tapping the bagged tassel or removing the tassel and shaking the pollen into the bag, as shown in Photo A. Pollination of individual ear shoots is facilitated by transferring the pollen to a more manageable glassine bag, removing the top of the glassine bag covering the ear shoot (Photo B), and dusting pollen onto the silks while at the same time holding the glassine bag covering the ear in place so it acts as a funnel (Photo C). Generally, maize plants produce an abundance of pollen. One pollen grain per silk is required to effect fertilization. For each silk on an ordinary ear, approximately 25,000 pollen grains are produced. Best seed set occurs when pollination is completed three to five days after the first silks have emerged, but pollination can be successful for up to eight or ten days.

Stripes of various colors on the bags covering the tassels identify parental materials and proper male and female combinations (Photo D). The bag is secured to the stalk by a paper clip or staple. At least one day should be allowed before bagged tassels are used for pollination to allow for anther exsertion and pollen dehiscence.

For self-fertilization, the pollen is taken from the tassel and placed on the silks of the same plant. For cross-fertilization, pollen from one tassel may be used for one or several ear shoots, or pollen from several tassels may be composited.

Figure 14.9 continues overleaf

Figure 14.9 continued

A.

B.

C.

D.

REFERENCES

Allard, R. W., and A. D. Bradshaw. 1964. Implications of Genotype: Environment Interactions in Applied Plant Breeding. *Crop Science* 4:503-507.

Burton, G. W. 1958. Cytoplasmic Male Sterility in Pearl Millet (*Pennisetum glaucum* [L.] R. Br.). *Agronomy Journal* 50:230.

Darrah, L. L., S. A. Eberhart, and L. H. Penny. 1972. A Maize Breeding Methods Study in Kenya. *Crop Science* 12:605-608.

Davis, R. L. 1929. Report of the Plant Breeder. *Annual Report of Puerto Rico Experimental Station for 1927*. Puerto Rico Experiment Station, Rio Piedras, pp. 14-15.

Douglas, A. G., J. W. Collier, M. F. El-Ebrashy, and J. S. Rogers. 1961. An Evaluation of Three Cycles of Reciprocal Recurrent Selection in a Corn Improvement Program. *Crop Science* 1:157-161.

Duvick, D. N. 1959. The Use of Cytoplasmic Male-Sterility in Hybrid Seed Production. *Economic Botany* 8:167-195.

————. 1965. Cytoplasmic Pollen Sterility in Corn. *Advances in Genetics* 13:1-56.

Eberhart, S. A., 1969. Yield Stability of Single Cross Genotypes. *Proceedings of the Annual Corn Sorghum Research Conference* 24:22-35.

Eberhart, S. A., and W. A. Russell. 1969. Yield and Stability for a 10-line Diallel of a Single-Cross and Double-Cross Maize Hybrids. *Crop Science* 9:357-361.

Eberhart, S. A., W. A. Russell, and L. H. Penny. 1964. Double Cross Hybrid Predication in Maize When Epistasis is Present. *Crop Science* 4:363-366.

Hallauer, A. R., and J. B. Miranda. 1981. *Quantitative Genetics in Maize Breeding*. Iowa State University Press, Ames.

Horner, E. S., W. H. Chapman, H. W. Lundy, and M. C. Lutrick. 1973. Comparsion of Three Methods of Recurrent Selection in Maize. *Crop Science* 13:485-489.

Hunter, R. B., T. B. Daynard, D. J. Hume, J. W. Tanner, J. D. Curtis, and L. W. Kannenberg. 1969. Effects of Tassel Removal on Grain Yield of Corn, *Zea mays* L. *Crop Science* 9:405-406.

Hunter, R. B., C. G. Mortimore, and L. W. Kannenberg. 1973. Inbred Maize Performance Following Tassel and Leaf Removal. *Agronomy Journal* 65:471-472.

Jenkins, M. T. 1934. Methods of Estimating the Performance of Double Crosses in Corn. *Journal of the American Society of Agronomy* 26:199-204.

Jenkins, M. T., and A. M. Brunson. 1932. Methods of Testing Inbred Lines of Maize in Crossbred Combinations. *Journal of the American Society of Agronomy.* 24:523-530.

Jones, D. F. 1918. *The Effects of Inbreeding and Crossbreeding Upon Development*. Connecticut Agricultural Experiment Station Bulletin 207. U.S. Department of Agriculture, Washington, DC. pp. 1-100.

————. 1939. Continued Inbreeding in Maize. *Genetics* 24:462-473.

————. 1958. Heterosis and Homeostasis in Evolution and Applied Genetics. *American Naturalist* 92:321-328.

Jones, H. A., and S. L. Emsweller. 1936. A Male Sterile Onion. *Proceedings of the American Society of Horticultural Science* 34:582-585.

Kiesselbach, T. A. 1945. The Detasseling Hazard of Hybrid Seed Corn Production. *Journal of the American Society of Agronomy* 37:806-811.

————. 1951. A Half-Century of Corn Research. *American Scientist* 39:629-655.

Lynch, P. J., R. B. Hunter, and L. W. Kannenberg. 1973. Relative Performance of Single Cross, Three-Way Cross and Double Cross Corn Hybrids Recommended in Ontario, 1968-72. *Canadian Journal of Plant Science* 53:805-810.

Richey, F. D. 1922. The Experimental Basis for the Present Status of Corn Breeding. *Journal of the American Society of Agronomy* 14:1-17.

Rogers, J. S., and J. R. Edwardson. 1952. The Utilization of Cytoplasmic Male-Sterile Inbreds in the Production of Corn Hybrids. *Agronomy Journal* 44:8-13.

Rojas, B. A., and G. F. Sprague. 1952. A Comparsion of Variance Components in Corn Yield Trials: III. General and Specific Combining Ability and Their Interaction with Location and Years. *Agronomy Journal* 44:462-466.

Russell, W. A., and S. A. Eberhart. 1975. Hybrid Performance of Selected Maize Lines from Reciprocal Recurrent and Testcross Selection Programs. *Crop Science* 15:1-4.

Russell, W. A., S. A. Eberhart, and U. A. Vega. 1973. Recurrent Selection for Specific Combining Ability for Yield in Two Maize Populations. *Crop Science* 13:257-261.

Shull, G. H. 1908. A Pure-Line Method in Corn Breeding. *American Breeders' Association* 5:51-59.

Sprague, G. F., and W. T. Federer. 1951. A comparsion of Variance Components in Corn Yield Trials: II. Error x Year and Variety, Location x Variety and Variety Components. *Agronomy Journal* 43:535-541.

Sprague, G. F., and L. A. Tatum. 1942. General vs. Specific Combining Ability in Single Crosses of Corn. *Journal of the American Society of Agronomy* 34:923-932.

Stephens, J. C., and R. F. Holland. 1954. Cytoplasmic Male-Sterility for Hybrid Sorghum Seed Production. *Agronomy Journal* 46:20-23.

Stephens, J. C., G. H. Kuykendall, and D. W. George. 1952. Experimental Production of Hybrid Sorghum Seed with a Three-Way Cross. *Agronomy Journal* 44:369-373.

Tatum, L. A. 1971. The Southern Corn Leaf Blight Epidemic. *Science* 171:1113-1116.

Ullstrup, A. J. 1972. The Impacts of the Southern Corn Leaf Blight Epidemic of 1970-71. *Annual Review of Phytopathology* 10:37-50.

Walejko, R. N., and W. A. Russell. 1977. Evaluation of Recurrent Selection for Specific Combining Ability in Two Open-Pollinated Maize Cultivars. *Crop Science* 17:647-651.

Weatherspoon, J. H. 1970. Comparative Yields of Single, Three-Way and Double Crosses in Maize. *Crop Science* 10:157-159.

Zambezi, B. T., E. S. Horner, and F. G. Martin. 1986. Inbred Lines and Testers for General Combining Ability in Maize. *Crop Science* 26:908-910.

CHAPTER 15

Synthetic Cultivars in Cross-fertilized Crops

Synthetic cultivars are advanced generations of cross-fertilized seed mixtures of a few selected parents. In maize and forage crops, these selected plants are maintained so that the synthetic cultivar can be reconstituted as desired. A synthetic cultivar is distinct from an open-pollinated cultivar and a pure-line cultivar. An open-pollinated cultivar is developed by mass selection, allowed to interpollinate, and maintained by further mass selection of parents. An open-pollinated cultivar is the same as a landrace cultivar or an unselected introduction. A pure-line cultivar generally originates from a single plant selection and is maintained by self-fertilization. Synthetic maize cultivars are of limited commercial value in the developed world, where F_1 hybrids dominate, but are commercially important in developing countries. Open-pollinated or synthetic maize cultivars are the common source of genetic variation used to improve maize populations for further selection of inbreds to be developed for F_1 hybrid production. Development of open-pollinated or synthetic maize cultivars is an important step in providing variability for hybrid breeding programs. In most forage species, synthetic cultivars are common, and the vast majority of cultivars sold are of the synthetic type.

BREEDING SYSTEM FOR SYNTHETIC CULTIVAR DEVELOPMENT

The phenomenal success enjoyed by maize breeders in the use of inbred lines produced by selfing individual plants naturally led breeders of other cross-fertilized crops to investigate inbred line development. Tysdal (1948) reported on efforts to inbreed selected plants of alfalfa for purposes of combining inbreds to produce F_1 resistance to bacterial wilt disease (Figure 15.1). Some of the lines were inbred (Figure 15.2) up to six generations, which was accompanied by a rapid decline in vigor. The reduction in vigor was so great in some lines that after four or five generations of selfing, difficulty was encountered in propagating the lines for seed.

When crosses were made between inbred lines, some vigorous F_1 hybrids were obtained, but the average yield of all the hybrids was about the same as for open-pollinated cultivars. The main difficulty, however, is that even if a desirable hybrid is obtained from a given combination, there might be no practical means of producing such a hybrid in quantity because of the low yield of seed and poor vegetative growth of the inbred lines. This situation may be resolved by the use of **artificial seeds**. Hybrids between two inbred lines often exhibit shortcomings.

It became obvious that for forages, an approach other than the one used in maize was required. Two fundamental differences exist between maize and forage crops. First, in maize it is necessary to have inbred lines in order to perpetuate the genotypes that produce superior

Figure 15.1   The flower being manipulated in this photo is that of alfalfa, a cross-fertilized species. Alfalfa is known as the "queen of the legumes" because of high yield, feeding value, and nitrogen-fixing ability. Alfalfa belongs to the genus *Medicago,* which includes an array of diverse species; *M. sativa* and *M. falcata* are the two main ones. *M. sativa* flowers usually are some shade of purple, and *M. falcata* flowers are yellow. Crosses can be made between the two species, resulting in purple, white, yellow, or variegated flower color in the progeny. Flower color is influenced by at least five genes, and the term variegated describes the shades of blue and green that may occur from the combined expression of genes for purple and yellow flower color.

Controlled pollination in alfalfa is achieved by hand-crossing emasculated flowers. This may be accomplished by collecting pollen on a toothpick and dabbing it on the stigma, as shown in this photograph.

Figure 15.2   The inflorescence of the alfalfa flower shown in Photo A consists of about ten flowers, each carried on its own pedicel and collectively known as a raceme. Self-fertilization can occur in alfalfa.

In critical genetic studies involving crosses of alfalfa plants, it may be necessary to emasculate each flower of a raceme prior to pollination to avoid selfing completely. One method of emasculating is to use a suction tube (Photo B). Suction is applied with glass tubing that has a 1 mm tip and is attached by a hose to a vacuum pump. Suction is applied to freshly tripped flowers to remove the anthers.

The alfalfa flower has ten stamens. Nine are fused to form a tubelike structure with alternately long and short filaments so that the anthers fit tightly around the stigma in a double ring. The tenth stamen is free.

Emasculation also may be accomplished with the use of alcohol, whereby the entire raceme is immersed for ten seconds in a solution of 57% ethyl alcohol and then washed in water for a few seconds. Even when immersing the raceme in alcohol is rapid, it may cause more injury to the stigma than the vacuum method.

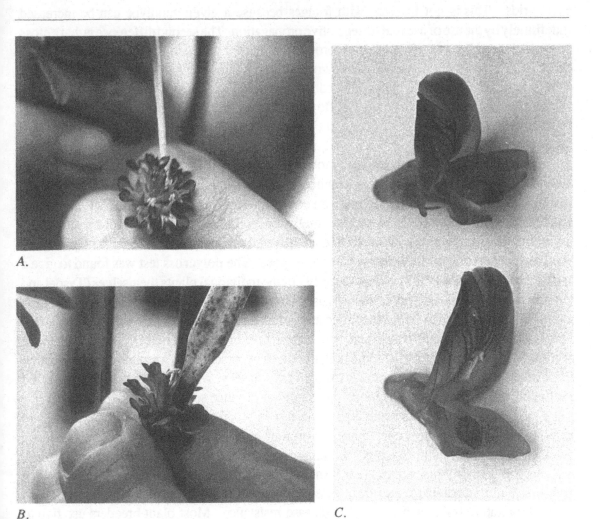

A.

B.   C.

Figure 15.2 continued

The flower of alfalfa exhibits a unique floral morphology and tripping mechanism that encourages cross-fertilization. Photo C shows a single alfalfa flower removed from the raceme, untripped (top) and tripped (bottom). The flower consists of five lobed sepals that compose the calyx, which holds the corolla consisting of five petals—the standard or banner petal, two lateral wing petals, and two fused petals that form the keel. The keel holds the sexual column by means of interlocking, hooklike projections of cutinized tissue on the closely fitting petal edges.

A cuticular membrane that forms a continuous film over the stigma may deter pollination prior to tripping. Generally, for pollination to occur, the flower must be tripped. This usually is accomplished by bees searching for nectar or pollen. Tripping occurs with an explosive force that ruptures the cuticular membrane as it strikes the standard petal or the bee. Barnes and Stephenson (1971) compared four techniques to self-pollinate alfalfa and found that rolling racemes between the fingers was about three times more efficient than tripping flowers with a toothpick, tripping flowers with a toothpick having an emery paper tip, or tripping flowers into a folded cardboard bag containing pollen.

Photo C courtesy Ontario Ministry of Agriculture and Food.

F_1 hybrids. This is not the case with forages because a given genotype can be increased indefinitely by the use of asexual or vegetative propagation. The second difference is that a cross between inbred lines is often more vigorous than a cross between noninbred forage plants. Selected genotypes of open-pollinated plants, however, appear capable of producing progeny as high-yielding as an F_1 hybrid produced from inbreds.

Unlike maize, forages have **perfect flowers**, so production of hybrid seed is difficult on a commercial scale. But the development of male-sterile lines has facilitated hybrid seed production.

If inbreeding, with its associated problems of low self-fertility and reduced vigor, is avoided, (Onokpise *et al.*, 1987), larger numbers of individual plants can be screened for desired features. A major problem in designing a breeding program was testing these individual plants and deciding how they could be used eventually for producing a commercial crop. Certainly, individual plants might be evaluated for qualitative traits, but quantitative traits might not be evaluated accurately from the individual phenotypes. The **polycross** test was found to give a satisfactory indication of the combining ability of the individual plants themselves (Tysdal and Crandall, 1948). The term polycross was used because any given clone might be crossed with a large number of pollen parents and because it helped to differentiate it from the top cross. Some selfing might occur in a polycross. The principle of the polycross is to have each clone pollinated by approximately the same pollen sources, represented by a randomized pollination from other clones in the same plot. A **polycross nursery**, where combining ability can be evaluated, may have 50 to 100 clones and is used to produce seed for progeny testing.

Tysdal (1948) noted that second-generation seed produced from a synthetic continues to exhibit, within a very narrow percentage, the original high yield. A synthetic cultivar produced by cross-fertilizing four to five high-combining, superior clones, identified by progeny tests, will result in a highly superior cultivar. An important feature of this system is that once such high-combining, desirable clones are selected, they can be improved by a backcrossing program to add qualitative traits such as specific disease resistance. Most plant breeders use five to twelve plants to produce a **synthetic** instead of the four to five initially suggested by Tysdal.

OUTLINE OF A PRACTICAL PROGRAM FOR DEVELOPING A SYNTHETIC ALFALFA CULTIVAR

As in any sound plant breeding program, the first step is to define the goal or objective. Consider the goal to be producing a synthetic alfalfa cultivar with phytophthora root rot (*Phytophthora megasperma* Drecks) resistance, for which the only known control is genetic resistance. The first consideration is a genetic source of resistance and agronomically suitable plants. Clones may be obtained from several sources, including old stands of alfalfa, introductions, landrace cultivars, synthetics, hybrid progenies, materials produced in a breeding program, cultivars with genes for specific attributes, or any other source capable of supplying desirable genes.

Step 1. **Clonal lines** are used to establish a source nursery with as many as 5,000 to 10,000 individual plants (Lackamp, 1966). Screening for specific disease resistance may be conducted in the greenhouse, but a field nursery is established to allow for cross-fertilization (Figure 15.3). The task of the plant breeder is to identify superior clones in the source nursery. Selection may be aided by subjecting plants to stress conditions such as severe clipping. This polycross nursery is used to produce seed for progeny testing.

Step 2.   Based entirely on phenotypic appearance, 50 to 150 agronomically desirable plants with satisfactory phytophthora resistance are selected. Once desired clones have been identified, seed is harvested from each of those plants. This seed is produced by cross-fertilization. Seed produced on replicated clones is bulked to produce enough seed for a yield trial.

Step 3.   Each of the selected clones is tested in a performance trial to determine which clones have general combining ability. The top 5 to 12 plants are identified and will be used as parents to produce a synthetic in a polycross nursery. A recombination nursery is used to produce Syn 1, or S_1, seed.

Step 4.   The 5 to 12 Syn 0 parents are vegetatively propagated and planted in isolation to effect cross-fertilization (Figure 15.4) and produce Syn 1 seed for commercial multiplication.

A.   B.

Figure 15.3   Source nurseries of forage legume clones (Photo A) and forage grasses (Photo B), which are repeatedly screened and evaluated for superior individuals on the basis of phenotypic features. A source nursery may be regarded as an assemblage of germplasm resources. The challenge to the plant breeder is to identify genetically superior clones in the population. Examination of the photos will reveal differences among individual clones. The plant breeder attempts to exploit this variation by selecting and combining desirable types to produce improved cultivars.

In order to identify superior genotypes, subjectively and phenotypically selected individuals are cloned and placed in a polycross nursery. A clearly defined objective will help with the identification of the initial selections. Seed from the polycross nursery is used to establish performance trials. From such trials, clones are selected to combine into a synthetic cultivar.

The perennial nature of the plants in this nursery means that the nursery will endure for years. For biennials or short-lived perennials, the nursery may be maintained by clones or by seed. Seed may be obtained in bulk or by bagging and selfing individually selected clones.

Figure 15.4 Isolation to produce synthetic cultivars of cross-fertilized crops can be achieved with insect-proof mesh tents supported by steel frames (Photo A). Cross-fertilization among all clones is achieved effectively by bees placed in the tent. Photos B and C are views of the cloned plants inside the tents.

A synthetic cultivar is developed by random mating of 5 to 12 clones selected on the basis of general combining ability. The seed harvested in bulk from all the clones is "synthesized" into a new cultivar. The parents or clones used to produce a synthetic cultivar are known as the S_0, and the offspring the S_1. S_2 and S_3 seed can be produced from synthetic cultivars with minimal loss of vigor and performance.

Photo D shows individual plants bagged to promote self-fertilization and small cages in which leafcutter bees are released to cross-pollinate six to ten selected clones to produce a synthetic. Photo E shows a box containing a small colony of leafcutter bees to serve as pollinating agents for the alfalfa plants in the foreground.

Cross-fertilized seed from the Syn 1 synthetic cultivar is known as Syn 2 seed. Performance of the synthetic at the Syn 2 or later generation is used to evaluate the performance and acceptability of the synthetic. A commercial seed production scheme is shown in Table 15.1.

The suggested scheme for the development of a synthetic cultivar should be regarded as a guide and may be subject to many large and small modifications. The scheme is applicable to cross-fertilized species in which vegetative propagation is readily possible and in which clones can be maintained for several years.

The concept of a synthetic cultivar was reinforced by the work of Tysdal, Kiesselbach, and Westover (1942), Frandsen (1940), and Wellensiek (1952). Synthetic cultivar development has application to cross-fertilized forage grasses. Schaepman (1952) outlined a scheme for forage grass breeding similar to that suggested for alfalfa and noted the need for a good and practical method to evaluate clones for producing a synthetic cultivar. Schaepman described seven methods of obtaining seed for progeny testing in order to evaluate clones for quantitatively inherited traits.

1. *Self-fertilization.* The quantity of seed produced from self-fertilization varies with the degree of self-fertility of the plant, but a common disadvantage is that comparatively little seed is obtained. Plant features associated with productivity are difficult to evaluate in inbred material.
2. *Pair crossings.* Two clone populations of alfalfa failed because the pollinating bees favored specific clones, leading to seed produced by self- or sib-pollination rather than by crosses between the two clones. Although it would not be a problem with wind-pollinated grasses, pair crossings give only specific combining ability. In forages, general combining ability is much more important.

Table 15.1 Seed Production of a Synthetic Cultivar Composed of 12 Parent Plants

Certification Level	Procedure	Synthetic Generation
Prebreeder	1. Twelve parent plants allowed to intermate at random	Syn 0
	2. Equal amount of seed taken from each parent to plant for S_1 generation. Random mating is assured by mixing seed together and then planting	Syn 1
Breeder	Seed planted and plants intermated	Syn 2
Foundation	Seed increase	Syn 3
Certified	Seed increase	Syn 4

Note: For alfalfa, in which parent plants can be cloned or asexually reproduced, the same mother plants as used in the polycross progeny test can be used to initiate seed production for seed of commerce. However, for annual plant species, asexual reproduction is not feasible and remnant seed of the inbred or partially inbred maternal parent must be used. For this reason, synthetic cultivars of maize are a result of combining completely or at least partially inbred parents. Component strains are maintained so that the synthetic may be reconstituted whenever desired.

3. *Diallel crossing*. This cumbersome and time-consuming procedure has the disadvantage, especially in grasses, that it is difficult to execute if a large number of clones are being tested.

4. *Free pollination among clones in close proximity*. Wit (1952) noted that in perennial ryegrass the distance of effective pollen distribution is very small. Under such circumstances, the progeny performance of each clone largely reflects the good or bad characters of the adjacent clones. Randomization of clones in a polycross nursery is important to assure that each clone is pollinated by approximately the same pollen sources.

5. *Top cross*. A top-cross paternal parent could be planted alternately between the clones under test. This would provide information on the specific combining ability of each plant with regard to that paternal clone. The choice of the top-cross paternal clone is critical, however. Progeny have the disadvantage of being half sisters and hence not suitable for further use.

6. *Top cross in cultivar improvement*. In this case, the paternal clone is replaced by a superior cultivar. This eliminates the problem of half-sister progeny.

7. *Polycross test*. This test can produce the same results as a top cross to a named cultivar. A plant breeder can find the clones with the best characters by selecting progenies showing the highest percentage of the desired characters. Cross-fertilized forage grasses may deteriorate through inbreeding, and so a synthetic cultivar should be developed with more than two clones. A relatively large number of clones is essential for adaptability, an important feature. Performance of a clone in combination with many clones consequently is more important. This results in general combining ability, which is of greater interest than specific combining ability. The polycross test closely approximates what happens in the advanced stages of multiplication of a Syn 1 and subsequent populations.

LAYOUT OF THE POLYCROSS IN THE FIELD

The polycross test is intended to provide information on the general combining ability of an assortment of clones. It is based on the principle that seed produced on each clone is pollinated by as many of the other individuals under test as possible. Seed of each clone is harvested separately. Two methods of laying out the field were suggested by Schaepman (1952). The first method proposes to propagate each selfed clone asexually into 12 plants. The resulting 144 plants are arranged in a square plot so that in every row, every clone is found once. This polycross test requires only a small area (Figure 15.5).

When a large number of clones is involved, a grid system may be used. Schaepman (1952) divided each of 50 clones into 120 individuals, arranged in eight grids or replications to allow all clones to be represented in any one plot. To prevent intercrossing between grids, buffer rows of another species may be planted to separate the grids. Clones within each grid can be systematically arranged so that each clone has two different adjacent clones to ensure the principle of securing a highly mixed pollination.

In a study with smooth bromegrass (*Bromus inermis*), Hittle (1954) emphasized the importance of minimizing pollen effects by producing polycross seed from a relatively large

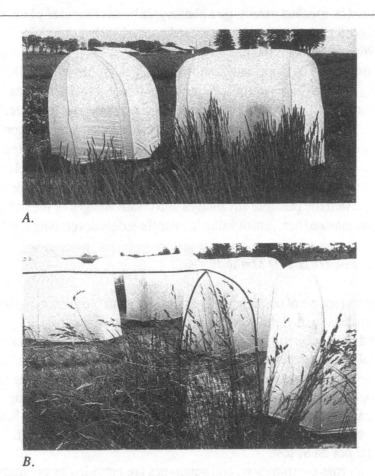

A.

B.

Figure 15.5   Synthetic cultivars of forage grasses can be produced from clonal propagules of selected clones. If similar flowering characteristics exist among the clones, seed of the first synthetic generation can be produced in a replicated polycross nursery under isolation. Isolation may be achieved by distance or by fine mesh tents (Photo A), which allow air currents to penetrate but restrict outside pollen. A steel frame (Photo B) is used to keep the cloth fabric taut. Once pollination is complete, the fabric can be removed so that photosynthesis is adequate to produce quality seeds. The cages shown in these photos are modifications of the system reported by Dunne (1959) and Knight (1966).

number of replications of single randomized plants. Hittle suggested ten or more replications as appropriate. Wassom and Kalton (1958) considered replicate number in a polycross and suggested that estimates of the cost per plot in the nursery would be helpful in estimating the optimum number of replications. These workers observed that eight replications in the nursery gave 75 to 87% of the information on panicle number and 97 to 100% on forage yield. Tripling the number of replications to 24 would also triple the cost for relatively little gain in information.

Unless attention is paid to designing a polycross nursery that ensures random interpollination, the term cross-fertilized progenies may be more appropriate. Perhaps because of inadequately designed polycross nurseries, questions have been raised as to the merits of the polycross test.

An alternative would be to select parents on a phenotypic basis without any kind of progeny test or to base selection on trials produced from inbred seed of each clone.

Lackamp (1966) questioned the merit of establishing 5,000 to 10,000 single plants in a nursery and selecting 60 to 100 clones based entirely on phenotype to perform the polycross test, thereby discarding 99% of the material. Using more than 60 to 100 entries in a polycross increases the difficulty of obtaining a theoretically good polycross scheme. Instead of a polycross, Lackamp (1966) suggested that S_1 (selfed) plants may reflect the combining ability of a grass clone for entry into the polycross. Suggestions of this nature prompted Carnahan and Miller (1968) to review the polycross in alfalfa. They cautioned that many polycross nurseries may possess too few replications to assure random pollination or may lack adequate isolation. These workers reviewed the possible advantages and disadvantages of the polycross method, and their points, summarized here, are of value for plant breeders developing synthetic cultivars.

Advantages of the Polycross Test Method

1. During the normal course of obtaining polycross seed, valuable phenotypic information can be obtained, such as seed yield, reaction to pests, and dormancy.
2. Sufficient seed can be obtained from a polycross nursery to derive performance data based on commercial seeding rates in fully replicated tests and at several locations. With seed produced by selfing clones, inadequate seed is often produced for testing programs.
3. Polycross progeny performance partially reflects genetic diversity among the individual clones to the extent that there is variation for genetic diversity and that it has an important impact on cross-fertilized progenies. This nonadditive genetic effect is obtained from neither a clonal nor an S_1 test.
4. A wide array of cross-combinations and genotypes are provided by polycrosses of a large group of clones, which can serve as additional source material during the process of normal evaluation.
5. A more efficient evaluation of general combining ability is obtained from a polycross than from a diallel crossing system. This is particularly true for a large group of clones to justify confidence in having important genetic differences among the set of clones.
6. Synthetic cultivars produced from a polycross nursery are less likely to suffer genetic shifts during the normal process of seed increase than would cultivars that are based on heterogeneous clones and are unselected for seed production.
7. Synthetic cultivars based on good polycross data may possess greater diversity than synthetics based on clonal or S_1 data from the same source of germplasm. This is the result of subjective judgment rather than objective selection. Yield in open-pollinated maize was reduced when the investigator selected some fixed ideal.

Disadvantages of the Polycross Test Method

1. Male parentage may be confounded among clones because of inadequate replication and isolation to ensure completely random pollination among entries. In crops cross-fertilized by insects, clonal preference by pollinators may exist. Clones may vary also in degree of self-fertility, and in rate of pollen tube growth and subsequent chance for fertilization.

2. The additively inherited characters of the parent are less sensitively measured in a polycross than in an evaluation of either the clone or its S_1 progeny.
3. Effort devoted to any kind of progeny test for certain highly heritable characters may not be justified.
4. The adoption of any progeny test prior to simple phenotypic selection for relatively unadapted populations or those possessing a low frequency of desirable genes for important agronomic features would be a serious waste of effort.

Top-cross testing in forages is also used. Schaepman (1952) indicated that either a single clone or a commercial cultivar could be employed as the common parent. A single clone would be used to evaluate specific combining ability and a commercial cultivar, general combining ability. When polycross, single-cross, and top-cross progenies of eight alfalfa clones were compared for efficiency in evaluating combining ability for forage yield, disease, and insect and cold resistance, all clones were ranked in a similar manner (Tysdal and Crandall, 1948). Wassom and Kalton (1958) concluded that top-cross progeny tests should be satisfactory for evaluating combining ability of orchard grass (*Dactylis glomerata* L.) clones.

PERFORMANCE OF SYNTHETIC CULTIVARS

Synthetic cultivars in the third or fourth generation from the original parents or inbred lines are not as high in performance as the Syn 1 generation. Like F_1 hybrid performance, the highest yield is found in the Syn 1 generation. The F_2 or Syn 2 generation would be expected to lose a portion of its heterozygosity and would not have the same level of performance.

Wright (1921, 1922) developed a formula to predict the F_2 yield of a group of inbred parents. This relationship is given by

$$F_2 \;=\; \frac{F_1 - (F_1 - P)}{n}$$

where F_2 is the expected performance of the F_2, F_1 equals the mean F_1 hybrid performance from combinations of inbred lines, P is the average performance of inbred lines, and n is the number of inbred lines.

A number of experiments in maize have shown that this relationship is remarkably close to theoretical predictions. The formula predicts that the F_2 yield can be increased by increasing the average F_1 yield, by increasing the yield of the parent lines, or by increasing the number of lines or parents that make up the synthetic cultivar. Kinman and Sprague (1945) studied the effect of these different factors on yield of maize synthetics and observed that the maximum yield was obtained when four to six parents were included in the synthetic. One of the reasons for this observation is illustrated in Figure 15.6. The results indicate that the assumption that as the number of parental lines increases the F_1 performance tends to decrease is more realistic than the assumption that F_1 yields are constant regardless of the number of parental lines involved.

Parental lines with lower combining ability will have lower F_1 yields than parents with high combining ability. In the Kinman and Sprague (1945) study, for example, parents with the highest yields in combining ability tests were put together first. Increasing the parent number

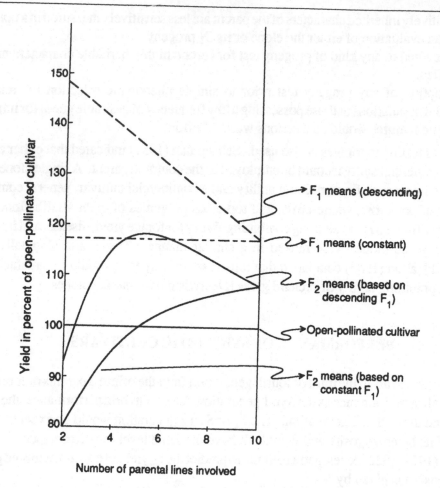

Figure 15.6   Curves showing the performance of synthetic cultivars based on constant and declining F_1 maize yields. Two contrasting assumptions were used to determine the F_2 performance. In the first case, the F_1 means were assumed to be constant regardless of the number of parental lines involved. In this situation, as the number of parent lines is increased, the yield of the F_2 generation also increases. This would be precisely as predicted by the formula of Sewall Wright (1921, 1922). The second assumption was that as the number of parents are increased, the F_1 yield will decrease. If this assumption is used to calculate F_2 yield, an optimum performance will be reached somewhere between four and six parents, after which F_2 yield will decline.

Source: Adapted from Kinman and Sprague, 1945, p. 347.

from four to six resulted in a decline in the F_1 yield because the combining ability of the parents was progressively less and less. The advantages related to increasing the number of parental lines resulted in an increase in the F_2 yield as long as the decline in the parental combining ability (lowering of the F_1 yield) did not override the beneficial effect of number of parents. It is a difficult task to find a large number of parents with excellent general combining ability to increase the number of parents in a synthetic. The effect of parental yield or synthetic performance is shown in Figure 15.7. As the yield of the inbred parents increases, the yield of the synthetic increases.

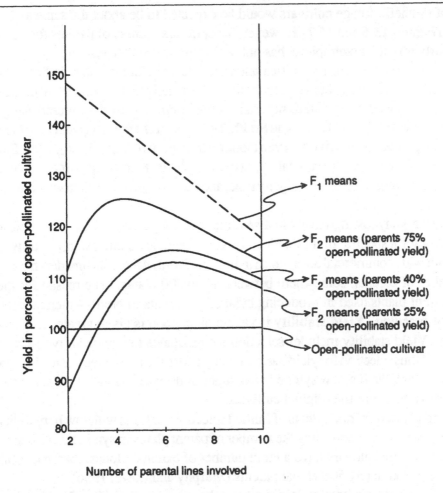

Figure 15.7 Curves showing the effect of inbred mean yield on the yield of synthetic cultivars. The assumption was made that F_1 yields decline as the number of parents is increased. This assumption is based on the performance of inbred lines used in the study. The synthetic yield increases as the yield of the inbred parent increases. Therefore, increasing inbred yield, and for noninbred parents increasing the parental yield, will likely result in an increase in the yield of the synthetic.
Source: Adapted from Kinman and Sprague, 1945, p. 349.

The formula of Wright (1921, 1922) is a useful tool for discussing the relative performance of a synthetic cultivar relative to the performance of the parents and the number of parents used to produce the synthetic. This relationship, however, only applies to inbred lines and diploid species. In practical terms, there is no simple relationship that is useful in predicting the yield of synthetic cultivars in cross-fertilized diploid and polyploid forage species.

Observations of forage breeders and geneticists indicate that the maximum yield decline in synthetic cultivars of cross-fertilized diploids and polyploids occurs between the Syn 1 and Syn 2 generations. This is roughly analogous to the situation described for maize synthetics based on inbred lines. As a general rule, a synthetic forage cultivar will yield approximately 10 to 12% higher in the Syn 1 generation than in the Syn 2 and Syn 3 generations. The optimum number of parents of synthetic forage cultivars would be expected to be about the same as in maize, suggested in Figures 15.6 and 15.7. However, the optimum number of parents for a synthetic, and particularly for an autotetraploid, has not been determined conclusively.

Theoretical studies in North America and particularly in Europe have indicated that the optimum number of parents is between four and six. The red clover synthetic Arlington is the result of intercrossing six lines, and Kenstar and Norlac were produced by intercrossing 10 and 11 initial clones, respectively (Taylor and Smith, 1979). In orchardgrass (*Dactylis glomerata*), synthetics were produced with five to seven clone combinations (Smith, 1956). Most European cultivars have relatively fewer parental genotypes or clones than North American cultivars. Europeans conduct more extensive progeny tests, and this may partly explain why fewer clones are used.

Extensive progeny testing results in a small number of parental plants that represent the desirable genes for a specific trait or series of traits. There are some examples of two-clone synthetics with wide commercial acceptance in Europe. North American breeders have favored a larger number of parents, somewhere between 12 to 200 plants. The reason for the large number of parent plants used in producing synthetic cultivars in North America rests in the philosophy of breeding. **Yield stability** is valued highly and is given greater emphasis than **yield ability**. Yield stability includes selection for resistance to a wide array of disease and insect pests, and thus selection for yield has been relegated to the final stages. Breeding progress is considered acceptable if a new synthetic has superior disease and insect resistance with the same yield performance as the original cultivars.

The potentially detrimental effects of limited selection for traits with low heritability, such as yield, are minimized by increasing the number of parent plants in synthetic cultivars. Some very successful alfalfa cultivars have a large number of parental clones. Saranac alfalfa, for example, has approximately 500 clonal parents (Murphy and Lowe, 1966).

The decline in yield of autotetraploid synthetics between Syn 1 and Syn 2 is widespread and is likely the result of loss of vigor associated with the reduction of triallelic and tetrallelic loci. Maximizing performance of synthetics in alfalfa should be, in part, a result of attempting to maximize the triallelic and tetrallelic genotypic structures in those synthetics.

The effect of tetrallelic loci on alfalfa performance has been demonstrated by Dunbier and Bingham (1975), who found that the higher the number of tetrallelic loci, the better the performance for forage yield, seed yield, height, longevity, and other agronomic traits. The effect of randomly mating four unrelated clones compared with mating four parent plants in a double-cross hybrid procedure is shown in Table 15.2. The results clearly demonstrate that if vigor of natural autotetraploids is decreased by inbreeding, then performance is improved by returning to a more heterozygous state. An increase in vigor in a single cross is termed heterosis, but these authors did not refer to the increase in the double cross as heterosis. Autotetraploids have the capacity for intra-locus **interaction**, and breeding programs might be designed to exploit these, as well as purely additive, forms of genetic variation.

Table 15.2 Proportion of Allelic Interactions When Four Unrelated Parent Clones Are Randomly Mated Compared to a Double Cross in an Autotetraploid

Allelic Class	Random Mated (percent)	Controlled Mating[a] (percent)
Monoallelic	0.19	0
Diallelic	9.6	1.2
Triallelic	49.2	19.8
Tetrallelic	41.0	79.0

[a] A x B, C x D and AB x CD

Note: If four clones are randomly mated for two generations, the proportion of tetrallelic loci is 41%. However, when four clones are first mated as a single cross and then the two single crosses are intermated, the proportion of tetrallelic loci is 79.0%. If a practical method of producing double-cross hybrid alfalfa could be found, agronomic performance likely could be increased.

Source: After Dunbier and Bingham, 1975, p. 529.

TYPE OF GENE ACTION

Additive gene action appears to be the most important aspect for optimum performance of synthetic cultivars and results in higher-performing parents capable of transmitting this superior performance to their offspring. Kinman and Sprague's results in Figures 15.6 and 15.7 illustrate the importance of parental performance for the performance of the synthetic. Dominance genotypic variance can be important, depending on the number of parents and the manner in which parents are combined for seed of commerce of a synthetic cultivar. Dominance might be relatively more important in a forage species in which only 20% or so of the seeds that are sown contribute to stand establishments, if genotypes showing dominance survive preferentially. In this situation, some specific parental combinations might result in a proportion of individuals in the population that have maximum performance.

It has been proposed that the best synthetic cultivar is one that has the best additive gene composition plus the best intralocus or allelic interactions. This is true for autotetraploids. There are two approaches to maximizing intralocus interactions. First, selecting parents for synthetic cultivars based on the selection of desirable alleles that can transmit their performance to their progeny. Second, increasing cultivar performance by maximizing good genotypic structure (i.e., tetrallelic loci). In an autotetraploid such as alfalfa or birdsfoot trefoil, selection of good alleles is difficult because the phenotypic performance of a plant will depend both on the alleles that the plant has and on the allelic structure (tetrallelic, triallelic, diallelic, etc.). For example, an individual that has a certain array of alleles would perform relatively better if those alleles were in a heterozygous genotypic structure rather than a less heterozygous genotypic structure.

Two approaches have been proposed to selecting effectively for better alleles among parent plants. The first approach is to select among inbred lines, and the second is to select among lines where heterozygosity is maximized. Both of these procedures appear to be feasible. For a diploid annual species, such as a maize S_1 or self progeny, testing has proved to be a rapid and efficient method to select and increase performance.

S_1 selection has been proposed for alfalfa and other autotetraploid species. The rapid loss of vigor and inbreeding depression are theorized to assist in identifying superior alleles among plants being screened. Examples can be found in which good progress has been made in improving the yield of a synthetic using a line selection method. Others have proposed selecting among heterozygous individuals that have maximum heterozygosity.

Superior performance among heterozygotes must occur because of superior alleles in addition to good allelic interactions. The use of inbred-line selection methods, such as the S_1 progeny test in alfalfa and trefoil, is difficult because of the decline in vigor associated with inbreeding depression and the difficulty of obtaining sufficient quantities of S_1 seed for testing.

ADVANTAGES AND DISADVANTAGES OF SYNTHETIC CULTIVARS

Synthetic cultivars are a compromise between maintaining a large proportion of the advantages of heterozygosity and heterosis and at the same time allowing economical seed production in many crop species. They preserve the performance due to additive gene action and make use of a sizable proportion of the heterozygosity. Synthetic cultivars of forages are relatively easy to use in producing commercial quantities of seed.

The primary disadvantage of synthetic cultivars is that potential performance based on hybrid vigor or heterosis is disregarded. Almost every forage species and cultivar would show much better performance if Syn 1 seed or its equivalent could be sold as the seed of commerce rather than the Syn 3 or Syn 4 generation of seed. In addition, the synthetic cultivar tends to lessen the emphasis on choosing parents that show heterosis.

Improvement of forage crops forms a very complex system, both from the standpoint of breeding systems and from a farm producer standpoint. The plant breeder must be concerned with longevity, hardiness, regrowth after cutting, quality, seed production, pest resistance, nitrogen-fixing ability, protein content, and other agronomic traits.

Forages are characterized by a large number of species, each with considerable diversity. Exchange of genetic stocks among breeders is necessarily restricted.

Forage breeding methods are in principle similar to those of maize, yet the practical value of inbreeding remains in question. In comparison with advances in maize breeding, progress in forage breeding has lagged. Few advances in techniques have been made since Tysdal (1948) advocated combining selected clones by means of a polycross to form a synthetic; yet technically appreciable progress has been made in yield performance.

REFERENCES

Barnes, D. K., and M. G. Stephenson. 1971. Relative Efficiencies of Four Self-pollination Techniques in Alfalfa. *Crop Science* 11:131-132.

Carnahan, H. L., and J. W. Miller. 1968. Effectiveness of Polycross Progeny Data for Determining Merit. *Report of the Twenty First Alfalfa Improvement Conference*, Reno, Nevada. United States Department of Agriculture, Washington, D.C., pp. 23-29.

Dunbier, M. W., and E. T. Bingham. 1975. Maximum Heterozygosity in Alfalfa: Results Using Haploid-Derived Autotetraploids. *Crop Science* 15:527-531.

Dunne, J. P. 1959. A System for the Control of Pollination in Insect and Wind Pollinated Plants. *Euphytica* 8:76-80.

Frandsen, H. N. 1940. Some Breeding Experiments with Timothy. *Imperial Agriculture Bureau Joint Publication* 3:80-92.

Hittle, C. N. 1954. A Study of the Polycross Progeny Testing Technique as Used in the Breeding of Smooth Bromegrass. *Agronomy Journal* 46:521-523.

Kinman, M. L., and G. F. Sprague. 1945. Relation Between Number of Parental Lines and Theoretical Performance of Synthetic Varieties of Corn. *Journal of the American Society of Agronomy* 37:341-351.

Knight, R. 1966. A Technique for Controlled Pollination in the Production of Grass Seed. *Euphytica* 15:374-376.

Lackamp, J. W. 1966. Some Remarks on the Polycross Test Method in Grasses. *Euphytica* 15:291-296.

Murphy, R. P., and C. C. Lowe. 1966. Registration of "Saranac" Alfalfa. *Crop Science* 6:611.

Onokpise, O. U., S. R. Bowley, D. T. Tomes, and B. E. Twamley. 1987. Evaluation of Self and Polycross Progeny Testing in Birdsfoot Trefoil (*Lotus corniculatus* L.) for Forage and Seed Yield. *Plant Breeding* 98:141-148.

Schaepman, H. 1952. Application of the Polycross Test to Grass Breeding. *Euphytica* 1:105-111.

Smith, D. C. 1956. Progress in Grass Breeding. *Advances in Agronomy* 8:127-162.

Taylor, N. L., and R. R. Smith. 1979. Red Clover Breeding and Genetics. *Advances in Agronomy* 31:125-154.

Tysdal, H. M. 1948. History and Development of the Polycross Technique in Alfalfa Breeding. *Report of the Eleventh Alfalfa Improvement Conference*, Lincoln, Nebraska. United States Department of Agriculture, Washington, D.C., pp. 36-39.

Tysdal, H. M., and B. H. Crandall. 1948. The Polycross Progeny Performance as an Index of the Combining Ability of Alfalfa Clones. *Journal of the American Society of Agronomy* 40:293-306.

Tysdal, H. M., T. A. Kiesselbach, and H. L. Westover. 1942. *Alfalfa Breeding*. College of Agriculture, University of Nebraska. Agricultural Experiment Station Research Bulletin, United States Department of Agriculture, Washington, D.C., 124:1-46.

Wassom, C. E., and R. R. Kalton. 1958. Evaluation of Combining Ability in *Dactylis glomerata* L. IV: Randomness of Pollination in Topcross and Polycross Nurseries. *Agronomy Journal* 50:640-643.

Wellensiek, S. J. 1952. The Theoretical Basis of the Polycross Test. *Euphytica* 1:15-19.

Wit, F. 1952. The Pollination of Perennial Ryegrass (*Lolium perenne* L.) in Clonal Plantations and Polycross Fields. *Euphytica* 1:95-104.

Wright, S. 1921. Systems of Mating. V: General Considerations. *Genetics* 6:167-178.

———. 1922. The Effects of Inbreeding and Cross Breeding on Guinea Pigs. United States Department of Agriculture, Washington, D.C., Bulletin 1121.

Recurrent Selection in Cross-fertilized Crops

Success in a crop improvement program depends on the available genetic variability and its utilization. In maize, for example, selection methods for identifying desirable inbreds for superior hybrid performance include some form of self-fertilization and selection from a cross-fertilized or heterozygous, heterogeneous population. Although desirable inbreds have been isolated, the frequency of superior genotypes is rare. Kiesselbach (1951) noted that in 50 years of maize breeding, only 60 inbreds out of 100,000 developed proved superior in hybrid combinations.

The frequency of desirable inbreds is low for at least three reasons, including inbreeding associated with selfing, the small population sizes in the selfing program, and linkage. A selfing series using a small number of plants in each generation reduces the probability of recombination of potentially desirable genotypes as heterozygosity is reduced, that is, effective recombination is reduced. Segregation also will be rapid during selfing generations and, in the case of small populations, could result in a lower probability of "fixing" desirable genes.

The proportion of outstanding inbred lines expected from a population depends on the heterogeneity in the original population, the number of heterozygous genotypes in the original population, and the effectiveness of selection in increasing gene frequency of desirable alleles. The evidence from second-cycle inbreds, that is, those developed by a selfing series from advanced generations of elite double-cross hybrids, indicates that there is no advantage in inbred-line yields obtained over those that give rise to the original double-cross hybrid. However, further selection from single crosses for several cycles of selection has been effective in increasing yielding ability.

Selection experiments using a selfing series with small population size, nevertheless, were generally ineffective in increasing the frequency of desirable genes. Two procedures are available to offset the disadvantage of these early selection programs. One procedure is to use much larger population sizes so that the probability of selecting superior inbred lines from a given source population is increased. A second procedure uses cumulative selection in which a gradual increase in desirable gene frequency is obtained through a cyclical breeding procedure.

The second method, that of cumulative selection, has been adopted by plant breeders in both self- and cross-fertilized crops. The procedure is called recurrent selection. Recurrent selection refers to reselection generation after generation, with intercrossing or interbreeding of the selected plants to provide for genetic recombination. The intercrossing of the selected plants allows for the occurrence of gene **recombination**, which often is important in obtaining free variability for rare genes.

Recurrent selection methods were developed to improve quantitatively inherited traits.

Quantitative traits involve a large number of genetic factors, each with a small effect, and these cannot be grouped into discrete classes. Confounding selection is the masking effect of environment. The aim of recurrent selection is to systematically increase desirable gene frequencies. Some form of recurrent selection is involved in plant improvement systems, but the methods of selection are not used systematically.

The objectives of recurrent selection are illustrated in Figure 16.1, which assumes finite population size. In practice, populations must, by necessity, be finite. One of the primary concerns in designing a recurrent selection program is the choice of an effective population size. The problem is to establish a balance between maximizing the short-term genetic progress within the constraints of resources available and keeping the effective population size large enough to avoid undue loss of genetic variance by random **drift**.

The effect of finiteness of population size on short-term recurrent selection success was considered by Rawlings (1980). The greatest impact is on the variance of the genetic change. As population size is restricted, the intensity of selection should be increased, that is, a smaller proportion of the entries should be saved.

A finite population places undue restriction on the genetic base for the next cycle. The effect for long-term selection is that many loci become fixed for the unfavorable alleles by random drift. The genetic variance within replicates is reduced by drift. The classical reference relating

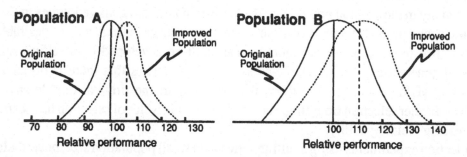

Figure 16.1   The objectives of recurrent selection are illustrated in this series of distributions of performance of single-cross maize hybrids. Success in plant breeding depends on the amount of genetic variability available in the original population and the initial gene frequencies in the population. Although recurrent selection may result in improvement of population A, genetic variability is lacking to meet intermediate and long-term goals.

The distribution curve B has the same mean value as curve A but a broader range of genetic variability. In both curves, the expected variabilities among single crosses from the original and improved populations are the same. The improved populations had the higher mean values and produced the best hybrids. This idealized situation assumes finite population size and ability to identify superior individuals.

Recurrent selection provides a means of cyclical selection by which the frequencies of favorable genes are increased systematically in a population so that the chance of developing superior genotypes is enhanced. The method is designed to (1) improve the mean of the populations by increasing the mean of the populations under selection and (2) maintain genetic variability to permit continued improvement and opportunity for selection of superior genotypes in any cycle.

Source: After Hallauer, 1981.

to long-term selection response is Robertson (1960), who considered the limits to selection, expected advances, and half-life of the recurrent selection process. Mathematical considerations by Rawlings (1980) suggest that the effective population size needed to retain the long-term genetic potential in a population is within the reach of any reasonable recurrent selection program and is about 30 selected individuals for many genetic systems. This figure may be increased slightly to allow for linkage depression. At a 10% selection intensity, a test population of 300 may be required.

RECURRENT SELECTION METHODS IN ANNUAL CROPS

Recurrent selection methods proposed for maize or other cross-fertilized annuals are shown in Figure 16.2 and have as their ultimate objective the production of inbred lines that show a superior performance in hybrid combinations. The recurrent selection procedures considered have three basic characteristics: (1) they are cyclical in nature, (2) they include some form of evaluation of selected plants, and (3) selected individuals are intercrossed in all combinations to facilitate recombination.

The effectiveness of a recurrent selection program is determined by: (1) the improvement of population performance after a particular cycle or cycles of selection, (2) the improved performance of inbreds selected from the population of hybrid combination or, in the case of synthetic cultivars, from parents of a population improvement program, and (3) the amount of genetic variability remaining in the population.

The most obvious indicator of the progress of selection is the change in the mean of the population. If the mean of the population increases in reference to the mean of the original population or a standard check cultivar, then the selection program can be judged as effective. Population improvement *per se*, as measured by an improvement in the mean of the population or by an improvement in the mean yield of inbred lines selected from a population, is only one gauge of effectiveness. In a hybrid breeding system, the most critical determinant of the improvement of a population is the ability to select parents from a population that in combination give superior hybrid performance.

In the case of an autotetraploid or diploid where pollen control is not feasible, the effectiveness of the selection in the population is determined by the performance of parents or synthetic cultivars where parents are derived from a population improvement program. The significance of genetic variability remaining in the population is in the expectations for continued successful selection. A population that has been improved but that has limited genetic variability is of limited potential, such as population A in Figure 16.1. Modest progress from cycle to cycle in a recurrent selection program is acceptable if genetic variability is maintained.

The two major types of recurrent selection in maize are phenotypic recurrent selection and recurrent selection based on some form of progeny testing. The major categories of recurrent selection using progeny testing are broad-base tester, narrow-base tester, self progeny, and reciprocal tester. In each of the progeny test methods, phenotypic selection for desirable qualitative traits is made in year one. Self-fertilization is performed on each individual phenotype that has the desired traits. In addition, pollen from the selected plants is transferred

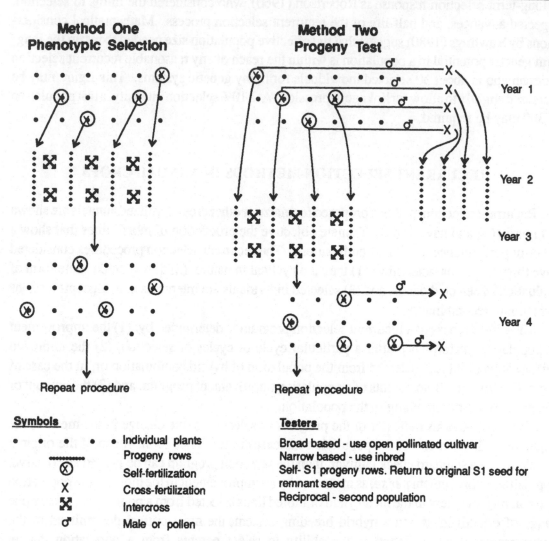

Figure 16.2 Recurrent selection methods in maize where pollen control is possible. Phenotypic recurrent selection requires two years to complete a cycle. In the first year, selections are made on the basis of phenotypically desirable characteristics and selected plants are self-fertilized. In year two, the self-fertilized progeny of each of the selections are grown and the progeny of the selected plants are intercrossed in diallel. Intercrossing in all possible combinations allows the recombination among the progeny of selected plants to form a population for the next cycle of selection. Recurrent selection procedures using progeny testing require three years for completion of a cycle.

to a tester parent. Selected plants are the male parent and the tester, the female parent in this combination.

The choice of the tester parent distinguishes the different recurrent selection methods. In year two, the progeny of the selected plants crossed onto the tester parent are grown. During the second year, the most desirable parents are chosen on the basis of their top-cross performance in the progeny row in combination with the tester parent. After the superior parents are chosen on the basis of their progeny performance, the progeny rows are discarded. In year three, remnant selfed seed from the desirable plants or parents chosen on the basis of their progeny performance is planted in progeny rows. Progeny from the selected parents are diallel crossed to give rise to the population for the next cycle of selection. The time it takes to complete a cycle can be reduced if a winter nursery is used.

TYPES OF TESTERS AND RESULTS OF SELECTION

The results from phenotypic recurrent selection indicate that the progress of this breeding procedure is dependent on the heritability of the trait under selection. For highly heritable traits, such as some forms of disease resistance and maturity, phenotypic recurrent selection has been shown to be effective. For quantitatively inherited traits with low heritability, the effectiveness of this method is low. In addition, the genetic variability of a population may be reduced by using the method. Yield and most phenotypic characteristics are not correlated, and selection for a trait with low heritability generally results in mediocrity and a loss of genetic variability associated with random drift.

Recurrent selection may be conducted using a tester with a wide genetic base, such as a double-cross hybrid, an open-pollinated cultivar, or a synthetic cultivar. Such broad-base tester parents act as a test for general combining ability in the population under improvement. This system tests for additive genetic variance. Population improvement *per se* with a broad-base tester has been fair to good. Some reduction in genetic variability has occurred using this method, although the reduction is not as severe as with phenotypic selection. Depending on the choice of the tester parent, broad-base testers for recurrent selection have increased general combining ability.

Recurrent selection using a narrow-base tester parent means that the test parent has a narrow genetic base. A typical test-cross parent in this situation is an inbred that is homozygous at most, if not all, genetic loci. Such a tester parent could have specific combining ability, although evidence was presented in Chapter 11 that it also is useful to have general combining ability.

The choice of the tester parent is a critical determinant of the success of recurrent selection. A good tester parent may be an inbred line that has poor performance itself. An individual with an abundance of recessive genes, often associated with poor performance, will be able to indicate parents with a high proportion of dominant and favorable alleles, but this view is not held universally. More often, a standard inbred with wide use as a parent in hybrid production is used as a tester.

Selected reports on recurrent selection are presented here to emphasize practical aspects. Johnson *et al.* (1986) used phenotypic recurrent selection to reduce plant height in tropical maize. Fifteen cycles of visual, full-sib recurrent selection for reduced height were conducted in one maize population. Selection resulted in a linear reduction in plant height from 282 to 179 cm (113 to 72 in.), or 2.4% per cycle.

Four cycles of full-sib selection on an open-pollinated maize cultivar were conducted for prolificacy under high and low plant densities (Singh, Khehra, and Dhillon, 1986). Linear response per cycle for number of ears per plant was significant and greater in the low-density selection environment.

Recurrent selection schemes for soybeans that use a genetic male-sterile factor to obtain natural outcrosses was outlined by Brim and Stuber (1973). Three generations per cycle—intermating, advancing, and testing—were used.

SELFING, OR S₁ RECURRENT SELECTION

The S_1 or S_2 progeny test has been employed to improve populations in maize. The progeny row used to evaluate the selected plants in year one is discarded, just as in the other method. Remnant seed from the original selected plants is then grown to intercross the progeny of the selected plants. The S_1 or S_2 progeny test used in conjunction with recurrent selection has shown good population improvement *per se*. Both the mean of the population and the performance of the inbred lines selected from the population have shown good improvement. In some studies, reduction in genetic variability has been less than expected. Selection for GCA occurs with this procedure.

As would be expected from a self-progeny test, inbred lines obtained from some populations have shown less inbreeding depression following a series of selfing generations. Data to indicate relative performance in hybrid combinations using the S_1 progeny test, however, are relatively rare. Although a population may be improved in regard to inbred line performance, the ultimate goal is improvement of performance in hybrid combination. A situation might arise, therefore, in which inbred lines of superior yielding ability are isolated, but in hybrid combination with other inbred lines they might give rise to less than superior performance.

Eight cycles of S_1 progeny and half-sib recurrent selection were conducted at Iowa State University in a strain of the open-pollinated maize cultivar Krug yellow dent to evaluate and compare the changes in grain yield, grain moisture, and lodging resistance (Tanner and Smith, 1987). The genetic gains per cycle for grain yield in the populations *per se* for cycles 0 to 4, adjusted for the effects of finite population size, were significantly larger for S_1 progeny than for half-sib family progeny. For cycles 4 to 8, genetic gains, adjusted for drift in S_1, showed no change, and the genetic gains for half-sib family progeny increased when compared with their respective rates in cycles 0 to 4. For each method, 100 families were evaluated in separate experiments in the same environments. Ten selected families for each method were recombined, each cycle using the bulk entry method outlined by Eberhart, Harrison, and Ogada (1967). The effects of genetic drift in such a finite population were significant for grain yield in both methods and confounded the observed response to selection, especially in cycles 4 to 8. These effects were larger in S_1 progeny.

S_1-line recurrent selection was conducted for five cycles on two maize populations of Iowa stiff stalk synthetic to improve cold tolerance (Hoard and Crosbie, 1985). Selection was based on percentage emergence, rate of emergence, and seedling dry weight in an unselected population under conditions of early seeding. An average of 164 lines were evaluated in each of the five cycles, and a 10% selection intensity was used. Five cycles of S_1-line recurrent selection were completed in nine years. Selection resulted in significant genetic gains of 2.1%/cycle for percentage emergence, 0.04 g/cycle for seedling dry weight, and 0.3 units/cycle for seedling vigor. Genetic drift associated with restricted population size was significant for seedling dry weight and seedling vigor, which emphasizes the need for larger population sizes to obtain long-term gains from selection.

Penny, Scott, and Guthrie (1967) manually infested five synthetic maize cultivars with European corn borer (*Ostrinia nubilalis* Hubner) and selected a group of S_1 lines from the original population. From each cycle of recurrent selection, S_1 lines were evaluated in single-row, nonreplicated, 25-plant plots. Two cycles of selection were sufficient to shift the frequencies of resistant genes, generally involving one to three or more loci, to high levels in all five cultivars. Three cycles of S_1 selection produced essentially borer-resistant cultivars. An average of 123 S_1 lines were rated over the three cycles, and a 12% selection pressure was used. Likewise, Jinahyon and Russell (1969) increased stalk rot (*Diplodia zeae*) resistance in three cycles of S_1 selection. Corn borer and stalk rot resistance were improved more rapidly than yield because the first two traits have higher heritabilities, and good artificial infestation and infection techniques aided selection and reduced the effects of environment.

RECIPROCAL RECURRENT SELECTION

To make maximum use of both general and specific combining ability, Comstock, Robinson, and Harvey (1949) proposed a method of **reciprocal recurrent selection** for improvement of commercial hybrids in diploid organisms. Two sources of foundation material, A and B, are used in this method. The hybrid or hybrids to be developed involve crossing material descended from the two sources, hence the sources should be as genetically divergent as possible. Source material may consist of two cultivars, two synthetics, or the F_2 generation of two single crosses involved in a successful double cross.

S_0 or S_1 plants from source A are self-pollinated and, at the same time, outcrossed to plants from source B. Selection is based on experimental comparisons among test-cross progenies, and selected plants are interbred the third year using their selfed seed produced the first year. In the fourth year, the cycle is reinitiated. Source B plants are tested against source A in the same way. The procedure is shown in Figure 16.3 in greater detail. By testing source A plants in crosses with source B plants, selection pressure for those specific genes that contribute most to the cross of material from the two sources is insured.

Reciprocal recurrent selection is an attempt to improve simultaneously two different populations, at the same time keeping each source population genetically distinct. Phenotypic selections are made in year one within both populations using criteria on which commercial hybrids are evaluated. Progenies, for example, that exhibit lodging or disease susceptibility should not be considered, even though their yield is high. In years that favor agronomic selection, such as for lodging or disease resistance, selection might well be based largely on such agronomic factors.

Reciprocal recurrent selection differs from other recurrent selection methods because the second population acts as the tester parent. After phenotypic selection is complete, plants from one population are used to fertilize plants of the second population. For example, if populations A and B are being improved, phenotypic selections in population A are selfed and pollen from A is used to pollinate plants in B. These progeny are used to test the performance of the selections in A. For population B, individual phenotypic selections are selfed and pollen from B is used to fertilize plants in population A. The progeny of this mating are then used to evaluate the performance of the parents in population B.

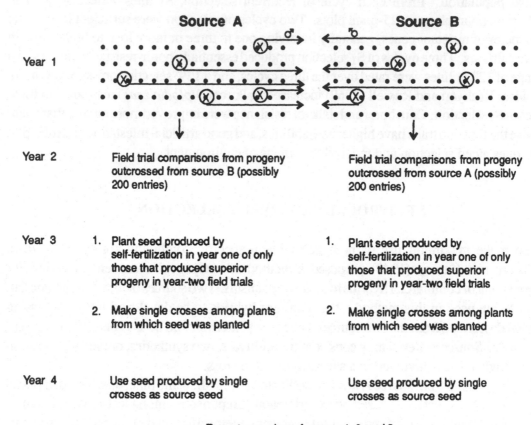

Figure 16.3 Detailed outline of a reciprocal recurrent selection program.

Year 1: Cross-pollinate each of about 200 plants from source A with four or five randomly selected plants from source B. Likewise, cross-pollinate each of about 200 plants from source B with four or five randomly selected plants from source A. Self-pollinate all plants used as pollen parents in these cross-fertilizations.

Year 2: Conduct yield trials on the progeny produced from crosses in year one. Seed from each plant, fertilized by one of the four or five selected pollen parents, constitutes one entry. At most, there would be 200 entries. Perhaps 169 progenies from each group could be tested in a 13 x 13 lattice design.

Year 3: Use the seed produced on the four or five pollen parents in year one produced by selfing these plants to plant a crossing nursery. Use only seed from those plants whose progenies were superior in the year two field trials. Within each source group, make all or as many as possible single crosses between plants from which seed was planted.

Year 4: Repeat the year one procedure using the group A and B seed produced in year three. Repeat procedures of years two and three.

In the third year, remnant selfed seed from those parents showing good progeny performance, as indicated by progeny tests, is planted in rows and single-crossed in all combinations within each population source.

Genetic gain per cycle of selection for maize yield improvement for all recurrent selection methods was about 3 to 5% (Sprague and Eberhart, 1977). For reciprocal recurrent selection, the average gain was lower at 2.9%, but this value included both populations A and B and population crosses. Populations had an average gain of 1.8%, and population crosses, 5.1%.

The primary advantage of reciprocal recurrent selection is the possible increase in heterotic response between two simultaneously improved populations. Other methods may be preferable if early output of improved hybrids is wanted or if inputs are restricted. Recurrent selection for specific combining ability to a single cross over a certain number of cycles may be a more acceptable method when both early output of improved hybrids and a long-term increase in yield are the goals.

Schnell (1961) reviewed reciprocal recurrent selection as a breeding method and concluded that it should not be regarded as a short-cut method but rather as a means to enhance the genetic diversity of future breeding stocks.

RECURRENT SELECTION METHODS IN PERENNIAL CROPS

Recurrent selection is an effective breeding method for population improvement in perennial cross-fertilized species. Methods and procedures for recurrent selection have evolved along different lines, partially because of the eventual end uses of populations and also because of the difficulties encountered in pollen control. The eventual end product in many perennial crops is an improved cultivar based on increased performance of the population or, more likely, a source of improved parental stocks for development of synthetic cultivars. For these reasons, progeny testing methods emphasize population improvement *per se*, as opposed to tests of general and specific combining ability that are associated with hybrid maize performance.

Recurrent selection for forage crops was first evaluated by Johnson (1952) and Johnson and Goforth (1953) using sweet clover (*Melilotus officinalis*). The techniques involved are worth reviewing as a guide. A selection nursery of 1,500 spaced and isolated plants was established by asexually propagating individual plant selections from the sweet clover cultivar Madrid. Of the 400 plants that remained after undesirable plants were cut prior to blooming, about 200 were self-pollinated; of these, 62 produced ten or more seeds. In addition, open-pollinated seed was obtained from all successfully selfed plants. S_1 progenies of the 62 S_0 plants were established, and based on top-cross performance tests on the open-pollinated seed of the 62 plants, the top ten were hand-crossed in all 45 possible combinations. Progeny were rated for vigor under field conditions.

Equal amounts of open-pollinated seed were harvested from each F_1 polycross and bulked. Subsequent Syn 2 plants were self-pollinated and progeny tested. Based on performance, the top ten Syn 2 plants were used to form a synthetic by intercrossing using honeybees. Syn 1 seed of the second cycle of recurrent selection was harvested separately from each plant and was field tested for vigor.

Large positive gains were measured in a single cycle of recurrent selection, suggesting recurrent selection is an effective method of breeding in forage plants. When S_1 progeny were contrasted with progress using mass selection relative to the original cultivar, it was found that two cycles of S_1 progeny selection increased yield 52%, compared to 11% for four cycles of mass selection. Four generations of visual selection for desired plants in the second year was not as effective as a single cycle of recurrent selection based on progeny performance.

Figure 16.4 illustrates four major types of recurrent selection that may be used in perennial crops, namely, individual plant selection, genotypic selection, family selection, and within-family selection. Breeding procedures associated with these selection schemes are presented in Figure 16.5. Although Hill and Haag (1974) outlined nine different recurrent selection procedures that could be used, not all have been tested and few have been tested long enough to indicate the superiority of one method over another.

Phenotypic recurrent selection is the most commonly used method in perennial forage species and is considered effective for improving qualitative traits such as disease resistance. As in maize, however, phenotypic recurrent selection in perennial crop species is ineffective for traits of low heritability or for quantitatively inherited traits involving many genes.

Theoretical studies conducted on autotetraploids by Hill and Haag (1974) indicate that genotypic selection using a selfed progeny test or selfed family selection should be the most effective breeding method in conjunction with recurrent selection. The polycross progeny test (genotypic method) is intermediate in its effectiveness, and the polycross method was thought to be the least effective breeding method in conjunction with recurrent selection. These theoretical studies indicate that phenotypic recurrent selection would be effective only if environmental variance is low.

Foliar disease resistance in alfalfa was used by Haag and Hill (1974) to compare the performance of selfed family, selfed progeny, polycross family, and polycross progeny for selection of resistance to rust fungus (*Uromyces striatus* Schroet. var. *medicaginis* Pass. Arth.) and black stem rust *(Phoma herbarum* West var. *medicaginis* Fckl.) in alfalfa. Each method of selection produced a significant increase in resistance to *U. striatus* (Table 16.1). As indicated by theory, the polycross family method was among the least effective of the breeding methods, but the results were not as bad as predicted. Genotypic selection methods from either the polycross or selfed progeny test were most efficient in increasing the gene frequency in the population. The polycross method would be expected to increase the mean performance of a population somewhat faster than the other methods. Family selection gave a slower response but may be a better method for obtaining a good response to selection and at the same time maintaining genetic variability for yield and other traits in the population.

Selecting for geno-phenotype is theoretically more productive than selecting for phenotype or genotype alone. Selection of individuals with outstanding appearance is phenotypic selection. Selection based on progeny performance is, of course, genotypic selection. When phenotypically outstanding individuals are chosen out of superior-performing lines, selection is by both phenotype and genotype and is termed geno-phenotype. This is analagous to family selection. Geno-phenotype was used by Fryer (1939) to substantially improve seed yield in alfalfa. This procedure was employed in a four-cycle recurrent selection program for the improvement of seed size in birdsfoot trefoil, with an average improvement of 6 and 20% per cycle for Empire and Viking cultivars, respectively (Draper and Wilsie, 1965).

Selection-by-genotype, selection-by-phenotype, and selection-by-geno-phenotype methods

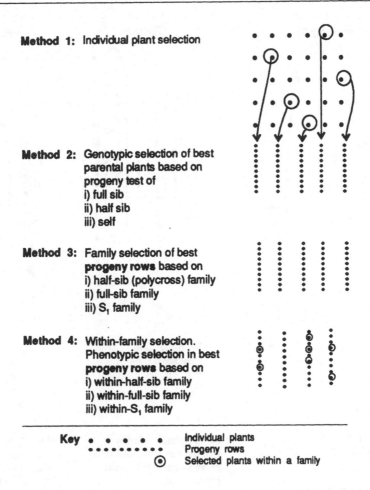

Method 1: Individual plant selection

Method 2: Genotypic selection of best parental plants based on progeny test of
i) full sib
ii) half sib
iii) self

Method 3: Family selection of best **progeny rows** based on
i) half-sib (polycross) family
ii) full-sib family
iii) S_1 family

Method 4: Within-family selection. Phenotypic selection in best **progeny rows** based on
i) within-half-sib family
ii) within-full-sib family
iii) within-S_1 family

Key
- • • • • • Individual plants
- .... Progeny rows
- ⊙ Selected plants within a family

Figure 16.4 Four selection schemes used in conjunction with recurrent selection in perennial cross-fertilized crop species.

Individual plant selection is based on the phenotypic performance of the individual plant.

Genotypic selection is based on the progeny performance of phenotypic selections. Progeny tests can be full sib, half sib, or selfed progeny. Once the progeny are used to identify superior parents, the progeny are discarded, as in recurrent selection methods in maize. The original parent, which can be maintained as an asexual propagate or clone in most perennial crops, is then further used in breeding.

In the family selection method, the progeny rows of phenotypic selections are evaluated. The basis of the progeny test in this method includes half-sib (polycross), full-sib, and S_1 families. The S_1 family is not as commonly used for family selection because of inbreeding depression. After superior parents have been identified on the basis of superior progeny row performance, all progeny of superior rows are used for further selection.

When using selection within the family, various types of progeny tests are used, as in the family selection method, to identify superior parents, including half-sib, full-sib, and S_1 family performance. After superior progeny rows have been identified, phenotypic selection is conducted within those rows. In this manner, phenotypic selection is restricted to superior progeny. The usual terminology with this selection scheme is within-half-sib family, within-full-sib family, or within-S_1 family selection.

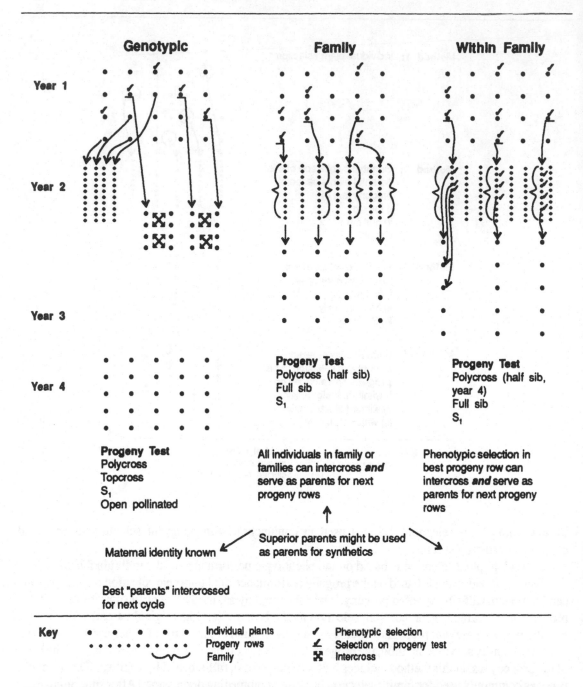

Figure 16.5 Major types of recurrent selection in perennial cross-fertilized crop species. Genotypic selection, as in the methods outlined for maize, involves some form of progeny testing of phenotypically selected plants. Progeny rows are evaluated in year two to pick the best parent plants. In year three, parent plants maintained asexually from year one are then intercrossed in diallel to give rise to a population for the second cycle of selection. The types of progeny testing are listed.

In the family selection method, phenotypic selections are progeny tested in year two. The best-performing progeny rows are intercrossed to give rise to the population in year three and, depending on the progeny testing method, can be intercrossed and used to evaluate parental performance. The polycross, or half-sib, progeny test could be used to complete a cycle in two years. However, since these breeding procedures are used in perennial crop species where persistence over winter is often important,

Table 16.1   Response of Alfalfa to Resistance for *Uromyces striatus* and *Phoma herbarum* with Five Different Breeding Methods

Disease	Breeding Method					
	Selfed Family	Selfed Progeny Test	Polycross Family	Polycross Progeny	Individual Phenotype	Check
U. striatus	0.40	0.45	0.20	0.18	0.30	3.67
P. herbarum	0.17	0.18	0.16	0.25	0.04	3.47

Note: Data are expressed as deviation from check mean; that is, a higher value indicates more successful selection. The selfed family and selfed progeny methods were much more effective in obtaining resistance to *U. striatus* than the polycross and polycross family methods. Individual selection in this situation was effective because the trait has high heritability. In breeding for resistance to *P. herbarum*, a trait of low heritability, the polycross family method should be only half as effective as the polycross progeny (genotypic) method. The two methods produced very nearly equal performance, however. Individual selection was ineffective because of the low heritability. The polycross progeny (genotypic) scheme was more effective than the selfed family, the selfed progeny test (genotypic scheme), or the polycross family. Recurrent phenotypic selection, a form of mass selection, proved effective for developing resistance in alfalfa for diseases that are highly heritable (Hanson *et al.*, 1972).

Source: After Haag and Hill, 1974, p. 592.

were compared for seedling vigor in birdsfoot trefoil (Twamley, 1971). Initially, 72 large-seeded lines were used to establish a seedling vigor test upon which the three selection schemes were based. Sixteen superior lines were identified, their maternal parents intercrossed, and their progeny genotypically selected. A second group consisted of phenotypically vigorous seedlings selected out of every fourth line from the original 72 lines. A third group was formed by phenotypically selecting four vigorous seedlings from the best 16 lines, which were later intercrossed for geno-phenotype selection. Progeny tests were used to evaluate the germplasm selected by the three methods. The geno-phenotype method gave more variable results but was potentially more rewarding than the straightforward and more effective genotypic selection. Phenotypic selection was the least desirable.

A recurrent selection program was employed to select for seed yield in the birdsfoot trefoil cultivar Leo (Sandha and Twamley, 1973) using both genotypic and geno-phenotypic selection methods. After two cycles of selection, seed yields for the genotypic and geno-phenotypic methods were 132 and 158%, respectively (Table 16.2).

In smooth bromegrass (*Bromus inermis* Leyss.), top-cross progenies of 50 plants selected for high seed weight and seed set from cycle 3 of a recurrent selection program were 45% higher

---

Figure 16.5 continued

most of the recurrent selection methods require more than two years if field evaluation is a part of the process.

For within-family selection, phenotypic selections in the original population are intercrossed and superior families identified by their performance. Phenotypic selection is restricted to superior progeny rows, and these selected plants form the basis of the population to start the second cycle of selection.

Table 16.2   Comparison of Genotypic and Geno-Phenotypic[a] Recurrent Selection for Seed Yield in the Birdsfoot Trefoil Cultivar Leo (in percentages)

	Genotypic Selection		Geno-Phenotypic Selection	
Characteristic	*Cycle 1*	*Cycle 2*	*Cycle 1*	*Cycle 2*
Seed yield	116	132	140	158
Heritability	48	9	61	61
Genotypic coefficient of variability	16	5	17	15

[a] Based on both genotypic and within-family selection using the polycross progeny test (half-sib).

Note: Geno-phenotypic or within-family selection showed very good progress and maintained better genetic variability in the population as measured by heritability estimates and the genotypic coefficient of variation. Although genotypic selection showed good progress, after two cycles of selection, heritability estimates and the coefficient of variation for genotypes practically were exhausted. Family selection can combine modest progress with the maintenance of genetic variability in the population.

Source: After Sandha and Twamley, 1973, p. 813.

in average weight per seedling one month after planting in the field than were five check cultivars (Trupp and Carlson, 1971).

USE OF POPULATIONS DEVELOPED BY RECURRENT SELECTION

In perennial crops, recurrent selection may result in a population that can be released as a cultivar in itself. The alfalfa cultivar Cherokee was released after seven cycles of recurrent selection (Dudley, Hill, and Hanson, 1963). Release of a cultivar might occur where a maximum amount of variability for traits other than those selected is desired or acceptable. Some recurrent selection procedures may result in sufficient loss of variability to prevent further successful selection for population improvement, although the improvement for the trait under selection may be sufficient to warrant use of the population as a cultivar *per se.*

The family selection method and the within-family selection method also allow a preliminary estimate of combining ability that could be used to select parents for synthetic cultivars. For example, in any cycle of selection in the family and within family methods, the parent plants are not used for the recurrent selection in the next cycle. These parents, however, could be used as potential parents of synthetic cultivars if they prove to have superior progeny performance.

Data from the birdsfoot trefoil breeding program at Guelph, Ontario, have indicated the superior performance of elite parents chosen on the basis of their performance in the recurrent selection program. Such parents would be likely candidates for forming an improved synthetic cultivar. Frequently, a population developed through recurrent selection for one specific trait may be used as the source population for one or two generations of intense selection to concentrate other characteristics that would be desirable in a commercial cultivar.

Recurrent selection has been shown to be a very effective breeding method for concentrating gene frequency in a relatively small number of individual plants. Although phenotypic recurrent selection has been used most frequently in both perennial crop species and the annual maize crop, its major advantage is likely in increasing rare genes for qualitatively inherited traits.

RECURRENT SELECTION IN HORTICULTURAL CROPS

The observation that interbreeding in cantaloupe (*Cucumis melo* L.) is achieved by natural populations of bees suggested the possibility of practicing mass and recurrent selection. Information obtained from breeding practices should be applicable to many crops that do not fall into the fully self-fertilized or fully cross-fertilized classifications (Andrus and Bohn, 1967). Nine cycles of selection were conducted based on an index of 16 fruit characteristics. Selection resulted in improvement in most traits without any evidence of loss of genetic variability. A feature of the program was that selection was conducted at alternating locations, with final evaluations made at both environments after nine generations of selection.

The original concept that persistent blending of germplasm selected for superior performance in widely divergent environments would lead to a population with a superior range of adaptation remains hypothetical. Apparently, mass selection in early generations from cantaloupe hybrids increases the favorable gene frequencies available for late-generation controlled inbreeding. Selection in later generations appears to have promise for high-quality cultivars with good adaptation.

Likewise, the sweet potato (*Ipomoea batatas* L. Lam.) is well suited to recurrent selection. It is a hexaploid, however, having 90 chromosomes with very complex inheritance that can be described as quantitative (Jones, Dukes, and Cuthbert, 1976). Sweet potato is propagated vegetatively, but no problems were encountered with lack of flower development, sterility, or incompatibility if population sizes were kept large. High seed set appeared to have no adverse effect on reductions of root yield. Phenotypic recurrent selection for resistance to a number of pests and for production and market qualities was conducted for six generations with notable success due to the high visibility and high heritability of the traits being selected.

Experience with sweet potato led Jones, Dukes, and Cuthbert (1976) to conclude that recurrent selection procedures offer the breeder flexibility. Selection priorities can be changed as the relative frequencies of the traits change, and new selection criteria can be added as the need or opportunity arises. Recurrent selection procedures can be altered or integrated with other parts of the program as necessary. A two-year selection cycle was suggested so that more traits can be added to those already included in the selection process.

In general, phenotypic recurrent selection has been used effectively in horticultural crops.

INTROGRESSION OF GERMPLASM

Most crop species have adequate genetic variability for reasonable progress to be achieved through plant breeding. After the variability in landrace cultivars was exploited, hybridization programs were established to increase genetic variability for use with conventional breeding systems.

Crosses with unadapted exotic and weedy germplasm require large populations to enhance the opportunity for new genetic combinations to occur, to avoid the loss of desirable genes, and to secure genes adapted to the environment. Recurrent selection provides a useful technique for the introgression of genes because recombination on a regular basis following the original cross is inherent in the system.

There have been expressions of concern that genetic variability is too limited in some crops (National Academy of Sciences, 1972; Miller, 1968). Brown (1975) noted that about 130 racial complexes make up the maize germplasm, but over 90% of the entire maize breeding program in the United States originates with three of the 130 races and 98% of the maize germplasm is ignored.

Recurrent selection can be a productive plant breeding method and can complement other systems of plant breeding, but it is not the answer to all breeding objectives (Hallauer, 1985). Recurrent selection methods have not been an integral part of most plant breeding programs, although they have received greater attention in recent years. Compared to the backcross and pedigree breeding systems, recurrent selection methods are relatively new.

REFERENCES

Andrus, C. F., and G. W. Bohn. 1967. Cantaloup Breeding: Shifts in Population Means and Variability Under Mass Selection. *Proceedings of the American Society of Horticultural Science* 90:209-222.

Brim, C. A., and C. W. Stuber. 1973. Application of Genetic Male Sterility to Recurrent Selection Schemes in Soybeans. *Crop Science* 13:528-530.

Brown, W. L. 1975. A Broader Germplasm Base in Corn and Sorghum. *Proceedings of the Corn Sorghum Research Conference* 30:81-89.

Comstock, R. E., H. F. Robinson, and P. H. Harvey. 1949. A Breeding Procedure Designed to Make Maximum Use of Both General and Specific Combining Ability. *Agronomy Journal* 41:360-367.

Draper, A. D., and C. P. Wilsie. 1965. Recurrent Selection for Seed Size in Birdsfoot Trefoil. *Crop Science* 5:313-315.

Dudley, J. W., R. R. Hill, Jr., and C. H. Hanson. 1963. Effects of Seven Cycles of Recurrent Selection on Means and Genetic Variances of Several Characters in Two Pools of Alfalfa Germ Plasm. *Crop Science* 3:543-546.

Eberhart, S. A., M. N. Harrison, and F. Ogada. 1967. A Comprehensive Breeding System. *Der Zuchter* 37:169-174.

Fryer, J. R. 1939. The Maternal-Line Selection Method of Breeding for Increased Seed-Setting in Alfalfa. *Scientific Agriculture* 20:131-139.

Haag, W. L., and R. R. Hill, Jr. 1974. Comparison of Selection Methods for Autotetraploids. II: Selection for Disease Resistance in Alfalfa. *Crop Science* 14:591-593.

Hallauer, A. R. 1981. Selection and Breeding Methods. In *Plant Breeding II*, edited by K. J. Frey. Iowa State University Press, Ames, pp. 3-55.

Hallauer, A. R. 1985. Compendium of Recurrent Selection Methods and Their Application. *CRC Critical Reviews in Plant Sciences* 3:1-33.

Hanson, C. H., T. H. Busbice, R. R. Hill, Jr., O. J. Hunt, and A. J. Oakes. 1972. Directed Mass Selection for Developing Multiple Pest Resistance and Conserving Germplasm in Alfalfa. *Journal of Environmental Quality* 1:106-111.

Hill, R. R., Jr., and W. L. Haag. 1974. Comparison of Selection Methods for Autotetraploids. I: Theoretical. *Crop Science* 14:587-590.

Hoard, K. G., and T. M. Crosbie. 1985. S_1-line Recurrent Selection for Cold Tolerance in Two Maize Populations. *Crop Science* 25:1041-1045.

Jinahyon, S., and W. A. Russell. 1969. Evaluation of Recurrent Selection for Stalk-Rot Resistance in an Open-pollinated Variety of Maize. *Iowa State Journal of Science* 43:229-239.

Johnson, E. C., K. S. Fischer, G. O. Edmeades, and A.F.E. Palmer. 1986. Recurrent Selection for Reduced Plant Height in Lowland Tropical Maize. *Crop Science* 26:253-260.

Johnson, I. J. 1952. Effectiveness of Recurrent Selection for General Combining Ability in Sweetclover *Melilotus officinalis*. *Agronomy Journal* 44:476-481.

Johnson, I. J., and F. Goforth. 1953. Comparison of Controlled Mass Selection and Recurrent Selection in Sweetclover, *Melilotus officinalis*. *Agronomy Journal* 45:535-539.

Jones, A., P. D. Dukes, and F. P. Cuthbert, Jr. 1976. Mass Selection in Sweet Potato: Breeding for Resistance to Insects and Diseases and for Horticultural Characteristics. *Journal of the American Society for Horticultural Science* 101:701-704.

Kiesselbach, T. A. 1951. A Half Century of Corn Research. *American Scientist* 39:629-655.

Miller, F. R. 1968. Genetic Diversity in the World Sorghum Collection. *Proceedings of the Corn Sorghum Research Conference* 23:120-128.

National Academy of Sciences. 1972. *Genetic Vulnerability of Major Crops*. Washington, D.C. 307 pp.

Penny, L. H., G. E. Scott, and W. D. Guthrie. 1967. Recurrent Selection for European Corn Borer Resistance in Maize. *Crop Science* 7:407-409.

Rawlings, J. O. 1980. Long- and Short-Term Recurrent Selection in Finite Populations: Choice of Population Size. In *World Soybean Research Conference II: Proceedings*, edited by F. T. Corbin. Westview Press, Boulder, Colorado. 897 pp.

Robertson, A. 1960. A Theory of Limits to Artificial Selection. *Proceedings of the Royal Society B*, Vol. 153:234-249.

Sandha, G. S., and B. E. Twamley. 1973. Recurrent Selection for Seed Yield Improvement in *Lotus corniculatus*, Cult. Leo. *Canadian Journal of Plant Science* 53:811-815.

Schnell, F. W. 1961. On Some Aspects of Reciprocal Recurrent Selection. *Euphytica* 10:24-30.

Singh, M., A. S. Khehra, and B. S. Dhillon. 1986. Direct and Correlated Response to Recurrent Full-Sib Selection for Proficiency in Maize. *Crop Science* 26:275-278.

Sprague, G. F., and S. A. Eberhart. 1977. Corn Breeding. In *Corn and Corn Improvement*, edited by G. F. Sprague. American Society of Agronomy, Madison, Wisconsin, pp. 305-362.

Tanner, A. H., and O. S. Smith. 1987. Comparison of Half-Sib and S_1 Recurrent Selection in the Krug Yellow Dent Maize Populations. *Crop Science* 27:509-513.

Trupp, C. R., and I. T. Carlson. 1971. Improvement of Seedling Vigor of Smooth Bromegrass (*Bromus inermis* Leyss.) by Recurrent Selection for High Seed Weight. *Crop Science* 11:225-228.

Twamley, B. E. 1971. Selection Methods for Seedling Vigor in Birdsfoot Trefoil. *Canadian Journal of Plant Science* 51:229-235.

CHAPTER 17

Interspecific and Intergeneric Hybridization

WIDE CROSSES

Limited genetic diversity in a crop species may restrict the amount of genetic improvement that can be achieved through plant breeding. Plant breeders characteristically use the most easily accessible sources of variation, namely, the ones that can be incorporated in the least amount of time, usually improved adapted cultivars. Plants of the same biological species generally cross readily to produce fertile progeny and present little or no hindrance to genetic recombination.

Plant introductions have served to provide plant breeders with valuable sources of germplasm that are useful for the development of cultivars that are responsive to improved cultural practices, that possess acceptable food, feed, or fiber quality, and that are resistant to pests and stress (Shands and Wiesner, 1991).

Hybridizing adapted cultivars to alien species often is considered when the variation among adapted cultivars or landraces appears to be exhausted. Sometimes a specific problem can be solved by obtaining specific genes from a wide cross, one involving representatives of different species or genera. Wide crosses may be difficult to make and, even if accomplished, may produce progeny that are sterile or **inviable**. Plant breeders who undertake wide crosses must have valid reasons to justify the extra effort required.

The chromosome number of plants to be considered for interspecific or intergeneric hybridization is useful information for the plant breeder (Cauderon, 1986). Different chromosome numbers in the parents or different ploidy levels may indicate that cross-incompatibility and hybrid inviability and sterility could be problems. Doubling the chromosome number of one or both parents and/or the F_1 hybrids may overcome the problem. The 2n chromosome number of nearly 400 plant species, along with the common name and the reference source, were listed by Armstrong (1987).

As a guide to the incorporation of wild germplasm, Bowley and Taylor (1987) compiled a comprehensive list of over 660 plant species that have been successfully introgressed with the cultivated counterpart. Introgression was considered successful if F_2 or backcross plants were produced or if progeny could be maintained through reproductive apomixis. Species in which one parent or the F_1 had to be artificially doubled to obtain the F_1 or a backcross progeny are indicated. The list, based on the literature, is intended to be complete through July 1982. Stalker (1980) reviewed the successful transfers of useful genes from wild species to related field crop plants.

Perseverance in meeting the challenges of wide crosses can be rewarding. Wide crosses to transfer qualitatively inherited traits such as disease or pest resistance in wheat (Knott and Dvorak, 1976; Cauderon, 1979) and sunflower (Laferriere, 1986) have been very successful. Internationally approximately 30 wheat cultivars carry disease resistant genes from alien sources (Metlin, Bluthner, and Schlegel, 1974; Brezhnev, 1978; Sharma and Gill, 1983; Zeller, 1974).

Quantitatively inherited traits generally are more complex in inheritance and are more difficult to manipulate in a breeding program. Species of *Agropyron* (*Elytrigia*) and *Triticum* (*Aegilops*) are crossable with wheat, and in many cases the material has been advanced to the state of addition lines that are available for screening (Fedak, 1985). Selections from these crosses promise enhanced levels of drought, cold, and salinity tolerance. In addition, flag leaves of primitive forms can have higher photosynthetic rates than those of cultivated hexaploid wheats (Evans and Dunstone, 1970; Khan and Tsunoda, 1970) and high protein content. In most cases, these features are products of quantitatively inherited complex metabolic pathways and are difficult to transfer to bread wheat. The success in extracting such traits from alien species has been minimal, despite laborious and time-consuming efforts (Fedak, 1985). The process requires adequate screening techniques. Recurrent selection was suggested as a possible means of transferring specific traits.

Despite the obstacles and barriers to wide crosses, the voluminous nature of the literature suggests that plant breeders view them as desirable. Wide crosses are made for the following reasons:

- To improve a species by transferring specific traits, such as disease, pest, and stress resistance.
- To achieve new character expressions, possibly through the action of complementary genes that could produce types beyond the range of either parent for such quantitative characters as yield, maturity, and winter hardiness.
- To produce new alloploid species.
- To satisfy an age-old human curiosity about characteristics produced when wide crosses are achieved.
- To investigate the relationships among species for phylogenetic studies where wide crosses are followed by a meiotic study of the hybrids to establish chromosome relationships between two parents.
- To increase genetic diversity as a safeguard against unpredictable biological and environmental hazards.
- As a source of favorable alleles for yield for increased genetic variation and for enhanced heterosis.

To understand the barriers to wide crosses, consider a hypothetical situation in the animal kingdom. Suppose that all but some very large and some very small dogs died from a worldwide epidemic. The two groups might become isolated from each other, and for various reasons, physical and geographic barriers might prevent crossing of the two types. Taxonomists might arbitrarily classify them as separate species. Artificial insemination could facilitate crossing, and the arbitrary classification would be modified. A true intergeneric cross is similar to crossing a dog and a cat.

Interspecific and intergeneric crosses have been by far the most important in ornamental plants. Rhododendrons, irises, orchids, cannas, dahlias, gladioli, roses, poppies, and violets were suggested by Emsweller *et al.* (1937) to have resulted from wide crosses.

Next in importance, tree crops such as apples, plums, cherries, filberts, grapes, and various kinds of berries have resulted from wide crosses. Crops such as wheat, tobacco, oat, cotton, and sugarcane, in which modern attempts to produce a wide cross have not been particularly successful, are allopolyploids that were derived originally from hybrids among different species.

Among the vegetables, only the potato and sweet potato have benefited from interspecific hybridization, whereas the common vegetables, such as tomato, bean, beet, carrot, asparagus, celery, and lettuce, are each derived from single diploid species, and interspecific hybridization has not been important.

Stebbins (1950, p. 292) suggested two reasons that could account for the range of importance of wide crosses among ornamentals, tree, field, and vegetable crops: Wide crosses may be more valuable in vegetatively propagated species than in those reproduced by seed; and wide crosses are of less value when quality aspects are more important than total quantity of yield.

ISOLATION MECHANISMS IN HYBRIDIZATION

The protective mechanisms or barriers that exist in plants serve to maintain the genetic integrity of species (Figure 17.1). These mechanisms will serve as the bases for consideration in breeding programs and are more fully described here.

External Barriers to Hybridization

Spatial Isolation. Isolation associated with geographic distances between two species is frequently an easy one to overcome. Provided isolation is the sole barrier, the free flow of genes can be facilitated by bringing the two populations together. Bates and Deyoe (1973, p. 401) described isolating barriers that "depend upon physical separations of time, distance, environment, or specific ecological niche" as only partial; all of these isolation barriers can be manipulated during cultivar production.

Sensitivity to photoperiod can result in spatial isolation and the free flow of genes. Under a relatively constant temperature, days to flowering can be manipulated by modifying the length of day or night (Figure 17.2).

The soybean genus (*Glycine*) includes extremely diverse species. The response of soybeans to photoperiod was a major factor in the development of a large number of local cultivars. Introductions of soybeans were adapted to latitudes similar to the latitudes from which they had been collected because of their photoperiodic response. Pronounced adaptation to photoperiod resulted in soybeans in North America being classed into twelve maturity groups designated by Roman numerals. Group 00 is the earliest to mature, 0 is the next earliest, and the groups extend south, with group VIII cultivars adapted to the southern parts of the United States, and end with groups adapted to the short days of tropical regions on either side of the equator.

The pedigree maturity grouping can represent an isolation mechanism, but one that can be overcome to allow for the free flow of genes.

A. EXTERNAL BARRIERS

I.   Spatial Isolation

II.   Prefertilization Reproductive Barriers
(barriers between the parental species)

    1.   Ecological isolation
    2.   Time and seasonal isolation
    3.   Mechanical isolation
    4.   Gametic incompatibility

B. INTERNAL BARRIERS

III.   Postfertilization Reproductive Barriers
(barriers between the hybrids)

    1.   Hybrid inviability or weakness
    2.   Failure of flowering in the progeny
    3.   Sterility of the progeny
    4.   Inviability and weakness of F_2 and later segregates

Figure 17.1 The classification of the different isolating mechanisms that form barriers between species shown in this figure was adopted from Stebbins (1950 p. 196) and modified according to the suggestions of Hadley and Openshaw (1980).

From the standpoint of organic evolution over the ages, the reproductive isolation mechanisms led to plants that can be grouped into landraces, species, genera, families, and higher orders. Such categories have been developed because groups of individuals resemble each other more than they do members of other groups. The inability to hybridize two plants may form the basis for speciation. Reproductive isolation barriers serve to maintain the integrity of species by restricting the flow of genes from one to another.

From a plant breeding viewpoint, these reproductive isolation mechanisms restrict the free flow of genes and the expanded use of germplasm. Plant breeders have little respect for species integrity and intuitively wish to break down these barriers temporarily, make crosses, and broaden the useful gene pool to meet ever-expanding human needs. Knowledge and understanding of the barriers to interspecific and intergeneric hybridization are essential to increasing the chances of success. Some barriers, such as that of spatial isolation, frequently are easy to overcome, but they may not be the sole barriers. If spatial isolation is the sole barrier, what was formerly considered a biological species may be more accurately described as a subspecies, ecotype, or race.

Figure 17.2 Isolation barriers of differences in flowering time resulting from a response to photoperiod can be readily overcome to facilitate hybrid production. The maize plants on the left in Photo A have reached and passed anthesis, whereas the maize on the right remains vegetative.

Maize is a short-day plant and if short-day-sensitive lines from equatorial regions are produced at high latitudes, reproductive development may not occur until late in the season. By placing artificial

Figure 17.2 continues overleaf

Ecological Isolation. Most crops, particularly the cereals, have spring and winter counterparts that can be freely hybridized but in which hybridization has been restricted and generally done in greenhouses because of the difficulty in the coincidence of anthesis. Spring and winter types in cereals belong to the same respective species, and although they may be grown side by side in any region, they are ecologically isolated.

For winter cereals to reach reproductive development, vernalization is required. Having spring and winter wheats flower at the same time can be achieved using elevation to provide the cold requirement for vernalization by November seeding of winter cereals and a January or February planting of spring cereals at the same site. This works well at the central Mexico station at Toluca (elevation 2,640 m, or 8,580 ft.). An alternative is to vernalize winter seedlings in pots for about six to seven weeks in cold chambers at 3 to 4° C (40° F) under an eight- to ten-hour photoperiod. In short-day regions, vernalized seedlings could be transplanted outdoors under artificial lights, as shown in Figure 17.2, to extend the day length an additional six hours (CIMMYT, 1980).

Following hybridization, seeds may be grown at sites to screen for winter or spring habit and for agronomic features. If fall-seeded under severe winter conditions, those lacking hardiness genes may be winter-killed. If spring-seeded, those winter genotypes requiring vernalization might not develop reproductive organs.

For example, a partially dominant allele for tolerance to barley yellow dwarf (*Rhopalosiphum padi*), Yd_2, was transferred from spring to winter barley (Bourne and Poehlman, 1987). Hybrids between spring and winter wheats produced significant levels of heterosis in F_1 yield trials (Grant and McKenzie, 1970). Kronstad *et al.* (1975) suggested that winter x spring wheat crosses have opened a new reserve of genetic germplasm. Pinthus (1967) indicated that winter wheat should be able to contribute to an increase in yield of progenies of its crosses with spring wheat.

Mechanical Isolation. It is believed that Midwestern dent maize of the United States originated from hybridization of two landraces, Northern flint and Southern dent, in the nineteenth century. This idea was verified by means of isozyme electrophoresis (Doebley *et al.*, 1988). The reason why deliberate crosses of these two contrasting types were not made earlier may be related to mechanical isolation. Northern flints are hard and vitreous and require considerable energy to be milled. The Southern dents are much softer and easy to mill or chew after minimal processing. Northern flints were adapted to the short-season areas of North

---

Figure 17.2 continued

lights above the crop (Photos B and C), day length can be modified to overcome isolation barriers and permit hybridization.

Plants respond to the length of dark or the night period. By interrupting the dark period, a long night becomes a short-night period. By extending the length of day, a short night can be achieved, depending on the crop and location. These techniques either delay or advance flower initiation and development.

Species that respond to long days (or short nights) can be induced to flower by a one-hour interruption with light in the middle of the night. In plants that require short days (or long nights), flowering can be delayed. Lighting the plants at sunset for about four hours will maintain the effect of long days in warm climates. In the latter case, lights are turned off about six to twelve weeks before the flowers are desired, depending on the species being grown.

America because of their early maturity but were resisted in areas where soft corn could be grown. The resistance against this genotype became a mechanical barrier, and crosses between the two were achieved in time purely by accident. Only when the superiority of the progeny from these crosses was recognized did more extensive crossing occur.

Another form of mechanical isolation was observed by Stebbins (1958) in species whose flowers have become adapted to insect pollination. Some plant species are visited by only one insect species. In such cases, two compatible plant species with exclusive insect vectors will not hybridize in nature because no insect will visit flowers of both species. Honeybees characteristically visit flowers of many species but may not visit flowers of more than one species during the same foraging flight. Discrimination of this type may lead to the conclusion that two species are sexually incompatible.

Gametic Incompatibility. Stebbins (1950) referred to gametic incompatibility as the prevention of fertilization. Gametic incompatibility is an incapacity of pollen to germinate on the style of the pistil or to form a viable zygote with an egg of another species (Solbrig, 1970). Barriers to hybridization that are beyond human control are considered "absolute" barriers and include gametic incompatibility and hybrid breakdown.

Gametic incompatibility is similar to sporophytic self-incompatibility, described in an earlier chapter, and may be associated with failure of the pollen to germinate on the stigma; failure of the pollen tube to grow down the length of the style; or failure of the male gamete to unite with the egg, even though the pollen tube reaches the ovary. Gametic incompatibility is governed by a complex multiple allelic system of *S*-genes that prevents **gametic union** (syngamy) of gametes carrying identical *S*-alleles.

Sporophytic incompatibility differs from gametic incompatibility in that the incompatibility reaction is imparted to the pollen by the plant upon which the pollen is borne.

S-alleles may have have two functions: (1) preventing self-fertilization and (2) preventing cross-fertilization. These possibilities are based on the fact that a cross between a self-compatible and a self-incompatible plant often results in fertilization, whereas the reciprocal cross does not (Hadley and Openshaw, 1980).

Other evidence (Abdalla and Hermsen, 1972; Hogenboom, 1972) suggests that self-incompatibility and interspecific incompatibility are two distinct phenomena. Stebbins (1958) suggested that gametic incompatibility seldom is the primary cause of reproductive isolation between closely related species.

Internal Barriers to Hybridization

Hybrid Inviability or Weakness. When fertilization has taken place, gametic incompatibility ends. In wide crosses, postfertilization abnormalities may result in internal barriers to hybridization, such as haploid formation (already observed in *Hordeum vulgare* x *H. bulbosum* crosses and associated embryo abortion). Progeny can be produced by removing young hybrid embryos and culturing them artificially through embryo rescue techniques.

In other cases, F_1 progeny from wide crosses are too weak to be of any use to the plant breeder. Hadley and Openshaw (1980) suggested possible causes for such weakness: (1) disharmony between the genomes of the parental species, (2) disharmony between the genome of one species and the cytoplasm of the other, or (3) disharmony between the genotype of the F_1

zygote and the genotypes of endosperm or the maternal tissue with which the developing F_1 embryo is associated.

Disharmony between the parental genomes may have a polygenic basis and is difficult to analyze. Hybrid weakness may be associated with improper timing of critical processes such as cell division, cell organization, and differentiation and may result from weak genetic messages from the two parental genomes.

The shriveling of grain has been a persistent problem in triticale (X *Triticosecale* Wittmack) and has limited the acceptance of the crop. Skovmand, Fox, and Villareal (1984) reviewed the problem and suggested the causes were associated with differences in the amount of terminal **heterochromatin** found on wheat chromosomes as compared to rye chromosomes. As the amount of heterochromatin slowly declined to about 50% of rye chromosomes, grain plumpness improved.

Double fertilization resulting in a triploid endosperm can result in serious disharmony. Two genomes from the female parent and one different genome from the male may produce unfavorable dosage effects. The endosperm could be *AAB* or *ABB*. Where disharmony occurs between the genes of one species and the cytoplasm of another, reciprocal crosses may avoid the problem. Reciprocal crosses may reduce seed shriveling as well, possibly as a result of improved chromosomal balance in the endosperm.

Failure of Flowering in the Progeny. Hybrid sterility due to genetic causes may result in failure of the progeny of wide crosses to produce flowers, failure of meiosis in either the anther or the ovary or both, or failure of the entire meiotic process.

Sterility of the Progeny. Sterility could have a genetic basis, as described previously, or could be chromosomal and result from differences in the chromosome numbers of the parents, partial or complete failure of the chromosomes to pair, or a number of abnormal chromosome associations. Genetic sterility generally produces completely sterile plants; chromosomal sterility may produce some fertile florets that can be useful to the plant breeder.

Inviability and Weakness of F_2 and Later Segregates, or Hybrid Breakdown. In some wide crosses, the F_1 progeny may show normal development and reasonable fertility, but the F_2 progeny may be less satisfactory and even sterile. The reason for this phenomenon suggested by Muller (1940) was segregation of complementary gene systems. Consider a cross between two parents with the genotypes *AABB* and *aabb* in which the homozygous *AA* or *aa* condition is complementary to *BB* or *bb*, respectively. It is further assumed that *AA* or *aa* is not complementary to *bb* or *BB*, respectively. The F_1 progeny with the *AaBb* genotype would be balanced and viable, but F_2 genotypes such as *A_bb* and *aaB_* would show degeneracy and sterility (Stebbins, 1958).

Deterioration of the F_2 generation of crosses occurs between the cultivated American Upland cotton (*Gossypium hirsutum*) and *G. barbadense* (Sea Island cotton) (Stephens, 1950). The cross is easy to make, and the F_1 progeny are vigorous and have normal fertility. In the F_2 generation, individuals are weak, sterile, or agronomically unsatisfactory.

OVERCOMING REPRODUCTIVE BARRIERS

Various techniques have evolved to overcome genetic and spatial barriers that restrict genetic interchanges in wide crosses. Some crosses can be made using conventional methods; however, others are difficult or impossible to achieve with conventional procedures and special techniques must be applied. Some of these techniques are described in the following sections.

Taxonomic Classification

A realistic approach to potential obstacles must be taken in choosing parents to combine in wide crosses. Taxonomic classifications based on morphological characteristics may serve as guides. Such classifications, however, do not reflect genetic relationships accurately, and data from previous hybridization studies are more useful. Of greatest value in determining which crosses may be possible are biosystematic studies made by cytotaxonomists, along with an appreciation of plant breeding problems and associated facilities.

Crosses attempted vary from those involving parental species that differ by several genes to those in which parental species differ in genomic constitution. Gene pools that might be used in wide crosses were grouped into three categories by Zohary (1973), according to degree of difficulty with which crosses could be made:

Gene pool I   Cultivars of the same crop
Gene pool II   Wild progenitors
Gene pool III   Alien or other species in the genus

Zohary suggested that hybridization with alien gene pools be attempted when the gene pools from groups I and II have been exhausted. In a similar manner, Harlan and de Wet (1971) ranked gene pools as primary, secondary, and tertiary. Although these authors did not rank the pools in order of ease or difficulty of gene transfer, such conclusions can be drawn, as shown for sunflower (*Helianthus annuus* L.) (Figure 17.3).

Use of Different Cultivars of Two Species

The considerable genetic variation that exists among cultivars and types within a wild relative should be sampled to avoid basing conclusions regarding crossability on too small a sample.

In interspecific crosses with tobacco, when the autotetraploid *Nicotiana repanda* Willd. ex. Lehm (2n = 4x = 96) was crossed with three cultivars of common tobacco (*Nicotiana tabacum* L., 2n = 2x = 48) (Pittarelli and Stavely, 1975), only one produced F_1 hybrid plants. A similar situation was reported in *Tripsacum dactyloides* L. (Figure 17.4). Likewise, Lambert and Leng (1965) found differences among five teosinte (*Zea mexicana*) strains in their ability to backcross to an inbred corn selection.

Reciprocal Hybridization

A cross may be successful if made in one direction, but the reciprocal cross may fail. If no *a priori* knowledge exists about a particular parental combination, plant breeders should

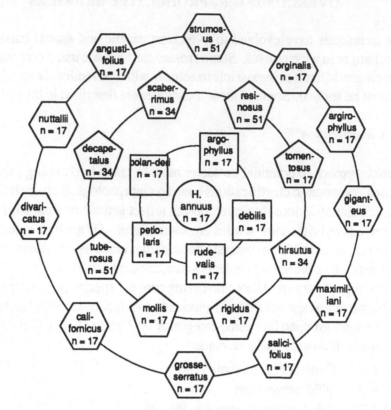

Figure 17.3 Diagram showing interspecific crossability with the cultivated sunflower *Helianthus annuus*. The genus *Helianthus* contains perhaps 49 species, all native to the Western Hemisphere, adapted to a wide range of environments, and exhibiting a great diversity of morphological and physiological characteristics.

The basic diploid has 17 chromosomes in its genome. There are three primary genomes, A, B, and C. Some are annuals, some are perennials. The cultivated sunflower, *H. annuus*, hybridizes most easily with other diploid annuals, less easily with the polyploid and perennial species. Those species that exhibit good compatibility are shown in the inner circle, those in which crosses are difficult but possible are found in the middle circle, those that exhibit complete incompatibility are shown in the outer circle, based on the findings of Georgieva-Todorova (1984).

Harlan and de Wet (1971) classed gene pools, in order of ease of making a cross, as primary (easiest), secondary, and tertiary. Secondary gene pools have serious barriers to hybridization, and plant breeders must persevere to make such crosses. The tertiary gene pools define the outermost limits of hybridization, which might be successful with highly specialized techniques.

It should be emphasized that as new techniques develop, crosses that once were considered very difficult might be achieved using genetic engineering or other techniques. Plant breeders should view taxonomic classifications with a skeptical eye.

Figure 17.4 The genus *Tripsacum*, shown in these photos, is a distant, wild relative of maize that has the potential to transfer drought resistance to maize. The successful completion of the intergeneric cross of *Zea mays* (maize) with *Tripsacum* may allow access to genetic variation not found in the maize breeding program.

Tripsacum consists of seven to nine species, of which the most widely known is *T. dactyloides,* or

Figure 17.4 caption continues on p. 356

A.

B.

C.

consider making reciprocal crosses. Usually the parent with the larger number of chromosomes should be used as the female, but female parents may be chosen based on ease of emasculation, or a plant with abundant pollen may be chosen as the male parent. If a species has a self-incompatibility system, pollination might be inhibited, but the reciprocal cross might not be affected.

If no prefertilization barriers are found in reciprocal crosses, postfertilization barriers caused by cytoplasmic-genomic disharmony may become evident in the F_1 progeny of the crosses. Likewise, one cross of a reciprocal may produce superior F_2 segregates.

In sunflower, Georgieva-Todorova (1984) crossed diploid *Helianthus annuus* (n = 17) as the female with other polyploid *Helianthus* species. When *H. decapetalus* (n = 34) and *H. hirsutum* (n = 34) were used as male parents, contrary to the popular practice of using the species with the larger number of chromosomes as the female, successful plant hybridization and full plant development were found. *H. annuus* and *H. scaberrimus* (n = 34) hybrids were obtained only when the tetraploid was used as the female parent. It is possible that reciprocal crosses may be unsuccessful because of incompatibility between pollen tubes and stylar tissue. It might be that pollen tubes from 4 x sporophytes are too thick for 2 x pollen tissue, but pollen tubes from 2 x sporophytes may penetrate readily 4 x stylar tissue.

Chromosome Doubling

A low level of compatibility was found between the diploid sunflower *Helianthus annuus* and the tetraploid species *H. decapetalus* (Georgieva-Todorova, 1984). The number of univalents was between 9 and 17, that of bivalents between 17 and 34. Doubling the chromosome number of the diploid species *H. annuus* before hybridization may have reduced these problems. Such was the case with trefoil. Crosses of narrow leaf trefoil *Lotus tenuis* (2n = 2x = 12) and broadleaf birdsfoot trefoil *L. corniculatus* (2n = 4x = 24) were not achieved until the chromosome number of *L. tenuis* was doubled (Wernsman, Davis, and Keim, 1965).

In another case, doubling the chromosomes of both diploid parents increased crossability between two rapeseed species (Olsson, 1963). Attempts to cross *Brassica campestris* L. (2n = 2x = 20) with *B. oleracea* L. (2n = 2x = 20) resulted in only 0.15% of 10,395 pollinations producing F_1 plants; the same cross at the tetraploid level resulted in 0.58% of pollinations producing F_1 plants.

Figure 17.4 continued

gama grass (Photo A). The plant bears several upright stems (Photo B) that may terminate in an inflorescence (Photo C). Flowers are borne either in single spikes or in a panicle of spikes. Staminate spikelets are arranged in pairs on the upper portion of each spike (Photo C) and pistillate flowers appear alternately along the axis. All species of *Tripsacum* are perennial.

Crosses with *Tripsacum* involving the diploid (2n = 36) and tetraploid (2n = 72) forms have been made. Harlan and de Wet (1977) found that only one strain from among many could be used effectively in crosses with *Tripsacum dactyloides* L. and *Zea mays*. Adequate sampling of potential parental populations is important in making wide crosses.

The *Tripsacum* spp shown in these photographs are maintained in a perennial garden at the CIMMYT research station at Tlaltizapan.

Doubling the chromosome number of the F_1 may increase fertility. If the chromosomes of the parents do not pair at meiosis, doubling the chromosome number gives each chromosome a partner. By doubling the chromosome number of the F_1 hybrids of a cross between the diploid species *Lolium multiflorum* (Italian ryegrass) and *L. perenne* (Perennial ryegrass), reasonable genetic stability is achieved over a number of seed generations (Figure 17.5).

Embryo Culture Techniques

Frequently, the embryo is inhibited in its development by incompatibility with the endosperm. To prevent abortion and loss, the young embryo can be excised and grown on a culture medium. Embryo rescue techniques were first used by Laibach (1929) and since then have been used extensively in many ways to achieve interspecific hybrids.

An adaptation of embryo culture techniques for use in producing hybrids of hard-to-cross sunflower species was reported by Chandler and Beard (1983). Embryos can be excised and cultured three to seven days after pollination and are developed initially on a solid medium containing inorganic components, vitamins, amino acids, and 120 g/L of sucrose. The cultures are transferred to a liquid medium containing only the inorganic components and 10 g/L of sucrose for embryo germination and seedling growth. Such techniques avoid embryo abortion and seed dormancy and decrease the generation time.

Interspecific hybrids were obtained from an otherwise incompatible cross between *Brassica juncea* and *Brassica hirta* (mustards) with $2n = 36$ and $2n = 64$, respectively (Mohapatra and Bajaj, 1987), using *in vitro* culture techniques for hybrid ovules and ovaries. The best response was obtained from ovaries cultured five to seven days after pollination.

Direct Genetic Transfers

Gill and Raupp (1987) reported on the methodology of direct genetic transfer from diploid *Aegilops squarrosa* L. ($2n = 14$) to cultivated wheat (*Triticum aestivum* L., $2n = 42$) as a possible applied plant breeding technique for rapid introgression of useful traits from one species to another. Chromosome pairing is used through recombination as a means to incorporate genes. In spite of the fact that *A. squarrosa* contains the *D* genome found in common wheat, interspecific hybridization between the two species is difficult, and only a few tetraploid hybrids have been reported. Major hurdles include the crossability of the parents, F_1 seed abortion, F_1 hybrid lethality, and high male and female sterility of F_1 hybrids.

Gill and Raupp (1987) suggested that *Aegilops squarrosa* may be the most suitable among the progenitors of wheat for direct introgression for various reasons. First, complete homology exists between hexaploid wheat *D*-genome chromosomes. Second, there is little adverse genetic interaction between the *D* genome of wheat and *A. squarrosa*. Third, *A. squarrosa* has greater useful genetic variability than is found in the other wheat progenitor species, and this genetic variability is readily accessible. Figure 17.6 illustrates the crossing scheme for direct genetic transfer.

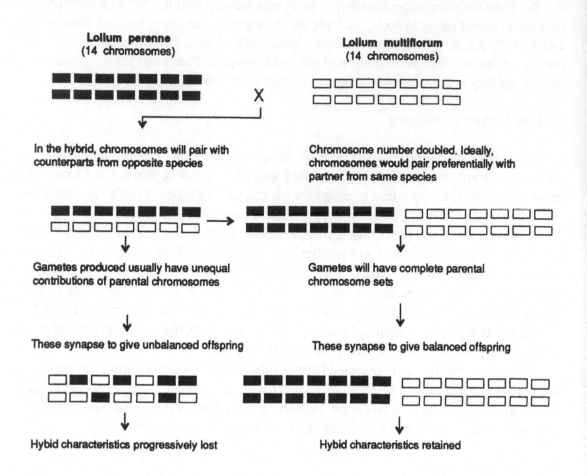

Figure 17.5 Schematic diagram showing chromosome segregation in diploid (left) and tetraploid (right) *Lolium* spp hybrids. Hybridization between the two diploid species *L. perenne* and *L. multiflorum* is achieved readily, and first generation hybrids are fertile. When F$_1$ hybrids form their own gametes, each chromosome is obliged to pair with its similar but nonidentical counterpart, or homologue, from the other species (interspecific pairing). Consequently, when the chromosomes separate again, the new sets of seven carry unequal contributions of chromosomal material from the parent species, leading to a loss of hybrid characteristics when the gametes fuse to produce the next generation.

By doubling the chromosomes to produce tetraploid F$_1$ hybrids using the drug colchicine, reasonable genetic stability is achieved over a number of seed generations. Doubling results in duplication of each chromosome set so that the chromosomes pair preferentially with identical partners. Each gamete receives a full set of seven chromosomes from each parent, allowing hybrid characteristics to be maintained and to be carried into the next generation.

The level of preferential pairing is not yet perfect, but improvements might be made by incorporating a gene or genes to enforce complete preferential pairing.

Combining the features of these two *Lolium* species has resulted in progeny that resemble Italian ryegrass and have good establishment vigor and early spring growth, together with the capacity to give good yields of high quality. Progeny are more persistent than Italian ryegrass, with greater tillering capacity and hence more grazing opportunity. A high degree of winter hardiness, drought tolerance, and disease resistance is found.

Source: Breese and Lewis, 1984, p. 22. Reprinted by permission.

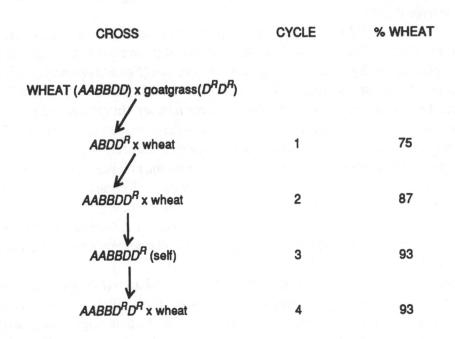

CROSS	CYCLE	% WHEAT

WHEAT (*AABBDD*) x goatgrass($D^R D^R$)

$ABDD^R$ x wheat   1   75

$AABBDD^R$ x wheat   2   87

$AABBDD^R$ (self)   3   93

$AABBD^R D^R$ x wheat   4   93

Figure 17.6 Illustrated in this figure is a crossing scheme for direct genetic transfer from *Aegilops squarrosa* (goatgrass) into hexaploid wheat. The symbol D^R designates the goatgrass genome carrying a resistance gene that is to be transferred to wheat.

In this sequence, *A. squarrosa* is the female parent, a crossing direction associated with hybrid seed abortion, so that it is necessary to rescue the embryos 10 to 12 days following pollination. Another major obstacle is high sterility of tetraploid F_1 hybrids *(ABDD)*. The probability of transferring a single gene to a BC_1 plant through a normal gamete is 0.5. Low BC_1 seed set could be a serious problem, but there is a 50% probability that the gamete may be of the **restitution type** *(ABDD)*, thereby raising the probability of the transfer of any gene to a BC_1 plant to 0.75. In this study, 15 to 40 seeds were produced per plant, and gene transfer to BC_2 plants was virtually assured.

The quick recovery of desirable agronomic types was associated with the fact that the *A* and *B* genomes of the adopted parent were unaffected because they are not present in *A. squarrosa*, and the *D* genome of the wheat parent shows almost complete homology with the *D* genome of *A. squarrosa*. As a result, there was little adverse genetic interaction in the hybrid populations.

Preferably, F_1 hybrids should be produced with wheat as the female parent to improve the viability of F_1 embryos. At least two BC_1 seeds should be obtained for each hybrid combination for the transfer of the desired gene with a probability of success greater than 90%. The partially fertile BC_1 plants should be used as males in a subsequent backcross with hexaploid wheat as the recurrent female parent to recover meiotically stable 42-chromosome BC_2 plants in the cytoplasmic background of hexaploid wheat. There is a strong selection pressure against aneuploid gametes in the pollen, and 21-chromosome gametes will be preferentially transmitted. The BC_2 plants should be allowed to self; the desired gene may be recovered in homozygous condition in subsequent selfed generations.

Source: Gill and Raupp, 1987. Reprinted by permission.

Bridging Crosses

When two species cannot be hybridized by sexual means because of barriers, an indirect approach may be adopted to circumvent the situation by making crosses with related species that can be hybridized.

Hybrids between diploid Italian ryegrass (*Lolium multiflorum*, 2n = 2x = 14) and the hexaploid tall fescue (*Festuca arundinacea*, 2n = 6x = 42) have not been achieved directly. Hybrids between *L. multiflorum* and diploid meadow fescue (*Festuca pratensis*, 2n = 2x = 14) can be achieved with embryo rescue techniques. This cross is sterile at the diploid level, but fertility can be restored by doubling the chromosome number. In general, ryegrass-meadow fescue tetraploid hybrids are not genetically stable but will hybridize with tall fescue. The initial ryegrass-meadow fescue hybrid serves as a genetic bridge to transfer specific characters into tall fescue. The procedure involves repeated backcrossing to the ryegrass-meadow fescue hybrid as the donor species onto tall fescue, the recipient species. Selection is concentrated on only one or a small number of key characteristics of the donor species, in this case the desired attributes of Italian ryegrass. By means of the bridging method, Buckner, Burrus, and Bush (1977) were able to stabilize a 42 chromosome cultivar of tall fescue with certain Italian ryegrass characteristics.

Resistance to tobacco mosaic virus was transferred to tobacco (*Nicotiana tabacum*) from *N. repanda* by means of a bridging cross with *N. sylvestris* (Burk, 1967). *N. sylvestris* will cross with both tobacco species, but a direct cross is not possible. Two subsequent backcrosses with *N. sylvestris* to progeny selected for mosaic resistance produced progeny sufficiently like *N. sylvestris* that they could be crossed with common tobacco. Subsequent backcrosses to common tobacco recovered an agronomic genotype with mosaic resistance.

Protoplast fusion was used to produce a somatic hybrid to serve as a bridge in the development of distant gene combinations (Evans, Flick, and Jensen, 1981) in order to achieve a cross with *N. repanda*. In this case, the somatic hybrid between *N. tabacum* and *N. nesophila* was the bridge cross with *N. repanda*. The bridge hybrid with genetic information from all three species was backcrossed to cultivated tobacco with selection in each backcross generation for disease resistance.

Additional Techniques

A number of other techniques may aid in making wide crosses:

1. Pollinate at different stages in the development of the stigma and style. This technique is based on the premise that early pollinations may succeed better than those made when the stigma normally is receptive because of slow and difficult growth of the pollen tube in foreign tissue.
2. Facilitate organelle transfer by using cytoplasts, incapable of mitosis, as organelle donors (Maliga *et al.* 1982) through somatic hybrids.
3. Rely on cross-pollination by use of male-sterile female plants. Bagging together heads of a male-sterile plant to act as a female parent and an adjacent plant of a different species may produce a chance fertilization.
4. Graft shoots of the female parent to the male species to be used as the male parent, which

may ease hybridization difficulties. Grafting the embryos from germinating seeds of one species onto the endosperms from germinating seeds of another species may help overcome cross-incompatibilities (Hall, 1954).

5. Increase genetic variation through **somaclonal variation** achieved by protoplast-derived plants or through somatic hybrids.

6. Shorten the length of the style to reduce the distance the pollen tube has to travel to effect fertilization. Attempts to hybridize maize and *Tripsacum* was aided when pollen was placed on a shortened silk (Mangelsdorf and Reeves, 1939).

7. Mix pollen from an incompatible parent with pollen from a compatible parent to facilitate interspecific mating (Knox, Willing, and Asford, 1972).

8. Apply growth-promoting substances to the pistil of the female parent to speed development of the pollen tubes and reduce the risk of egg degeneration or flower drop before fertilization can occur. Gibberellic acid applied to autotetraploid barley culms increased pollen tube growth when pollinated with rye (Larter and Chaubey, 1965).

PRODUCING NEW SPECIES AND NEW PRODUCTS

Triticale (X *Triticosecale* Wittmack) is the product of a cross between the genera *Triticum* and *Secale*. This cross was produced initially in 1876, but the progeny were sterile. The first fertile triticale was produced in 1891. The problems, progress, and promise of this crop are well documented (Skovmand, Fox, and Villareal, 1984; Forsberg, 1985). Perhaps triticale (Figure 17.7) symbolizes the challenge to plant breeders to produce new crop species.

Unlike other cereals, which emerged through natural hybridization processes and through selection pressure by evolutionary forces over thousands of years, triticale was synthesized by humans, and despite the protective field laboratory conditions, it took plant breeders approximately one hundred years to produce it. Triticale development does not represent the work of a single scientist or even a single laboratory. Rather, progress in triticale represents the collective contributions from worldwide efforts.

Triticale is a predominantly self-fertilizing crop, but because the rye parent is a near-obligate cross-fertilized crop, some differences in the reproductive behavior compared with wheat may be expected. Initially, fertility, grain shriveling and test weight, preharvest sprouting, late maturity, and excessive height were problems. All of these have been overcome to a considerable degree.

Triticale represents a wide cross, and severe limitations had to be overcome, including cross-incompatibility, embryo starvation, hybrid necrosis, sterility, cytogenetic malfunction, and associated endosperm malformation. Steps in the development of triticale are illustrated in Figure 17.8. The value of chromosome substitutions must be emphasized.

Centuries of evolutionary pressure and years of plant breeding and selection pressure have resulted in wide adaptation among the major cereal crops of the world. International testing programs reveal that triticale is notoriously poor in adaptation. When triticale is subjected to changes in latitude, elevation, temperature, day length, and soil fertility and variations in moisture levels, it exhibits a crippling intolerance to these changes. This poor adaptation generally is believed to result from a lack of evolutionary adaptation.

Figure 17.7   The three cereal heads, from left to right, are wheat, rye, and triticale. The evolution of triticale has been a topic of keen interest, considerable research, and intense effort by plant breeders. Over the years, triticale has raised considerable excitement and publicity as a potentially valuable economic crop, but such enthusiasm has exceeded actual biological and economic development. Lack of markets for triticale as a grain crop has meant that it has yet to find its niche, despite yield performance that is competitive with wheat and adaptation to a diversity of environments, including sandy, droughty, and acid soils and high elevations. As a forage producer and animal feed crop, triticale has potential, but that role has not yet been established.

The importance of triticale may rest with the academic knowledge it has contributed as a vehicle to transfer desirable characteristics to wheat following crosses between hexaploid wheat and hexaploid triticale. Such crosses provide an opportunity for chromosome substitutions to occur whereby the *R*-genome chromosomes from the rye component of triticale are replaced or substituted for by *D*-genome chromosomes from hexaploid wheat. Improvement in the substitutional triticales is due to a combination of factors, including genic, cytoplasmic, and chromosomal interactions.

Photo courtesy Ontario Ministry of Agriculture and Food.

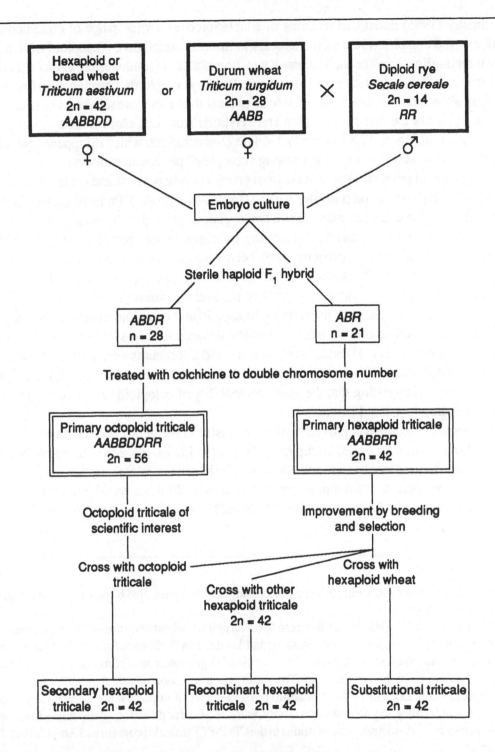

Figure 17.8 Schematic presentation of the steps in the development of triticale. Research emphasis initially directed to octoploid triticale gradually shifted to work on primary hexaploid triticale. A further shift in research emphasis occurred that combined efforts on primary hexaploid and secondary hexaploid forms, including recombined hexaploid forms and substitutional triticale cultivars.

Populations arising from bread wheat-triticale hybridization provide excellent opportunities for

Figure 17.8 continues overleaf

Zillinsky (1985) noted that in terms of adaptation over a wide range of conditions, the presence of the complete rye genome appears to increase adaptation. Octoploid triticale has good winter hardiness, and on the Yunnan-Kweichow plateau in southwestern China at 2,000 m (6,562 ft.) has proved superior to wheat and rye. Primary octoploids, carrying dwarfing genes from both wheat and rye, are more suitable as parents than either bread wheats or dwarf ryes since they possess the full rye complement. Hybrids from octoploid x hexaploid triticale crosses, shown in Figure 17.8 have only a single *D*-genome set, which disappears quickly in the next few generations when grown among "complete" hexaploid triticales.

The successful production of triticale from crosses between wheat and rye has encouraged attempts to combine wheat and cultivated barley (*Hordeum vulgare* L.). Promising allohexaploids have been derived from durum wheat (*Triticum turgidum*) crossed with a South American wild barley (*Hordeum chilense*), and by duplicating the chromosome complement of the hybrid (Cubers *et al.*, 1986), fertile progeny have been obtained. The new alloploid is named tritordeum. Primary tritordeums have small but well-filled, plump seed and, compared with advanced wheat cultivars, yielded 20 to 40% of the average wheat yield.

Based on the good results of hybridizing hexaploid and octoploid triticales, an octoploid tritordeum was synthesized to hybridize with hexaploid forms (Martin, Padilla, and Fernandez-Escobar, 1987). Hybrids between octoploid tritordeums were sterile. Hexaploid x octoploid tritordeum hybrids were fertile, and secondary tritordeums were extracted with increased fertility, suggesting that the genetic variability of octoploid tritordeum can be very useful in breeding hexaploid tritordeum.

Interspecific crossing has proved highly successful in the genus *Musa* (bananas), which contains about 30 to 40 species, all diploids (2n = 2x = 14, 18, 20, 22). The edible bananas commonly are divided into banana and plantain. The banana is palatable when eaten raw, but plantain must be cooked. Two diploid species, both with 2n = 22, are of importance: *Musa acuminata*, carrying the *AA* genomes, and *Musa balbisiana*, with the *BB* genomes.

Figure 17.8 continued

chromosome substitutions to occur. Natural outcrossing could be promoted by planting both wheat and triticale close together.

Crosses of triticale x wheat resulted in zero to seven pairs of wheat chromosomes being substituted for chromosomes from rye (Gustafson, Bennett, and Larter, 1985). Some of these substitutions have been very beneficial, such as the substitution of a pair of *D*-genome wheat chromosomes for a pair of chromosomes (2*R*) from the rye genome. The result was a selection named Armadillo, a hexaploid triticale selected in Mexico in 1968. Armadillo possesses several important characteristics, the most significant being fertility, day-length insensitivity, and one gene for dwarfing. Armadillo proved to be an excellent parent, and most productive strains in the CIMMYT triticale program had Armadillo in their parentage.

Results of chromosome substitutions have made it clear that crosses between hexaploid triticale and hexaploid wheat can be of considerable value to both triticale and wheat by transferring various characteristics from rye into wheat and from wheat into triticale. These transfers are relatively easy to obtain.

Source: Stoskopf, 1985, p. 417. Copyright © 1985; reprinted by permission of Prentice-Hall, Englewood Cliffs, New Jersey.

Dessert bananas have traditionally been triploid, produced by crossing tetraploids with diploids. In the diploid state, thousands of seeds are produced, but in the triploid state, few if any seeds are produced. Once the seedless banana was developed, vegetative propagation made banana production possible, but the scarcity of seeds makes breeding new cultivars a challenging process.

A first step in banana breeding is to understand the inflorescence (Figure 17.9). The banana breeding process is illustrated in Figure 17.10. Interspecific crossing had a beneficial impact on banana breeding in that it introduced much-needed genetic variability. *Musa balbisiana* is a hardier and more drought-tolerant species than *M. acuminata*, so the hybrid groups not only extended the range of plant characteristics and quality features but also helped to extend the geographical distribution of banana out of wet into seasonably drier zones. The *B* genomes from *M. balbisiana* give starchiness and acidity to the fruit, characteristics that are associated with cooking quality, or plantain, bananas. Starchiness, however, depends on ripeness, so potentially sweet (*AAA*) bananas are starchy and nonsweet if cooked green.

Musa balbisiana provided much-needed genes for resistance to Panama disease, leaf spot, and nematodes. Cultivars containing *B* genomes are not as hardy or as resistant as the wild species itself but are superior to the *AA* or *AAA* cultivars. The *AAB* and *ABB* groups are important in India and provide both diversity to diets and hardy genotypes.

Seedless triploid watermelons are a product of different ploidy levels and were first described by Kihara (1951). To create triploid seeds, a tetraploid line is produced by doubling the $2n = 2x = 22$ chromosomes with colchicine. The cross of a diploid pollinator with a tetraploid female will yield triploid seed, but the reciprocal cross with the tetraploid as the pollinator is not successful. Production of seedless fruit by triploid plants in a producer's field is dependent upon the triploid flowers being pollinated by diploid plants. The diploid pollen stimulates **parthenocarpy**, but the ovules fail to develop because of the sterility accompanying the triploid condition. Triploid watermelons have not become widely adapted because of problems in germinating triploid seed and in providing a pollen source.

ALIEN CHROMOSOME ADDITION

The production of triticale has provided encouragement to plant breeders to attempt other alien chromosome additions to wheat. A cross between durum wheat, *Triticum turgidum* (*AABB*) and diploid *Agropyron elongatum* (*EE*) produced hexaploid *Agropyron* of the constitution *AABBEE*, which exhibited good vigor. Additional possible advantages of transferring *Agropyron* genes to wheat are disease resistance, winter hardiness, and semidwarfness.

Restoration backcrossing was proposed by Schulz-Schaeffer and McNeal (1977) using *Agropyron intermedium* (*II*) to produce *AABBII*. Kimber (1984) proposed that numerical measures be derived of relative affinity between the genomes of alien species and cultivated wheat to provide a basis for the logical introduction of variation.

Intergeneric hybrids were attempted between *Hordeum* (barley) and *Secale* (rye) (Gupta and Fedak, 1985). A total of seven hybrids were produced that were vegetatively vigorous,

Figure 17.9 The inflorescence shown in this photo is that of the edible banana, *Musa acuminata*. Banana flowers are produced in nodal clusters that lie under protective bracts that are compressed one upon the other at the apex of the inflorescence. The protective bracts lift successively and fall off as the inflorescence develops.

Three types of flowers are borne along the inflorescence stem. The first clusters are female, normally with only rudimentary anthers, which develop into the familiar hands of bananas. Fruit on these flowers develop without fertilization, a process called parthenocarpy, and are seedless. The inflorescence abruptly changes to imperfect hermaphroditic flowers, which show rudimentary male and female organs. Male flowers are produced as more bracts fall. By this time, most of the female flowers have begun to shrivel, perhaps a means by which nature prevents self-fertilization. Male flowers normally have only a rudimentary stigma and ovary, and may continue to be produced for several months, which helps ensure that male and female flower development coincide. Pollen grains are large and sticky and must be used within a few hours for controlled pollination. Natural pollination is by nectar-seeking insects or birds.

To ensure controlled pollination, reproductive structures can be covered with calico bags or cages, which can be lifted each morning to allow receptive stigmas to be daubed with pollen.

Thousands of seeds are produced when a seedy, wild diploid banana plant is pollinated. Although cultivated bananas have many ovules, seed set is rare and ranges from one to two seeds per bunch. Banana seeds are hard and difficult to germinate and are undesirable in edible bananas.

Bananas are one of the few tropical fruit crops to have become a major item of world trade. Edibility is determined by the capacity of fruit to grow and develop edible parenchymatous pulp without pollination.

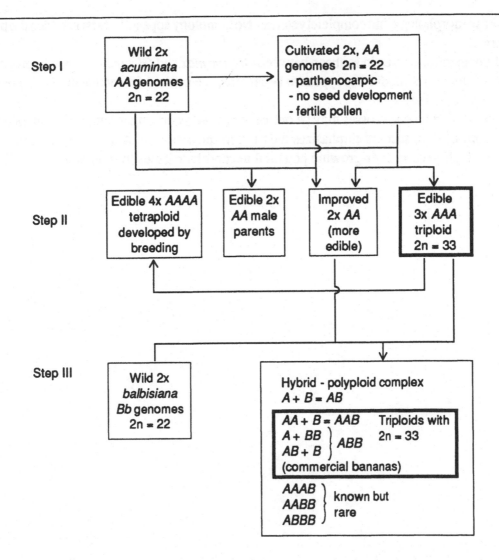

Figure 17.10 The development of dessert bananas is shown in this diagram. Only two species of the 30 to 40 known species of *Musa* are of importance in banana breeding to date, *Musa acuminata* with the *AA* genomes and *Musa balbisiana* with the *BB* genomes. Both are diploids with 2n = 22.

Step I involved the crucial development of parthenocarpy and seed sterility in *M. acuminata* to produce the edible, but economically unimportant, diploid cultivars, *AA*. Parthenocarpy is the capacity of the fruits to grow and fill with edible parenchymatous pulp without pollination. Seed sterility is due to cytogenetic factors and is very important because banana seeds are stony and most unpleasant to encounter unawares. Diploid bananas normally are full of seeds and contain little edible pulp.

From the *AA* cultivars, the commercially important triploids (*AAA*), which are seedless and incapable of sexual reproduction, were produced. A few unreduced triploid egg cells became fertilized to give tetraploid embryos. The value of this phenomenon was that genes from the mother plant did not segregate, and therefore the maternal characteristics that made the fruit so marketable were retained. Triploids can be produced by crossing these tetraploids with seed-producing diploids, using the diploids as the female parent. Tetraploid bananas are seedless.

Musa acuminata banana production was limited, however, by susceptibility to disease. They lacked

Figure 17.10 continues overleaf

ryelike in morphology, but completely sterile. No homology appears to exist between the parent genomes.

Interspecific crosses in soybean between *Glycine max* and *Glycine soja*, a wild progenitor of cultivated soybean, did not appear to be an effective method for increasing yield potential (Ertl and Fehr, 1985).

Success with wide crosses may be one of the greatest challenges tempting plant breeders, but the promise of success emphasizes the need for preservation of wild genotypes and alien genes, which have an ever-growing potential to contribute genes for human use.

Figure 17.10 continued

hardiness and wide adaptation. The introduction of triploids by crossing *Musa balbisiana* was an important step in forming the interspecific allopolyploid cultivars, which are hardier and more drought tolerant than *M. acuminata*.

The level of ploidy normally is established by chromosome counting, but the experienced observer usually can diagnose ploidy level by eye.

The principal value of the *AA* cultivars lies in their breeding potential as male parent breeding stock.
Source: After Simmonds, 1983.

REFERENCES

Abdalla, M.M.F., and J. G. Hermsen. 1972. Unilateral Incompatibility: Hypotheses, Debate and Its Implications for Plant Breeding. *Euphytica* 21:32-47.

Armstrong, K. C. 1987. Chromosome Numbers of Crop Species. In *CRC Handbook of Plant Science in Agriculture*, edited by B. R. Christie. Vol. 1. CRC Press, Boca Raton, Florida, pp. 3-13.

Bates, L. S., and C. W. Deyoe. 1973. Wide Hybridization and Cereal Improvement. *Economic Botany* 27:401-412.

Bourne, T. F., and J. M. Poehlman. 1987. Evaluation of Barley Yellow Dwarf Tolerance from Plant Traits in Two Winter x Spring Barley Crosses. *Euphytica* 36:585-589.

Bowley, S. R., and N. L. Taylor. 1987. Introgressive Hybridization. In *CRC Handbook of Plant Science in Agriculture*, edited by B. R. Christie. Vol. 1. CRC Press, Boca Raton, Florida, pp. 23-59.

Breese, E. L., and E. J. Lewis. 1984. Breeding Versatile Hybrid Grasses. *Span* 27:21-23.

Brezhnev, D. D. 1978. The Utilization of the World Plant Gene Pool of the U.S.S.R. in Distant Hybridization. In *Interspecific Hybridization in Plant Breeding*. Proceedings of the Eighth Congress of Eucarpia, Madrid, Spain. Universidad Politecnica de Madrid, Madrid, pp. 34-50.

Buckner, R. C., P. B. Burrus II, and L. B. Bush. 1977. Registration of Kenhy Tall Fescue. *Crop Science* 17:672-673.

Burk, L. G. 1967. An Interspecific Bridge: Cross-*Nicotiana repanda* Through *N. sylvestris* to *N. tabacum. Journal of Heredity* 58:215-218.

Cauderon, Y. 1979. Use of *Agropyron* Species for Wheat Improvement. In *Proceedings of a Conference on Broadening the Genetic Base of Crops*. Pudoc, Wageningen, pp. 175-186.

———. 1986. Cytogenetics in Breeding Programmes Dealing with Polyploidy, Interspecific Hybridization and Introgression. In *Genetic Manipulation in Plant Breeding*, edited by W. Horn, C. J. Jensen, W. Odenbach, and O. Schieder. Proceedings International Symposium, Sept. 8-13, 1985, Berlin (West), Germany. Walter de Grupter, Berlin, New York, pp. 83-104.

Chandler, J. M., and B. H. Beard. 1983. Embryo Culture of Helianthus Hybrids. *Crop Science* 23:1004-1007.

CIMMYT. 1980. Probing the Gene Pools: Spring x Winter Crosses in Bread Wheat. *CIMMYT Today*, No. 12. 11 pp.

Cubers, J. I., A. Martin, T. Millar, A. Gomez-Cabrera, and A. de Haro. 1986. Tritordeum: A New Alloploid of Potential Importance as a Protein Source Crop. *Crop Science* 26:1186-1190.

Doebley, J., J. D. Wendel, J.S.C. Smith, C. W. Stuber, and M. M. Goodman. 1988. The Origin of Cornbelt Maize: The Isozyme Evidence. *Economic Botany* 42:120-131.

Emsweller, S. L., P. Brierley, D. V. Lumsden, and F. L. Mulford. 1937. Improvement of Flowers by Breeding. *United States Department of Agriculture Yearbook*. Washington, D.C., pp. 890-998.

Ertl, D. S., and W. R. Fehr. 1985. Agronomic Performance of Soybean Genotypes from *Glycina max* x *Glycina soja* crosses. *Crop Science* 25:589-592.

Evans, D. A., C. E. Flick, and R. A. Jensen. 1981. Somatic Hybrid Plants Between Sexually Incompatible Species of the Genus *Nicotiana. Science* 213:907-909.

Evans, L. T., and R. L. Dunstone. 1970. Some Physiological Aspects of Evolution in Wheat. *Australian Journal of Biological Science* 23:725-741.

Fedak, G. 1985. Alien Species as Sources of Physiological Traits for Wheat Improvement. *Euphytica* 34:673-680.

Forsberg, R. A., ed. 1985. *Triticale*. Crop Science of America Special Publication No. 9. American Society of Agronomy, Madison, Wisconsin. 82 pp.

Georgieva-Todorova, J. 1984. Interspecific Hybridization in the Genus *Helianthus* L. *Zeitschrift für Pflanzenzuchtung* 93:265-279.

Gill, B. S., and W. J. Raupp. 1987. Direct Genetic Transfers from *Aegilops squarrosa* L. to Hexaploid Wheat. *Crop Science* 27:445-450.

Grant, M. N., and H. McKenzie. 1970. Heterosis in F_1 Hybrids Between Spring and Winter Wheats. *Canadian Journal of Plant Science* 50:137-140.

Gupta, P. K., and G. Fedak. 1985. Meiosis in Seven Intergeneric Hybrids Between *Hordum* and *Secale. Zeitschrift für Pflanzenzuchtung* 95:262-273.

Gustafson, J. P., M. D. Bennett, and E. N. Larter. 1985. Cytological Studies in Triticale. In *Triticale,* edited by R.A. Forsberg. Crop Science of America Special Publication No. 9. American Society of Agronomy, Madison, Wisconsin, pp. 9-24.

Hadley, H. H., and S. J. Openshaw. 1980. Interspecific and Intergeneric Hybridization. In *Hybridization of Crop Plants*, edited by W. R. Fehr, and H. H. Hadley. American Society of Agronomy-Crop Science Society of America, Madison, Wisconsin, pp. 133-159.

Hall, O. L. 1954. Hybridization of Wheat and Rye After Embryo Transplantation. *Hereditas* 40:453-458.

Harlan, J. R., and J.M.J. de Wet. 1971. Toward a Rational Classification of Cultivated Plants. *Taxon* 20:509-517.

———. 1977. Pathways of Genetic Transfer from *Tripsacum* to *Zea mays. Proceedings of the National Academy of Science USA* 74:3494-3497.

Hogenboom, N. G. 1972. Breaking Breeding Barriers in *Lycopersicum*. 5: The Inheritance of the Unilateral Incompatibility Between *L. peruvianum* (L.) Mill and *L. esculentum* Mill and the Genetics of Its Breakdown. *Euphytica* 21:405-414.

Khan, M. A., and S. Tsunoda. 1970. Evolutionary Trends in Leaf Photosynthesis and Related Leaf Characters Among Cultivated Wheat Species and its Wild Relatives. *Japanese Journal of Breeding* 20:133-140.

Kihara, H. 1951. Triploid Watermelons. *American Society for Horticultural Science* 58:217-230.

Kimber, G. 1984. Technique Selection for the Introduction of Alien Variation in Wheat. *Zeitschrift für Pflanzenzuchtung* 92:15-21.

Knott, D. R., and J. Dvorak. 1976. Alien Germplasm as a Source of Resistance to Disease. *Annual Review of Phytopathology* 14:211-235.

Knox, R. B., R. R. Willing, and A. E. Asford. 1972. Role of Pollen-Wall Proteins as Recognition Substances in Interspecific Incompatibility in Poplars. *Nature* 237:381-383.

Kronstad, W., A. Klatt, A. Bayraktar, and S. Rajaram. 1975. Spring x Winter Crosses for Winter and Spring Wheat Improvement. *Proceedings of the Second International Winter Wheat Conference*, Zagreb, Yugoslavia. Agricultural Institute, Zagreb; Institute for Breeding and Production of Field Crops, Zagreb; Agricultural Experimental Station, University of Nebraska, Lincoln; and Agricultural Research Service, U.S. Department of Agriculture, Lincoln, pp. 105-107.

Laferriere, J. E. 1986. Interspecific Hybridization in Sunflowers: An Illustration of the Importance of Wild Genetic Resources in Plant Breeding. *Outlook on Agriculture* 15:104-109.

Laibach, R. 1929. Ectogenesis in Plants. *Journal of Heredity* 20:200-208.

Lambert, R. J., and E. R. Leng. 1965. Backcross Response of Two Mature Plant Traits for Certain Corn-Teosinte Hybrids. *Crop Science* 5:239-241.

Larter, E. N., and C. Chaubey. 1965. Use of Exogenous Growth Substances in Promoting Pollen Tube Growth and Fertilization in Barley-Rye Crosses. *Canadian Journal of Genetics and Cytology* 7:511-518.

Maliga, P., H. Lorz, G. Lazar, and F. Nagy. 1982. Cytoplast-Protoplast Fusion for Interspecific Chloroplast Transfer in *Nicotiana. Molecular and General Genetics* 185:211-215.

Mangelsdorf, P. C., and R. G. Reeves. 1939. *The Origin of Indian Corn and Its Relatives.* Texas Agricultural Experiment Station Bulletin No. 574. Texas Agricultural Experimental Station, College Station.

Martin, A., J. A. Padilla, and J. Fernandez-Escobar. 1987. The Amphiploid *Hordeum chilense* x *Triticum aestivum* spp. *sphaerococcum*: Variability in Octoploid Tritordeum. *Plant Breeding* 99:336-339.

Metlin, D., W. D. Bluthner, and G. Schlegel. 1974. Additional Evidence on Spontaneous IB/IR Wheat-Rye Substitutions and Translocations. In *Proceedings of the Fourth International Wheat Genetics Symposium*, University of Missouri, Columbia, Missouri, pp. 179-184.

Mohapatra, D., and Y.P.S. Bajaj. 1987. Interspecific Hybridization in *Brassica juncea* x *Brassica hirta* Using Embryo Rescue. *Euphytica* 36:321-326.

Muller, H. J. 1940. *The New Systematics*. Edited by J. Haxby. Oxford University Press, New York, pp. 185-268.

Olsson, G. 1963. Induced Polyploids in Brassica. In *Recent Plant Breeding Research:* Svalof 1946-1961, edited by E. Akerberg, A. Hagberg, G. Olsson, and O. Tedin. John Wiley and Sons, New York, pp. 179-192.

Pinthus, N. 1967. Evaluation of Winter Wheat as a Source of High Yield Potential for the Breeding of Spring Wheat. *Euphytica* 16:231-251.

Pittarelli, G. W., and J. R. Stavely. 1975. Direct Hybridization of *Nicotiana repanda* x *N. tabacum*. *Journal of Heredity* 66:281-284.

Schulz-Schaeffer, J., and F. H. McNeal. 1977. Alien Chromosome Addition in Wheat. *Crop Science* 17:891-896.

Shands, H. L., and L. E. Wiesner. 1991. *Cultivar-Development Part I*. Crop Science Society of America Special Publication No. 17. American Society of Agronomy, Madison, Wisconsin. 164 pp.

Sharma, H. C., and B. S. Gill. 1983. Current Status of Wide Hybridization in Wheat. *Euphytica* 32:17-23.

Simmonds, N. W. 1983. Classification and Breeding of the Bananas. In *Banana Production and Research in Eastern and Central Africa*, Proceedings of a Regional Workshop Held in Bujumbura, Burundi. International Development Research Centre, Ibadan, Nigeria, pp. 62-73.

Skovmand, B., P. N. Fox, and R. L. Villareal. 1984. Triticale in Commercial Agriculture: Progress and Promise. *Advances in Agronomy* 37:1-45.

Solbrig, O. T. 1970. *Principles and Methods of Plant Biosystematics*. Macmillan Co., Collier-Macmillan Co., Toronto, pp. 108-109.

Stalker, H. T. 1980. Utilization of Wild Species for Crop Improvement. *Advances in Agronomy* 33:111-147.

Stebbins, G. L., Jr. 1950. *Variation and Evolution in Plants*. Columbia University Press, New York. 643 pp.

―――. 1958. The Inviability, Weakness, and Sterility of Interspecific Hybrids. *Advances in Genetics* 9:147-215.

Stephens, S. G. 1950. The Internal Mechanisms of Speciation in *Gossypium*. *Botanical Review* 16:115-149.

Stoskopf, N. C. 1985. *Cereal Grain Crop*. Prentice Hall, Englewood Cliffs, New Jersey. 417 pp.

Wernsman, E. A., R. L. Davis, and W. F. Keim. 1965. Interspecific Fertility of Two *Lotus* Species and Their F_1 Hybrids. *Crop Science* 5:452-454.

Zeller, F. J. 1974. IB/IR Wheat Rye Chromosome Substitutions and Translocations. *Proceedings of the Fourth International Wheat Genetics Symposium*. Columbia, Missouri, pp. 209-222.

Zillinsky, F. J. 1985. Triticale: An Update on Yield, Adaptation, and World Production. In *Triticale*, edited by R. A. Forsberg. Crop Science Society of America Special Publication No. 9. American Society of Agronomy, Madison, Wisconsin, pp. 1-7.

Zohary, D. 1973. Gene-Pools for Plant Breeding. In *Agricultural Genetics*, edited by R. Moav. John Wiley and Sons, New York, pp. 177-183.

CHAPTER 18

Mutation Breeding

Spontaneous **mutations** and subsequent recombination through hybridization followed by natural selection have provided the sources of genetic variability whereby the forces of nature and eventually human effort directed crop evolution. It was through spontaneous mutations that progress toward crop domestication took place, and subsequent mutations have provided valuable new variation to improve crops over the centuries.

A myriad of variants arising from mutations have occurred and are still occurring, some desirable and some undesirable. From among this mass of variants, those that appear to be of potential value are tested or hybridized to adapted cultivars for further selection and evaluation. Many mutations are deleterious and must be discarded.

Variation always has been important for plant improvement and will continue to be important in the future. Modern agricultural systems continue to demand higher yields and improved adaptation to changing production practices. As crops are forced into the fringe areas of production, wide adaptation and new sources of variability are not only critical but take on an increasing sense of urgency. Awareness of nutritional aspects, and changes in industrial processing techniques and storage and transportation systems also demand new genetic variation.

Ecologists concerned about **genetic erosion** associated with intense production of a single crop of the highest-yielding cultivar over a major food-producing region scour the earth for new genotypes. Recognition of the value of genetic variation has given new meaning to the preservation of existing genotypes. These factors have contributed to the concern that spontaneous mutation may not provide variation fast enough to keep pace with accelerating human needs. The ability to induce mutations was therefore embraced as a possible new and productive means of crop improvement.

INDUCED MUTATIONS

Physical Mutagens

The ability to induce mutations using **x-rays** was demonstrated by Muller (1927) in the fruit fly (*Drosophila melanogaster*) and by Stadler (1928, 1929) in crop plants. The possibilities inherent in the ability to produce mutations artificially were recognized quickly, and a wave of optimism developed that new crop cultivars would be produced more rapidly and more economically than by conventional breeding methods involving hybridization and selection. But within a decade of the release of the knowledge about mutation-causing x-rays, plant breeders abandoned their interest in induced mutations. One exception was a group of Swedish scientists under Gustafsson (1975) studying barley.

The initial work with x-rays was directed equally at biophysical studies of the effects of radiation exposure upon biological systems and the production of useful new variants. Although x-rays were capable of inducing mutations and of increasing mutation rates to about one thousand times the spontaneous rate in barley at near lethal levels, many undesirable, lethal, or repetitious types of mutations arose (Figure 18.1). Some mutations are useful, including positive increases in grain yield, seed size, plant height, maturity, leaf morphology, and malting

Figure 18.1   The white barley plants in this photo exhibit mutations in which chlorophyll and carotenoid pigments are lacking. These albino plants cannot use light energy in photosynthesis and die once the reserve food in the endosperm is depleted.

Chlorophyll-deficient mutants in barley have been studied extensively (Gustafsson, 1974) and, based on visual differences, have been divided into at least seven classes, as follows:

1. Albino plants, which are shown in the photo, are white, lack carotenoids and chlorophyll, and generally have small plastids.
2. Xantha forms generally have little or no chlorophyll but have carotenoid pigmentation and are yellow. Plastid morphology varies.
3. Alboviridis mutants have different rates of plastid development at the base and tips of each leaf and exhibit different colors in these leaf areas.
4. Viridis forms are light green in color.
5. Tigrina mutants exhibit transverse stripes along the leaf with alternating narrow bands of green and yellow or brown where pigment destruction has occurred.
6. Striata forms have yellow or white longitudinal bands alternating with green.
7. Maculata mutants have spots where chlorophyll and/or carotene has been destroyed.

The various chlorophyll-deficient mutants generally can be accounted for by single, recessive mutated genes. Their phenotypic expression can be influenced by the environment, and each class is highly heterogeneous. Irregularities in segregating populations from crosses indicate that gene complementation, gamete or zygote elimination, and maternal inheritance also occur.

Photo courtesy Ontario Ministry of Agriculture and Food.

quality, but many are hard to detect; large populations are required for screening, and most mutant plants cannot be used directly for cultivar release. One barley mutant designated "erectoides" has compact spikes and very stiff straw, useful agricultural features, but when field tested, only a few plants exhibited any superiority over the mother cultivar (Gustafsson, 1965). In addition, seemingly desirable mutations are associated with poor genotypes, possibly as a result of additional mutation. Despite considerable effort at mutation breeding with numerous crops, few practical results have emerged. One researcher likened mutation breeding to firing a rifle bullet into an automobile engine with the hope of improving performance.

Some positive results were produced, however. The barley cultivar Jutta, introduced in 1944, was, according to Sigurbjornsson (1975), the first to be developed by x-ray irradiation. The cultivar was released by the German Democratic Republic in 1955. The barley cultivar Pallas, an x-ray-induced mutant released by the Swedish team in 1960, has been exploited as a parent for six other cultivars (Konzak, Kleinhofs, and Ullrich, 1984).

Following World War II (1939-1945), knowledge of atomic energy opened the way for the use of several ionizing mutagenic agents besides x-rays (Table 18.1). The next step was to determine effective dosage rates and handling.

Optimum dosage rates of x-rays and gamma irradiation were reported by Micke, Hsieh, and Sigurbjornsson (1972) to be 15 to 30 k rad for indica rice cultivars and 15 to 25 k rad for japonica rice cultivars. The dose range of neutrons suitable for rice was approximately one-tenth the gamma dose at 1 to 3.5 k rad. Low dosage rates of 5 k rad were suggested by Yonezawa and Yamagata (1977) to minimize chromosomal damage and drastic mutations.

At least 200 to 300 seeds should be used for treatment and about 3,000 to 5,000 seeds should be harvested in the M_1 generation. From each M_1 panicle, 15 to 25 seeds may be planted in rows for identifying mutants on a single-plant basis in the M_2 generation. The M_3 generation may be grown again in progeny rows and rechecked for mutants. Each promising and uniform row can be harvested in bulk as lines and subjected to preliminary yield trials.

Fifteen to 19 k rad of gamma rays were used to treat soybean seeds (Wang *et al.*, 1984). The frequency of qualitative character mutations was low, whereas the frequency of exceeding the parental population was high for main quantitative characters. Thermal neutrons were considered more effective than gamma and beta rays. Radiation injury was apparent for such characteristics as seed weight and number of seeds per pod. Segregation of characters was mainly in the M_3 generation.

X-ray dosage rates recommended for seeds of five crops were given by Gustafsson and von Wettstein (1958) as follows:

Barley	10 to 15 k rad
Flax	40 to 50 k rad
Mustard	greater than 100 k rad
Pea	5 to 15 k rad
Wheat	15 to 20 k rad

Briggs and Knowles (1967) suggested that different cultivars react differently to mutagens and advised the treatment of several different cultivars. If dosage rates for a crop are unknown, it was proposed that treatment kill 50 to 70% of the seeds or developing seedlings.

Mutagens can be applied not only to seeds but also to buds, pollen, somatic tissue and cells, tubers, or bulbs.

Table 18.1   Types of Ionizing Radiations Most Commonly Used and Their Characteristics

Type of Radiation	Source	Hazard	Penetration into Tissues
X-rays	X-ray units	Dangerous	Penetrating (up to many cm)
Gamma rays	Radioisotypes and nuclear reactors	Dangerous	Very penetrating
Neutrons (fast, slow, and thermal)	Nuclear reactors or accelerators	Very hazardous	Up to many cm
Beta particles	Radioactive isotypes or accelerators	May be dangerous	Up to several mm
Alpha particles	Radioisotypes	Very dangerous internally	Small fraction of a mm
Protons or deuterons[a]	Nuclear reactors or accelerators	Very dangerous	Up to many cm

[a] A proton is the nucleus of the common isotype of hydrogen; a deuteron is the nucleus of the heavy isotype of hydrogen.

Note: The effect of radiation generally is proportional to the energy absorbed by a particular tissue, and so depth of penetration is important. The effect of ionizing radiation results from excitation, the transfer of energy, and ionization when radiation interacts with matter. The rad (radiation absorbed dose) is a unit of absorbed dose of ionizing radiation and is equal to 100 ergs/g. The oldest unit, the Roentgen (abbreviated as r or R), is a unit of exposure and not a unit of absorption. Great care must be exercised in the use of ionizing radiation because exposure can result in gene mutation, chromosomal aberration, change in chromosome number, inhibition of cell division, sterility, cell death, destruction of the nuclei, and retardation or stimulation of growth rates. A high dose rate may increase mutation rate but does not necessarily result in the increase of practical, desirable mutants. An optimum mutation rate and optimum mutagen dose based on theoretical considerations was suggested by Yonezawa and Yamagata (1977). For cereal crops such as rice and barley, mutagen doses lower than 5,000 r or 5 rad for gamma or x-rays are appropriate for obtaining directly applicable mutants. This level produces a low mutation rate. High mutation rates seem to be limited to the production of some rare breeding stocks. The availability, ease of operation, and versatility of x-ray machines means that they are the most widely used source of ionizing radiation.

Chemical Mutagens

A number of chemical mutagens have been identified, such as myleran, which like x-rays, gamma rays, and neutrons can induce high ratios of chromosome aberrations and mutations. Ethylene-imine can induce about equal frequencies of chromosome aberrations to mutations. Mutagens such as ethyl methanesulphonate, diethyl sulphate, sodium azide, and certain base substitution and nitroso compounds appear to induce higher proportions of mutations than chromosome aberrations (Konzak, Nilan, and Kleinhofs, 1977).

Sodium azide was developed in the late 1960s and is recognized as one of the most potent mutagens available for higher plants (Kleinhofs, Owais, and Nilan, 1978). Azide is unique in that it induces very high frequencies of gene mutations but is ineffective in producing major chromosomal changes in plants (Nilan, 1981). It appears to induce point mutations, but whether these mutations are small, deleterious, or true base changes has to be resolved. Owais *et al.* (1978, 1979) determined that it is not the inorganic azide *per se* that acts as the mutagenic agent but rather an organic metabolite synthesized in azide-treated barley, bacteria cells, and probably other plants. The azide metabolite is only one of three mutagenic metabolites, the others being atrazine and 1, 2-dibromo-ethane.

Treatment to induce mutations in plants involves soaking seeds, buds, roots, and dormant cuttings in a solution of the chemical mutagen. Most work has been conducted on ethyl methanesulphonate because it has been available and recognized longest. Concentrations of this compound normally are in the 0.1 to 2.0% range. Treatment should last long enough to ensure infusion of the mutagen into the plant tissue. Wang *et al.* (1984) treated soybeans with 0.2 to 0.4% ethyl methanesulphonate and diethyl sulfate. Micke, Hsieh, and Sigurbjornsson (1972) treated rice with rates of various compounds ranging from 0.01 to 2.0%. All mutagenic agents should be used with extreme caution because many are powerful **carcinogens**.

The concept of preventing gene expression by blocking **translation** of messenger RNA may provide a new method of studying gene action in the cells of higher organisms and hence of studying mutation (Marx, 1984). Procedures aimed at blocking gene expression by inhibiting gene translation, the second major step in gene expression, may provide a new method of inducing mutations. The first step in gene expression is the copying of the gene into messenger RNA, after which the nucleic acid message is translated and the protein is produced. Translation can be prevented by introducing into cells RNA complementary in structure to the messenger RNA for the protein in question.

The ability to induce mutations has progressed to the point where it provides a viable additional option to plant breeders for creating useful genetic variability. Mutation breeding initially was interpreted as selecting mutants for direct release as commercial cultivars. More aptly, inducing mutations may be regarded as a technique for creating useful genetic variability, which in combination with hybridization and selection can be used to achieve desired plant breeding goals. Mutation breeding, therefore, is the use of mutants, either directly or indirectly, for plant cultivar development.

Mutation breeding is a complementary method of plant improvement whereby existing germplasm variability can be supplemented by induced mutations. Variation resulting from induced mutations can provide genes at previously unknown loci as well as a new series of alleles with similar phenotypic expressions but in an improved genetic background.

Because the application of mutagenic agents cannot be directed to force a desired mutation, screening mutated populations is not in keeping with the recommendation that plant breeding programs be oriented toward specific objectives. It is highly advisable that mutation induction programs be directed toward well-defined objectives. To do this, a desired mutant trait is selected and a program of screening developed within the limits of the resources available that will provide the maximum probability of identifying the desired mutant.

The need for a clear objective for mutation induction programs becomes apparent when the breeder realizes that the most frequently observed mutations are those that have no practical value and are of least interest. In sorting out mutations, it is imperative that a clear goal be stated in advance. The need for an effective and efficient screening procedure is also emphasized by this fact.

Redei (1974) and Yoshida (1962) suggested that it is more efficient to grow one offspring from each of a large number of M_1 plants grown close together to reduce the number of tillers than it is to grow a large number of offspring from fewer M_1 plants. This practice is especially important for qualitative traits; and may hold true for quantitative traits, but for the latter, selections in subsequent generations should be grown at several locations and over several years to verify the quantitative trait.

INDUCED VERSUS SPONTANEOUS MUTATIONS

Plant breeders working with induced mutations have raised the question whether induced mutations differ from spontaneous mutations. Research has convincingly shown that no major detectable differences exist between induced and spontaneous gene and chromosome mutants (Brock, 1980; Konzak, Nilan, and Kleinhofs, 1977; Nilan, Kleinhofs, and Konzak, 1977: Nilan and Vig, 1976). The only difference between the two mutant groups is that older spontaneous mutants have gone through an evolutionary process of coadaptation through recombination and natural selective forces. Newly arisen spontaneous or induced mutations have not had time to be shaped into such complexes. It is the task of the plant breeder, through mutation breeding, to find a "happy home" for new mutations, that is, the trait must be incorporated into a plant combining other genetically acceptable traits.

The equal status of induced and spontaneous mutations has caused plant breeders to question the value of induced mutations. Sigurbjornsson (1975) suggested that existing mutants are being used for plant improvement more frequently than new mutants are being induced. Based on the numerous reports of mutation experiments in the literature, Gustafsson (1974) concluded that the blind induction of mutations in higher plants is producing "worthless junk" instead of meaningful selections (Figure 18.2).

On the other hand, although induced mutations can result in practically any genetic changes that occur naturally, the advantage is that the desired changes can be incorporated directly into an agronomically desirable genotype with a good chance of leaving the residual genotype reasonably intact. Sigurbjornsson (1975) pointed out that the gene pool of barley, possibly one of the most highly bred crops for the temperate zone, has been screened thoroughly for useful genes. In designing the best ideotype barley cultivar, the breeder inevitably and repeatedly comes up against barriers that cannot be penetrated with existing and usable genotypes. If new ideotypes with advances in yield are to be achieved, induced mutations must become an even more important component of barley development.

Figure 18.2   The heads of triticale and barley (Photos A and B, respectively) exhibit mutations. These mutations are visible and easily identified, but their practical value is questionable. Generally, 90% of induced mutations are deleterious, and careful testing is required to evaluate each one.

Wheat with branched heads, illustrated by the two heads on the left in Photo A, was described by Sharman (1944) as "miracle"-headed wheat and was produced through hybridization. Years later, Koric (1973) produced similar branching in wheat and noted that branching is controlled by several genes, the phenotypic manifestation of which is repressed by an inhibitor. Branched heads may be of value in increasing grains per head only if adequate provision is made to increase photosynthesis to fill the larger sink.

Figure 18.2 continued

Photo B shows barley heads with normal awns (left), no awns (center) and multiple awns (right). Multiple awns appear to result from the development of the rudimentary outer glumes. Awns are an important source of photosynthate, and under certain conditions with other things being equal, awned heads normally outyield awnless genotypes. The multiple-awned head has not proved to be of commercial value, possibly because of associated deleterious mutations or because of excess shading and lack of free carbon dioxide movement.

Photo courtesy Ontario Ministry of Agriculture and Food.

In view of the success of induced mutations in barley, Sigurbjornsson (1975) expressed surprise that mutation breeding is not used as a major method of barley improvement anywhere and that very few breeders are making use of mutagens for the purpose of barley breeding. The majority of released mutant barley cultivars are byproducts of academic studies on mutations.

Another question to be answered by plant breeders who are considering the use of induced mutations is whether entirely new mutant forms can be produced that have not been observed among the spontaneous genetic variability. Undoubtedly, mutagenic agents can produce mutants that have not arisen in recent evolutionary history but which arose during the plant's evolution and which have since been lost and hence not encountered by the plant breeder.

The extent to which induced mutations can be developed is illustrated by the variants for "waxless" plant surface. The microscopic morphology of the leaf surface wax is altered in the mutants and its quantity and chemical composition changed. The wax cover on barley leaves may be reduced 17 to 48%. Natural variability for this trait was confined to a few spontaneous mutants at six controlling loci. By using a wide array of mutagens, which appear to vary in the frequency with which they induce mutations at different loci (Sogaard, 1974), a total of 1,302 induced mutations have been produced (Lundquist, 1976). Several of these loci have been located on chromosomes 1, 2, 3, 4, 5, and 7.

Appropriate genetic tests for induced chlorophyll-deficient mutants (Figure 18.1) have revealed 600 to 700 loci controlling chlorophyll development (Nilan, 1981). Induced mutations play an important role in identifying specific genes that may be transferred eventually by genetic engineering.

THE NATURE OF MUTATIONS

Despite plant breeders' strong desire for a mutagenic agent that can be directed to mutate a specific gene, the induction of mutations is essentially a random process that cannot be directed. The majority of mutations are recessive (A to a) or partially recessive. Improvement in barley by means of mutation breeding has been due to the selection for and retention of recessive traits. Recessive traits are expressed in the homozygous condition achieved by self-fertilization.

The interpretation of numerical studies on mutation frequencies is a challenging task. Mutation frequencies may be reported as per plant, per spike, or per 100 M_2 plants. Mutation frequencies as high as 40 to 57% of M_1 barley spikes were reported following treatment with ethyl methylsulfonate and 17% per spike for x-rays (Nilan, 1964). A mutation is seldom induced in a single gene without some other genetic changes occurring in the treated genotype. Often, relatively high doses of a mutagen are applied to obtain maximum mutation frequencies or with the objective of mutating a specific gene or character. At such dosage levels, the possibility is high that other genes will be mutated or a group of genes simultaneously deleted. Selection for a specific mutant genotype through the use of high dosage levels may result in the desired mutant, but one accompanied by a number of additional attributes that may not be desirable. This is the reason Yonezawa and Yamagata (1977) recommended moderate to low dosage rates. Recognizing that many induced mutations are deleterious and that an apparently desirable mutation may be accompanied by other undesirable mutations, plant breeders should select several of the desired mutant phenotypes to ensure that the mutant being sought is obtained. Perhaps only 10% of the total number of mutations recorded in rice and barley can be regarded as favorable genetic changes.

Mutations may occur in both qualitatively and quantitatively inherited traits. A recessive point mutation can be identified in the M_2 by selfing, but quantitative mutations can be subtle and less apparent. Induced mutations, however, can generate useful variation in polygenic features if appropriate selection is conducted. A third category, that of dominant mutants, is difficult to identify.

MUTATION BREEDING SYSTEMS

Successful mutation breeding programs require large plant populations for two reasons. First, only 10% of induced mutations can be expected to be useful. Second, minor but useful mutations are difficult to identify by visual means. These factors mean that a large number of seeds must be treated and several thousand M_1 plants are needed to build up an M_2 population with sufficient variability for successful selection opportunity.

Population size depends on the mutation frequency, whether selection is aimed at mutations that result from chromosome changes, from several recessive genes, from single recessive genes, or from single dominant genes. Mikaelsen (1980) suggested that insufficient population size may be a major reason for many failures in mutation breeding.

In crops that tiller, all the primary tillers can be used to identify mutations because they represent the maximum potential for induced genetic variability. Secondary tillers may repeat the mutations found in the primary tillers.

Bulk breeding methods are suggested as suitable for mutation breeding because they provide the highest statistical probability for the appearance of a single mutant in the plant population sampled. Population size and breeding system may be influenced by other considerations, such as resources available, economics, and breeding objectives.

In addition to gene or point mutations, chromosome mutations occur also. Mutagenic agents, particularly ionizing radiations, may break chromosomes, resulting in many structural arrangements. Chromosome mutations include **translocation, inversion, duplication**, and **deficiency**.

A mutation breeding method called recurrent mutation induction, designed to overcome sterility associated with multivalent chromosome associations in autoploids, was suggested by Gaul (1964). Autoploid populations from widely different sources are given a mutagenic treatment at least once per generation to induce gene mutations and chromosome changes. Then the treated autoploids and the F_1 plants are crossed in as many combinations as possible for the purpose of inducing as much structural and genetic variation as possible to discourage multivalent associations. Finally, selection is made for fertile, desirable types.

PRACTICAL RESULTS FROM MUTATION BREEDING IN SEED CROPS

The ability to induce mutations in plants has been enthusiastically received by plant breeders, and sweeping statements have been made about large numbers of useful genetic changes that could be induced. In addition, generalized comparisons have been made with conventional breeding methods in regard to the relative ease, speed, and economic advantages of induced mutations. Now, sixty years after the discovery that mutations can be induced, we might well ask what has been achieved with mutation breeding.

Mutation breeding appears to have matured at last. It is now recognized more clearly as a supplement to conventional hybridization methods, to be used in particular circumstances and not as an alternative to traditional systems.

Mutation breeding generally has been more successful in self-fertilized than in cross-fertilized crops because of the recessive nature of most mutations. Mutations are more easily recognized in diploid than in polyploid species, but examples of success with hexaploid wheat and oat crops are found. Verticillium wilt resistance, for example, was obtained in the polyploid peppermint (*Mentha piperita* L.) following screening of over 100,000 stolons (Murray, 1969). Verticillium wilt resistance is regarded as a most difficult achievement using conventional breeding methods.

The development of a new conceptual model, or **ideotype**, was achieved from mutant genetic variability (Adams, 1982; Blixt and Vose, 1984; Davies, 1977). The early determinate bush habit trait from an induced mutant of a dry bean cultivar (*Phaseolus vulgaris* L.) provided germplasm necessary for cultivar improvement through hybridization. Other mutations and crosses provided the necessary genetic variability to combine traits to produce a bean with a tall, erect, narrow plant profile and with a strong central axis and a few erect basal branches, superior lodging resistance, and excellent pod development. A drastically new ideotype comparable to that produced in beans has been found in pea (*Pisum sativum*) as a result of induced mutations (Figure 18.3). Modifications to nodulation through mutations have exciting implications (Jacobsen, 1984). The success in ideotype development with mutations offers yield advances when coupled with production practices.

Forty-nine new grain legume cultivars produced either directly or indirectly from physical irradiation treatments were listed by Micke (1984). Nilan (1981) reported that 465 cultivars of sexually and vegetatively reproducing crops have been released. These have considerable economic impact and owe some of their production advantage to an induced gene or chromosome mutant.

Mutation breeding for disease resistance in sexually propagated crops has not been particularly successful perhaps because of the dominant nature of resistance genes and because vertical resistance lacks stability. Identification of mutant genes offering horizontal resistance has not been possible. Nevertheless, a number of examples of disease resistance achieved through mutation were described by Konzak, Kleinhofs, and Ullrich (1984). Notable instances include resistance to Victoria blight (*Helminthosporium victoriae*) in the hexaploid oat crop, powdery mildew resistance (*Erysiphe graminis*) in barley, and downy mildew resistance (*Sclerospora graminicola*) in pearl millet.

Mutations for early maturity have been induced frequently in a wide range of crops. Mutants for earliness provide an example of mutants affecting physiological traits. Using

Figure 18.3   This leafless type of pea arose as an induced mutation and has been released as a cultivar in Poland (Jaranowski, 1977). The mutation changed leaflets to tendrils, thereby facilitating sunlight penetration to cause cell differentiation in the stem, lodging resistance, and suitability for combine harvesting. Photosynthesis is conducted in the green stems and tendrils, sepals and pods, resulting in a high harvest index and high potential yield per unit area of soil.

Micke (1984) reported on mutation breeding work with grain legumes, and through a number of mutation breeding projects in different countries involving grain legume species, it has been demonstrated that mutation induction is a good tool for supplementing available genetic variation. High-yielding cultivars have been developed from induced mutants.

Jacobsen (1984) demonstrated that the symbiotic behavior of a host plant can be modified by means of induced mutations. Out of a population of 250 M_2 families, one mutant was found that could form nodules at high nitrate concentrations and nodulated abundantly with several *Rhizobia* strains, both in the absence and presence of nitrate. Root morphology was changed also, possibly as the result of a pleiotropic effect. This work shows that pea mutants with modified nodulation capacities can be obtained readily by treatment with ethyl methylsulphonate, which has implications for breeders.

Photo courtesy Ontario Ministry of Agriculture and Food.

sodium azide, Kivi (1981) was able to induce 22 early six-row barley mutants in an already early-maturing cultivar by screening 350,000 M_2 plants. Two of the 22 mutants proved worthy of advancement in testing beyond M_5 for possible direct use as cultivars. Earliness has been induced in rice, soybeans, maize, cotton, peppers, beans, peas, castor beans, and wheat (Konzak, Kleinhofs, and Ullrich, 1984).

The discovery of high lysine levels associated with visible opaque and floury endosperms produced by gene mutations in maize led to discoveries of other mutants with changed protein content and/or composition. Other quality improvements associated with mutations include barley free of proanthocyanidin, which prevents the formation of chill-haze in beer produced from barley malt (von Wettstein *et al.*, 1977). Chemical stabilization is not required for beer manufactured from barley malt of this mutant type, resulting in economic savings to brewers and improved purity. Mutants free of proanthocyanidin are induced readily and are identified in an M_2 population by a lack of the purple anthocyanin pigment most evident at the heading stage.

Low-erucic-acid rapeseed mutants (*Brassica napus* L. and *B. campestris* L.), either induced or spontaneous, have given rise to a subspecies named canola that is low in erucic acid and glucosinolates. Canola has revitalized these species as edible oil crops adapted to cool northern climates where other oilseed crops, such as soybean, cannot be produced (Axtell, 1979). The rapeseed oil meal from low-glucosinolate mutants can be used for animal feed (Robbelen and Nitsch, 1975). Other breeders are searching for rapeseed mutants with high erucic acid, in contrast to canola, for production of industrial oil (Axtell, 1979).

Interest in commercial F_1 hybrid seed production has raised interest in mutants affecting cross- or self-fertility. Evidence of nuclear and cytoplasmic forms of male sterility suggests that mutations for these forms can be induced in almost any crop if large enough populations are screened.

MUTATIONS IN VEGETATIVELY PROPAGATED CROPS

The large number of released crop cultivars based on mutated characters is a reflection of the increasingly important role induced mutations have played in barley, rice, wheat, beans, peas, and oilseed crops. In vegetatively propagated ornamental plants, the potential of induced mutations also can be demonstrated. In contrast, however, success with vegetatively propagated crop plants and woody plants has been rather limited. Reasons for this include the difficulty of producing large populations of mutagen-treated plants from which the breeder may select, the lack of efficient screening techniques applicable in early stages of development, and the long time required to assess clearly the superiority of a selected genotype under field conditions, especially in woody plants (Figure 18.4).

One of the problems of applying mutagens to multicellular **explant** material from vegetatively propagated species is the formation of a **chimera**. Buds, petioles, apices, tubers, or bulbs may be treated. Chimera formation has a number of unfavorable consequences, such as the reduced possibility of recognizing mutations, which prevents early selection and obliges the breeder to propagate all the material once or twice to arrive at complete chimeras (Figure 18.5) and true mutants before selection can be applied.

A.   B.

Figure 18.4   The genus *Populus*, a fast-growing woody tree species used for biomass production for fuel, is in theory an ideal subject for a mutation breeding program. Poplar trees are highly heterozygous, and therefore induced mutants may show phenotypically without the necessity of a seed-propagated generation. They normally are propagated vegetatively, so that desirable new mutants can be easily multiplied, and in some regions schemes exist for the dissemination of new clones to the forest industry.

Hybridization of woody plant parents, each having different desirable traits, is an attempt to synthesize new genotypes that combine traits from both parents. When hybridization is undertaken in fruit trees, new cultivars are produced that in most cases differ from both parents with respect to several characteristics. Desirable attributes of both parents may be lost through random segregation.

Mutation breeding of vegetatively propagated crops aims at the direct improvement of a given cultivar for a limited number of traits. Objectives need to be defined clearly. Photo A, for example, shows one-year-old poplar plants. Asexual propagation of clones leads to extremely uniform rows of plants in the nursery, making the selection of plants with vegetative characteristics possible, such as the vigorous plant in Photo A. The phenotypic expression of leaf size in Photo B provides increased photosynthate area and may be a valuable attribute on which to select.

Photo courtesy Ontario Tree Improvement and Forest Biomass Institute, Ministry of Natural Resources, Maple, Ontario.

The number of initial cells present in a growing point subjected to mutagens cannot be determined reliably but can be estimated indirectly from the segregating seeds in the M_2 and later generations. The segregating ratio of example A in Figure 18.5 with a heterozygous *Cc* genotype is 3:1. In example B, the segregating ratio is approximately 8:1, but the amount of **diplontic selection** in the M_1 plants will determine the ratio. Often the ratio is smaller than

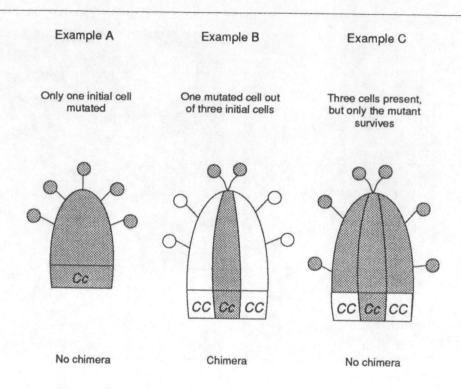

Figure 18.5   These schematic drawings, presented to explain a chimera, represent the growing point of a vegetatively growing plant that has been treated with a mutagen to induce mutation. Most plant species contain several initial cells in their embryonic growing points. Treatment with a mutagen may cause a distinct mutation, such as that for flower color (C-), in one of these cells, and if this cell is propagated into a plant, all flowers will exhibit the same flower color (example A). Plants subsequently asexually propagated all bear the mutated gene and all will exhibit the same flower color.

In example B, three initial cells are present, and a distinct mutation in one of these cells has produced a chimeral plant. If Cc has dark flowers, CC has light-colored flowers. Both types of flower color may appear on a single plant.

Competition may occur between different cell types in a chimera. This is known as diplontic selection and is illustrated in example C.

Mutant cells very often are not fully competitive with the nonmutant cells, and this can result in a low frequency of mutants and a narrow mutation spectrum. Simply stated, chimeras are plants that are composed of two genetically different tissues. Chimeras may affect flower color and pattern, which results from two different color genes forming segments of the flower tissue.

Source: Adapted from Gottschalk and Wolff, 1983.

expected because of reduced activity of the heterozygous meristematic cells, resulting in an additional deficit of recessives in the M_2 families. **Zertation** of the pollen tubes containing the mutant genes in relation to the style with the nonmutant dominant alleles can cause variation in the segregating ratios. Reduced speed of pollen tube growth can result in a reduction of fertilization chances for the mutant germ cells. The plant in example C of Figure 18.5 will produce seed with a 3:1 segregating ratio.

The African violet (*Saintpaulia ionantha*) has been chosen for use in studying the effect of various mutagenic agents because its unusual propagation by means of leaf cuttings results in the production of a number of young plants at the base of the petiole. Each plantlet apparently arises from only one epidermal cell. After irradiation, a very high percentage of plantlets turn out to be nonchimeral, and since diplontic selection is reduced to a minimum, the mutation frequency is high. Broertjes (1968b) reported that about 30% of the plantlets show visible mutants after an optimal, acute x-ray dose of 5 k rad. The mutation spectrum is very wide and is expressed in flower color, flower design, leaf pattern, and leaf shape (Figures 18.6 and 18.7).

To evaluate the effects of x-ray dosage and rate in African violet, Broertjes (1968b) used three criteria: (1) the percent survival in terms of the control of irradiated leaves, (2) the production of young plants, and (3) the mutation frequency. The critical dose rate appears to be 5 to 7 k rad/minute.

Dahlia (*Dahlia variabilis*) is regarded as an excellent species for mutation breeding because of the extremely high degree of heterozygosity. As a result of numerous cross-breeding programs since 1900 and considering the fact that dahlia is an octoploid that usually is cross-fertilized and is vegetatively propagated (Broertjes, 1968b), considerable progress has been made. Dahlia flower color is influenced by four different pigments, two anthocyanins and two flavones in simplex, duplex, triplex, and quadruplex condition, with cumulative or partially cumulative action. Interaction among genes, as well as inhibitors for flower color, result in wide variability.

Dahlia tubers are irradiated soon after harvest when buds are in the youngest possible stage of development, with minimal cells and with less chance of chimera production. X-ray doses range from 1 to 4 k rad with 4 k rad being almost lethal. The 1-, 2- and 3-k rad treatments produced 11.6% mutations that varied in flower form, flower size, and color, as well as in growth habit and plant morphology. Four of the 21 mutations formed in this study were registered for commercial use.

Induced somatic mutations in most higher plants are chimeric in nature. Since a chimera is composed of two generally distinct tissues, the mutation may or may not be transmitted in the seed obtained from a mutant sector. To produce an asexually propagated plant that will exhibit the same features as the mother plant, both genetic portions of the explant material must be included in the new plantlet. For these reasons, chimeras are not favored for commercial purposes.

The eye-catching and commercially desirable pinwheel, or star-flowered, African violets (shown in Figure 18.8) are the result of a chimera. Careful selection of explant material has meant that this pinwheel effect can be asexually propagated but not sexually propagated. Specialized techniques to propagate chimera plants have provided mutation breeders with a degree of protection against pirating of genetic stocks.

A chimera may exhibit two distinctly different flowers on one plant, as shown in Figure 18.9. A chimeral plant of this nature is of academic interest only because it cannot be propagated to maintain dual flower color by asexual or sexual means.

To overcome the problems of high polyploidy heterozygosity and intricate flowering behavior in sugarcane, a mutation breeding program was begun in 1975 in India (Jagathesan, 1982). Six commercial cultivars under cultivation in India were subjected to mutagenic treatment to induce mutations for specific characters. More than 50 mutants for various morphological characters, disease resistance, and increased sugar content were obtained in

Figure 18.6   Treating African violet tissue with x-rays produces a wide spectrum of mutations. The six flowers in this photo are indicative of some of those variations.

The star design on the top left has a white central strip the length of each petal. Each flower has a smooth edge trimmed in fuchsia (bright purplish red).

The double flower in the middle of the top row is a solid purple color with a smooth petal edge.

The flower at the top right is white with a fuchsia edge that spreads into the central part of the petal, thereby providing a variegated pattern with the most intense color concentrated along the edges. The pronounced wrinkled petal edge results in a sawtooth effect.

In contrast to the variably edged petals on the previous flower, the flower at the bottom left is neatly edged in a light green color. The frilled edge adds an attractive touch. The flower is violet (bluish purple) in color, with the most intense color in the central portion.

The lavender (pale bluish purple) flower color mutant in the middle of the bottom row represents a common mutation that arises spontaneously or by radiation induction. The mutation may occur at a color inhibitor locus.

The crimped edge trimmed in white on the sparkling magenta (reddish purple) flower at the bottom right provides a distinctive brilliance and a reflective property.

All six flowers are accented by five yellow stamens around a protruding but inconspicuous stigma and style. The flowers in this photo are indicative of the color combinations and flower designs possible through mutations. Hybridization of these mutant types may result in additional color and design combinations that could have aesthetic and economic value.

The ease of asexual propagation of African violets allows new variants to be utilized commercially very quickly.

Photo courtesy Ontario Ministry of Agriculture and Food and the grower, M. Epp.

Figure 18.7   The African violet leaves in this photograph illustrate some mutations in leaf morphology. The leaves in the upper row all exhibit a crenate margin. The leaf on the top left has an intensely dark green color with a pebbled or rough surface. The center top leaf has a light-colored, yellow-green area at the base of the leaf known as "semi-girl" marking. Overall, the leaf is pale green in color. The leaf at the top right has an orbicular shape and a plicate, or quilted, leaf blade and exhibits the "boy," or loss of the dominant semi-girl locus.

In the bottom row, the left leaf is a normal, or "boy," orbicular-shaped leaf with an entire margin. The center leaf is less crenate than leaves in the upper row and slightly more elongated. The leaf on the right exhibits a variable pattern of variegation that may range from extensive to flecked. Individual leaves contain patches of dark green, light green, or white. Some mutations produce a white edge on the margin. White areas represent a loss of chlorophyll and photosynthetic activity and frequently are discarded by growers because they are more difficult to grow.

Other mutants include a curling of the leaf blade to give a spooning effect, miniature leaves and plants, and pubescent or glabrous leaf surface. Many induced mutations are deleterious or lack aesthetic appeal.

Photo courtesy Ontario Ministry of Agriculture and Food and the grower, M. Epp.

these cultivars. Mutant characters of economic value include glabrous leaf sheath (smooth and free of hair or pubescence), nonflowering, vigor and high yield, and early maturity. Two mutant selections were advanced for commercial consideration.

In Pakistan, 1.5 to 3.0 k rad of gamma rays were found optimum for sugarcane, and a broad spectrum of variability was obtained (Siddiqui and Javed, 1982). To avoid problems with chimeras, mutants are grown for four to five years to test for stability.

Chimerism can occur in mutated seeds. The numerous meristematic cells in embryos of higher plants each have the potential to contribute to the formation of various plant parts as the mutant plant develops in the M_1 generation. Chimerism can occur in the plant, the spike, spikelet, and anther of cereal plants (Lindgren, Eriksson, and Sulovska, 1970). Chimerism at the plant level is identified easily since spikes from different tillers will segregate for mutants.

Figure 18.8   The attractive flower on this African violet plant is the result of a chimera. A chimera is produced when two or more cells that are used to propagate new plantlets mutate. Each of the mutated cells differs from the other and from the mother cells. Diplontic competition between any two mutants may eventually eliminate one mutation, and the resulting plant will produce a stable mutation type.

A second possibility is that the two genetically different tissues will grow together to produce a unique effect, such as the pinwheel flower in this photo. As the flower develops, the outer edges of each petal grow from certain tissue layers and the center of each flower petal from a different layer. Because of the unique effects produced, chimeras are sought by some producers (Coleman, 1987).

If such a plant is propagated by the normal method of leaf cutting, resulting plantlets will usually arise from a single cell and the flowers produced will be of one color—either red or white. A chimera that will produce the flowers in this photo can be propagated by taking side shoots or by taking cuttings from the flower stalks just below the flower. This explant material will perpetuate the pinwheel flower.

Photo courtesy Ontario Ministry of Agriculture and Food and the grower, M. Epp.

At the spike level, chimerism can be identified by planting intact spikes, followed by a visual scoring of mutant spikes for the number and location of chlorophyll-deficient mutant seedlings (Constantin, 1975). Microscopic examination of pollen grains for differential staining by iodine will identify mutations at the waxy locus. This technique has been used to demonstrate chimerism at the spikelet and anther levels (Lindgren, Eriksson, and Sulovska, 1970).

INDUCED MUTATIONS IN FRUIT TREES

The world apple industry is dominated by a limited number of apple cultivars, the majority of which arose from spontaneous mutations. The genotype of a well-known apple cultivar with

Figure 18.9   A chimeric African violet exhibiting two distinctly different flowers on one plant. One group of flowers, to the right of center, is lavender, and the other flowers, to the left of center, are much lighter in color and show only a tinge of lavender with darker lavender flecks.

Asexual propagation of this plant will result in plants with either flower color but not both on the same plant. The plant in this photo, therefore, is of academic interest only and of no commercial value. If such a plant is hybridized, the chimeric nature will not be perpetuated.

Plants in which single cells are **totipotent** and can be developed into new plants are useful to avoid chimeras. In plants where tissue culture techniques permit the growth of whole plants from single cells, induced mutations offer promise as a breeding tool.

Photo courtesy Ontario Ministry of Agriculture and Food and the grower, M. Epp.

consumer acceptance may be lost in hybridizing programs but may be improved in one or more aspects by a useful mutant without altering most of the cultivar's desirable characteristics. Mutation breeding for fruit trees is becoming an integral part of some research programs, but much effort has been devoted to treatment and screening techniques (Lapins, 1973; Visser, 1973).

Lacy and Campbell (1982) irradiated over 4,000 scions of fifteen apple cultivars and reported that some mutants are performing well in large-scale trials. Chimeral makeup is a problem in treating scions, and about half of the selected mutants must be discarded. Because some of this material appears to have potential, efforts are being made to resolve the chimeral makeup.

BIOTECHNOLOGY AND MUTATION BREEDING

The possibility that a single gene may be transferred from one organism to another through genetic engineering demands that each gene or set of genes contributing to a trait be pinpointed on the chromosome. Induced chromosome mutations and improved banding techniques could prove to be very real assets in this effort.

Gene mapping is a first step toward success in gene transfers by genetic engineering. Locating genes on a specific chromosome is facilitated by **trisomics** $(2n + 1)$, monosomics $(2n - 1)$, and translocation break points, and although trisomic series are well-developed in wheat and a few other crops, many crops lack well-developed gene maps. For example, gene mapping in red clover, an important forage legume $(2n = 2x = 14)$, has never been reported except for a few fragmentary data. A few trisomics have been reported in red clover but no detailed investigations into trisomic frequency have been conducted.

Translocation break points may be recognized and used in mapping through partial sterility or cytological techniques. Translocations can be induced by mutagenic agents, and in barley, over 300 translocations are available for cytologically locating genes, most of which have been induced (Nilan, 1974). Induced translocation break points may provide cytological markers in regions of chromosomes lacking suitable genes.

The use of cell cultures, protoplasts, and pollen to induce mutations is a new field of endeavor about which little is known. Single-cell suspensions and the protoplasts derived from them are a true single-cell system for mutagenesis and were considered by Carlson (1973) to be the best method for screening variant phenotypes in culture. Routine screening of million of cells is possible, but for cereals, no reliable and well-established system for regenerating plants from protoplasts is known. In tobacco (*Nicotiana tabacum*), mutant nitrate-reductase-deficient lines occurring at a frequency of one in 10 million were identified (Muller, 1983).

Genetic variation occurs more frequently, or may be more readily detectable, in tissue culture regenerates than in seed-grown plants. Plants regenerated from callus and suspension cultures grown for extended periods show more mutants than new cultivars. Such variation is known as **somaclonal variation**. Some of the tissue-culture-induced changes may be valuable (Larkin and Scowcroft, 1981). No tissue-culture-selected mutation has been used to produce a commercial cultivar (Maliga, 1985), perhaps because insufficient time has passed for useful mutations to be isolated. The value of somaclonal variation and cell culture isolation of mutant types has yet to be realized.

THE PLANT BREEDER'S VIEWPOINT ON MUTATIONS

Mutation breeding methods can contribute to crop improvement when they are correctly and intelligently applied and may be regarded as effective additional tools in ordinary, conventional breeding systems. There is no need to develop mutation breeding specialists to carry out this work.

In developing countries that have limited adapted genetic resources for building up good breeding materials, induced mutations may be a useful source of variability. This statement recognizes that mutation breeding techniques are sophisticated, but despite limited resources, developing countries can benefit from mutation breeding programs.

Mutation breeding is an extensive subject as indicated by the numerous references in the literature. Close to 1,300 publications were cited by Gottschalk and Wolfe (1983) in their review article. In keeping within the constraints of this text, this chapter presented an overview and represents a first step in educating researchers about mutation breeding.

REFERENCES

Adams, M. W. 1982. Plant Architecture and Yield Breeding. *Iowa State Journal of Research* 56:225-254.

Axtell, J. D. 1979. Breeding for Improved Nutritional Quality. In *Plant Breeding II*, edited by K. J. Frey. Iowa State University Press, Ames, pp. 365-432.

Blixt, S., and P. B. Vose. 1984. Breeding Towards an Ideotype: Aiming at a Moving Target? In *Crop Breeding: A Contemporary Basis,* edited by P. B. Vose and S. Blixt. Pergamon Press, London and New York, pp. 414-426.

Briggs, F. N., and P. F. Knowles. 1967. *Introduction to Plant Breeding.* Reinhold Publishing Corporation, Davis, California, p. 305.

Brock, R. D. 1980. Mutagenesis and Crop Improvement. In *The Biology of Crop Productivity,* edited by P. S. Carlson. Academic Press, New York, pp. 383-409.

Broertjes, C. 1968a. Dose-Rate Effects in *Saintpaulia.* In *Mutations in Plant Breeding II,* Proceedings of a Research Coordination Meeting on the Use of Induced Mutations in Plant Breeding. International Atomic Energy Agency, Vienna, pp. 63-71.

————. 1968b. Mutation Breeding in Vegetatively Propagated Crops. In *Mutations in Plant Breeding II,* Proceedings of a Research Co-ordination Meeting on the Use of Induced Mutations in Plant Breeding. International Atomic Energy Agency, Vienna. pp. 59-62.

Carlson, P. S. 1973. The Use of Protoplasts for Genetic Research. *Proceedings of the National Academy of Science USA* 70:598-602.

Coleman, K. S. 1987. What is a Chimera? *African Violet Magazine,* March-April, 9-11.

Constantin, M. J. 1975. Mutations for Chlorophyll-Deficiency in Barley: Comparative Effects of Physical and Chemical Mutagens. In *Barley Genetics II,* Proceedings of the Third International Barley Genetics Symposium, Garching. Verlag Karl Thiemig, Munchen, pp. 96-112.

Davies, D. R. 1977. Creation of New Models for Crop Plants and Their Use in Plant Breeding. *Applied Biology* 2:87-127.

Gaul, H. 1964. Mutations in Plant Breeding. *Radiation Botany* 4:155-232.

Gottschalk, W., and G. Wolff. 1983. *Induced Mutations in Plant Breeding.* Monographs on Theoretical and Applied Genetics 7. Springer-Verlag, Berlin, Heidelberg, New York, Tokyo. pp. 238.

Gustafsson, A. 1965. Characteristics and Rates of High-Productive Mutants in Diploid Barley. In *The Use of Induced Mutations in Plant Breeding.* Pergamon Press, Oxford, pp. 323-327.

————. 1974. *Mutations in Plant Breeding: A Glance Back and a Look Forward.* Lecture presented at the 5th International Congress of Radiation Research, Seattle, Academic Press, New York, July 18, 1974.

————. 1975. Mutations in Plant Breeding: A Glance Back and a Look Forward. In *Radiation Research,* edited by O. F. Nygaard, H. I. Adler, and W. K. Sinclair. Proceedings of the Fifth International Congress on Radiation Research. Academic Press, New York, pp. 81-95.

Gustafsson, A., and D. von Wettstein. 1958. *Handbuch der Pflanzenzuchtung* 2. Auflage, Band 1. Paul Parey, Berlin, pp. 612-699.

Jacobsen, E. 1984. Modification of Symbiotic Interaction of Pea (*Pisum sativum* L.) and *Rhizobium leguminosarum* by Induced Mutations. *Plant and Soil* 82:427-438.

Jagathesan, D. 1982. Improvement of Sugarcane Through Induced Mutations. In *Induced Mutations in Vegetatively Propagated Plants II*. International Atomic Energy Agency, Vienna, pp. 139-154.

Jaranowski, J. K. 1977. New Genotypes of Pisum Derived from Hybridization of Mutants and Cultivars. *Genetica Polonica* 18:337-355.

Kivi, E. I. 1981. Earliness Mutants from Sodium Azide Treated Six-Row Barley. In *Barley Genetics IV*, Proceedings of the Fourth International Barley Genetics Symposium, Edinburgh. Edinburgh University Press, Edinburgh, pp. 855-857.

Kleinhofs, A., W. M. Owais, and R. A. Nilan. 1978. Azide. *Mutation Research* 55:165-195.

Konzak, C. F., A. Kleinhofs, and S. E. Ullrich. 1984. Induced Mutations in Seed-Propagated Crops. In *Plant Breeding Reviews*, edited by J. Janick, vol. 2. AVI Publishing Company, Westport, Connecticut, pp. 13-72.

Konzak, C. F., R. A. Nilan, and A. Kleinhofs. 1977. Artificial Mutagenesis as an Aid in Overcoming Genetic Vulnerability of Crop Plants. In *Genetic Diversity in Plants*, edited by A. Muhammed, R. Askel, and R. C. von Borstel. Plenum Publishing Corporation, New York, pp. 163-177.

Koric, S. 1973. Branching Genes in *Triticum aestivum*. In *Proceedings of the Fourth International Wheat Genetics Symposium*, edited by E. R. Sears and L.M.S. Sears. University of Missouri, Columbia, Missouri, pp. 283-288.

Lacy, C.N.D., and A. I. Campbell. 1982. Progress in Mutation Breeding of Apples (*Malus pumilla* Mill.) at Long Ashton Research Station, Bristol, United Kingdom. In *Induced Mutations in Vegetatively Propagated Plants II*. International Atomic Energy Agency, Vienna, pp. 11-28.

Lapins, K. O. 1973. Induced Mutations in Fruit Trees. In *Induced Mutations in Vegetatively Propagated Plants*, Proceedings of a Panel on Mutation Breeding of Vegetatively Propagated and Perennial Crops. International Atomic Energy Agency, Vienna, pp. 1-20.

Larkin, P. J., and W. R. Scowcroft. 1981. Somaclonal Variation: A Novel Source of Variability from Cell Cultures. *Theoretical and Applied Genetics* 60:197-214.

Lindgren, D., G. Eriksson, and K. Sulovska. 1970. The Size and Appearance of the Mutated Sector in Barley Spikes. *Hereditas* 65:107-132.

Lundquist, U. 1976. Locus, Distribution of Induced *eceriferum* Mutants in Barley. In *Barley Genetics III*, edited by H. Gaul. Karl Thiessig, Munich, pp. 162-163.

Maliga, P. 1985. Cell Culture Isolation and Characterization of Agronomically Useful Mutants of Higher Plants. In *Biotechnology in International Agricultural Research*. International Rice Research Institute, Manila, Philippines, pp. 111-120.

Marx, J. L. 1984. New Ways to "Mutate" Genes. *Science* 225:819.

Micke, A. 1984. Mutation Breeding of Grain Legumes. *Plant and Soil* 82:337-357.

Micke, A., S. C. Hsieh, and B. Sigurbjornsson. 1972. Rice Breeding with Induced Mutations. In *Rice Breeding*. International Rice Research Institute, Los Banos, Philippines, pp. 573-580.

Mikaelsen, K. 1980. Mutation Breeding in Rice. In *Innovative Approaches to Rice Breeding*. International Rice Research Institute, Los Banos, Philippines, pp. 67-79.

Muller, A. J. 1983. Genetic Analysis of Nitrate Reductase Deficient Tobacco Plants Regenerated from Mutant Cells: Evidence for Duplicate Structural Genes. *Molecular and General Genetics* 192:275-281.

Muller, H. J. 1927. Artificial Transmutation of the Gene. *Science* 66:84-87.

Murray, M. J. 1969. Successful Use of Irradiation Breeding to Obtain *Verticillium*-resistant Strains of Peppermint, *Mentha piperita* L. In *Induced Mutations in Plants*. International Atomic Energy Agency, Vienna, pp. 345-371.

Nilan, R. A. 1964. *The Cytology and Genetics of Barley, 1951-1962*. Monographic Supplement No. 3, Research Studies. Vol. 32, no. 1. Washington State University, Pullman, p. 278.

———. 1974. Barley (*Hordeum vulgare*). In *Handbook of Genetics*, edited by R. C. King, vol. 2. Plenum Press, New York, pp. 93-100.

————. 1981. Induced Gene and Chromosome Mutants. In *Philosophical Transactions of the Royal Society of London.* Series B, vol. 292 (no. 1062). Royal Society, London, pp. 457-466.

Nilan, R. A., and B. K. Vig. 1976. Plant Test Systems for Detection of Chemical Mutagens. In *Chemical Mutagens: Principles and Methods for their Detection,* edited by A. Hollaender, vol. 4. Plenum Press, New York, pp. 143-170.

Nilan, R. A., A. Kleinhofs, and C. F. Konzak. 1977. The Role of Induced Mutation in Supplementary Natural Genetic Variability. *Annals of the New York Academy of Science* 287:367-384.

Owais, W. M., A. Kleinhofs, and R. A. Nilan. 1979. In Vivo Conversion of Sodium Azide to a Stable Mutagenic Metabolite in *Salmonella typhimurum. Mutation Research* 68:15-22.

Owais, W. M., M. A. Zarowitz, R. A. Gunovich, A. L. Hodgdon, A. Kleinhofs, and R. A. Nilan. 1978. A Mutagenic In Vivo Metabolite of Sodium Azide. *Mutation Research* 50:67-75.

Redei, G. P. 1974. Economy in Mutation Experiments. *Zeitschrift für Pflanzenzuchtung* 73:87-96.

Robbelen, G., and A. Nitsch. 1975. Genetical and Physiological Investigations on Mutants for Polyenic Fatty Acids in Rape Seed *Brassica napus* L. I: Selection and Description of New Mutants. *Zeitschrift für Pflanzenzuchtung* 75:93-105.

Sharman, B. C. 1944. Branched Heads in Wheat and Wheat Hybrids. *Nature* 153:497-498.

Siddiqui, S. H., and M. Javed. 1982. Mutation Breeding in Sugarcane (*Saccharum* spp. Hybrid) by Gamma Irradiation of Cuttings and Tissue Cultures. In *Induced Mutations in Vegetatively Propagated Plants II.* International Atomic Energy Agency, Vienna, pp. 155-166.

Sigurbjornsson, B. 1975. The Improvement of Barley Through Induced Mutation. In *Barley Genetics III,* Proceedings of the Third International Barley Genetics Symposium, Garching. Verlag Karl Thiessig, Munchen, pp. 84-95.

Sogaard, B. 1974. The Localization of *Eceriferum* Loci in Barley. *Hereditas* 76:41-47.

Stadler, L. J. 1928. Mutations in Barley Induced by X-rays and Radium. *Science* 68:186-187.

————. 1929. Chromosome Number and the Mutation Rate in *Avena* and *Triticum. Proceedings of the National Academy of Science* USA 15:876-881.

Visser, T. 1973. Methods and Results of Mutation Breeding in Deciduous Fruits, With Special Reference to the Induction of Compact and Fruit Mutations in Apples. In *Induced Mutations in Vegetatively Propagated Plants,* Proceedings of a Panel on Mutation Breeding of Vegetatively Propagated and Perennial Crops. International Atomic Energy Agency, Vienna, pp. 21-34.

von Wettstein, D., B. Jende-Strid, B. Ahrenst-Larsen, and J. A. Sorensen. 1977. Biochemical Mutant in Barley Renders Chemical Stabilization of Beer Superfluous. *Carlsberg Research Communication* 42:341-351.

Wang, Y., H. Yuan, S. Gao, and G. Lu. 1984. Character Mutation and Mutation Breeding in Soybean. In *Proceedings of the Second US-China Soybean Symposium,* edited by S. Wong *et al.* Published in Cooperation with the Office of International Cooperation and Development. United States Department of Agriculture, Washington, D.C. pp. 228-231.

Yonezawa, K., and H. Yamagata. 1977. On the Optimum Mutation Rate and Optimum Dose for Practical Mutation Breeding. *Euphytica* 26:413-426.

Yoshida, Y. 1962. Theoretical Studies on the Methodological Procedures of Radiation Breeding. *Euphytica* 11:95-111.

CHAPTER 19

Haploid Breeding Systems

Haploid is a general term that refers to a plant containing the gametic, or half, chromosome number. When produced from a diploid species, haploid plants are more accurately described as **monoploids**. The term *haploid*, however, remains in common usage. By doubling the haploid chromosome number, the normal number of chromosomes is restored in the resulting **dihaploid**. Doubled haploids, or dihaploidy, offer the quickest possible approach to homozygosity. Duplication of the haploid chromosome number results in the most homozygous genotypes possible for research purposes or cultivar release.

Ever since A. D. Bergner discovered haploid plants in Jimson weed (*Datura stramonium*) in 1921 (Blakeslee *et al.*, 1922), interest has grown in developing techniques for inducing haploidy. Three years later, utilization in breeding programs was suggested (Nitzsche and Wenzel, 1977) but remained a theoretical consideration until the early 1970s, when it was discovered that haploids can be produced *in vitro* by anther or pollen culture, by unpollinated ovary culture, and by chromosome elimination using the **bulbosum method** (Figure 19.1). Spontaneous haploids occur *in vivo* from **polyembryony, semigamy,** and **androgenesis**.

Hu (1985) suggested that anther culture techniques are the simplest and most efficient method of haploid production. The system is limited to specific crops until appropriate techniques can be developed for all crops. Haploids have been induced in 247 species from 88 genera and 34 families through anther or pollen (Hu, 1985).

For plant breeding purposes, haploids have potential only if large numbers can be produced efficiently in many diverse parents. It appears that no one method of producing haploids is effective across many species. In maize, spontaneous haploids have been found (Chase, 1949); in potato, haploids occur after crosses involving different levels of ploidy (Hougas, Peloquin, and Ross, 1958; Iwanaga, 1985); in cotton as a result of semigamy (Turcotte and Feaster, 1967); in tobacco and rice by anther culture (Nitsch and Nitsch, 1969; Zapata, 1985); in wheat through the use of alien cytoplasm (Tsunewaki, Noda, and Fujisawa, 1968); in pepper as a result of twinning (polyembryony) (Morgan and Rappleye, 1950); and in barley by means of the interspecific cross with *Hordeum bulbosum* (Kasha and Kao, 1970).

The frequency of haploid production is very high in barley with the bulbosum technique. In potato, the source of pollen influences haploid frequency. The desirable traits of haploids and the many 2x relatives of the potato are combined at the 2x level and can be efficiently transferred into the 4x progeny by 4x x 2x crosses using diploid male parents. Ploidy manipulation in the potato takes advantage of two cytological tools—haploids and first-division, restitution 2n pollen. Manipulating ploidy involves three steps:

1. Extracting haploids from selected tetraploids.
2. Hybridizing haploids with useful 2x species and subsequently selecting for disease resistance, agronomic adaptation, and 2n pollen production. Population breeding at the

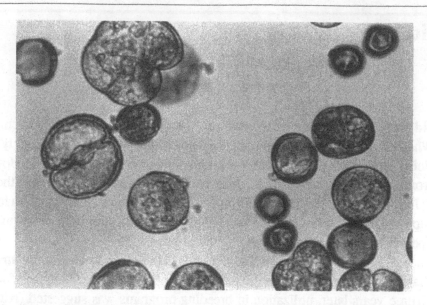

Figure 19.1   Photo of haploid cells dividing. Haploid plants contain half the normal number of chromosomes, and where haploids can be induced in crop plants, they provide the most rapid technique for producing homozygous lines.

Kasperbauer and Eizenga (1985) used the haploid and doubled-haploid technique to speed up the improvement of tall fescue (*Festuca arundinacea*) over that achieved by conventional breeding methods. Androgenic haploid lines were cultured *in vitro*. This method offers the possibility of more rapid identification of superior genotypes among the haploids and then culturing fertile doubled haploids with the characteristics identified in the haploids.

Photo courtesy K. P. Pauls.

2x level involves **disomic** inheritance, which is more efficient than traditional breeding at the 4x level and involves tetrasomic inheritance.

3. Transferring the improved 2x germplasm into 4x through first-division, restitution 2n pollen.

The 4x progeny produced by 4x x 2x crosses would be excellent not only in a clonal selection program but also as true potato seed cultivars because the genetic homogeneity of first-division, restitution 2n pollen results in a highly heterozygous and high-yielding 4x progeny. Multiple resistance combined at the 2x level is efficiently transmitted to the progeny by first-division, restitution 2n pollen.

In cotton, superior pollinators increased haploid frequency 5 to 20 times that of some seed parents (Peloquin, 1981).

Haploids have provided a major contribution to basic genetic knowledge through the development of **trisomics** and **monosomics**, which was achieved with the use of Chinese Spring wheat (Sears, 1954). In time, in addition to trisomics and monosomics, the complete series of **nullisomics** and **tetrasomics** were produced. These aneuploids have provided basic information on wheat evolution, genetics, cytogenetics, and plant breeding. For comparable understanding in other crops, trisomic series need to be developed.

One fundamental piece of information obtained from a haploid wheat deficient for chromosome 5B was that particular genes control the pairing specificity among selected chromosomes. Homoeologous pairing was prevented by the activity of a gene(s) or the 5B chromosome (Riley and Chapman, 1958), and this realization revolutionized cytogenetic thinking about chromosome pairing in polyploids.

The use of haploids in cultivar breeding can be quite variable depending on whether a species is a diploid, a **disomic polyploid**, or a **polysomic polyploid**. Polyploids can be classed as either disomic or polysomic; this is a superior classification to that of allopolyploid and autopolyploid because the terms disomic and polysomic give a more precise conceptual picture of the nature of polyploids from the viewpoint of cytogenetics and plant breeding (MacKey, 1970).

Barley cultivars have been released as a result of barley haploid breeding systems, and the bulbosum method is used in selected laboratories. Doubled haploids of maize have been used successfully in commercial hybrids (Chase, 1974), but until maize haploids can be produced more readily, it is doubtful their use will replace conventional inbreeding systems. Cultivars were produced by the haploid technique in oilseed rape (*Brassica napus*) (Thompson, 1972) and in tobacco (*Nicotiana tabacum*) (Nakamura *et al.*, 1974).

In the dioecious asparagus (*Asparagus officinalis*), haploids have been developed from polyembryonic seeds and by anther culture. In the asparagus, male plants (*YY*) outyield female plants (*XX*). Haploids produced by anther culture resulted in 50% homozygous *YY* male plants, which when crossed with any female plant produced entirely male progenies (Thevenin, 1974).

In polysomic polyploids, haploids have a tremendous potential, which is enhanced if the polyploid has wild and cultivated relatives at lower ploidy levels because crosses can be achieved. This situation is found in potatoes, alfalfa, strawberries (*Fragaria virginiana*), coffee (*Coffea arabica*), blueberries (*Vaccinium* spp), sweet potato (*Ipomoea batatas*), yams (*Dioscorea sativa*), as well as in ornamentals, many tropical grasses, and subtropical trees and shrubs (Peloquin, 1981).

Haploid research has contributed cultivars in many species and will continue to do so in the future. Hu (1985) reported 81 rice and 20 wheat cultivars and strains produced through anther culture. Essentially, haploidy is becoming an increasingly potent tool in the plant breeders tool kit but, like other methods of plant breeding, must be put in the proper perspective. It would be unfortunate if one system of plant breeding dominated the scene.

The key to using dihaploids in commercial plant breeding programs rests with the availability and reliability of methods for their production. Success with methods for obtaining haploids in cultivated plants since the early 1970s has made the haploid method a viable breeding system for some crops and has resulted in the release of commercial cultivars (Figure 19.2).

ONTARIO
REGIONAL
BARLEY
TEST

Figure 19.2   Some of the barley cultivars under test in this photo have been developed by the doubled-haploid technique. The first barley cultivar to be released commercially using this method was named Mingo, licensed in 1979 by Ciba-Geigy Seeds, Ltd. (Ho and Jones, 1980).

The use of doubled haploids as commercial cultivars raises questions about their agronomic performance under field conditions because of their high level of homogeneity. In contrast to cultivars produced by the doubled-haploid technique, landrace cultivars are heterogeneous, a feature that is felt to offer yield stability over environments and years. Most conventionally produced inbred cultivars are heterogeneous for complex characteristics, even though they satisfy basic uniformity requirements for plant height, maturity, and morphological features.   Allard and Bradshaw (1964) suggested that heterogeneity is valuable to buffer a cultivar.

The use of genetic mixtures rather than homogeneous or pure-line cultivars was suggested as an alternative means to reduce genotype x environment interactions.   Jensen (1952) suggested that multiline cultivars would possess greater stability of production and broader adaptation. There is no reason to believe that a similar buffer effect cannot be achieved by homozygous genotypes, since the characteristics of an organism on which its adaptation depends and all its complex adaptive characters are generally admitted to be determined by a combination of genes.

The first tobacco cultivar to be produced by the doubled-haploid breeding method was designated MC 1610 and released in 1974 (Nakamura, Yamada, Nadotani, and Itagaki, 1974).

USING HAPLOID PROCEDURES

Efficient use of the haploid technique in crop breeding programs requires (1) a dependable method of producing haploids, (2) a method in which haploid production represents a random sample of gametes, (3) a reliable method of doubling haploid chromosome number, and (4) adequate technical competence and suitable physical facilities.

Dependability of Haploid Production

Haploid plants can be identified and reproduced with a low frequency in any plant species. The first attempt to isolate haploids for practical breeding purposes was made by Chase (1952) in the cross-fertilized maize crop. He observed the occurrence of maternal haploids in the progenies of single- and double-cross hybrids and in inbred lines when fertilized by four unrelated pollen stocks. Haploid frequencies were low and ranged from none in 4,500 to 1 in 145 plants examined. Chase observed that when certain stocks were used as seed parents, a higher frequency of haploids were found than when other stocks were used, regardless of the pollen parent.

Such observations helped emphasize the requirements for haploid production. Specifically, a high frequency must be produced with a minimal amount of labor, and the production of haploids must not be limited to particular genotypes.

Symko (1969) and Kasha and Kao (1970) reported on a technique for obtaining haploids in barley using an interspecific cross between a 28-chromosome, autotetraploid cultivated barley (*Hordeum vulgare*) and a wild tetraploid barley from *Hordeum bulbosum* (Figure 19.3). Progeny from reciprocal crosses were nearly all 14-chromosome "diploid" plants resembling cultivated barley.

When *Hordeum bulbosum* was crossed to diploid ($2 = x = 14$) forms, haploid plants were obtained (Kasha and Kao, 1970; Kao and Kasha, 1970). Although a hybrid zygote is formed following this interspecific cross, subsequent **embryogenesis** and endosperm development undergo chromosome elimination (Subrahmanyam and Kasha, 1973). The elimination process is preferential for the loss of the chromosomes in the *Hordeum bulbosum* genome. The chromosome instability of the hybrid cells of the developing embryo probably causes both physiological and developmental instability, which, in time, causes the embryo to abort. The embryo must be removed eight to ten days after hybridization. Following excision, *in vitro* culturing of the embryo removed from the endosperm provides better conditions for survival; many embryos will differentiate slowly into plants, most of which are haploid (Figure 19.4). The chromosomes of the haploid plant can be doubled by colchicine treatment. The resulting dihaploid plant is homozygous for all gene loci in the haploid gamete of the *Hordeum vulgare* parent used in the initial interspecific cross.

Kasha and Kao (1970) made some unusual and important observations, namely, that (1) haploid embryos can be obtained with a very high frequency; (2) in crosses between haploid and dihaploid barley, only haploids have been obtained; (3) any cultivated barley genotype can be used to produce haploids; and (4) the haploid plants always carry the cultivated barley genome.

The formation of gametes in *Hordeum vulgare* following pollination by *Hordeum bulbosum* is a result of fertilization and not **parthenogenesis**, as suggested initially by Chase (1949). Although fertilization occurs the *Hordeum bulbosum* chromosomes are lost during the first few mitotic divisions of the embryo.

Haploid plants can also be produced directly from pollen or anthers (Guka and Maheshwari, 1964), and this played a major role in developing interest in haploidy. Anther culture has been applied successfully in tobacco (Sunderland, 1974) and rice (Collins, Legg, and Kasperbauer, 1974), which has stimulated intensive research on other species of economic importance.

Figure 19.3   Spike of the wild barley grass *Hordeum bulbosum*, the pollen parent used for the production of haploids in barley. The plant is at the stage of anthesis, and anthers are extruded fully. *Hordeum bulbosum* became the model system for haploid barley production because the interspecific cross with *Hordeum vulgare* leads to the development of haploid embryos through chromosome elimination.

Hordeum bulbosum is characterized by prostrate growth habit, fine stems, narrow leaves and prolific tillering, light pubescence on the stems and leaves, perennial and winter growth habits, resistance to powdery mildew, spikes that are two-rowed with fine, short awns, and large anthers that are exserted from the florets at the time of anthesis, as shown in the photo.

Production of *Hordeum bulbosum* plants with a 16 hour photoperiod at 21° C (70° F), 17° C (63° F) during the 8-hour dark period, and 60 to 80% relative humidity produces vigorous plants and will help circumvent partial incompatibility that may exist in certain parents.

Pollen release from *Hordeum bulbosum* must coincide with receptivity in the female *H. vulgare* flower. To induce flowering in *Hordeum bulbosum,* prior vernalization is required. Prompt application of the *bulbosum* pollen is suggested. Twenty-four hours after pollination, tillers bearing fertilized heads are removed from the *H. vulgare* plant and maintained in aerated Hoagland's solution. Seed set can be enhanced by the application of gibberellic acid (Jensen, 1977; Kasha, Subrahmanyam, and Ali, 1978).

Sunderland (1974) described anther culture as offering simplicity of operation, ease of induction, and high induction of frequencies in some instances.

Using anther culture, immature anthers are excised from the flower bud and nurtured on an *in vitro* culture medium, which induces the haploid microspores to go through an embryonic development into a haploid plant.

Haploids can be obtained in many economically important crops, but the frequency of production is much lower than with *bulbosum*-induced barley or another-cultured tobacco. Anther culture has been successful in both wheat and rice, but the frequency of haploids from cultured anthers has been either low (about 0.1% of cultured wheat anthers) or, in the case of

Figure 19.4 Photo of a haploid barley plant cultured *in vitro* and developed from an embryo that was removed from the mother plant about ten days after fertilization to prevent abortion.

Following hybridization of a diploid barley (*Hordeum vulgare* L.), and a diploid wild barley (*Hordeum bulbosum* L.), haploid embryos result from the rapid elimination of the *Hordeum bulbosum* genome from dividing cells in the developing embryo and endosperm. The haploid embryo bears the maternal genome only. The growth of the haploid embryo tends to be slow while the developing endosperm turns watery and aborts, leaving a small, flat, immature seed. Such embryos are unfit to maintain themselves under these conditions and must be "rescued" approximately fourteen days after pollination. This requires that the embryos be aseptically removed from the haploid seed and placed in culture medium in more amenable conditions, where they have the potential to survive and grow into haploid plants. The following steps are involved in embryo rescue:

1. Pluck haploid seeds from the florets with fine, sterile tweezers.
2. Remove the embryo from the endosperm under a stereoscope. The haploid seed is flat and elongate, with the stigma remnants at one end and the haploid embryo at the other. With fine tweezers, tear the seed along the middle downward toward the embryo and gently dislodge the opaque, white embryo, which can be up to 3 mm in size.
3. Transfer each embryo to a plate or vial with medium. Seal with paraffin. Avoid contamination by flame-sterilizing tweezers and equipment.
4. Plates, each containing a number of well-spaced embryos, or vials are placed in a dark growth cabinet at 23° C (73° F) for seven days.
5. After seven to ten days, when shoot apices have formed, move the plantlets to an illuminated growth cabinet to promote chlorophyll development.

Figure 19.4 continues overleaf

rice, associated with **albino** seedling production (Maheshwari *et al.*, 1980). A similar albino problem in anther-induced barley haploids led Huang *et al.* (1984) to conclude that until the problems of poor regeneration and high numbers of albinos can be overcome, the method chosen for haploid production in barley will remain the bulbosum technique.

Huang *et al.* (1984) compared the relative merits of haploid production in barley using interspecific hybridization involving *Hordeum bulbosum* with anther culture. Although the most successful culture methods enabled more than 50% of barley anthers to produce microspore callus, the overall level of plant regeneration from this callus was very low. The majority of regenerants were albino, and the highest yield of green plants was 0.5%. The *Hordeum bulbosum* technique was much more reliable, with a high of a 15.4% haploid success rate.

Progress in developing wheat haploids from anthers resulted in a success rate of 7.31% in one cultivar (Liang, Xu, and Hoang-Tang, 1987).

Maheshwari *et al.* (1980) observed that in the 16 years prior to 1980, the anther culture technique was extended to over 150 different species, and that in spite of intense research activity, real progress was very limited. The most important reason for this is that the anther culture technique works well on plants belonging to only a few families, such as *Solanaceae* (potato). Even where response has been obtained, haploid frequency often is very low. In the case of wheat, rice, and oat crops, cultivar effects are significant (Rines, 1983) in the production of haploids.

A genotypic response among wheat cultivars was observed when haploid plants were produced from callus culture (Ahloowalia, 1982) and may be attributed to differences in cytokinin and auxin. Genotypic responses limit the value of the doubled haploid breeding system.

Barley genotype was observed to be a factor in haploid production when five *Hordeum vulgare* and three *Hordeum bulbosum* genotypes were used in a specific environment (Pickering, 1983).

A success rate of near 10% is necessary in order for the haploid technique to be a useful plant breeding tool. Some cultivar or genotypic specificity for haploid development is acceptable as long as a wide range of genotypes is capable of giving rise to a reasonably high frequency of haploids.

Necessity of Random Samples of Gametes

To determine whether or not haploids from either the *Hordeum bulbosum* or the anther

---

Figure 19.4 continued

6. When the root systems are sturdy two or three weeks later, as seen in the photo, transplant the plantlets to soil. After washing off the medium, plant seedlings in small pots and cover with clear plastic to maintain high humidity. Within a week, the plants will have **acclimatized** to the growth room conditions and the plastic covering will be unnecessary.

Although haploid embryos can often be quite irregular in shape, well formed embryos have a differentiated scutellum, root, and shoot apex.

---

culture method represent a random sample of the gametes produced by an individual genotype, two approaches might be used. One is to study the segregation of qualitatively inherited traits, and the second is to study the mean and variance of quantitatively inherited traits from heterozygous genotypes.

Johns (1974) studied the expected segregation in barley genotypes heterozygous for a number of different marker genes. He observed that the haploids obtained conformed to the 1:1 genetic ratio expected of a heterozygous individual. In a comparable study, Legg and Collins (1973, 1978) studied the inheritance of alkaloid level in heterozygous populations of cultivated tobacco. They observed the ratios of low- and high-alkaloid types expected from the two gene system for alkaloid level. We may conclude from these studies that gametes sampled by the haploid techniques conform to the expected normal distribution of the gametic output.

Reliable Chromosome Doubling

The production of haploid plants represents a first step in the use of haploidy as a breeding method. Haploid number must be doubled to produce a homozygous plant (Figure 19.5). Two procedures have been used to achieve doubling of the chromosome number in barley and tobacco, namely, **colchicine** treatment and culturing stem and root sections.

The most common procedure used for chromosome doubling is to apply colchicine. This treatment has proved especially successful with barley where careful consideration has been given to the length of colchicine treatment and to additional compounds that improve and facilitate the doubling effect. Colchicine can be used also to double haploid plants obtained from the anther culture technique.

Essentially, chromosome doubling requires the treatment of the apical meristems for five hours with a 0.1% colchicine solution mixed with a penetrating agent, such as 2% dimethylsulfate, and a surfactant. Colchicine also can be applied by capping the plant, by the immersion method, or by first splitting the base of each stem with a razor blade before immersion (Pickering and Morgan, 1979). The addition of gibberellic acid and benzyladenine to the colchicine mixture is recommended (Thiebaut, Kasha, and Tsai, 1979) to increase plant survival and the number of doubled sectors per plant. After doubling, repotted plants are rogued for *Hordeum bulbosum* x *Hordeum vulgare* hybrids. Nonhybrids are indicated by excessive tillering and prostrate growth habit. In the first generation of doubled haploids, seed set is reduced by the colchicine treatment. But subsequent generations recover from the effects, and the use of doubled haploids in a breeding program is not hampered (Finch and Bennett, 1979).

Another method of chromosome doubling is to take stem and root sections of haploid plants and culture them *in vitro* on a medium that initiates shoot differentiation. Because some cells in the stem and root segments are normally **diploidized**, diploid plants are obtained when such cells give rise to shoots.

Studies of barley by Walsh, Reinbergs, and Kasha (1973) indicated that an extra generation of seed increase following colchicine doubling should occur to eliminate seed source effects on seedling vigor. The suggestion is that doubled haploid lines should be advanced one generation before their performance relative to conventional lines is evaluated. It is potentially erroneous to discard agronomically undesirable lines when evaluation is based on their performance in plots grown from original seed. Walsh and his coworkers found that in the first generation some

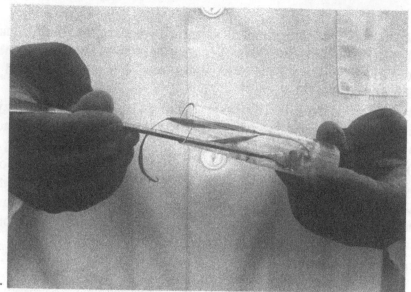

Figure 19.5 The barley seedling in the vial in Photo A is haploid and has been produced by the *Hordeum bulbosum* technique. The seedling is ready to be transplanted into small pots (Photo B) containing light soil. If the 2n chromosome number is not restored, the plant will be distinct from diploids in that it will be smaller, have finer leaves, have a bushy appearance due to abundant tillering, and will produce sterile spikes. Without chromosome doubling, haploid plants are useless.

Haploid plants are ready for doubling at the three- to five-leaf stage. This may be accomplished by depotting the plants and immersing the washed roots and crown of each plant in small vials of a 0.1% colchicine solution for five hours at 32° C (90° F). Remove the plants from the solution, rinse the colchicine from the roots, clip the shoots to 10 cm and the roots to 3 cm, and repot. This procedure is highly effective in producing fertile, homozygous diploids from haploid plants. Chromosome-doubled sectors of the original haploid plant produce seeds that are all identical in genotype and that produce completely homozygous plants in subsequent generations.

Colchicine is a compound extracted from the autumn crocus and was discovered in the mid-1930s (Kostoff, 1938). The compound acts by preventing spindle fibers from being effective during cell division. Colchicine is a potent compound, and direct contact with it should be avoided.

colchicine-doubled haploids show an increase in hard seeds, which are less vigorous. In the next generation, vigor is restored. The time required, however, for a generation of multiplication following colchicine treatment and prior to agronomic assessment reduces the time-efficiency factor of the system. Work by Powell *et al.* (1985b) showed no differences in ranking between the first and second generations of self-fertilization following colchicine treatment.

Technical Competence and Physical Plant Facilities

Suitable physical plant facilities are more critical for haploid production than for traditional breeding programs. Adequate environmental control in greenhouses or growth rooms is a prerequisite for successful haploid production using bulbosum and anther culture techniques. In addition, the level of technical expertise required for the haploid procedure is greater than that normally required for more traditional plant breeding systems.

These extra requirements translate into extra costs for privately or publicly funded breeding organizations that specialize in using haploid procedures. Additional costs may be offset to a large measure by the improved efficiency that results because less time is required to complete a breeding program. This will be considered later in this chapter.

HAPLOID BREEDING PROCEDURES

Although the haploid breeding system using *Hordeum bulbosum* and presented in Figure 19.6 is focused on barley, it serves as a model to illustrate comparative efficiencies and basic principles of this method. Selected parents are hybridized in a normal fashion, but a major difference between this and other breeding systems is that F_1 hybrid plants are crossed with *Hordeum bulbosum* to obtain haploid progeny. Since F_1 plants are uniformly heterozygous the only opportunity for selection is to observe F_1 plants to be sure they are hybrids. There is no opportunity to select superior F_1 gametes, and the breeder must hybridize sufficient plants with an adequate representation of gametes to ensure that the chance occurrence of favorable genes from both parents has resulted from **crossing over** and **recombination**. Haploids could be extracted from later generations, but the F_1 gametes represent a sample of the genetic variability that would be present in the F_2 generation. The number of haploids necessary to sample a particular cross is less than would be required by the pedigree and bulk methods because F_1 gametes are sampled rather than F_2 individuals. The number of gametes required for sampling is the square root of the population size that would normally be required in the F_2. A population size of 10,000 in the F_2 generation in the pedigree or bulk breeding method could be sampled by 100 haploids from the same cross:

$$\sqrt{10,000} = 100$$

Note that the doubled-haploid breeding system is distinguished from the pedigree, single-seed-descent, bulk breeding, and other methods in that no segregating populations exist because they are not needed to reach homozygosity (Figure 19.6).

Calculations of the means and variance components of doubled-haploid barley populations (Snape and Simpson, 1981), the mathematical formulation of Choo (1981) of the frequency of

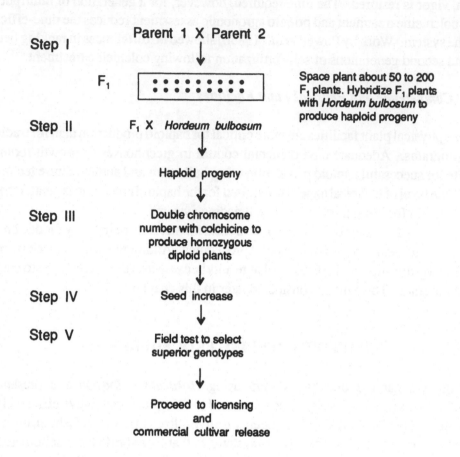

Step I

Parent 1 X Parent 2

F_1

Space plant about 50 to 200 F_1 plants. Hybridize F_1 plants with *Hordeum bulbosum* to produce haploid progeny

Step II F_1 X   *Hordeum bulbosum*

Haploid progeny

Step III   Double chromosome number with colchicine to produce homozygous diploid plants

Step IV   Seed increase

Step V   Field test to select superior genotypes

Proceed to licensing and commercial cultivar release

Figure 19.6   Schematic procedure of a haploid barley breeding system using interspecific hybridization with *Hordeum bulbosum*. Five basic steps are involved:

Step I   Define objective(s), select appropriate parents, and hybridize parents to produce 15 to 20 F_1 seeds. This procedure may be completed under controlled environmental conditions over a three- or four-month period.

Step II   Grow F_1 progeny in environmentally controlled facilities. Hybridize F_1 plants with the wild barley *Hordeum bulbosum* to produce haploid embryos. This procedure may require six months.

Step III   Grow the embryos into seedlings. Treat plants at three- to five-leaf stage with colchicine to cause chromosome doubling, and grow plants to maturity. Requires four to six months.

Step IV   Increase seed from doubled, homozygous haploids for replicated field trial evaluation. Requires a growing season.

Step V   Begin replicated yield trials on station. Proceed to cooperative and regional testing to determine genotype x environment interaction. Proceed to licensing and release of commercial cultivar.

The *Hordeum bulbosum* method has become a well-defined procedure and for barley breeding is a standard procedure. Haploid breeding in other crops may use systems of androgenesis involving anther or pollen culture. Pollen culture is preferred over anther culture because plants may arise from various parts of the anther, with the result that a mixed population of plants with various ploidy levels might be obtained in a given culture.

desirable recombinants, and the computer simulation study of Yonezawa, Nomura, and Sasaki (1987), indicated that the doubled-haploid method could be more efficient if applied to the gametes produced by F_2 plants. Desirable F_2 plants could be selected for haploid production.

COMPARISONS OF THE HAPLOID METHOD WITH OTHER TECHNIQUES FOR HANDLING SEGREGATING POPULATIONS

A major feature of the haploid procedure is the time saved in achieving homozygosity compared with most other breeding methods. With the bulk and pedigree methods, normally seven years are required before relatively homozygous selections can be field tested in replicated trials (Reinbergs, Park, and Kasha, 1975; Kasha and Reinbergs, 1972). Assuming that the F_3, F_4, and F_5 segregating generations can be grown in one calendar year under controlled environmental conditions using the single-seed-descent system, approximately five years are required before field testing. Two years are required to reach an equivalent stage in the haploid method, assuming that the parents are planted, the crosses made, the haploids extracted, and the chromosome number doubled within the first 12 months. A second assumption is that within the next 12 months seed is increased from the doubled haploids and replicated field tests are established. The haploid production rate is assumed to be approximately 100 haploids per week, with a 60% doubling efficiency (Kasha, 1976; Kasha and Reinbergs, 1975).

The results of field comparisons of haploid-derived lines with selections from conventional breeding methods were reported by Park *et al.* (1976). No deleterious results from complete homozygosity and homogeneity were detected in the haploid progeny. It was concluded that the doubled-haploid technique is a useful tool for producing homozygous barley lines in a relatively short time.

In a subsequent study using the same barley crosses, the doubled-haploid method was compared with the bulk plot method (Song *et al.*, 1978). The results (Table 19.1) suggest that the time-saving, doubled-haploid method can produce barley genotypes with the same yield potential as those developed by the bulk plot method. In addition, doubled haploids had greater environment variability and yield stability.

Doubled-haploid breeding techniques have given positive results in many institutions, yet Yonezawa, Nomura, and Sasaki (1987) noted that the techniques are not routinely applied in practical breeding projects. The main reason appears to be technical production difficulties in producing haploid and doubled-haploid plants, although the bulbosum technique in barley has a better record than anther or pollen culture techniques. Perhaps another reason is the relative inefficiency of doubled-haploid techniques. Mathematical calculations (Nei, 1963) have shown conventional breeding methods to be much more efficient than the doubled-haploid method when a large number of loci are involved for the breeding objective chosen.

A computer simulation study by Walsh (1974) using a model of six linked loci proved the doubled-haploid method to be undesireable under these conditions. The single-seed-descent method was shown by Riggs and Snape (1977) to produce a population with a larger genetic variation than that derived by diploidization of F_1 gametes. The doubled-haploid technique was not viewed as useful when linkage is important. Yonezawa, Nomura, and Sasaki (1987) used

Table 19.1   Grain Yield Comparison of 70 Barley Entries Derived from the Doubled-Haploid and Bulk Plot Methods and Representing Two Crosses

	Breeding Method	Number of Entries	Grain Yield Over Three Environments (g/hill)	(g/row)
Cross I	Doubled haploid	70	25.7a[a]	190a
	Bulk plot	70	27.0b	200a
Cross II	Doubled haploid	70	30.2a	244a
	Bulk plot	70	32.4b	232b

[a] Means with different letters are significant at the 5% level of probability.

Note: In the bulk plot method, the F_2, F_3, and F_4 generations were grown in unselected plots. F_4 seed was harvested from 150 randomly selected plants to produce F_5 lines. Seventy of the 150 were chosen at random for further study. Seventy doubled-haploid lines were obtained from F_1 plants. Evaluations were conducted in hill plots with ten replications in each of three environments. The ten highest-yielding lines were selected from each cross for each method and evaluated in four-row plots using three replications at each of three locations. In both crosses, the bulk population had a higher mean grain yield but less variation among lines than the doubled-haploid population. In one cross, the bulk population headed earlier and resulted in taller plants than the doubled-haploid population. No yield differences were found among the four groups of top ten lines, but selections from the bulk population showed less variation in stability for grain yield than selections from doubled-haploid population. Selections for evaluation were chosen at random. Conscious selection pressure on doubled haploids can produce yields comparable to other breeding systems.

Source: Adapted from Song *et al.*, 1978, p. 275.

computer simulation to investigate the conditions favoring doubled-haploid breeding over conventional breeding of self-fertilized crops. Three conventional breeding systems, namely, postponed selection, early selection, and accelerated generation advances were compared to first and second generation doubled haploids. The probability of obtaining desired genotypes and the expected genetic advance of selected lines were examined. An efficiency factor whereby the same level of success could be achieved with 50% or fewer plants was determined. With this factor, the doubled-haploid breeding method can be used efficiently when one or more of the following conditions are met:

• A relatively small number of loci, possibly ten or less, is involved with the stated breeding objective.
• Desirable alleles are recessive to undesirable ones at most, if not all, the segregating loci.
• The genes are not strongly linked.

The overall conclusion of Yonezawa, Nomura, and Sasaki (1987) was that large-scale tests of doubled-haploid lines need to be conducted for various crop plants over a wide range of environmental conditions.

STUDIES OF GENETIC VARIANCE IN SELF-FERTILIZED CROPS

Haploid breeding procedures have been used to study the extent of variation and the type of genotypic variance (additive, dominance, or epistasis) present in segregating populations. Considerable discussion has evolved concerning the breeding methodology, and perhaps some controversy as to which procedure for obtaining haploids is suited to particular crop species. Superior agronomic performance, for example, may be possible through F_1 hybrids. In essence, the distinction between using a pure-line breeding approach and F_1 hybrids is that the former uses mostly additive genetic variance, whereas the latter uses both additive and dominance genetic variance.

To assess the feasibility of combining the increased yield effects observed in F_1 hybrids into a homozygous line, Kasha *et al.* (1977) considered doubled haploids, which provide assurance that the lines are pure. Results indicated that the amount of recombination in the F_1 gametes should be sufficient to allow superior gene combinations in the resultant homozygous diploid lines. Additive genetic variances might make it possible to fix heterotic effects, a conclusion reached by Aastveit (1964) and Busch, Lucken, and Frohberg (1971). Homozygous lines in F_5 to F_8 yielded as well as the F_1 hybrids of barley and spring wheat crosses, which showed marked heterosis. Similar conclusions regarding barley were reached by Yap and Harvey (1971) and by Upadhaya and Rasmussen (1967). In contrast, Powell *et al.* (1985a) found evidence that the haploidization process does not appear to be generating random changes. Instead, it is sampling the variation present in parental material.

The hypothesis that doubled haploids may be a viable alternative to hybrid barley development needs further investigation.

A fundamental step in any breeding program is the wise selection of parental material. Parents that combine well are very important to successful cultivar development using doubled-haploid techniques. The bulbosum method in barley may be particularly advantageous in determining parental combinations that have the potential to produce high-yielding progeny.

The question of how many doubled-haploid progeny are needed to evaluate both the mean performance of a cross and, more important, the variability available from a particular cross needs to be resolved. A study to evaluate the potential of barley crosses by testing a relatively small number of completely homozygous and randomly selected doubled-haploid lines was carried out by Reinbergs, Park, and Song (1976); its results are shown in Table 19.2. The bulbosum method of doubled haploids is a useful technique for the early identification of superior barley crosses. The method shows the combining ability of selected parents, and hence parental selection is aided.

Cross evaluations based on seven barley cultivars used to produce five crosses, 20 doubled haploids, 20 F_3 families, and 40 single-seed-descent lines grown in a replicated experiment for one year were reported by Caligari, Powell, and Jinks (1985). All three methods seemed effective, but the doubled-haploid system appeared superior because it gave the closest correspondence when the crosses were ranked according to their ability to better a standard cultivar. These workers cautioned that genotype x environment interactions need to be considered if cross predictions are to be of value.

Both these studies suggest that doubled haploids can be used successfully for univariate-cross prediction purposes, but breeding programs are rarely based on a single characteristic.

Table 19.2   Comparison of the Effectiveness of Two Population Sizes of Doubled-Haploid Lines to Analyze the Value of a Cross

| | Mean Grain Yield (g/plot) Based on: | | | |
| | 100 Lines Extracted per Cross | | 20 Lines Extracted per Cross | |
Cross	Mean	Range	Mean	Range
1	17a [a]	6-29	17a	8-29
2	21b	5-36	23b	14-36
3	21b	6-33	19b	6-30

[a] Means with different letters indicate significance at the 0.05% level.

Note: The study is based on three crosses. The data indicate that sampling of 20 doubled-haploid lines is as effective as sampling of 100 doubled-haploid lines in identifying the yield potential of crosses. Similar ranges of performance were also found. The small population size required and the fact that selections of doubled haploids are completely homozygous means that many crosses could be tested simultaneously.

Source: Adapted from Reinbergs, Park, and Song, 1976, p. 217.

Multivariate-cross prediction methods, based on five crosses, to compare doubled-haploid and F[3] single-seed-descent progenies were evaluated by Powell *et al*. (1985a). The single-seed-descent method was superior because it gave the closest agreement between observed and predicted rankings.

Monogenic mildew resistance genes in barley were found to be sampled by the bulbosum technique (Powell *et al.*, 1984). Of significance is the observation that haploid barley plants carrying mildew resistance genes in the **hemizygous** condition exhibited disease reactions comparable to their diploid counterparts. This fact should allow identification of mildew resistance prior to agronomic testing.

APPLICATIONS

The techniques for and results of producing haploids from higher plants were summarized by Nitzsche and Wenzel (1977). Included in this review are rapeseed, tobacco, potato, tomato, pepper, sugarbeet, maize, barley, oat, wheat, rye, triticale, rice, sorghum, various forage crops, asparagus, cucumber, flax, hemp, cotton, various trees, grape, geranium, African violet, petunia, and a number of pharmaceutical plants. This list demonstrates the widespread interest in producing haploids and the range of crops under investigation.

The use of haploids in barley breeding has been thoroughly documented (Choo, Reinbergs, and Kasha, 1985). Instead of using *Hordeum bulbosum* as an interspecific cross with *Hordeum vulgare*, Sitch and Snape (1986a) conducted intergeneric crosses with *Hordeum bulbosum* and *Triticum aestivum* (spring wheat) and X *Triticosecale* Wittmack (triticale) in an effort to produce haploids in these crops. Chromosome elimination and haploid production in wheat following pollination by tetraploid *Hordeum bulbosum* was observed initially by Barclay (1975). Within the tetraploid *H. bulbosum* species, considerable variation exists for seed setting potential on crossable wheat genotypes (Sitch and Snape, 1986b). *Hordeum bulbosum* genotypes selected for high crossability were hybridized to 16 winter wheat and 6 spring wheat and winter triticale genotypes. A low frequency of seed set, 0.20% with wheat and 0.27% with triticale, was achieved, but subsequent doubled-haploid production was further limited by poor embryo differentiation and regeneration (Sitch and Snape, 1986a). Although several steps need to be taken before the bulbosum technique can be applied to other cereals, the potential does exist for expanded opportunities.

The doubled-haploid wheat cultivar Florin, produced by anther culture, demonstrated that complete homozygosity in a hexaploid wheat is not detrimental and that high-yielding, doubled-haploid cultivars can be obtained from *in vitro* anther culture techniques (de Buyser, Henry, and Taleb, 1985; de Buyser *et al.*, 1987).

Using techniques of anther culture, Kasperbauer and Eizenga (1985) developed a limited number of haploids in tall fescue (*Festuca arundinacea* Schieb.). Although fescue represents a cross-fertilized species, the study is of interest because it demonstrates an extension of haploid production followed by asexual propagation of superior haploids.

The ability to produce haploids in cross-fertilized and highly heterozygous maize plants would greatly speed up inbred and subsequent hybrid development. Genovesi and Collins (1982) noted that efficient chromosome doubling techniques are needed in maize to convert haploids to diploids because separate systems determine the development of ear and tassel. Treatment of haploid maize seedlings must result in chromosome doubling of both structures.

The view of haploid breeding techniques stated by Chase (1974 p. 227) remains valid: "When the problem of developing large numbers of haploids and derivative homozygous diploid **sporophytes,** dependably, at low cost from species of economic interest is solved through the controlled culture of sporophytes from microspore and **gametophyte** initials, the tremendous leverage of the haploid method of breeding diploids and autopolyploids and **amphidiploids,** will overcome the reluctance of plant breeders to utilize this powerful tool. It is, however, my opinion that the method will be most successful in practice if those who are masters of the haploid technique are also masters of plant breeding."

OBSERVATIONS ON DOUBLED-HAPLOID METHODS

1. *Haploid* is a general term referring to a plant containing the gametic chromosome number, or half the somatic number. Such plants are more accurately described as monoploids when produced from a diploid species. When the chromosome number of a haploid is doubled it becomes a dihaploid. The system of plant breeding associated with dihaploids is known as the doubled-haploid method. Many technical steps have been resolved, and success in plant breeding using haploids is becoming evident.

2. Doubled-haploid generations provide a method of estimating the population mean and additive and genetical variances, revealing the distribution of inbred lines that would be expected from a cross.

3. Chromosome doubled haploids offer the quickest possible approach to homozygosity and provide the most homozygous genotypes possible for research purposes and cultivar release.

4. The key to the utilization of chromosome doubled haploids is the availability of methods for the production of haploids. In barley, haploids can be produced reliably using interspecific hybridization with *Hordeum bulbosum* and subsequently doubling the chromosome number with colchicine. A limited number of crop species have sufficiently reliable techniques for obtaining haploids to exploit commercially anther and pollen culture procedures. Refinement of techniques involving cell fusion, transfer of specific genes, and wide crosses with *H. bulbosum* offer possibilities for the future.

5. A huge gap exists between speculation and practice, but a number of commercial cereal and tobacco cultivars have been produced using haploid techniques. From the time of hybridization until a cultivar was released, five years elapsed, including three years in licensing trials for the development of the first doubled-haploid barley cultivar, Mingo (Ho and Jones, 1980). Despite success in cultivar development in Canada and the United Kingdom (Kasha and Reinbergs, 1981), the bulbosum technique has not become a common breeding method.

6. Every stage of the haploid procedure, including hybridization, haploid production, chromosome doubling, and seed increase, can be accomplished at a site other than one where a new cultivar will be used commercially. Only the final testing procedure need be accomplished in the region of adaptation.

7. A much higher level of technical expertise is required for the haploid procedure than for other breeding techniques. Suitable technical and physical facilities are also needed.

8. Haploid breeding methods are useful for meeting specific objectives, both qualitative and quantitative, of a breeding program, especially if time is limited.

9. For specific objectives such as maturity, vertical pest resistance, and standability, where repeated selection is not necessary, doubled-haploid techniques are most useful.

10. Doubled-haploid techniques can be used for obtaining either **vertical** or **horizontal resistance**.

11. For many crops, lack of genetic variability is not the limiting factor in their further improvement. The major difficulty is identifying and isolating desired recombinants. This important step may be simpler with doubled-haploid techniques than with any other breeding system.

12. The two major methods of isolating haploids are **microspore culture** and wide crosses with *Hordeum bulbosum*.

13. Doubled-haploid techniques had to await the discovery and use of colchicine, the development of controlled environment facilities, and cytological knowledge. For the systems to be used in routine, commercial practice, a number of further evolutionary steps may need to be completed.

14. Doubled-haploid lines have a greater additive genetic variance than diploid lines, and the response to selection should be greater than in diploid populations.

15. A random sample of gametes from F_1 hybrids using the doubled-haploid techniques can be employed for quantitative genetic studies. Genetic models of doubled haploids do not require the unrealistic assumption of no epistasis or no linkage or both.

REFERENCES

Aastveit, K. 1964. Heterosis and Selection in Barley. *Genetics* 49:159-164.

Ahloowalia, B. S. 1982. Plant Regeneration from Callus Culture in Wheat. *Crop Science* 22:405-410.

Allard, R. W., and A. D. Bradshaw. 1964. Implications of Genotypic Environmental Interactions in Applied Plant Breeding. *Crop Science* 4:503-508.

Barclay, I. R. 1975. High Frequencies of Haploid Production in Wheat (*Triticum aestivum*) by Chromosome Elimination. *Nature* 256:410-411.

Blakeslee, A. F., J. Bellina, M. E. Farnham, and A. D. Bergner. 1922. A Haploid Mutant in the Jimson Weed, *Datura stramonium. Science* 55:646-647.

Busch, R. A., K. A. Lucken, and R. C. Frohberg. 1971. F_1 Hybrids Versus Random F_5 Line Performance and Estimates of Genetic Effects in Spring Wheat. *Crop Science* 11:357-361.

Caligari, P.D.S., W. Powell, and J. L. Jinks. 1985. The Use of Doubled Haploids in Barley Breeding. 2: An Assessment of Univariate Cross Prediction Methods. *Heredity* 54:353-358.

Chase, S. S. 1949. Monoploid Frequencies in a Commercial Double Cross Hybrid Maize and its Component Single Cross Hybrids and Inbred Lines. *Genetics* 34:328-332.

————. 1952. Production of Homozygous Diploids of Maize from Monoploids. *Agronomy Journal* 44:263-267.

————. 1974. Utilization of Haploids in Plant Breeding: Breeding Diploid Species. In *Haploids in Higher Plants: Advances and Potential,* edited by K. J. Kasha. University of Guelph, pp. 211-230.

Choo, T. M. 1981. Doubled Haploids for Studying the Inheritance of Quantitative Characters. *Genetics* 99:525-540.

Choo, T. M., E. Reinbergs, and K. J. Kasha. 1985. Use of Haploids in Breeding Barley. In *Plant Breeding Reviews,* edited by J. Janick, vol. 3. Avi Publishing Company, Westport, Connecticut, pp. 219-252.

Collins, G. B., P. D. Legg, and M. J. Kasperbauer. 1974. Use of Anther-Derived Haploids in *Nicotiana.* I: Isolation of Breeding Lines Differing in Total Alkaloid Content. *Crop Science* 14:77-80.

de Buyser, J., Y. Henry, and G. Taleb. 1985. Wheat Androgenesis: Cytogenetical Analysis and Agronomic Performance of Doubled Haploids. *Zeitschrift für Pflanzenzuchtung* 95:23-34.

de Buyser, J., Y. Henry, P. Lonnet, R. Hertzog, and A. Hespel. 1987. "Florin": A Doubled Haploid Wheat Variety Developed by the Anther Culture Method. *Plant Breeding* 98:53-56.

Finch, R. A., and M. D. Bennett. 1979. Meiotic Stability in Control and Newly Colchicine-Induced Dihaploid Barley. *Canadian Journal Genetics and Cytology* 21:33-35.

Genovesi, A. D., and G. B. Collins. 1982. *In Vitro* Production of Haploid Plants of Corn via Anther Culture. *Crop Science* 22:1137-1144.

Guka, S., and S. C. Maheshwari. 1964. *In Vitro* Production of Embryo as from Anthers of *Datura. Nature* 204:497.

Ho, K. M., and G. E. Jones. 1980. Mingo Barley. *Canadian Journal of Plant Science* 60:279-280.

Hougas, R. W., S. J. Peloquin, and R. W. Ross. 1958. Haploids of the Common Potato. *Journal of Heredity* 47:103-107.

Hu, H. 1985. Use of Haploids in Crop Improvement. In *Biotechnology in International Agricultural Research.* International Rice Research Institute, Manila, Philippines, pp. 75-84.

Huang, B., J. M. Dunwell, W. Powell, A. M. Hayter, and W. Wood. 1984. The Relative Efficiency of Microspore Culture and Chromosome Elimination as Methods of Haploid Production in *Hordeum vulgare* L. *Zeitschrift für Pflanzenzuchtung* 92:22-29.

Iwanaga, M. 1985. Haploids Ploidy Manipulation and Meiotic Mutants in Potato Breeding. In *Biotechnology in Agricultural Research.* International Rice Research Institute, Manila, Philippines, pp. 139-148.

Jensen, C. J. 1977. Monoploid Production by Chromosome Elimination. In *Applied and Fundamental Aspects of Plant Cell, Tissue and Organ Culture,* edited by J. Reinhart and Y.P.S. Bajaj. Springer Verlag, Berlin, pp. 299-340.

Jensen, N. F. 1952. Intra-varietal Diversification in Oat Breeding. *Agronomy Journal* 44:30-34.

Johns, W. A. 1974. A Preliminary Evaluation of Haploid as a Breeding Technique in Barley (*Hordeum vulgare* L.). Ph.D. Thesis, University of Guelph. Dissertation Abstract 35:4750B.

Kao, K. N., and K. J. Kasha. 1970. Haploidy from Interspecific Crosses with Tetraploid Barley. In *Barley Genetics II,* Proceedings of the Second International Barley Genetics Symposium, edited by R. A. Nilan. Washington State University Press, Pullman, Washington, pp. 82-88.

Kasha, K. J. 1976. Utilization of Haploidy in Plant Breeding and Mutation. *Acta Biologica Series F of Genetika* 8:101-110.

Kasha, K. J., and K. N. Kao. 1970. High Frequency Haploid Production in Barley (*Hordeum vulgare* L.). *Nature* 225:874-876.

Kasha, K. J., and E. Reinbergs. 1972. The Haploid Technique in Barley Breeding. *Technical Quarterly* 9:128-130.

———. 1975. Haploidy and Polyploidy and Its Application in Breeding Techniques. In *Barley Genetics III,* Proceedings of the Third International Barley Genetics Symposium, Garching. Verlag Karl Thieming, Munich, pp. 307-315.

———. 1981. Recent Developments in the Production and Utilization of Haploids in Barley. In *Barley Genetics IV,* Proceedings of the Fourth International Barley Genetics Symposium, Edinburgh. Edinburgh University Press, Edinburgh, pp. 655-665.

Kasha, K. J., L.S.P. Song, S. J. Park, and E. Reinbergs. 1977. Fixation of Heterosis: Comparison of F_1 Hybrids with Their Respective Homozygous Lines Developed Using Doubled Haploid Procedures. *Cereal Research Communications* 5(3):205-214.

Kasha, K. J., N. C. Subrahmanyam, and A. Ali. 1978. Effect of Gibberellic Acid Treatment and Nutrient Supply Through Detached Tillers Upon Haploid Frequency in Barley. *Theoretical and Applied Genetics* 51:169-175.

Kasperbauer, M. J., and G. C. Eizenga. 1985. Tall Fescue Doubled Haploids via Tissue Culture and Plant Regeneration. *Crop Science* 25:1091-1095.

Kostoff, D. 1938. Colchicine and Acenaphthene as Polyploidizing Agents. *Nature* 142:753.

Legg, P. D., and G. B. Collins. 1973, 1978. The Potential Value of Doubled Haploids in Quantitative Genetics Research. *Horticultural Science* 8:270; *Plant Breeding Abstracts* 44:1891.

Liang, G. H., A. Xu, and Hoang-Tang. 1987. Direct Generation of Wheat Haploids via Anther Culture. *Crop Science* 27:336-339.

MacKey, J. 1970. Significance of Mating Systems for Chromosomes and Gametes in Polyploids. *Hereditas* 66:165-176.

Maheshwari, S. C., A. K. Tyagi, K. Malhotra, and S. K. Sopory. 1980. Induction of Haploidy from Pollen Grains in Angiosperms: The Current Status. *Theoretical and Applied Genetics* 58:193-206.

Morgan, D. T., and R. D. Rappleye. 1950. Twin and Triplet Pepper Seedlings: A Study of Polyembryony in *Capsicum frutescens. Journal of Heredity* 41:91-95.

Nakamura, A., T. Yamada, N. Nadotani, and R. Itagaki. 1974. Improvement of Flue-Cured Tobacco Variety MC 1610 by Means of Haploid Breeding Method and Some Problems of this Method. In *Haploids in Higher Plants: Advances and Potential,* Proceedings of the First International Symposium. University of Guelph, pp. 277-278.

Nei, N. 1963. The Efficiency of Haploid Method of Plant Breeding. *Heredity* 18:95-100.

Nitsch, J. P., and C. Nitsch. 1969. Haploid Plants from Pollen Grains. *Science* 163:85-87.

Nitzsche, W., and G. Wenzel. 1977. *Haploids In Plant Breeding.* Advances in Plant Breeding, Supplement 8 to Journal of Plant Breeding. P. Parey, Hamburg. 101 pp.

Park, S. J., E. J. Walsh, E. Reinbergs, L.S.P. Song, and K. J. Kasha. 1976. Field Performance of Doubled Haploid Barley Lines in Comparison with Lines Developed by the Pedigree and Single Seed Descent Methods. *Canadian Journal of Plant Science* 56:467-474.

Peloquin, S. J. 1981. Chromosomal and Cytoplasmic Manipulations. In *Plant Breeding II*, edited by K. J. Frey. Iowa State University Press, Ames, pp. 117-150.

Pickering, R. A. 1983. The Influence of Genotype on Doubled Haploid Barley Production. *Euphytica* 32:863-876.

Pickering, R. A., and P. W. Morgan. 1979. Progress in Doubled Haploid Production at the Welsh Plant Breeding Station. *Barley Newsletter* 23:141-143.

Powell, W., M.J.C. Asher, W. Wood, and A. M. Hayter. 1984. The Manipulation of Mildew Resistance Genes in a Barley Breeding Programme by the Use of Doubled Haploids. *Zeitschrift für Pflanzenzuchtung* 93:43-48.

Powell, W., P.D.S. Caligari, J.W. McNicol, and J. L. Jinks. 1985a. The Use of Doubled Haploids in Barley Breeding. 3: An Assessment of Multivariate Cross Prediction Methods. *Heredity* 55:249-254.

Powell, W., P.D.S. Caligari, J. L. Jinks, and A. M. Hayter. 1985b. The Use of Doubled Haploids in Barley Breeding. I: Comparison of H_1 and H_2 Generations. *Heredity* 54:261-266.

Reinbergs, E., S. J. Park, and K. J. Kasha. 1975. The Haploid Technique in Comparison with Conventional Methods in Barley Breeding. In *Barley Genetics III*, Proceedings of the Third International Barley Genetics Symposium, Garching. Verlag Karl Thieming, Munich, pp. 346-350.

Reinbergs, E., S. J. Park, and L.S.P. Song. 1976. Early Identification of Superior Barley Crosses by the Doubled Haploid Technique. *Zeitschrift für Pflanzenzuchtung* 76:215-224.

Riggs, T. J., and J. W. Snape. 1977. Effects of Linkage and Interaction in a Comparison of Theoretical Populations Derived by Diploidized and Single Seed Descent Methods. *Theoretical and Applied Genetics* 49:111-115.

Riley, R., and V. Chapman. 1958. Genetic Control of Cytologically Diploid Behavior of Hexaploid Wheat. *Nature* 182:713-715.

Rines, H. W. 1983. Oat Anther Culture: Genotype Effects on Callus Initiation and the Production of a Haploid Plant. *Crop Science* 23:268-272.

Sears, E. R. 1954. *The Aneuploids of Common Wheat*. Missouri Agriculture Experiment Station Research Bulletin 572. pp.1-58.

Sitch, L. A., and J. W. Snape. 1986a. Doubled Haploid Production in Winter Wheat and Triticale Genotypes, Using the *Hordeum bulbosum* System. *Euphytica* 35:1045-1051.

————. 1986b. The Influence of the *Hordeum bulbosum* and the Wheat Genotype on Haploid Production in Wheat (*Triticum aestivum*). *Zeitschrift für Pflanzenzuchtung* 96:304-319.

Snape, J. W., and E. Simpson. 1981. The Genetical Expectations of Doubled Haploid Lines Derived from Different Filial Generations. *Theoretical and Applied Genetics* 60:123-128.

Song, L.S.P., S. J. Park, E. Reinbergs, T. M. Choo, and K. J. Kasha. 1978. Doubled Haploid vs the Bulk Plot Method for Production of Homozygous Lines in Barley. *Zeitschrift für Pflanzenzuchtung* 81:271-280.

Subrahmanyam, N. C., and K. J. Kasha. 1973. Selective Chromosomal Elimination During Haploid Formation in Barley Following Interspecific Hybridization. *Chromosoma* (Berlin) 42:111-125.

Sunderland, N. 1974. Anther Culture as a Means of Haploid Induction. In *Haploids in Higher Plants: Advances and Potential*, Proceedings of the First International Symposium, Guelph, edited by K.J. Kasha. University of Guelph, pp. 91-122.

Symko, S. 1969. Haploid Barley from Crosses of *Hordeum bulbosum* (2x) x *H. vulgare* (2x). *Canadian Journal of Genetics and Cytology* 11:602-608.

Thevenin, L. 1974. Haploids in Asparagus Breeding. In *Haploids in Higher Plants: Advances and Potential*, Proceedings of the First International Symposium, Guelph, edited by K. J. Kasha. University of Guelph, p. 279.

Thiebaut, J., K. J. Kasha, and A. Tsai. 1979. Influence of Plant Development Stage, Temperature and Plant Hormones on Chromosome Doubling of Barley Haploids Using Colchicine. *Canadian Journal of Botany* 57: 480-483.

Thompson, K. F. 1972. *Oilseed Rape.* Report of the Plant Breeding Institute, Cambridge. Plant Breeding Institute, Cambridge, pp. 94-96.

Tsunewaki, K., K. Noda, and T. Fujisawa. 1968. Haploid and Twin Formation in a Wheat Strain Salmon with Alien Cytoplasms. *Cytologia* 33:526-538.

Turcotte, E. L., and C. V. Feaster. 1967. Semigamy in Pima Cotton. *Journal of Heredity* 58:54-57.

Upadhaya, B. R., and D. C. Rasmussen. 1967. Heterosis and Combining Ability in Barley. *Crop Science* 7:644-647.

Walsh, E. 1974. Efficiency of the Haploid Method of Breeding Autogamous Diploid Species: A Computer Simulation Study. In *Haploids in Higher Plants: Advances and Potential,* Proceedings of the First International Symposium. University of Guelph, pp. 195-209.

Walsh, E. J., E. Reinbergs, and K. J. Kasha. 1973. Importance of Seed Source in Preliminary Evaluations of Doubled Haploids in Barley. *Canadian Journal of Plant Science* 53:257-260.

Yap, T. C., and B. L. Harvey. 1971. Heterosis and Combining Ability of Barley Hybrids in Densely and Widely Seeded Conditions. *Canadian Journal of Plant Science* 51:115-122.

Yonezawa, K., T. Nomura, and Y. Sasaki. 1987. Conditions Favouring Doubled Haploid Breeding Over Conventional Breeding of Self-fertilizing Crops. *Euphytica* 36:441-453.

Zapata, F. J. 1985. Rice Anther Culture at IRRI. In *Biotechnology in International Agricultural Research.* International Rice Research Institute, Manila, Philippines, pp. 85-95.

CHAPTER 20

Asexual Breeding Systems

SEXUAL VERSUS ASEXUAL REPRODUCTION

Plants can be divided into two principal groups: (1) those that are mainly reproduced from seed (sexual reproduction) and (2) those that are propagated vegetatively by runners, stems, rhizomes, tubers, bulbs, and corms, as well as root suckers (asexual reproduction). Humans have added asexual forms such as grafting, layering, cuttings, division, and, more recently, rapid *in vitro* techniques.

The method of propagation of various plant species presents a unique set of challenges and opportunities for the plant breeder or the commercial grower. It is largely on the basis of these opportunities that humans have arbitrarily divided plants into the two categories because nature never intended plants to be restricted in their evolution by a lack of sexual reproductive organs. By careful manipulation, all higher plants could be induced to develop sexual reproductive organs that could produce highly variable progeny. Methods of asexual reproduction were adopted by humans to avoid variable progeny and to maintain specific genotypes.

The promise of producing commercially successful, new and attractive garden ornamentals, attracted attention and stimulated horticultural plant breeders to hybridize plants sexually in the 1800s. At that time, lacking a knowledge of genetic laws, humans did not understand how to develop a pure line. Thus if a winning progeny was produced by sexual means, asexual propagation was used to maintain the specific genotype. Asexual reproduction assumed major importance in many crops before the time of Mendel. Following Mendel, sexual hybridization took on a new dimension, but the convenient asexual means of reproduction persists to this day.

The arbitrary classification of plants as either sexually or asexually reproduced is not rigid. A few perennial ornamentals, including *Begonia, Delphinium, Lilium,* and certain *Lupinus, Pelargonium, Primula, Viola, Hibiscus,* and African violets (*Saintpaulia*), are regularly propagated both by seed and by vegetative means. Among vegetable crops, asparagus, chicory, and sea kale are propagated both ways.

Most perennial horticultural plants are asexually propagated, and annuals and biennials are sexually propagated; again this is determined mainly by humans and based largely on economic considerations. A few perennials such as *Anemone coronaria, Cyclamen,* and runner bean are almost invariably seed-propagated, despite the fact that they all produce storage organs and could be grown as perennials. Seed propagation is preferred when rapid, inexpensive vegetative propagation is difficult.

Grapes (Figure 20.1) are propagated by a number of vegetative means, even though seeds can be germinated readily. Seedless cultivars of grapes and other crops are likely the product of a diploid and a tetraploid and would be lost under natural evolutionary forces if asexual

Figure 20.1   This is a photo of grape (*Vitis* spp) plants. Grapevines are propagated by seeds, cuttings, layering, budding, or grafting. Seeds are used only in sexual breeding programs to produce new cultivars.

Most commercial asexual propagation is by dormant hardwood cuttings. Budding or grafting on rootstocks is used occasionally to increase vine life, plant vigor, and yield. Where cultivars such as *Vitis vinifera* are to be grown in soil containing noxious soil organisms, such as phylloxera (*Dactylosphaera vitifoliae*) or root-knot nematodes (*Meloidogyne* spp), it is necessary to graft or bud onto a resistant rootstock. For grape cultivars that are difficult to root, such as the Muscadine grape (*Vitis rotundifolia*), layering or the use of leaf cuttings under mist is necessary.

For hybridization purposes, grape seeds are used to produce plants. Grape seeds are not difficult to germinate. Germination can be enhanced by a moist stratification period at 0.5 to 4.0° C (33 to 40° F) for about 12 weeks before planting (Harmon and Weinberger, 1959).

Asexual reproduction, without the union of gametes, is a common means by which many plants with high commercial value are reproduced. Without the benefit of vegetative propagation, many valuable cultivars of crops, such as seedless grapes, would be lost.

reproduction was not employed. Seedless forms could be maintained, however, through the parental genotypes.

In experiments with methods aimed at countering deforestation, Leakey (1986) reported success with vegetative propagation and clonal selection of hardwood trees to produce a large and continuous supply of planting stock, and with the capture and multiplication of genotypic variation. The identification of superior genotypes among clonal selections can result in large genetic gains in both yield and quality. For hardwood tree species that root with difficulty, rooting was promoted by dipping single-node cuttings into a 0.4% solution of indole-3-butyric acid, maintaining humid conditions and warm air (30° C or 86° F), and keeping the cuttings out of direct sunlight. The rooting medium should be at a temperature of 35 to 40° C (95 to 104° F).

CHARACTERISTICS OF ASEXUAL REPRODUCTION

Although sexual reproduction has occurred in crops at some time in their development, asexual reproduction is the current normal means of their commercial propagation and has the following features:

1. A major advantage of vegetative reproduction is that only one superior plant need be developed. The time-consuming and expensive task of progeny testing can be eliminated. A superior genetic combination resulting in high yield, disease resistance, heterosis, high quality, agronomic adaptation and stability, and unique flower color or design can be maintained.

2. A disadvantage of vegetative reproduction is that a number of small, undetected mutations may alter the genetic constitution. If this occurs, the original genetic constitution may be restored only by chance from the original parents. Mutations may result in the loss of vigor, disease resistance, or brilliance of an ornamental.

3. If the economically valuable component is a plant part such as a tuber, root, stem, or forage, sterile plants can be employed. The assurance that good hybrids can be used commercially raises the possibility of using wide crosses to combine the desired traits of different species or genera. Sterile cultivars can be eradicated easily and are less likely to become weed problems than cultivars that produce seeds.

4. The approach to breeding vegetatively propagated plants can be based on a different philosophy than that for seed-propagated plants. Once a good genetic combination has been obtained, it can be propagated immediately as a new cultivar, whereas with seed-propagated plants, it is necessary to "fix" the cultivar so that it breeds true. This can be a lengthy and difficult process. In certain respects, it is easier to achieve success with vegetatively propagated crops.

5. Vegetative propagation often carries diseases, especially viruses or viroids, to the new plants, whereas seed propagation usually screens out non-seed-borne pathogens. Precautions taken to produce virus-free stock, which will be described later, can reduce this problem.

6. Large fields or regions of vegetatively propagated crops such as sugarcane may be subject to attack from new, highly virulent disease strains. The problem may be no more severe, however, than for large regions of genetically uniform seed-produced crops.

7. Compared to sexual reproduction in seed-bearing crops, vegetative reproduction is a

slow means of multiplying new cultivars or disease-free stocks of established cultivars. In maize, a relatively standard yield of 6,200 kg/ha (100 bu./ac.) could produce enough seed to plant 245 ha (600 ac.). In orchids, clones can be multiplied vegetatively only very slowly.

8. Hybrid vigor can be exploited by combining two unrelated parents and finding a superior F_1 progeny. It is difficult in self-fertilized, seed-producing crops to obtain all the desired growth factors in any one plant in a homozygous condition. Asexual propagation maintains the highly heterozygous condition of an F_1 hybrid.

9. The recognition that forages can be propagated readily by asexual means has opened up new vistas in forage breeding (Figure 20.2). Chief among these is the opportunity to clone superior plants to form a polycross nursery for the development of a synthetic cultivar.

10. Compared to compact, concentrated dry seeds, which can be stored for long periods of time in controlled environments or in cryopreservation storage facilities, vegetatively propagated germplasm is difficult to maintain in a viable state for either a short or a long time period. Potato tubers, sugarcane cuttings, or forage explant materials cannot be stored for extended periods as readily as seeds. Tubers and rhizomes are bulky and require considerable space. Production of seed through selfing has been practiced to alleviate the vegetative storage problem, but loss of heterozygosity may not be desirable.

In vitro conservation has been given a high research priority by the International Board for Plant Genetic Resources (Withers and Williams, 1985). Short-term storage is available for shoot cultures and cryopreservation (Kartha, 1984) and can meet both short- and long-term requirements for cell cultures. But methods for storing protoplasts and anther and callus cultures are lacking. Shoot-tip cryopreservation has progressed considerably in recent years, but cannot be applied yet to a wide range of crops. Often, poor recovery results from stored shoot-tip tissue.

Perhaps artificial seeds, produced by encasing large numbers of asexually propagated plantlets from **callus tissue** in a medium, may reduce this problem. Artificial seeds are small in size, can facilitate high rates of clonal multiplication, and can be kept free of diseases.

BREEDING VEGETATIVELY PROPAGATED PLANTS

To produce new genetic combinations of plants that are normally produced asexually, it is necessary to use sexual methods of hybridization. Sexual methods of breeding plants belonging to the asexually propagated group are not widely different from those used with other seed-producing crop plants, but precautions must be taken to ensure the development of sexual reproductive organs in certain crops such as sugarcane (*Saccharum* spp) (Figure 20.3).

Flowering in commercial sugarcane production fields in the first year of a two-year crop has a detrimental effect on production, and the use of chemicals to retard flowering is standard practice. The induction of flowering for sexual breeding is a critical task in sugarcane. This crop has been grown for centuries by means of vegetative production with an accompanying selection away from reproductive development by means of seeds. Flowering can be induced by photoperiodic manipulation.

To ease the physical movement of plants, to facilitate coincidence of male and female

A.   *B.*

Figure 20.2   The realization that superior heterozygous individual and phenotypically superior forage plants can be cloned asexually, as evident in the source nursery in Photo A, has opened the way to new aspects of forage crop improvement through breeding. Vegetative propagation is used by the forage breeder to

- Establish clones from superior plants for clonal evaluation (Photo B).
- Establish clones for synthetic cultivar development in a polycross nursery.
- Maintain original genotype used in synthetic cultivar production.
- Propagate strains or cultivars that are poor seed producers.
- Propagate genotypes that have unusual chromosome numbers and could not be propagated by seed because of sterility.

Most forage grasses and legumes can be propagated by stem cuttings, stolons, or rhizomes or through division of the crown. Stem cuttings should consist of at least one node and a portion of an internode. When placed in moist sand or vermiculite with the node exposed, top growth develops from a nodal bud and roots from the cut end of the internode. A high success rate with these asexual cuttings generally means that treatment with a growth promoting substance such as indolebutyric acid is unnecessary.

Obtaining sufficient plants for a polycross nursery represented by four to six parental plants presents no problem for the plant breeder. Use of vegetative cuttings for commercial hybrid seed production was suggested by Tysdal, Kiesselbach, and Westover (1942), but this would be too great an undertaking for commercial seed production on the scale required in major forage-producing areas. Success with artificial seeds produced from callus tissue may make hybrids a feasible venture.

organs, and to facilitate the isolation of male-sterile female plants by transporting them to crossing rooms, a system of marcotting (Figure 20.4) is used. The marcotting method has greatly facilitated isolation of field-grown male and female flowering stalks and has eliminated the need for elaborate techniques to control crossing, described by Blackburn (1984). Marcotting allows stems to be moved into a controlled environment for photoperiodic, temperature-induced floral production.

Where inflorescences are formed in the field, it is possible to cut such stalks and keep them alive for several days or weeks by immersing them in 500 ppm of sulphurous acid solution.

Figure 20.3 The sugarcane shown in this photo is in a vegetative stage. Sugarcane is an asexually produced crop. The crop normally is propagated by stem cuttings, each containing one or more buds, that are dropped into furrows at the rate of about 14 buds/m of row (39 in.) in rows 1.0 to 1.5 m (about 39 to 58 in.) apart. Buds develop into a series of underground stems that produce a number of erect tillers or stalks. At harvest one to two years after planting, a dense stand results, although many tillers will die out from competition.

Sugarcane is vegetatively propagated so that the genetic constitution of the derived hybrids is fixed. Asexual breeding techniques involving mutation breeding and callus tissue culture are useful adjuncts. This latter method can be used to obtain disease resistance. Hybridization and sexual reproduction have provided substantial genetic improvement because of an enormous range of genetic variability within the species *Saccharum* and high chromosome number (*S. officinarum*, 2n = 80).

Natural flowering at high latitudes rarely occurs because sugarcane generally is regarded as a short-day plant that requires an uninterrupted dark period in order to flower. The optimum conditions for floral induction, outlined by Clements and Awada (1965) and Coleman (1968), include a photoperiod of just over 12 hours at high light level and a day-night temperature regime of 27 to 31° C (81 to 88° F) and 22 to 23° C (72 to 73° F), respectively, when the plant has two to four elongated internodes exposed. Photoperiod is the major influence, but reproductive development is also influenced by latitude, temperature, physiological age, soil nitrogen status, and water supply.

Figure 20.4   A method of asexual reproduction known as marcotting, or air layering, is illustrated in this photo, which shows the marcotted stem of the woody shrub hibiscus. Root primordia occur at each bud, or node, and can be induced to grow if placed in a suitable environment. Marcotting can be done in the field by surrounding the tissue with an aluminum or plastic container filled with a rooting medium such as moist soil, peat, or coconut fiber. The stem is cut below the makeshift container, the cover removed, and the marcotted upper part placed in a pot of soil.

Marcotting may be required in breeding sugarcane in areas where low temperature affects fertility. By transferring marcotted stems to a heated greenhouse with appropriate day length, the inflorescence, called an arrow or tassel, emerges and controlled crosses can be made.

Sugarcane is propagated vegetatively, and seeds produced by sexual hybridization can be grown in a selection nursery for evaluation. Superior individuals can be vegetatively propagated to maintain the superior genotype without further segregation. F_1 heterosis is fixed in subsequent populations as long as asexual reproduction is practiced.

In sugarcane breeding, a clear objective is required because the plant breeder is faced with selecting desirable individuals from among several thousand to two million or more seeds each year.

James (1980) presented modifications to the solution method that have become widely adopted. The flowering stalks are simply cut at the proper time after flower emergence and moved to the hybridization area.

In the development of potato cultivars, the traditional breeding process is to apply pollen from the male parent to the stigma of the emasculated female parent. From 50 to 100 seeds per fruit can develop in seedballs resembling immature tomatoes. Each seed can produce a new plant, and the genetic makeup of superior F_1 selections can be fixed for future generations by subsequent asexual propagation.

Seed tubers have become the conventional means of tuber production, despite the fact that tubers are bulky, expensive, and not always free of disease. In developing countries, these are becoming limiting factors in the production of potatoes.

Where locally produced seed tubers are used, accumulation of tuber-borne diseases in successive generations can result in significant yield reductions. Exacerbating this problem are storage facilities that lack environmental control. And high temperatures may cause a further decrease in quality through physiological deterioration of the tubers. Importation of seed potatoes from areas relatively free of disease is a solution to the disease and physiological quality problems, but at a substantial increase in production costs.

To overcome these problems, asexual propagation may be abandoned in favor of propagation using true potato seed. Major advantages of the use of true seed include low disease transmission, ease of transport and storage, and the release of tubers normally saved for planting for human consumption.

One difficulty with true potato seed is the wide range in seed set on various potato cultivars. A plant yield of 40 berries capable of producing 200 to 400 seeds is considered good, but cultivars range from those that produce several hundred berries per plant to those that never flower. Cultivars fail to set seed for reasons ranging from pollen sterility to a large and strong tuber sink where competition for photosynthate leaves little or nothing for seed development.

True seed has been found to transmit only four viruses and a viroid (Jones, 1982), of which only potato spindle tuber viroid is considered to pose a potential problem. In contrast, potato tubers can transmit most of the major potato diseases, including late blight, bacterial wilt, and about 50 viruses.

Seedlings from true potato seed appear to be less vigorous than those from tubers, perhaps because of the differences in nutrient reserves. Several vigorous shoots from tubers provide rapid development of a photosynthetic factory, whereas seedlings from true seed normally produce only one main shoot. Experimental procedures designed to overcome this problem involve transplanting four- to six-week-old true potato seed seedlings from nursery beds to the field.

An alternative to using true potato seed is to produce small disease-free tubers from true potato seed. This is achieved by dense plantings in sterilized soil with high nutrient fertility (White and Sadik, 1983).

The potential benefits of true potato seed are great but there is a need for much work to adapt present knowledge to the needs and resources of producers in areas where this potential may be realized. For true potato seed to become a commercial reality, changes in breeding would be required. Conventional potato breeding methods are dictated by the asexual and cross-fertilized nature of potato propagation. Self-fertilization to produce homozygous parents for F_1 hybrid seed production may be necessary, but commercial production of F_1 hybrids may prove too expensive and impractical for developing regions. An alternative is to produce open-pollinated cultivars, which so far have shown considerable promise.

Until true potato seed problems are resolved, disease-free seed tubers could be supplied by culturing potatoes under sterile laboratory conditions. **Biotechnology** may offer the tools necessary for the recovery of disease-free plants (Roca, 1985). Methods are available to grow plants to maturity from shoot-tip explants comprising the apical meristem and at least one primordial leaf (Quak, 1977). This explant material is not necessarily pathogen free, and so *in vitro* culture techniques should include

- identifying and eliminating disease organisms
- testing regenerated plants for freedom from pathogens
- propagating healthy plants under conditions that prevent reinfection

To produce disease-free stock, a heat treatment can be applied, and although it does not eliminate most viruses, it reduces virus multiplication and translocation in the plant. A heat treatment of 35° C (95° F) suppresses Caribbean mosaic symptoms in cassava (*Manihot esculenta*) (Figure 20.5).

The production of disease-free stock by chemical treatment has had limited success, and many chemicals are phytotoxic (Kartha, 1984). Viruses and viroids cannot be controlled by applying chemical compounds to infected plants. Breeding for resistance to viruses is generally difficult, and valuable cultivars cannot be restored to their original agronomic value with chemical treatment. *In vitro* culture techniques to produce clean planting material could be a short-term approach to restoring yield and quality in local cultivars of vegetatively propagated crops.

APOMIXIS

Apomixis is an asexual method of reproduction through the seed that occurs at low levels in some cultivated species and may be found in closely related species in some cultivated crops. Apomixis provides unique opportunities for developing superior cultivars in the future (Hanna and Bashaw, 1987). Transfer of apomixis to major food and feed crops would make possible the development of true-breeding hybrids and the commercial production of hybrids without the need for cytoplasmic male sterility and high cost, labor-intensive processes associated with F_1 hybrid seed production. If plant breeders could induce apomixis in plants, the results would be a system with the efficiency of reproduction through vegetative organs but with the convenience of propagation through asexually developed seed.

In plants that reproduce by sexual means, the egg is produced in an embryo sac derived from meiotic division in the megaspore mother cell. Meiosis and fertilization are involved in the development of the embryo in sexually reproducing plants and allow for genetic recombination, segregation, and production of a diversity of genotypes.

In apomictic reproduction, the embryo is derived from mitotic division of a megaspore mother cell or a somatic cell of the ovule. Meiosis and fertilization are not involved in development of the embryo, and the progeny of apomictic plants are exact replicas of the female parent. Apomixis therefore is a means of cloning plants through seed. Apomixis has important implications for use as a plant breeding tool. Mechanisms of apomixis were outlined in Chapter 2.

Some species, such as bluegrass (*Poa pratensis* L.), are classed as **facultative apomicts** that can produce both sexual and apomictic seeds. Plants that reproduce sexually can be mated with facultative apomictic plants to produce new hybrids. **Obligate apomicts**, such as Bahia grass (*Paspalum notatum*), rarely produce sexual types. Without sexual types for gene

Figure 20.5   In asexually propagated plants, diseases caused by viruses and viroids are particularly significant because once the crop is infected its propagates transmit the disease indefinitely. Severe yield and quality losses can occur. Shoot-tip culture techniques have been used to recover healthy cassava clones and have resulted in dramatic yield increases (Roca, 1985). The disease-resistant and high-yielding cassava plants in Photo A and the cassava tubers in Photo B are free of disease.

Disease-free clones can be regenerated through tissue culture techniques, which can be applied to all known infectious agents but are especially valuable in eliminating viruses and viroids from vegetatively propagated crops. Producing clean planting material through *in vitro* methods can significantly raise crop yield and quality. Roca (1985) reported on three cassava diseases that have been controlled by shoot-tip culture.

Clean planting material obtained through thermotherapy followed by shoot-tip culture eliminated cassava Caribbean mosaic disease and increased fresh weight yield by 70%. After three years, yields remained stable, but healthy plants must be propagated under conditions that prevent reinfection.

Cassava frog skin disease, endemic to an area in southern Colombia, can reduce yield up to 80%. Yield in root fresh weight and starch content of two local cultivars, cleaned of the disease by *in vitro* methods, more than doubled. In the second year, yield of treated material decreased slightly.

recombination, new genetic variation cannot be produced. Until the discovery of a plant that reproduces by sexual means in the obligate apomictic buffelgrass (*Pennisetum ciliare*), no genetic recombination was possible.

Apomictic species or plants cannot be identified by their phenotype but only through their progeny. Obligate apomicts can be identified in a cross-fertilized, heterozygous species by complete uniformity of progeny, identical appearance to the maternal parent, and complete lack of segregation following hybridization. Obligate apomicts are unlikely to appear suddenly in a plant breeding program involving sexually reproducing species. The tendency to produce spontaneous haploid offspring appears to be genetically controlled in some species, such as maize and cotton (Bashaw, 1980).

Apomixis is suspected if unusually high seed fertility is found in **aneuploids, triploids**, wide crosses, or other plants expected to be sterile. Other indicators of apomixis include multiple seedling development per seed, multiple stigmas, multiple ovules per floret, and double or fused ovaries (Hanna *et al.*, 1973; Hanna, Schertz, and Bashaw, 1970). If the aneuploid chromosome number or other structural heterozygosity remains constant from parent to progeny, apomixis may be present.

It is a more challenging task to identify a facultative apomict than an obligate apomict. The primary indicator is a disproportionately high number of identical homozygous progeny that resemble the maternal parent, along with some progeny that are distinctly different in appearance and presumably of hybrid origin. Facultative apomicts seldom produce obligate apomictic offspring. Bashaw (1980) suggested that facultative apomicts may produce up to 98% of their offspring as apomictic plants.

In self-fertilized crop species, indicators of apomixis include the failure of F_1 progeny to segregate normally following a cross and the sudden appearance of true-breeding lines in early-generation progenies.

ASPECTS OF BREEDING IN APOMICTIC PLANTS

Apomixis occurs at low levels in some cultivated species and can be found in closely related species of many crops. Hybridization of cultivated crops with an apomictic species can be either a useful tool for the plant breeder or a hindrance to breeding progress. Plant breeders should recognize the following features of apomixis:

1. The major advantage of asexual reproduction through seed is that it permits the development of hybrids or genotypes that breed true regardless of heterozygosity for an unlimited number of generations without loss of vigor or change in genotype. Apomictic plants usually produce normal reduced male gametes, and therefore new apomictic

---

Figure 20.5 continued

A third case deals with cassava cultivars that showed continued decrease in performance with no apparent disease. *In vitro* propagation greatly increased root yields, and healthy plants showed large leaf lobes and branching morphology.

Photo courtesy International Institute of Tropical Agriculture, Ibadan, Nigeria.

genotypes could be produced readily by hybridizing apomictic plants with plants that reproduce by sexual means.

2. Rapid production of true-breeding apomictic genotypes might be a strong force in reducing the risks associated with genetic vulnerability due to the widespread use of a limited number of genotypes.

3. Apomixis provides a rapid means of producing and fixing superior gene combinations and a means for the rapid incorporation of desirable genes. Apomictic plants can be performance tested immediately following hybridization.

4. Apomixis provides the means for the commercial production of F_1 hybrids in crops where cytoplasmic male sterility and/or fertility restoration systems are not available.

5. The germplasm base of existing maize, sorghum, and pearl millet hybrids would be broadened if apomictic forms could be identified because plant breeders would not be limited to females with the male-sterility-inducing system for commercial hybrid production.

6. Isolation is not necessary to produce commercial F_1 hybrid seed or to maintain and increase parental genotypes. In obligate apomicts, outcross contamination is eliminated. Hybrid seed production costs are reduced and the chances of mechanical mixtures are lessened because fewer lines need to be handled to produce each hybrid.

7. Problems in breeding apomictic plants include facultative apomictic behavior and environmental factors influencing apomixis. A better understanding is needed of associated factors such as the number of genes involved with apomixis, modifier genes, ploidy, and seed sterility before apomixis can be successfully used to produce new cultivars.

8. Apomixis could mean the potential loss of control of commercial hybrids and reduced seed sales because producers could process and plant their own seed. Quality commercial seed may continue to be purchased by producers in developed countries who recognize the benefits of high-quality treated, graded, and inspected seed. In developing countries, apomixis would make superior cultivars available for increased crop productivity.

9. A commonly accepted view is that asexual species tend to lose out because of their lesser ability to keep up with changes in the environment. An asexual population is limited to the most extreme individual in the population, and selection can go no further without mutation. Selection in an asexual population decreases the variance, whereas the variance of a sexual population quickly reaches a steady state (Crow, 1992). In a stable environment, however, there are advantages to asexual reproduction.

To convert crops that normally reproduce by sexual means into apomicts, it is necessary to find genes that govern apomictic development in the species or in a cross-compatible wild relative. In existing apomictic species, it is necessary to identify plants that have sexual or partially sexual means of reproduction. Discovery of a buffelgrass plant that reproduces by sexual means (Bashaw, 1962) has permitted the rapid genetic improvement of this crop and the release of a true-breeding apomictic cultivar (Taliaferro and Bashaw, 1966). New obligate apomictic cultivars can be produced by selfing sexually reproducing plants or by using them as female parents for hybridization to apomictic male parents (Figure 20.6).

Although it is easy to utilize obligate apomixis in breeding programs, provided an asexually reproducing counterpart exists, the use of facultative apomixis is associated with some difficulties because of sexual and apomictic plants produced. New cultivars of facultative apomicts have been successfully used only in forage grasses (Funk and Han, 1967). The Kentucky bluegrass cultivar Bonnieblue is a highly apomictic, first-generation hybrid produced when an unreduced egg of one parent was fertilized by a reduced gamete from the other parent, which resulted in a hybrid possessing approximately 94 chromosomes (Funk *et al.*, 1974). The Kentucky bluegrass cultivar Adelphi is a first-generation hybrid produced when an unreduced egg was fertilized by a reduced gamete, resulting in a facultative apomict possessing approximately 80 chromosomes. The mode of reproduction of Adelphi was determined by examination of field-grown progenies (Funk *et al.*, 1973).

The reproductive behavior of four facultative apomictic sorghum lines (*Sorghum bicolor* L. Moench) was studied in crosses with sexually reproducing lines. Crosses with facultative apomicts yield some hybrids that reproduce by apomictic means (Murty *et al.*, 1984). Ideally, obligate apomicts should be used for the fixation of heterozygosity, but since obligate apomixis does not exist in sorghum at present, facultative apomixis can be exploited for the production of what Murty *et al.* (1984) termed **vybrids**. In each generation, heterozygotes identical to the "hybrid" of the facultative apomicts occur with about equal frequency. The concept of vybrids leads to a useful and practical method for partial fixation of heterosis. If the vybrid is perpetuated from the seed of these genotypes only, then the heterozygote is perpetuated indefinitely. The method of producing vybrids was outlined as follows:

1. Hybridize selected sorghum cultivars and one or more of the available facultative apomicts. Variation will occur in the F_1 because the facultative apomict will be heterozygous.
2. Carry preferred F_1 plants to F_2.
3. Test desirable individual plants from F_2 for their cross-sterility reaction. Fertilization is prevented in facultative apomicts by the occurrence of cross-sterility.
4. Mate plants identified as cross-sterile. The progeny again have to be screened first for cross-sterility and then for apomixis. Desirable plants can be tested in field trials.

Apomixis is controlled genetically by a few genes that can be either recessive or dominant (Asker, 1979; Nogler, 1984). On the basis of the segregating ratios presented in Figure 20.6, the genotype of the parent that reproduced by sexual means was hypothesized to be *AaBb*, where dominant gene *B* conditions sexual reproduction and is epistatic to dominant gene *A*, which

---

Figure 20.6   The development of a new apomictic cultivar of buffelgrass (*Pennisetum ciliare* L.), an obligate apomictic species, is shown in this figure. In 1958, a variant of buffelgrass was discovered that reproduces by sexual means (Bashaw, 1962). This sexually reproducing plant is heterozygous for method of reproduction, and therefore its selfed progeny also provide a source of apomictic plants for selection. Two discrete types are produced: completely sexually reproduced plants and obligate apomicts (shown under A).

In the sequence shown in B, a heterozygous sexually producing plant is used as the female and pollinated by an apomictic male. The F_1 progeny are obligate apomicts, which produce three apomicts for every five sexually reproducing plants. Primary emphasis is placed on the selection of superior apomictic F_1 hybrids to take advantage of maximum hybrid vigor. Selection of apomictic plants in the F_2 or F_3 generation is possible, but considerable loss in vigor was observed.

Figure 20.6 continues overleaf

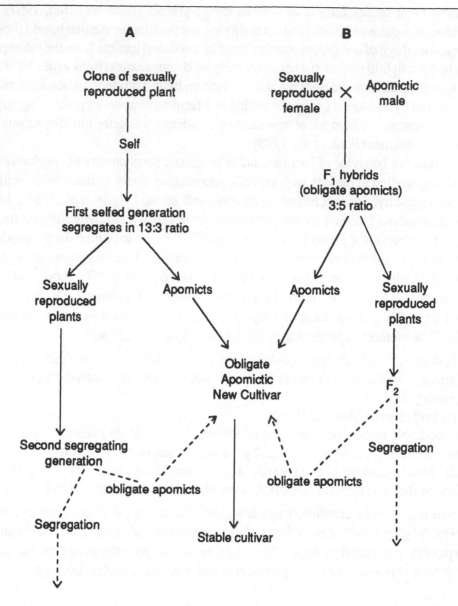

A

Clone of sexually
reproduced plant

↓ Self

First selfed generation
segregates in 13:3 ratio

Sexually
reproduced
plants   Apomicts

Second segregating
generation

obligate apomicts

Segregation

B

Sexually
reproduced ✕ Apomictic
female   male

F₁ hybrids
(obligate apomicts)
3:5 ratio

Apomicts   Sexually
reproduced
plants

Obligate
Apomictic
New Cultivar   F₂

obligate apomicts   Segregation

Stable cultivar

Figure 20.6 continued

Selection of superior apomictic plants among the first or second generation of selfed plants (in A) offers the most rapid means of developing new cultivars. Hybridization, however, not only provides extensive new gene combinations but substantially increases the proportion of apomictic progeny compared with selfed populations.

Buffelgrass is a vigorous, warm-season bunchgrass with an extensive native range from Africa to India. It is highly drought resistant and well adapted to arid and semiarid areas of the southwestern United States. The polymorphic species lacks cold tolerance. Progress in developing superior cultivars was limited by apomixis to selection of the better plant introductions. The discovery of a sexually reproducing plant opened opportunities for hybridization. The superiority of initial segregates led Taliaferro and Bashaw (1966) to conclude that apomixis can serve as a useful tool in breeding buffelgrass.

Source: After Taliaferro and Bashaw, 1966.

conditions **apospory** (Bashaw, 1980). On the basis of this hypothesis, the genotype of the two apomictic male parents would be *Aabb*. In the absence of dominant gene *A*, the double recessive *aabb* would reproduce sexually and produce the observed ratios.

Concentrated efforts to find apomixis in additional cultivated species or in their wild relatives could lead to success. The transfer of genes controlling apomixis from wild to related cultivated forms is possible but will not be done easily or rapidly. Dujardin and Hanna (1985a) hybridized pearl millet (*Pennisetum americanum*, 2n = 14) (Figure 20.7) with *P. squamulatum* (2n = 54), which produced 2n = 41 progeny that exhibited apomictic and sexual reproduction systems (Dujardin and Hanna, 1985b). Efforts to produce apomictic cultivars in crop plants are just beginning, as suggested by the work on pearl millet. Prospects for more extensive use of apomixis in plant breeding are promising.

Figure 20.7   Pearl millet plants (*Pennisetum americanum*), shown in this photo, normally reproduce from sexually developed seeds. Although apomixis is of little evolutionary value, plant scientists now realize that apomixis is widespread and offers promising potential to plant breeders (Bashaw, 1980).

To meet this challenge, *Pennisetum squamulatum* (2n = 54), an obligate apomictic perennial grass, was hybridized to pearl millet for purposes of transferring genes for apomixis (Dujardin and Hanna, 1985a, 1985b). Progeny from this interspecific cross showed conventional sexual reproduction and apomictic hybrids. *P. squamulatum* was used as the maternal parent, and progeny (2n = 41) were backcrossed to pearl millet. Backcrosses to diploid pearl millet (2n = 14) resulted in 12-, 28- or 27-chromosome male and female-sterile plants from more than 600 pollinated inflorescences. Backcrosses to tetraploid pearl millet (2n = 4x = 28) produced progenies with somatic chromosome numbers (2n) ranging from 36 to 39. The task of transferring apomixis into an agronomically desirable cultivar is a challenging one, but the production of partially male-fertile apomictic BC$_1$ plants might indicate that it is possible to continue gene transfer from *P. squamulatum* to pearl millet in subsequent generations. Apomixis was observed in backcross plants only where apomictic F$_1$ hybrids were used as the pollen parent. This suggests that apomixis is dominant to sexuality in *P. squamulatum*.

Photo courtesy International Crops Research Institute for the Semi-Arid Tropics (ICRISAT), India.

REFERENCES

Asker, S. 1979. Progress in Apomixis Research. *Heredity* 91:231-240.

Bashaw, E. C. 1962. Apomixis and Sexuality in Buffelgrass. *Crop Science* 2:412-415.

———. 1980. Apomixis and its Application in Crop Improvement. In *Hybridization of Crop Plants,* edited by W. R. Fehr and H. H. Hadley. American Society of Agronomy and Crop Science Society of America, Madison, Wisconsin, pp. 45-63.

Blackburn, F. 1984. Varieties of Sugar-cane. In *Sugar-Cane.* Longman, London and New York, pp. 101-135.

Clements, H. F., and M. Awada. 1965. Experiments on the Artificial Induction of Flowering in Sugarcane. *Proceedings of the International Society of Sugar Cane Technology* 12:795-812.

Coleman, R. E. 1968. Physiology of Flowering in Sugarcane. *Proceedings of the International Society of Sugar Cane Technology* 13:992-1000.

Crow, J. F. 1992. An Advantage of Sexual Reproduction in a Rapidly Changing Environment. *Journal of Heredity* 83:169-173.

Dujardin, M., and W. W. Hanna. 1985a. Cytology and Reproduction of Reciprocal Backcrosses Between Pearl Millet and Sexual and Apomictic Hybrids of Pearl Millet x *Pennisetum squamulatum. Crop Science* 25:59-62.

———. 1985b. Apomictic and Sexual Pearl Millet x *Pennisetum squamulatum. Journal of Heredity* 74:277-279.

Funk, C. R., R. E. Engel, G. W. Pepin, A. M. Radko, and R. J. Peterson. 1974. Registration of Bonnieblue Kentucky Bluegrass. *Crop Science* 14:906.

Funk, C. R., R. E. Engel, G. W. Pepin, and R. A. Russell. 1973. Registration of Adelphi Kentucky Bluegrass. *Crop Science* 13:580.

Funk, C. R., and S. J. Han. 1967. Recurrent Intraspecific Hybridization: A Proposed Method of Breeding Kentucky Bluegrass, *Poa pratensis. New Jersey Agricultural Experiment Station Bulletin* 818:3-14.

Hanna, W. W., and E. C. Bashaw. 1987. Apomixis: Its Identification and Use in Plant Breeding. *Crop Science* 27:1136-1139.

Hanna, W. W., J. B. Powell, J. C. Millot, and G. W. Burton. 1973. Cytology of Obligate Sexual Plants in *Panicum maximum* Jacq. and Their Use in Controlled Hybrids. *Crop Science* 13:695-697.

Hanna, W. W., K. F. Schertz, and E. C. Bashaw. 1970. Apospory in *Sorghum bicolor* L. Moench. *Science* 170:338-339.

Harmon, F. N., and J. H. Weinberger. 1959. Effects of Storage and Stratification on Germination of *Vinifera* Grape Seeds. *Proceedings of the Society of Horticulture Science* 73:147-150.

James, N. I. 1980. Sugarcane. In *Hybridization of Crop Plants,* edited by W. R. Fehr and H. H. Hadley. American Society of Agronomy and Crop Science Society of America, Madison, Wisconsin, pp. 617-629.

Jones, R.A.C. 1982. Tests for Transmission for Four Potato Viruses Through Potato True Seed. *Annals of Applied Biology* 100:315-320.

Kartha, K. K. 1984. Tissue Culture Techniques for Virus Elimination and Germplasm Preservation. In *Plant Breeding Reviews,* edited by J. Janick, vol. 2. Avi Publishing Company, Westport, Connecticut, pp. 265-282.

Leakey, R.R.B. 1986. Cloned Tropical Hardwoods: Quicker Genetic Gain. *Span* 29(1):35.

Murty, U. R., P. B. Kirti, M. Bharathi, and N.G.P. Rao. 1984. The Nature of Apomixis and Its Utilization in the Production of Hybrids ("Vybrids") in *Sorghum bicolor* L. Moench. *Zeitschrift für Pflanzenzuchtung* 92:30-39.

Nogler, G. A. 1984. Gametophytic Apomixis. In *Embryology of Angiosperms,* edited by B. M. Johri. Spring-Verlag, New York, pp. 475-518.

Quak, F. 1977. Meristem Culture and Virus-free Plants. In *Applied and Fundamental Aspects of Plant Cell, Tissue and Organ Culture*, edited by J. Reinert and Y.P.S. Bajaj. Springer-Verlag, Berlin, pp. 598-615.

Roca, W. M. 1985. In Vitro Clonal Propagation to Eliminate Crop Diseases. In *Biotechnology in International Agricultural Research*. International Rice Research Institute, Manila, Philippines, pp. 3-10.

Taliaferro, C. M., and E. C. Bashaw. 1966. Inheritance and Control of Obligate Apomixis in Breeding Buffelgrass, *Pennisetum ciliare*. *Crop Science* 6:473-476.

Tysdal, H. M., T. A. Kiesselbach, and H. L. Westover. 1942. *Alfalfa Breeding*. College of Agriculture, University of Nebraska Agricultural Experiment Station Research Bulletin 124. Lincoln, Nebraska. 46 pp.

White J. W., and S. Sadik. 1983. Potatoes from True Seed: A Promising Alternative. *Span* 26:23-25.

Withers, L. A., and J. T. Williams. 1985. Research on Long-term Storage and Exchange of In Vitro Plant Germplasm. In *Biotechnology in International Agricultural Research*. International Rice Research Institute, Manila, Phillipines, pp. 11-24.

CHAPTER 21

Cell and Molecular Biology Tools for Plant Breeding

Plant breeders have adopted new technologies when a consistent advantage in ability to provide new and novel germplasm is demonstrated. Examples of adopted technologies are the widespread use of computers in analyzing field data and small plot mechanization, which increases the number of plots that can be evaluated at any time or location. Various aspects of cellular and molecular biology have been used in both public and private sectors to produce new germplasm and cultivars (Tomes, 1990b). The accessibility of technology is a very important component of acceptance, which explains why **embryo culture** (Figure 21.1) **and anther culture** have a longer history of use in some breeding programs than more recent technologies such as **genetic engineering** and **genetic transformation**. Nonetheless, enough examples have been tested in laboratory and field trials to demonstrate the potential impact of these basic biological tools.

SOMACLONAL VARIATION

Fertile plants can be recovered from callus, suspension, or **protoplast** cultures in all of the major crop species (Figures 21.2 and 21.3) (Tomes, 1990b). Initially, it was assumed that such plants would be exact duplicates of the parent plant that gave rise to the *in vitro* culture. However, as plant regeneration became routine and plants were carefully analyzed, some plants were observed that differed from the original donor plant. Somaclonal variation describes the variation, often of genetic origin, that occurs as a result of the tissue culture process (Larkin and Scowcroft, 1981; Evans and Sharp, 1983). This technically simple procedure offered the possibility of producing new cultivars with little effort. However, the lack of control over traits that are modified and the absence of critical comparisons of alternative germplasm sources offering similar variability have dampened earlier more optimistic predictions (Bingham and McCoy, 1986; Vasil, 1990). Examples of cultivar traits that originated from somaclonal variation include potato virus Y resistance in tobacco following anther culture (Witherspoon *et al.*, 1991) and army worm resistance in sorghum recovered from callus culture selection against salt (Isenhour *et al.*, 1991).

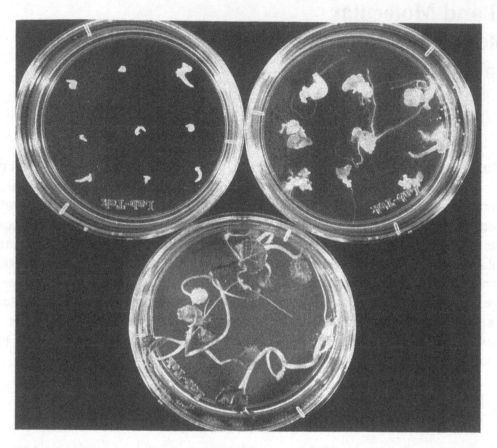

Figure 21.1   Photo showing embryo culture sequence, a useful technique in interspecific hybridization of otherwise incompatible crosses. Mohapatra and Bajaj (1987) reported hybridizing *Brassica juncea* (2n = 36) x *Brassica hirta* (2n = 24), a cross that is normally incompatible using conventional breeding methods. The researchers, using embryo culture techniques, excised fertilized ovaries four to nine days after pollination to avoid abortion associated with abnormal endosperm development. Excised ovaries were sterilized in freshly prepared chlorine water or mercuric chloride (0.25%) and rinsed twice with sterilized distilled water. The base of the pedicel was cut with a blade and the ovaries were cultured on a prescribed medium supplemented with various concentrations and combinations of growth regulators, including indoleacetic acid, kinetin, and casein hydrolysate. The resulting hybrids were sterile and had an intermediate number of chromosomes (2n = 30).

Photo courtesy K. P. Pauls.

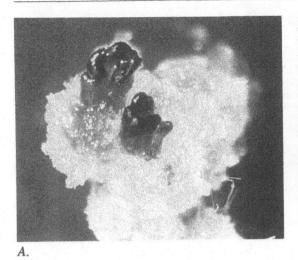
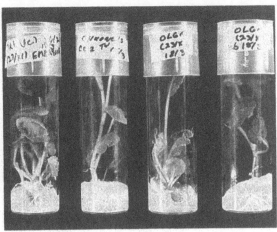

A.   B.

Figure 21.2   Photo A shows callus tissue propagated *in vitro* and displaying the development of leaf primordia. In crops such as potato, tomato, and rapeseed, **totipotency** is a routine procedure, but in the major monocotyledonous cereal crops, regeneration of whole plants from single cells and protoplasts has been reported in only a few cases with certain limited genotypes. No routine regeneration system exists for cereals.

In vitro propagation of plants from tissue is not a new concept. Espinoza *et al.* (1986) noted that in 1939, the German worker T. Schwann expressed the view that each living cell of a multicellular organism should be capable of independent development if provided with the appropriate external conditions. Asexual propagation using small pieces of stem or root has been practiced for years. The question of how small a piece could actually be used started researchers thinking about micropropagation. The use of micropropagation has been extended from asexual multiplication to maintain desired genotypes to extensive genetic recombination above and beyond that achieved through natural pollination and fertilization.

Photo B shows rapeseed plantlets regenerated from callus tissue ready for multiplication.
Photos courtesy K. P. Pauls.

DIRECTED SELECTION

In vitro selection is a process of selection at the cellular level in which chemicals such as herbicides or toxins are added to culture medium, allowing only cells with the desired tolerance traits (i.e., herbicide resistance) to grow and thus "capturing" genetic variation among cells. Traits are conditioned by well-characterized biochemical pathways and associated genes have given rise to mutants that express alternative traits. Herbicide-resistant mutants have been obtained with clear levels of resistance at both the cellular and the plant level (Chaleff and Ray, 1984). Ten- to 100-fold increases in tolerance to the imidazilinone herbicides in maize are commercially available from Pioneer Hi-Bred International (Tomes, 1990b; see also the discussion of haploidy in the next section and Swanson *et al.*, 1988). Imidazilinone-resistant corn allows the use of herbicides that give better control of grass weeds. Corn hybrids without the mutant gene would be killed by these herbicides. Selection strategies for other herbicides (Swanson and Tomes, 1983) and disease resistance have not been as successful and have not

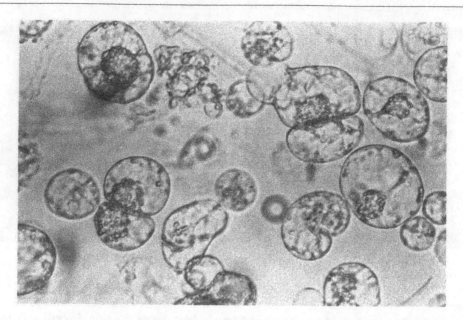

Figure 21.3   Photo showing plant protoplasts, or cells from which the surrounding rigid cellulose cell walls have been removed. Protoplast production is a first step in manipulating the contents of a cell. Cell walls can be removed by incubating plant tissue in a mixture of cell-wall-degrading enzymes for several hours. Cell wall removal allows the protoplast, the living part of the cell, to escape intact. Each protoplast contains a nucleus, where the majority of DNA is found, along with organelles such as mitochondria and chloroplasts, which also contain some DNA. The DNA in the nucleus is the primary determinant of the characteristics of the plant, but the **organelle DNA** also determines some agronomically important traits, such as male sterility and herbicide resistance.

Transformation of protoplasts can be achieved following cell wall removal by fusion of two unrelated protoplasts or by inserting pieces of DNA directly into the protoplasts. Fusion results in major changes in the genetic makeup of the resulting plants because the entire contents of two protoplasts are combined. Insertion of specific pieces of DNA can result in specific gene modifications.

Protoplast fusion does not occur spontaneously because the protoplasts carry a negative charge. Fusion can be induced by adding Ca^{++} in a polyethylene glycol solution or by subjecting the cells to a strong electric field.

Photo courtesy K. P. Pauls.

produced germplasm or cultivars that have proceeded into advanced field trials (Vasil, 1990). In a recent evaluation in barley, wheat, and potato, the progeny of plants regenerated following *in vitro* selection against *Fusarium*, *Helminthosporium*, and *Phytophthora* showed no detectable differences in disease reaction (Wenzel and Foroughi-Wehr, 1990). Traits that are poorly understood at the biochemical or molecular level are less likely candidates for trait improvement using *in vitro* selection.

HAPLOIDY

Haploid breeding procedures have been successful because the culture technology has been available for a relatively wide germplasm base in many of the cereals and because plant breeders could readily anticipate the advantages of haploids. Many techniques are available to produce haploids, including anther culture (Figure 21.4), **chromosome elimination** in interspecific hybrids (i.e., bulbosum technique), and **microspore culture**. Seven barley cultivars have been released in Canada using the bulbosum maternal haploid method. Cultivars are either in test or have been released in all of the major cereals, including barley, wheat, rice, and corn (Tomes, 1990b). Examples include Texmont, which is an early-maturing, high-yielding rice cultivar released in Texas (Bollich, 1990); Florin, a wheat cultivar released in 1985 in France because of its improved agronomic characteristics (deBuyser *et al.*, 1987); and a maize inbred designated as AC4115, which progressed to the precommercial stage from an anther-derived haploid in maize (Genovisi, 1990). Microspore culture is possible in *Brassica*, *Hordeum*, and *Zea*. The ability to culture individual haploid microspores has made possible the isolation of mutants for several traits, including chlorsulfuron and imidazilinone herbicide resistance in *Brassica* (Swanson *et al.*, 1988).

GENE TRANSFER

Somatic Hybridization

Regeneration of plants from protoplasts (plant cells with their cell walls removed) provided new possibilities for the transfer of discrete traits and/or genes between plant species that are sexually incompatible. Nonsexual gene transfer was first reported using **somatic hybridization** in *Nicotiana* (Carlson, Smith, and Dearing 1972), in which protoplasts of two species were first fused and somatic hybrid plants preferentially recovered because of their auxin-autotrophy (i.e., ability of cells to grow without auxin or cytokinin being added to the medium). Subsequently, somatic hybrids between both distant and more closely related species were produced (Figure 21.5). Although wide somatic hybrids are more striking, hybrids between closely related species have been of more immediate use to commercial plant breeding (Vasil, 1990). A notable example is the cultivar Delfield (R. S. Pandeya, personal communication, 1990), which originated from somatic hybrids between *N. tabacum* and *N. rustica* (Pandeya *et al.*, 1986). These somatic hybrids had unique traits (genes) for nicotine content and blue mold and black root rot disease resistance. The cultivars only resulted after multiple backcrosses to an *N. tabacum* parent. *Brassica* somatic hybrids have been recovered that contain unique cytoplasmic organelles (Figure 21.6), including cytoplasmic male sterility from mitochondria and triazine tolerance from chloroplasts with the nuclear genome of *Brassica napus* (Barsby *et al.*, 1987).

The major factors limiting more widespread use of somatic hybridization are restrictions imposed by the technical difficulties of protoplast regeneration and the inability to selectively use portions of the genome from one species in a recipient species (Vasil, 1990).

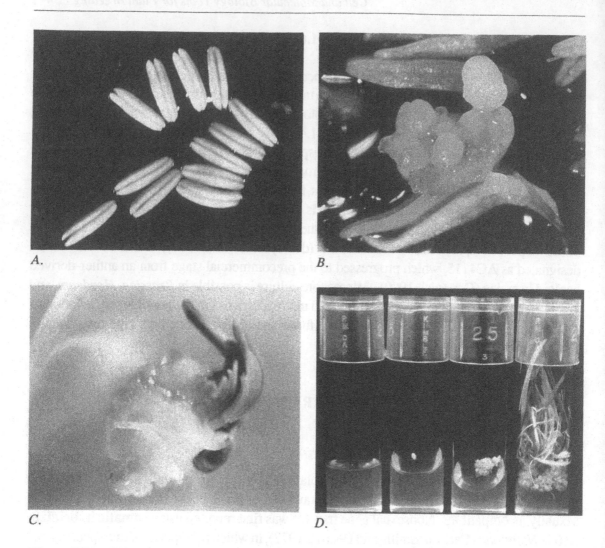

A.

B.

C.

D.

Figure 21.4   This sequence of photos illustrates anther culture in barley that results in haploid plants. Photo A shows barley anthers containing immature pollen or microspores. When placed on a culture medium with plant hormones, continued cell division is induced rather than normal development into mature pollen. These dividing cells may form either embryos (Photo B) or a mass of callus cells (Photo C) from which roots and shoots may be induced to grow.

Callus tissue may be transferred to a culture medium (Photo D) solidified with agar. In an appropriate culture growth medium, which must be customized for each species, such plant cells will grow and divide and can be kept alive indefinitely. Slowly, the nondescript callus cells start to form recognizable plant structures, which may be embryolike or leaflike. With further medium manipulation, roots can be induced to form at the base of the shoots and complete plantlets will result (right vial in Photo D). This is **micropropagation**. The entire micropropagation process takes place within the sterile confines of a laboratory flask. The resulting plantlets are haploid. Doubled haploids are completely uniform because they are homozygous. Anther culture is one means of producing a pure line.

This process of pure-line production is not feasible for all crops because some, like alfalfa and potato, contain four or more sets of chromosomes. Haploid plants normally contain two or more sets of chromosomes and when doubled will still have genetic variability, although much reduced compared to the parental plant.

Photos courtesy J. Lettre and L. Kott.

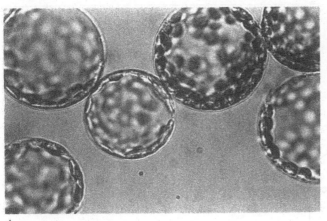 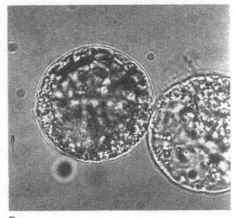

A.   B.

Figure 21.5 The rapeseed (*Brassica napus*) leaf protoplasts in Photo A contain green chloroplasts that can be used to distinguish such cells microscopically. Protoplasts obtained from the hypocotyl lack chloroplasts and also can be distinguished.

Of the two protoplasts in Photo B, the one on the left is hybrid and can be identified by the fact that the top half is green and the lower half colorless. The hybrid protoplast can be isolated, and because rapeseed protoplasts are totipotent, a hybrid plant can be produced.

Fusions between potato and tomato protoplasts have been used to produce potato or tomato plants that displayed characteristics of both parent plants. Unfortunately, the hybrid plants produced neither tomatoes nor potatoes, but the experiments illustrate that the genetic makeup of plant cells can be modified considerably without destroying their ability to regenerate into a plant (Pauls, 1987).

Triazine-resistant, female-sterile plants could serve as the maternal parent to produce hybrid rapeseed or canola. The paternal parent would be a male-fertile plant carrying genes that would restore fertility (Pauls and Beversdorf, 1988). After fertilization, an application of a triazine compound would be used to eliminate the pollinators, and the hybrid seed could be harvested from the field.

Photos courtesy K. P. Pauls.

Molecular Biology

Gene transfer technology at the molecular level has either reduced or eliminated barriers that prevented transferring defined genes into plants from diverse species. Genes, once isolated and characterized, can be transferred into plants regardless of source. The three components of genetic transformation are a gene or genes composed of a specific DNA sequence from a donor organism, a recipient cell in the plant to accept the DNA, and a DNA delivery technique. Cell and tissue culture procedures have allowed manipulation and increase of cells that are both receptive to inserted DNA and can then be regenerated into plants (Maniatis, Fritsch, and Sambrook, 1982). Three major methods used for gene transfer are (1) direct transfer through protoplasts, (2) *Agrobacterium*, and (3) direct transfer into intact cells using accelerated microprojectiles coated with DNA (Tomes, 1990a) (Figure 21.7). Genetic transformation has been achieved in many economically important species, such as tomato, potato, and soybean (Hinchee *et al.*, 1988), using a modification of the *Agrobacterium* leaf disc technique (Horsch *et al.*, 1985). The microprojectile bombardment procedure has been successfully used to obtain

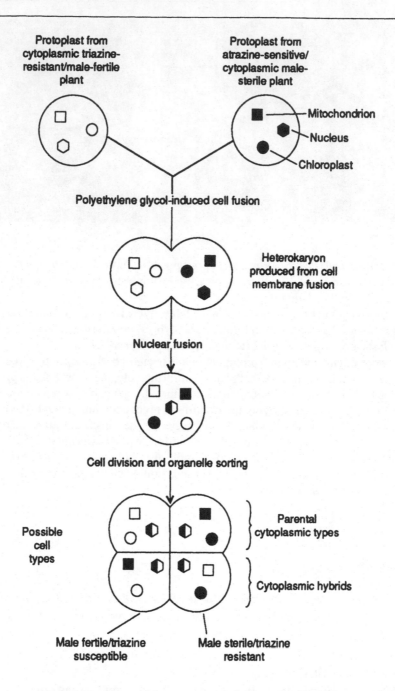

Figure 21.6 A specialized application for protoplast fusion is the creation of new combinations of organelles within a plant cell. Mitochondria and chloroplasts are inherited through the female plant, which means that progeny from a conventional cross will be the same as the female parent with regard to any traits that are determined by the organellar DNA. Protoplast fusion offers a means to combine organellar traits from both parents.

This diagram illustrates how protoplast fusion was used to achieve the transfer of triazine resistance and male sterility in a rapeseed fusion product. The triazine-resistance trait is carried by the chloroplasts of one parent and the male-sterility trait by mitochondria of the other parent. The fate of the organelles after protoplast fusion is illustrated.

During the process of protoplast fusion, the contents of one cell are combined with those of another

transgenic plants in maize (Gordon-Kamm *et al.*, 1990). Monocots such as maize have been much more difficult to transform compared with dicots such as tobacco, petunia, and tomato.

Genes for several different traits have been transferred into plants and have been extensively field tested. Virus resistance in tobacco, tomato (Tumer *et al.*, 1987), and alfalfa (Loesch-Fries *et al.*, 1987) have been tested. The gene for resistance to a virus in many cases comes from the coat protein gene of the virus. For example, the coat protein gene from alfalfa mosaic virus gives resistance to alfalfa mosaic virus in alfalfa, tobacco, and tomato when the gene is transferred into the plants that are normally susceptible to this virus.

Insect resistance is also a possibility, primarily by transferring the toxin gene from *Bacillus thuringiensis* into plants such as tomato and cotton (Delannay *et al.*, 1989). Recent field tests in cotton indicate that field levels of resistance comparable to insecticide treatments are observed against the *Lepidopteran* insects, cabbage looper (*Trichoplusia ni*) beet armyworm (*Spodoptera exigua*), and cotton boll worm (*Heliothis zea*) (Perlak *et al.*, 1990).

Another trait that has been genetically modified by gene transfer is tomato fruit softening (Kramer, Sheehy, and Hiatt, 1989). One gene particularly important in fruit softening controls production of the enzyme polygalacturonase, a pectin degrading enzyme. A modified form of the polygalacturonase gene when transferred into tomato disrupts normal gene function, which in turn substantially inhibits fruit softening.

GERMPLASM AND PEDIGREE IDENTIFICATION

Precise identification of cultivars, breeding lines, and germplasm for plant breeding is very important because of legal requirements for plant patents and because the choice of parents in developing new cultivars and inbreds is crucial to success. Historically, the pedigree records maintained by breeders have been used to determine the breeding lines that will be employed for further breeding research and also to establish ownership for patents and/or commercialization. Mature breeding programs in many of the cereals, including wheat, barley, and maize, have pedigree records that span more than one hundred years.

The degree of relatedness is one determinant in choosing parents in both self- and cross-pollinated species. Breeding for new inbred parents or cultivars often relies on using closely related breeding lines, whereas the best hybrids come from crossing two inbred parents that are the least related. The degree of relatedness can be determined by inspecting pedigree

---

Figure 21.6 continued

cell. The resulting **heterokaryon** contains six different types of DNA, including two nuclei, two types of mitochondria, and two chloroplasts.

During subsequent cell division, organelles in the heterokaryon sort out randomly so that the plant that develops has only one type of chloroplast and one type of mitochondrion. The nuclei of the heterokaryon fuse or one nucleus degenerates so that the uninucleate condition is reestablished. In the callus that results from division of the hybrid cell, one type of cell ultimately predominates and gives rise to a plant with those characteristics. The cytoplasmic hybrid plants produced are an important step toward the production of hybrid rapeseed.

Source: After Pauls, 1987.

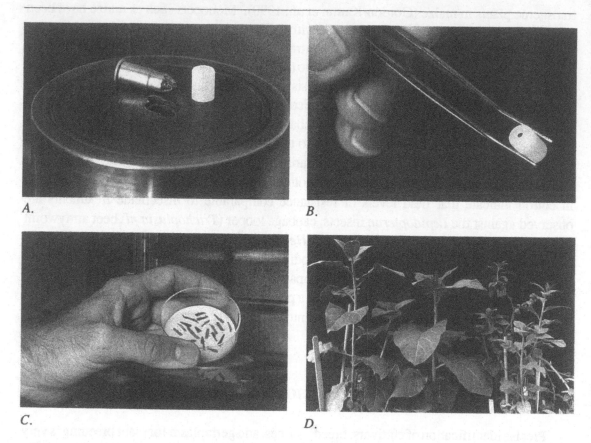

Figure 21.7 Sequence showing the production of transgenic tobacco plants using microprojectile bombardment. A .22 calibre blank is used to accelerate a macroprojectile in the barrel of the device (Photo A). In Photo B, DNA is carried into plant cells through a mixture of DNA and tungsten or gold microprojectiles with an average diameter of 1.2 microns (drop on macroprojectile). In Photo C, a sample of tobacco leaves is removed from the chamber beneath the barrel (see top). During treatment, the chamber is evacuated to about 1/10 atmosphere to ensure that microprojectiles reach sufficient velocity to penetrate the cells. Tobacco plants that have been treated to carry resistance to the antibiotic kanamycin are shown in Photo D (left and right plants). Leaf pieces are cultured on a medium that contains kanamycin. Only transformed cells are capable of growing, which first produce callus and then regenerate plants. Only transgenic plants show root growth in the presence of the antibiotic. Genes that have specific agronomic uses, such as disease, insect, or herbicide resistance, can be attached to the antibiotic-resistance gene to obtain plants that express both genes.

Photos courtesy Pioneer Hi-Bred International, Inc., and D. Tomes.

records, performance from hybrid combination of parents, and morphological attributes. Each of these methods suffers from imprecision—specific lines may not reflect expected performance based on pedigree or may be modified by the environment chosen to assess particular traits. The tools of **enzyme electrophoresis** and restriction fragment length polymorphism (RFLP) largely circumvent problems of inaccuracies in pedigree records and environmental variation (Smith *et al.*, 1987; Walton and Helentjaris, 1987).

Enzyme electrophoresis relies on quantifying a series of enzymes that are present in a specific tissue, such as germinating seedlings. Within each enzyme measured, different alleles (**allozymes** or isozymes) can be measured by their differential migration on a starch or polyacrylamide gel (Figure 21.8). Enzyme electrophoresis refers to the migration of proteins (enzymes) from a starting point at the base of the gel and across an electric field. The amount of migration is dependent on the molecular weight of the enzyme, charge differences, and three-dimensional structure. Early studies in maize could resolve approximately 85% of a sample of inbreds with known pedigrees (Stuber and Goodman, 1983). Smith *et al.* (1987) were able to distinguish 94% of 62 inbred lines of known pedigree. Furthermore, these inbreds could be identified in hybrid combinations and hybrid yield could be predicted based on their enzyme profile. Biochemical data is generally accepted as one method of identifying germplasm and in at least one legal case in the United States has been used to verify ownership of a maize inbred (United States District Court, 1987).

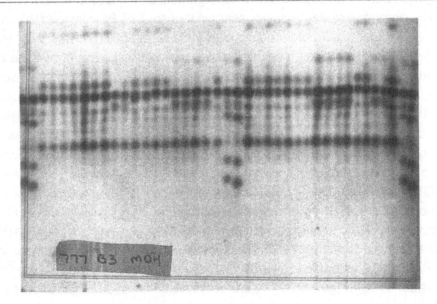

Figure 21.8   Isoenzyme patterns produced from polyacrylamide gel electrophoresis offer a reliable method of identifying a genotype, in this case verifying the hybridity of a somatic hybrid plant. Each of the approximately 40 vertical columns in this photo represents a different genotype.

Gene action is expressed through enzymes that influence chemical reactions in plants. Enzymes can be divided into subunits called allozymes that perform a similar function but exhibit different molecular forms. The different mobilities of these allozymes following electrophoresis in standard polyacrylamide or starch gels produce banding patterns, as is evident in this photo. By comparing potential new plant hybrids with parental materials, evidence of hybridity can be found. A somatic hybrid may exhibit the sum of parental isoenzyme bands.

Gel plates can be examined visually and differences noted, or they can be scanned with a densitometer to indicate both location and intensity of the bands.

It is extremely important to test samples for isoenzyme analysis from identical plant tissues at the same stage of development because isoenzyme expression may differ depending on the plant tissue used and stage of growth of the plant.

Photo courtesy D. Mather and B. Rennie.

RFLPs are DNA fragments created when plant or other DNA is digested by restriction endonucleases (Figure 21.9). RFLPs can identify specific inbreds, hybrids, or other germplasm, provided that there are sufficient known sequences (probes) and that the restriction enzymes employed produce different fragment lengths (polymorphisms) when used in different genotypes (Helentjaris, 1987) (Figure 21.9). Walton and Helentjaris (1987) described the applications for RFLP, which include identification of cultivars, inbreds, and hybrids, and the recovery of donor-parent genotypes in a backcross program. The RFLP technique offers more precision than electrophoresis because a much larger sample of polymorphic genetic markers are spread throughout the genome as compared with allozymes. A comparison of 78 U.S. maize hybrids using 38 probes revealed 288 RFLP variants. These data cluster most hybrids into groupings similar to those determined on the basis of enzyme electrophoresis and known pedigree information. Furthermore, field performance data gave the identical groups shown by RFLP designations (Smith and Smith, 1991).

Young and Tanksley (1989) used RFLPs to significantly reduce the number of backcrosses required to incorporate disease-resistance genes in tomato. RFLPs were used to predict inbred combinations in maize that would produce high-yield F_1 hybrids (Smith *et al.*, 1990).

Because RFLP markers can be genetically mapped just like morphological or enzyme markers, traits of interest can be potentially correlated to an RFLP linkage map. Such linkage data requires the use of segregating populations (either F_2 or backcross) for the traits of interest, a linkage map based on RFLPs and a series of probes (Walton and Helentjaris, 1987). Potentially, not only single-gene traits but also quantitative trait loci (QTLs), such as yield, might be mapped to specific chromosome locations. Three RFLP loci on three different linkage groups were associated with insect resistance in tomato (Nïenhuis *et al.*, 1987). Analysis of a cross of CO159 and Tx303 (Stuber, Edwards, and Wendel, 1987) revealed that several RFLP loci were correlated with height and were on chromosomes 4 and 9 of maize (Walton and Helentjaris, 1987) (Figure 21.9). It is of interest to note that chromosome 9 is also the location of the *d3* dwarfing gene, which suggests that other allelic forms of this gene may be responsible for height differences in maize (Edwards, Stuber, and Wendel, 1987; W. Beavis, personal communication, 1990) (Figure 21.9). Similar correlations have been made for other yield

---

Figure 21.9 RFLP refers to a fragment of DNA that is created when plant or other DNA is digested by one of a series of DNA-digesting enzymes called restriction endonucleases. These nucleases cleave DNA at specific locations depending on the base sequence. For example, the restriction enzyme *BAM HI* cleaves wherever the DNA sequence guanine-guanine-adenine-thymidine-cytosine-cytosine (GGATCC) is located. The genome of maize produces over 10^6 DNA fragments based on cleaving or cutting at each site that has the DNA sequence recognized by the enzyme. These fragments of different molecular weights can then be separated by agarose gel electrophoresis (see Figure 21.8). Different genotypes can produce a different series of RFLPs when digested with the same restriction enzyme, which forms the basis to distinguish different individuals based on their RFLPs. Finally, these differences are visualized by transferring the sized DNA produced by the restriction endonucleases onto a membrane and exposing the membrane to a 32p-labelled probe by a technique called a Southern blot (see Maniatis, Fritsch, and Sambrook, 1982, for a description). A probe is a single-stranded DNA of known sequence that is used for identification purposes (Walton and Helentjaris, 1987).

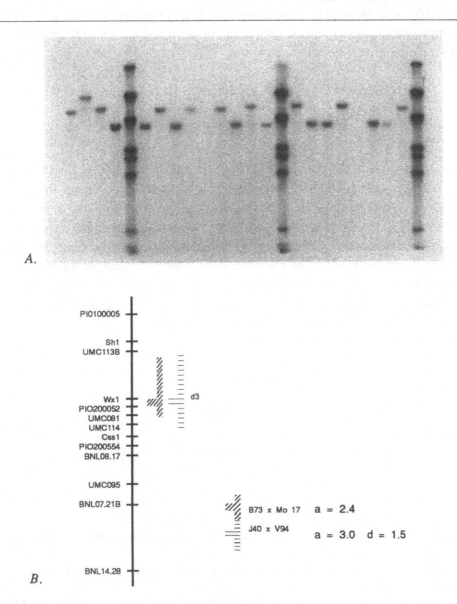

A.

B.

B73 x Mo 17   a = 2.4

J40 x V94   a = 3.0   d = 1.5

Figure 21.9 continued

Photo A shows RFLPs of 19 different inbred lines using a *BAM* HI endonuclease digest of the DNA and using a specific probe that has been mapped to chromosome 5L in maize. Columns or lanes with multiple bands are molecular weight markers for the DNA. The location of the single band in each column distinguishes each inbred, with three distinct types shown. Multiple probes can then be used to separate inbreds or hybrids. (Courtesy Pioneer Hi-Bred International, Inc., Stephen Smith)

Figure B reproduces a portion of a maize linkage map based on RFLP analysis that correlates a QTL (quantitative trait locus), height, to a specific chromosome location. Chromosome 9 is shown, with probes displayed on the left side. Analysis of two segregating populations from B73 x Mo17 and J40 x V94 for plant height were mapped at the approximate location d_3, a dwarf allele. "a" is a dominance effect arising from the second parent.

Photo A courtesy Pioneer Hi-Bred International, Inc.; Figure B courtesy W. D. Beavis.

components, which may be used to more accurately predict combinations of lines that might give rise to high-yielding, single-cross hybrids in maize and other crops (Smith *et al.*, 1990; Walton and Helentjaris, 1987; Stuber, Edwards, and Wendel, 1987; Edwards, Stuber, and Wendel, 1987).

CONCLUSION

The number of cultivars and/or hybrids that can be directly traced in part to either cellular or molecular biology tools is relatively small at the present time. In addition, the techniques of cellular and molecular biology are only one component of the integrated disciplines of modern plant breeding. Even those who are actively involved in these new technologies are quick to point out the necessity of combining new technologies with proved classical breeding methods. The best combination occurs when the skills of people trained in cell and molecular biology are fully integrated into the breeding operation. The cultivars and hybrids produced in the next five to ten years offer the possibility of dramatically improving the efficiency of crop production and lessening the environmental impact of intensive agriculture.

REFERENCES

Barsby, T. L., P. V. Chuong, S. A. Yarrow, Sau-Ching Wu, M. Coumans, R. J. Kemble, A. D. Powell, W. D. Beversdorf, and K. P. Pauls. 1987. The Combination of Polima cms and Cytoplasmic Triazine Resistance in *Brassica napus*. *Theoretical and Applied Genetics* 73:809-814.

Bingham, E. T., and T. J. McCoy. 1986. Somaclonal Variation in Alfalfa. *Plant Breeding Reviews* 4:123-152.

Bollich, C. N. 1990. Report of Technical Committee on Seed Release and Increase: Texmont Rice. *Texas Agricultural Experiment Station Form 97-72*, Beaumont, Texas.

Carlson, P. S., H. H. Smith, and R. D. Dearing. 1972. Parasexual Interspecific Plant Hybridization. *Proceedings of the National Academy of Sciences* 69:2292-2294.

Chaleff, R. S., and T. B. Ray. 1984. Herbicide-Resistant Mutants from Tobacco Cell Cultures. *Science* 223:1148-1151.

deBuyser, J., Y. Henry, P. Lonnet, R. Hertzog, and A. Hespel. 1987. "Florin": A Doubled Haploid Wheat Variety Developed by the Anther Culture Method. *Plant Breeding* 98:53-56.

Delannay, X., B. J. LaVallee, R. K. Proksch, R. L. Fuchs, S. R. Sims, J. T. Greenplate, P. G. Marrone, R. B. Dodson, J. J. Augustine, J. G. Layton, and D. A. Fischhoff. 1989. Yield Performance of Transgenic Tomato Plants Expressing the *Bacillus thuringiensis* var. Kurstaki. *Bio/Technology* 7:1265-1269.

Edwards, M. D., C. W. Stuber, and J. F. Wendel. 1987. Molecular-Marker-Facilitated Investigations of Quantitative-Trait Loci in Maize. I: Numbers, Genomic Distribution and Types of Gene Action. *Genetics* 116:113-125.

Espinoza, N. O., R. Estrada, D. Silva-Rodriguez, P. Tovar, R. Lizarraga, and J. H. Dodds. 1986. The Potato: A Model Crop Plant for Tissue Culture. *Outlook on Agriculture* 15:21-26.

Evans, D. A., and W. R. Sharp. 1983. Single Gene Mutations in Tomato Plants Regenerated from Tissue Culture. *Science* 221:949-951.

Genovisi, A. D. 1990. Maize (*Zea mays* L.): In-Vitro Production of Haploids. *Biotechnology in Agriculture and Forestry*, vol. 12, Haploids in Crop Improvement I, edited by Y.P.S. Bajaj. Springer-Verlag, Heidelberg, pp. 176-203.

Gordon-Kamm, W. J., T. M. Spencer, M. L. Mangano, T. R. Adams, R. J. Draines, W. G. Start, J. V. O'Brien, S. A. Chamabers, W. R. Adams, Jr., N. G. Willetts, T. B. Rice, C. J. Mackey, R. W. Krueger, A. P. Kausch, and P. G. Lemaux. 1990. Transformation of Maize Cells and Regeneration of Fertile Transgenic Plants. *Plant Cell* 2:603-618.

Helentjaris, T. 1987. A Genetic Linkage Map for Maize Based on RFLP's. *Trends in Genetics* 3:217-221.

Hinchee, M.A.W., D. V. Connor-Ward, C. A. Newell, R. E. McDonnell, S. J. Sato, C. S. Gasser, D. A. Fischhoff, D. B. Re, R. T. Fraley, and R. B. Horsch. 1988. Production of Transgenic Soybean Plants Using Agrobacterium-Mediated DNA Transfer. *Bio/Technology* 6:915-922.

Horsch, R. B., J. E. Frey, N. L. Hoffmann, D. Eichholtz, S. G. Rogers, and R. T. Fraley. 1985. A Simple and General Method for Transferring Genes into Plants. *Science* 227:1229-1231.

Isenhour, D. J., R. R. Duncan, D. R. Miller, R. M. Waskom, G. E. Hanning, B. R. Wiseman, and M. W. Nabors. 1991. Resistance to Leaf-Feeding by the Fall Armyworm (Lepidoptera: Noctuidae) in Tissue Culture Derived Sorghums. *Journal Economic Entomology* 84:680-684.

Kramer, M., R. E. Sheehy, and W. R. Hiatt. 1989. Progress Towards the Genetic Engineering of Tomato Fruit Softening. *Tibtech* 7:191-194.

Larkin, P. J., and W. R. Scowcroft. 1981. Somaclonal Variation: A Novel Source of Variability from Cell Cultures for Plant Improvement. *Theoretical and Applied Genetics* 60:197-214.

Loesch-Fries, L. S., D. Merlo, T. Zinnen, L. Burhop, K. Hill, K. Krahn, N. Jarvis, R. S. Nelson, and E. Halk. 1987. Expression of Alfalfa Mosaic Virus RNA 4 in Transgenic Plants Confers Virus Resistance. *EMBO Journal* 6:1845-1851.

Maniatis, T., E. F. Fritsch, and J. Sambrook. 1982. *Molecular Cloning: A Laboratory Manual.* Cold Spring Harbor Laboratory, Cold Spring Harbor, New York. 545 pp.

Mohapatra, D., and Y.P.S. Bajaj. 1987. Interspecific Hybridization in *Brassica juncea* x *Brassica hirta* Using Embryo Rescue. *Euphytica* 36:321-326.

Nienhuis, J., T. Helentjaris, M. Slocum, B. Ruggero, and A. Schaefer. 1987. Restriction Fragment Length Polymorphism Analysis of Loci Associated with Insect Resistance in Tomato. *Crop Science* 27:797-803.

Pandeya, R. S., G. C. Douglas, W. A. Keller, G. Setterfield, and Z. A. Patrick. 1986. Somatic Hybridization Between *Nicotiana rustica* and *N. tabacum*: Development of Tobacco Breeding Strains with Disease Resistance and Elevated Nicotine Content. *Zeitschrift für Pflanzenzuchtung* 96:346-352.

Pauls, K. P. 1987. Plant Protoplasts. In Plant Biotechnology: Applications to Agriculture. *Notes on Agriculture.* University of Guelph, pp. 20-21.

Pauls, K. P., and W. D. Beversdorf. 1988. From the Lab to the Field: The Progress of Hybrid Canola. *Highlights of Agricultural Research in Ontario.* II(i): 23-26.

Perlak, F. J., R. W. Deaton, T. A. Armstrong, R. L. Fuchs, S. R. Sims, J. T. Greenplate, and D.A. Fischhoff. 1990. Insect Resistant Cotton Plants. *Bio/Technology* 8:939-943.

Smith, J.S.C., S. Paszkiewics, O. S. Smith, and J. Schaffer. 1987. Electrophoretic, Chromatographic, and Genetic Techniques for Identifying Associations and Measuring Genetic Diversity Among Corn Hybrids. *Proceedings of 42nd Annual Corn and Sorghum Industry Research Conference.* American Seed Trade Association, Washington, D.C., pp. 187-203.

Smith, J.S.C., and O. S. Smith. 1991. Restriction Fragment Length Polymorphisms Can Differentiate Among U.S. Maize Hybrids. *Crop Science.* 31:893-899.

Smith, O. S., J.S.C. Smith, S. L. Bowen, R. A. Tenborg, and S. J. Wall. 1990. Similarities Among a Group of Elite Maize Inbreds as Measured by Pedigree, F_1 Grain Yield, Grain Yield, Heterosis, and RFLP's. *Theoretical and Applied Genetics* 80:833-840.

Stuber, C. W., M. D. Edwards, and J. F. Wendel. 1987. Molecular Marker-Facilitated Investigations of Quantitative Trait Loci in Maize. II: Factors Influencing Yield and Its Component Traits. *Crop Science* 27:639-648.

Stuber, C. W., and M. M. Goodman. 1983. *Allozyme Genotypes for Popular and Historically Important Inbred Lines of Corn, Zea mays L.* USDA-ARS, ARR-S-16. U.S. Govt. Printing Office, Washington, D.C.

Swanson, E. B., and D. T. Tomes. 1983. Evaluation of Birdsfoot Trefoil Regenerated Plants and Their Progeny After In-Vitro Selection for 2,4-Dichlorophenoxyacetic Acid. *Plant Science Letters* 29:19-24.

Swanson, E. B., M. P. Coumans, G. L. Brown, J. D. Patel, and W. D. Beversdorf. 1988. The Characterization of Herbicide Tolerant Plants in *Brassica napus* L. After In-Vitro Selection of Microspores and Protoplasts. *Plant Cell Reports* 7:83-87.

Tomes, D. T. 1990a. Transformation in Corn: Non-Sexual Gene Transfer. *26th Annual Illinois Corn Breeders School.* pp. 1-13.

————. 1990b. Current Research in Biotechnology with Application to Breeding. *Progress in Plant Cellular and Molecular Biology*, edited by H.J.J. Nijkamp, L.H.W. VanderApage, and J. VanArtrijk. Springer Verlag, Berlin, pp. 23-32.

Tumer, N. E., K. M. O'Connell, R. S. Nelson, P. R. Sanders, R. N. Beachy, R. T. Farley, and D. M. Shah. 1987. Expression of Alfalfa Mosaic Virus Coat Protein Gene Confers Crossprotection in Transgenic Tobacco and Tomato Plants. *EMBO* 6:1181-1188.

United States District Court. 1987. Pioneer Hi-Bred International Inc., an Iowa Corporation, Plaintiff, versus Holden Foundation Seed, Inc., et al., Defendants. U.S. District Court, Southern District of Iowa. Case No. 81-60-E. 108 pp.

Vasil, I. K. 1990. The Realities and Challenge of Plant Biotechnology. *Bio/Technology* 8:296-301.

Walton, M., and T. Helentjaris. 1987. Application of Restriction Fragment Length Polymorphism (RFLP) Technology to Maize Breeding. *Proceedings of the 42nd Annual Corn and Sorghum Industry Research Conference.* America Seed Trade Association, Washington, D.C., pp. 48-75.

Wenzel, G., and B. Foroughi-Wehr. 1990. Progeny Tests of Barley, Wheat, and Potato Regenerated From Cell Cultures After In-Vitro Selection for Disease Resistance. *Theoretical and Applied Genetics* 80:359-365.

Witherspoon, W. D., Jr., E. A. Wernsman, G. V. Gooding Jr., and R. C. Rufty. 1991. Characterization of a Gametoclonal Variant Controlling Virus Resistance in Tobacco. *Theoretical and Applied Genetics* 81:1-5.

Young, N. D., and S. D. Tanksley. 1989. RFLP Analysis of the Size of Chromosomal Segments Retained Around the *TM*-2 Locus of Tomato During Backcross Breeding. *Theoretical and Applied Genetics* 77:353-359.

CHAPTER 22

Commercial Hybrids in Self-fertilized Crops

In the quest to increase yields of self-fertilized crops, much attention is being directed toward the production of F_1 **hybrids**. The phenomenon of hybrid vigor and the commercial success of hybrids in many naturally cross-fertilized crops have added impetus to extend hybrids to self-fertilized crops. Indeed, the production of plants from hybrid seed has been one of the outstanding scientific advances in agricultural history.

The rationale, procedures, and model for F_1 hybrids in self-fertilized crops are based on success with maize and sorghum. Hybridization in maize is relatively simple because the male (tassel) and female (silk) flowers are on different parts of the same plant. To produce the hybrid, the tassel is removed mechanically from the female seed parent and cross-pollination occurs from adjacent male rows. Hybrid systems for self-fertilized crops are much more difficult because flowers are small and often contain both male and female parts in the same floret. Hand emasculation and pollination generally are impractical on a commercial scale, except where labor is abundant and inexpensive and/or the crop is of high value and where many seeds are produced per pollination. Nevertheless, the differences and similarities between cross- and self-fertilized crops have guided perceptions on hybrid development.

Other methods for achieving cross-fertilization in normally self-fertilized crops include the following:

- The use of male sterility, whereby such plants produce no viable pollen and cross-fertilization is promoted.
- The application of chemical hybridizing agents, which interferes with pollen production and self-fertilization.
- The use of genetic incompatibility, which results in self-fertilization. This system has not been employed on a commercial scale and will not be considered further.

A major barrier to successful hybridization in many seed-bearing plants is the transformation from easy, if not obligate, self-fertilization to cross-fertilization.

Hybrid systems basically involve, first, a form of genetic or cytoplasmic pollen control, **cytoplasmic inheritance,** in the female parent, with alternating strips or rows of the male parent, and, second, harvesting hybridized seed from the female parent rows for commercial F_1 production. Accordingly, a three-line system was developed. In this method, the first line is the cytoplasmic male-sterile female, or **A-line,** plant. The second is the maintainer, or **B-line,** which is used to pollinate the male- sterile line and make it productive, although progeny are still male sterile. The third is the fertility-restorer line, which acts as the male parent to pollinate the male-sterile line in order to restore the ability to produce F_1 seeds (Figure 22.1).

The objective of hybrid breeding programs is to make the improved performance from the immediate F_1 progeny of two selected parents commercially available. In addition to **heterosis**

STEP ONE: Select parents 1 and 2 for F_1 hybrid seed production

STEP TWO: Backcross cytoplasmic male sterility (ms) into parent 1 (donor) to produce an ms BC_4F_1-A parent. Self-pollinate parent 1 to produce maintainer B-line. Simultaneously backcross a genetic source of fertility restoration into parent C to form a BC_4F_1R parent. Lines A & B are isogenic lines

STEP THREE: Increase seed of female parent and restorer lines (field scale)

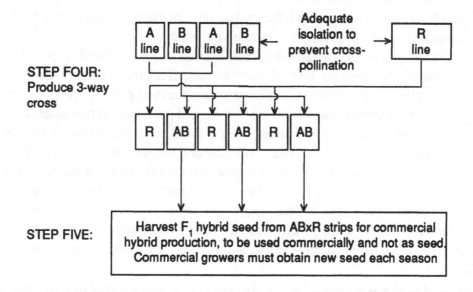

STEP FOUR: Produce 3-way cross

STEP FIVE: Harvest F_1 hybrid seed from ABxR strips for commercial hybrid production, to be used commercially and not as seed. Commercial growers must obtain new seed each season

Figure 22.1   The cytoplasmic genetic system for commercial hybrid seed production involves the production of three types of seed: the A-, B-, and R-lines. The A, or female, parent is cytoplasmically male sterile because pollen production is blocked and it will not set seed until cross-fertilized. To perpetuate the genotype of the A parent, a maintainer B parent is used to fertilize the A parent. The B parent is genetically similar to the A parent but is lacking male sterility. R or **restorer lines** contain genes that restore fertility. The basic procedure involved in creating A- and R-lines is the time-honored backcross procedure.

The production of commercial hybrid seed involves several steps, as follows:

1. Cytoplasmic male sterility is backcrossed into a selected hybrid parent. Since a cytoplasmic factor is involved, the genotype bearing the cytoplasmic factor acts as the donor and the recurrent parent is the chosen parent.
2. The cytoplasmic A-line is maintained by hybridizing to the B- or maintainer, line.
3. Genetic fertility restoration is backcrossed into parent R.
4. Commercial seed is produced from the cross AB x R.

Cross-pollination is accomplished on a field scale by seeding alternate rows or strips, depending on the crop of parents to be crossed. At maturity, the parental strips must be harvested separately. Superior restorers are agronomically desirable and are characterized by high anther extrusion and nick with the male-sterile parent.

and yield performance, advantages include advances in disease resistance, stress tolerance, uniformity, and adaptation. From a plant breeding viewpoint, three factors influence commercial hybrid success: (1) breeding method efficiency or the rate-of-progress factor compared to other breeding systems, (2) the negative or positive effects of male sterility-fertility restoration system(s) used to produce the hybrid, and (3) the seed multiplication ratio, commercial seeding rate, and cost of seed production relative to the inherent heterosis.

Hybrids in any self-fertilized crop are biologically possible, but commercial production of hybrids is considerably more challenging. A prerequisite to commercial production is control of the pollination process. Plants designated as the female parents must be fertilized by a known and controlled pollen source on a field scale rather than by their own pollen or by a random pollen source. Hybrid seed produced on the female plant must be harvested separately from that on the male parent and must be fertile in the F_1 population. For the commercial producer, hybrid crop production means purchasing F_1 seed for each crop because F_2 seed will segregate and lose its yield advantage.

Commercial hybrid seed production involves successfully integrating a number of factors, including:

1. A means of producing male-sterile, female parent plants.
2. A source of viable pollen from a male parent that can outcross with the male-sterile female parent.
3. A male to female planting configuration that maximizes outcrossing, such as a 1:1, 2:1, or 3:1 male-sterile to restorer ratio.
4. Agents such as wind or insects to move pollen from the male to the female parent.
5. Female and male parents that are in synchrony, or **nick**, so that pollen from the male parent is shed when the florets of the male-sterile female are receptive.
6. An abundance of pollen from the male parent to help ensure cross-pollination.
7. Good flower-opening characteristics in the female parent so that pollen can make contact with the stigma.
8. A complete and reliable fertility-restoration system so that commercial producers are not confronted with sterile progeny.
9. A reasonable yield advantage.

The amount of hybrid vigor required to make hybrids in self-fertilizing crops a viable economic reality remains uncertain. McRae (1985) suggested that in wheat, yields should exceed those of the best commercial cultivar by 10 to 20%. Carver and Nash (1984) suggested a minimum yield advantage of 10% over the most productive commercial cultivar available. Generally, yield advantages of this magnitude are possible. Yield trials for soft and hard winter wheats in the United States have established a 20 to 30% yield superiority for F_1 hybrids (Lucken, 1986), but the consistency with which such yield increases can be sustained across years and environments remains uncertain. Under Australian dryland conditions, F_1 hybrids from adapted Australian wheats grown in hill plots at six environments over two years outyielded the mid and high parents by 31.5 and 26.8%, respectively (Uddin *et al.*, 1992).

The phenomenal success of hybrid maize has produced an aura of glamour around the concept of hybrid production. Much of the glamour has proved true for some cross-fertilized crops, such as onion, pearl millet, and sorghum, and in many horticultural crops, both self- and cross-fertilized. However, in the major field crops of the world, such as wheat, rice, barley, oat, soybean, and cotton, the possibilities for profitable commercial hybridization remain to be demonstrated on a broad scale.

MALE STERILITY IN SELF-FERTILIZED CROPS

Several mechanisms exist to induce male sterility in self-fertilized crops. We explore these mechanisms in the following sections.

Genetic Male Sterility

At least one male-sterility locus has been found in virtually every diploid and polyploid plant species (Duvick, 1966). Male sterility may result from pollen abortion, **pistillody,** or other causes. Genetic male sterility is conditioned frequently by recessive genes (*ms*) in the homozygous state, which makes this form of male sterility difficult to manage from a breeding standpoint. A high frequency of the dominant *Ms* gene in a population of recessive male-sterile plants produces progenies that are 100% heterozygous (*Msms*). In crossing fertile heterozygous plants to homozygous, recessive, sterile plants in each generation, segregates will be produced in a 1:1 *Msms* to *msms* ratio. Since a strain will never be completely sterile with this genetic system, if it is used as a seed parent, 50% of the male-sterile plants must be rogued from the seed production rows prior to pollen shedding. Unless male sterile plants can be detected before pollen is shed, commercial production of hybrid seed through genetic male sterility is nearly impossible. Wilson (1984) observed that a genetic male-sterile method in wheat that requires much roguing is likely to fail. A system of genetic markers (Figure 22.2) may facilitate **roguing** in a male-sterile system.

Use of a single, recessive genetic male-sterile gene for hybrid development has been explored in many crops but has not been utilized successfully in cereals. Chromosomal genes for male sterility still seem practical.

The possibility of using balanced tertiary trisomics to provide a basis for commercial production of hybrid barley was reported by Ramage (1963), who described a method of employing them to produce female parents of hybrids (1965). Balanced tertiary trisomics are defined as tertiary trisomics set up in such a way that the dominant allele of a marker gene, closely linked with the translocation breakpoint, is carried on the extra chromosome and the recessive allele of the marker gene is carried on the two normal chromosomes that constitute the diploid complement. Trisomic indicates the presence of an extra chromosome; tertiary indicates that the extra chromosome is a translocated chromosome, that is, it consists of segments of two nonhomologous chromosomes; and balanced refers to breeding behavior, that is, because the extra chromosome is rarely transmitted through the pollen, the selfed progeny of a balanced tertiary trisomic consist mostly of dominant trisomic and recessive diploid individuals. In the selfed progeny of a balanced tertiary trisomic, all plants exhibiting the dominant character should be of the same gene and chromosome constitution because the parental trisomic and all of the functional pollen grains produced by such plants should carry the recessive marker allele.

To produce F_1 hybrid barley, alternate strips of male and female rows are planted. The male

Figure 22.2 If genetic male sterility is to be used to produce F_2 hybrids on a commercial scale, a marker gene closely linked to male-sterility genes may prove advantageous to detect male-sterile progeny prior to anthesis. The diversity of barley plants portrayed in this photo provides opportunity to search for useful markers. The problem is to find a completely reliable marker.

Recessive genetic male sterility is valuable in the female parent for hybrid seed production because any normal fertile cultivar will carry the dominant fertile allele, resulting in F_1 hybrids that are completely fertile. The problem is that male-sterile plants cannot be selfed to maintain the male-sterile line in the homozygous condition. Plants heterozygous for male sterility, when selfed, produce 3 fertile:1 male sterile.

Falk, Kasha, and Reinbergs (1981) reported the coupling of a male-sterile gene to a xenia-expressing shrunken endosperm gene, which makes it possible to select seeds, before planting, that will produce male-sterile plants. In populations with the shrunken endosperm gene tightly coupled to the male-sterile gene, the male sterility can be maintained in or eliminated from the population by selection for or against the linked endosperm trait. Selection of shrunken endosperm seeds will produce a population consisting almost entirely of male-sterile plants that can be used as the female line for hybridization with any normal male parent barley cultivar. The problems associated with such a proposal include the possibility that such a linkage may be broken, that incomplete linkage will cause some plants to escape, and that screening is incomplete and less than 100% of the F_1 population will be hybrid. Roguing plant populations may limit the scale of commercial seed production.

parent is a regular barley cultivar that can serve as an adequate pollinator. The female parent is a balanced tertiary trisomic. The selfed progeny of trisomic plants comprise about 30% tertiary trisomics and 70% male-sterile diploids. The trisomic plants are shorter, later in flowering, and much weaker in competition than the male-sterile diploids. Advantage can be taken of these characteristics to produce nearly pure stands of male-sterile diploid plants. Under commercial conditions, seed from balanced tertiary trisomic plants is sown in the female rows at a rate of 25 to 30 kg/ha (21 to 25 lb./ac.), which almost entirely eliminates the trisomic plants by competition, leaving a nearly pure stand of male-sterile plants.

Seed of balanced tertiary trisomics is produced at a seeding rate of 5 to 7 kg/ha (4 to 6 lb./ac.), which produces little competition from male-sterile diploids and produces about 30% trisomics and 70% diploids. Diploid seedlings have normal leaves; trisomics have long, narrow leaves. On this basis, diploid plants can be removed. Some of the seed from the trisomic increase field is used to produce more trisomics, and the rest of the seed is used as the female parent of a hybrid (Ramage, 1975).

The first hybrid barley using the balanced tertiary trisomic system was released in the United States in 1968. Because of problems of cross-fertilization, susceptibility to infection by ergot (*Claviceps purpurea*), economic considerations, and the complexity of the system, the balanced tertiary trisomic method has not proved practicable for implementing hybrid barley on a commercial scale. Progress, however, is encouraging, and problems may be overcome.

Cytoplasmic Male Sterility

The most widely accepted means of producing controllable male sterility in many crops is through cytoplasmic male sterility (CMS). In wheat, male sterility was reported in the early 1950s and was produced by hybridizing modern cultivars with *Aegilops caudata* or *Triticum timopheevi* Zhuk. (Kihara, 1951; Fukasawa, 1953; Wilson and Ross, 1962). Progeny produced by using these wild strains as mother plants and modern cultivars as pollinators were male sterile as a result of an imbalance between the cytoplasm of the wild strains and the paternal contribution derived from the pollen of the modern cultivar. Cytoplasmic male sterility has been reported in many crops. Male sterility prevents the maturation of functional pollen and ranges from complete pistillody to degeneration of pollen grains after meiosis.

Cytoplasmic male sterility is due, in part, to factors carried only through the female and is not diluted or lost in successive generations of reproduction. Nuclear genes generally influence the expression of this trait, and environmental conditions may alter its expression in many, but not all, gene-cytoplasm combinations.

F_1 hybrid production using CMS methods involves a long and complex breeding program with repetitive backcrossing of critical test crosses and reciprocal crosses. The male-sterile (*ms*) and restorer (*rf*) genes often must be transferred from one cultivar to another. Linkage may be a problem, and the time needed for several backcross generations can prove unprofitable in the dynamic and competitive field of commercial plant breeding.

To ensure adequate fertility in the F_1 offspring, restorer genes must be transferred from the male pollinator. In wheat, restoration to normal fertility requires two or more genes in hybrids having *Triticum timopheevi* cytoplasm. The number of genes required for restoration is dependent on the genetic background and the environment (Wilson, 1968). Some genotypes carry suppressor genes, and their presence tends to make restoration unstable in the hybrid under cool temperature conditions during the booting and flowering stages. A plant breeder must attempt to accumulate the necessary restorer genes, a situation complicated by the fact that restorer genes are more effective in the homozygous than in the heterozygous state. Self-fertility of the restorer lines (in *Triticum timopheevi* cytoplasm) provides only a partial indication of restoration capacity in F_1 hybrids (Lucken, 1986). Hybrid wheat development has been conducted primarily with hexaploid rather than tetraploid wheat, and restoration genes are scattered throughout the *ABD* genomes on seven different chromosomes (Wilson, 1984).

Pollen restoration for *Triticum timopheevi* cytoplasm may be obtained from a number of sources, including *Triticum timopheevi*, *T. boeoticum*, *T. Zhukovskyi*, *T. araraticum*, *T. spelta* L., *T. dicoccoides*, *Secale cereale* L., and common wheat cultivars. Pollen restoration from common wheat cultivars results from natural genetic crossovers from alien genes during introgression of specific genes.

The law of homologous series suggests that cytoplasms that cause male sterility and genes for fertility restoration can be found in any crop if a wide enough search is conducted. The use of cytoplasmic male sterility to produce F_1 hybrids is therefore biologically applicable to any crop species.

Chemical Hybridizing Agents

The possibility of a **chemical hybridizing agent** to induce male sterility has considerable appeal because it would facilitate hybrid production in the following ways:

- The parent selected to be the male-sterile female could be any cultivar or line, thereby saving the time required to incorporate male sterility by backcrossing.
- The incorporation and testing of fertility restorers would be eliminated.
- The maintenance of male-sterile stocks would no longer be necessary.
- The scale of production of hybrid seed could be adjusted readily to match demand.
- Plant breeders would be able to evaluate many parental combinations, easily discarding the poorest and keeping only the best without a great deal of backcrossing effort.

Chemical induction of sterility in plants was first demonstrated by Moore (1950) and Naylor (1950). Various terms, such as male sterilant, selective male sterilant, pollen suppressant, pollenocide, androcide, and chemical gametocide, have been used, but such terms may not embrace all conceivable modes or sites of action. The term chemical hybridizing agent (CHA) was suggested as appropriate (McRae, 1985).

The purpose of a chemical hybridizing agent is to facilitate hybrid seed production. A conventional female parent is treated at some stage before anthesis, and the resulting male-sterile plants are pollinated by windborne pollen from the alternating untreated male-parent drill strips to produce hybrid seed. Since the effect of the chemical is not carried to the next generation, fertility restoration is of no concern.

Chemical hybridizing agents must fulfill rigid requirements, suggested as:

1. The capacity to induce complete or near complete male sterility without causing a reduction in female fertility. The proximity of the pistil and stamens within a perfect floret and the coordinated timing of their development make a differential response a challenge. On the other hand, there is both gametophytic and sporophytic control of male sterility, suggesting selective responses are possible.
2. From a practical viewpoint, chemical hybridizing agents must be effective if applied over a range of crop growth stages, on different genotypes, and in different environments. In other words, genotype x chemical environment, environment x chemical, and genotype x chemical x environment interactions cannot be large. Latitude in time of application is required in case of inclement weather and to allow commercial seed producers time to treat relatively large areas.
3. A suitable chemical hybridizing agent should effectively sterilize both early- and late-formed flowers on tillers or branches. This may require a degree of persistence or even **systemic** action.

4. A chemical should be in a form that can be safely sprayed on the designated female strips without danger of drift to the adjacent male rows.
5. Chemical hybridizing agents should have a minimal impact on such factors as spike development and emergence from the boot of cereal crops, leaf chlorosis, and heading date or productivity. There were problems associated with the first chemicals tested.

A number of compounds can induce male sterility, including simple halogenated aliphatic acids such as sodium 2, 2-dichloroproprionate (dalapon), plant growth regulators such as triiodobenzoic acid (TIBA), gibberellic acid, and ethylene generators such as ethephon (2-chloro-ethylphosphoric acid). Several patented compounds have been produced by Rohm and Haas and Shell Chemical Company (Alper, 1983). These foliar sprays are applied at a premeiotic stage, but there is the possibility of a compound that would be applied as a seed treatment (McRae, 1985).

Evidence of the actions of chemical hybridizing agents suggests effects that can inhibit the early stages of sporogenous cell formation to the inhibition of another dehiscence. McRae (1985) listed four specific effects:

1. Disruption of meiosis, degeneration of the pollen mother cells or early microspores resulting in pollen abortion, and the arrest of anther development at an early stage.
2. Disruption of **exine** formation, resulting in thin-walled microspores with nonviable pollen in abnormally developed anthers.
3. Nonviable pollen development associated with abnormalities in microspore vacuoles, decreased starch deposition and **tapetum** persistence.
4. Production of viable pollen. Anthers appear normal but either do not dehisce or show delayed dehiscence.

A compound that causes any of these effects can be classed as a chemical hybridizing agent.

MULTIPLIER EFFECT

Hand hybridization to produce F_1 hybrid seed on a commercial basis is possible with a few high-value crops, especially when the yield of seed per individual hand operation is large, as in tomato, cucumber, and snapdragon. Nevertheless, F_1 seed produced in this manner is expensive, and progeny may be sold as bedding plants rather than in seed lots. The seed multiplication ratio and the plant population in commercial plantings are functional constraints that strongly affect whether commercial hybrid seed production is possible and how it is conducted. The seed multiplication ratio and the commercial seeding rate together make up the multiplier effect (Figure 22.3).

A comparison of seeding rates and multiplication ratios for selected crops will help demonstrate the multiplier effect. In hybrid-seed-production fields of maize, each inbred female seed may produce an ear bearing about 215 seeds (Figure 22.4). At a female plant population of 20,000 plants/ha (8,100 plants/ac.), a total of 4,300,000 seeds would be produced. This is sufficient to seed 87 hectacres at a commercial seeding rate of 49,4000 seeds/ha (20,000 seeds/ac.). In wheat, each female parent seed may produce 25 F_1 seeds, which with alternating male and female strips may produce 25 million seeds capable of seeding 15 hectares at 120 kg/ha (37 ac. at 100 lb./ac.) under North American prairie conditions. In sunflower, a favorable seed

A.

B.

Figure 22.3   The seedlings in Photo A are rice plants that have been transplanted by hand, a practice that drastically reduces the seed requirement of this crop. Seeding rates of 15 to 22 kg/ha (13 to 20 lb./ac.) are common in the rice-growing areas of the world.

In contrast to the relatively low seeding rate for rice, the population of spring wheat seedlings in Photo B are seeded at a much higher rate. Seeding rates for wheat vary from 40 to 160 kg/ha (36 to 140 lb./ac.). Wheat seeding rates are based on tradition and are high because growers tend to plant excess seed to offset possible losses due to low moisture, poor seedbed preparation, and the competitive edge of weeds, as well as to reduce soil erosion from water or wind. Most cereal cultivars are capable of a high tillering rate, and lower seeding rates may be possible. At high seeding rates, seed costs are too high.

Seeding rate is an important factor in determining the economic feasibility of hybrid crop production because the cost of producing F_1 seed is high. In the very successful cross-fertilized hybrid crops of sorghum, sunflower, and corn, seeding rates are 2 to 12 (2 to 10), 3 to 8 (3 to 7), and 11 to 20 kg/ha (10 to 18 lb./ac.), respectively.

A high level of efficiency must be achieved in hybrid seed production, especially in crops such as wheat, to minimize costs to commercial grain growers. Grain growers producing commercial crops from hybrid seed must obtain seed for each F_1 crop and break with the traditional system of producing crop after crop from farm-stored seed.

Photo B courtesy Ontario Ministry of Agriculture and Food.

A.   B.

Figure 22.4   The multiplier effect profoundly influences economic aspects of commercial seed production in F_1 hybrids. The ear of maize in Photo A may carry 500 or more seeds and is produced by a single pollination of the female component. This is a large F_1 ear, not an inbred. In the early stages of hybrid maize production, seed produced on inbred ears was small in size, high in cost, and lacked appeal for producers. This problem was overcome by double-cross hybrid production in which two F_1 hybrids were crossed. In contrast to ears of corn that produce a large number of seeds, wheat spikes (Photo B) produce considerably fewer seeds.

Another component of the multiplier effect is the seeding rate. The multiplier effect is one factor that influences the economics of hybrid seed production. An important axiom for commercial utilization of F_1 hybrids is to produce large quantities of seed with minimum expense.

Photo Courtesy Ontario Ministry of Agriculture and Food.

multiplication ratio is found. Each female parent seed may produce 450 F_1 seeds. A hectare of hybrid seed production may produce sufficient seed for 220 commercial hectares at a 4 kg/ha seeding rate (543 ac. at a 3.3 lb./ac. seeding rate).

The magnitudes of commercial seeding rates and multiplication ratios are important factors affecting the commercial feasibility of a selected crop. Economically feasible production of hybrid cereals is a challenging task because of a poor multiplier ratio and a high seeding rate. This situation impacts on plant breeding programs. Hand pollinating a single ear of an experimental line of maize produces sufficient seed to conduct yield trials at several locations.

Pollinating a single cereal head produces 25 to 30 seeds, increasing the labor and cost of testing and limiting the number and diversity of hybrids tested. For large-scale cereal testing, the female parent must be converted to its male-sterile counterpart by backcrossing, both male and female parents must be increased on a substantial scale because of a low multiplier effect, and hybrid seed must be produced in isolated crossing blocks. These factors require a long lead time, demanding care, isolation, and patience and involve considerable delays and expense.

Crops with a low multiplier effect require large areas to produce hybrid seed on a commercial scale, the management and inspection of massive seed areas, and a comprehensive marketing, handling, and distribution system.

The economics of commercial F_1 hybrid seed production depends to a large extent on the price of hybrid seed relative to the hybrid yield advantage. Seed production costs are determined by the female:male ratio required in seed production fields and depend on seed set on the male-sterile parent through cross-fertilization. Seed set is a function of environment and genotype, and a major challenge and responsibility of the plant breeder is to improve genetic aspects affecting cross-pollination in normally self-pollinated crops.

Thompson (1971) observed differences in the ability to cross-pollinate among barley cultivars, which appear to be related to favorable pollinating conditions as well as to morphological features of the flower (Figure 22.5). Likewise, Lucken (1986) noted that seed set is consistently higher from some male-steriles than from others. Seed set variation associated with genetic differences among male-sterile and restorer lines can be as great as two- or threefold. Some of the plant or floral features that contribute to cross-pollination include extrusion of the anthers on the male parent, extrusion of the stigmas on the female parent, action of lodicules in forcing open the floret, and lack of barriers such as awns, as illustrated in Figure 22.6. Proper matching, or nick, of male-sterile and pollinator flower development is critical.

Three main problems to be solved for hybrid barley development, as specified by Scholz and Kunzel (1981), are adequate heterotic performance of F_1 hybrids, sufficient cross-pollination, and efficient genetic systems. The future of F_1 hybrid barley in the United Kingdom, according to Foster and Fothergill (1981), depends upon three major factors: (1) the development of an efficient and reliable system of producing male-sterile stocks of female parents, (2) the attainment of high levels of seed set in the production of F_1 hybrid seed, and (3) economically significant increases in grain yield by F_1 hybrid cultivars.

Lucken (1986) suggested that environmental effects on cross-pollenation are large. He observed a sixfold difference in grain yield in a male-sterile parent seeded at a 1:1 ratio among locations in North Dakota. In Europe, Keydel (1978) suggested prolonged rainfall during pollination can reduce cross-pollination. Drought and high temperatures can also reduce cross-pollination. In areas where high stable yields are consistent, hybrid seed production can be successful.

In response to the problem of insufficient pollen dispersal in naturally self-fertilized cereals, Rutger and Carnahan (1981) suggested that pollinator parents with increased height would enhance pollen dispersal by wind and gravity onto short maternal plants. To overcome the fact that the tall character is usually dominant to the short in cereals, a recessive tall rice plant was reported. Incorporating this gene for elongated upper internode into pollen fertility-restoring parents in hybrid seed production in situations where a semidwarf F_1 generation is desired would enhance cross-pollination and reduce the problems of lodging and a poor harvest index.

A.

B.

Figure 22.5   Normally self-fertilized crops such as wheat (Photo A) and barley (Photo B) must be transformed to cross-fertilized crops for F_1 hybrid production. Awns, extensions of the lemma, may serve as physical barriers to cross-pollination if such lines serve as female parents. If awns restrict cross-pollination, the problem may be avoided by using these lines as the male parent. This requires close attention to parents prior to male-sterile conversion to predict which will have high cross-pollination potential. In the male parent, anther extrusion is an important parameter influencing cross-pollination. Parents noted for anther extrusion should be used as the male parent.

More research is needed to assess the various floral traits and cross-pollination capacity. De Vries (1971, 1973) gives a complete review of the topic.

Photo courtesy Ontario Ministry of Agriculture and Food.

Figure 22.6   The structures shown in this photograph are the reproductive organs of barley (*Hordeum vulgare* L.) and include the pistil with two feathery stigmas, three stamens each with two lobes prior to pollen release, and one of two lodicules on the bottom left of the ovary, which has partially been dislodged to help distinguish it from the ovary. These structures are normally contained within a floret, and self-pollination usually occurs when anthers release pollen onto the feathery stigmas.

For hybrid seed to be developed, cross-fertilization must occur. Mechanisms for male sterility render the anthers ineffective in producing viable pollen. Lack of self-fertilization is associated with swelling of the lodicules, forcing the lemma and palea of the floret apart and thereby exposing the feathery stigma. Breeding and selection of genotypes with large lodicules and associated open florets, extruded anthers or filaments that extend and push the anthers out of the floret, anthers that shed large quantities of pollen, and large, feathery and exposed stigmas may be useful in converting a self-fertilized crop to one that is cross-fertilized.

Thompson (1971) used male-sterile barley plants to measure the degree of outcrossing from the wild barley collection and found seed set ranged from near 0 to close to 100%. It is clear that selection of genotypes can be used to transform self-fertilized to cross-fertilized crops.

Photo courtesy Ontario Ministry of Agriculture and Food.

HYBRID PERFORMANCE

Despite 25 years of hybrid wheat research, the question of F_1 hybrid yield advantage has not been answered clearly. A number of wheat hybrids have been developed since 1970 by private seed companies and released for farm production in the United States and Australia (Wilson, 1984). Most of the hybrids released have not been continued because of various economic problems and competition from newly improved cultivars. The first commercial hybrid barley released in the United States in 1968 was not accepted by producers because of a lack of lodging resistance and the greater economic viability of other competitive crops (Lehmann, 1981).

There are several standards against which the performance of hybrids can be compared. By definition, a measure of hybrid vigor is obtained by comparison with the high-yielding parent or with the parental mean of midparent yield. Commercial producers, however, are interested in F_1 performance relative to the best cultivars available. In horticulture crops, especially ornamental crops, quality is of paramount importance.

In plants where yield is of primary importance, F_1 hybrid performance may be viewed in terms of **photorespiration**. Plants can be classed according to specific photosynthetic pathways as C_3 or C_4 plants. C_4 plants that respond to warm, dry, and high light conditions are more efficient than C_3 plants. Corn, sorghum, and millet are C_4 plants; cereals, cotton, and soybeans are C_3 plants. The question of the impact of heterosis on photorespiration remains unresolved.

In any consideration of hybrid performance, actual hybrid performance must be compared with potential hybrid performance (Lucken, 1986). Perhaps plant breeders have failed to hybridize appropriate parents. Lucken (1986) suggested seven variables that can affect hybrids produced by a genetic system.

- agronomic merit of male-sterile and restorer lines
- genetic diversity of parents
- combining ability of parents
- genetic system used to produce hybrids, including *Triticum timopheevi* cytoplasm, introgressed restorer genes, and fertility restoration
- quality aspects of hybrids
- purity of hybrid seed
- inherent hybrid vigor

F_1 HYBRIDS IN SPECIFIC CROPS

Cotton

The potential advantages of hybrid cotton (*Gossypium* spp.) have been considered since early in the twentieth century (Cook, 1909; Kearney, 1924), but large-scale commercial utilization of natural crossing to produce hybrid cotton remains a problem. Despite the fact that interest in hybrid cotton has existed for over eight decades, the technology required to produce commercial hybrid cotton was developed only in the 1970s. During the 1980s, production of commercial hybrid cotton seed moved hesitatingly forward, and the first commercial cotton hybrid was grown in Texas in 1984. The problems of seed cost, yield, and quality advantages remain unresolved.

The necessary cytoplasmic male-sterility system was described for cotton by Meyer (1975). Prior to cytoplasmic male sterility, 800,000 ha (1.75 million ac.) of hybrid cotton were grown in India annually using hand-produced hybrid seed and hand labor for pollination (Srinivasan, Santhanan, and Rajasekaran, 1972).

The effectiveness of the honeybee (*Apis mellifera* L.) as an agent for cross-pollination has been evaluated (Waller *et al.*, 1985). Satisfactory seed yields were obtained when the area was saturated with honeybees. Five colonies/ha (2/ac.) with 28,000 bees per colony were required to provide adequate pollen transfer. Although these results are promising, further work is needed to establish recommendations for economic use of bees in producing hybrid cotton seed commercially (Figure 22.7).

Four cotton hybrids from the interspecific cross of *Gossypium hirsutum* x *G. barbadense* produced an average yield increase of 325 kg/ha (290 lb./ac.) over four check cultivars that averaged 851 kg/ha (760 lb./ac.) or an increase of 37%. In the intraspecific *G. hirsutum* hybrids, seven check cultivars averaged 1,046 kg/ha (934 lb./ac.) of lint, and F_1 hybrids produced an additional 280 kg/ha (250 lb./ac.) for an average increase of 33% (Davis, 1978).

Several genetic and cytoplasmic male-sterile systems exist in cotton (Davis, 1978). Two male-sterile systems controlled by recessive genes produce complete sterility. A third

Figure 22.7 The honeybee (*Apis mellifera* L.), shown in this photograph, is one of several insects that are effective in cross-pollination of some plants. Cotton (*Gossypium* spp) is a generally self-fertilized crop. The stigmas of most cultivars are not exserted clear of the uppermost anthers, and selfing usually occurs immediately after anther dehiscence. Cross-pollination can be promoted by insect vectors when the stigma is exserted well above the anthers, a situation found in primitive forms of *Gossypium hirsutum* and *Gossypium barbadense*.

Species that rely on insect vectors for cross-fertilization have evolved mechanisms for attracting bees by floral nectar. Studies by Waller *et al.* (1985) showed that saturation of hybrid cotton seed-producing fields with honeybees, amounting to 5 colonies/ha with 28,000 bees per colony, was sufficient to pollinate male-sterile cotton under the conditions of this test. The planting pattern used was 2 rows of male-sterile female seed parent alternately planted with 2 rows of male-fertile maintainer in a 56 ha (138 ac.) seed production field. Floral nectar was collected, and sugar concentration of the nectar was generally above the 30% level preferred by honeybees (Waller, 1972).

Cotton plants bloom over a three- to five-week period, and matching flower dates for male and female plants is rarely a problem.

A problem associated with using insect vectors for cross-pollination is that insecticides are needed to control insect pests in cotton fields but must not be allowed to damage the pollen vectors.

Photo courtesy Ontario Ministry of Agriculture and Food.

cytoplasmic system is less satisfactory because it is influenced by the environment. Three separate cytoplasmic male-sterile systems are known: two are from *G. anomalum* and *G. arboreum*, but that from *G. harknessii* is the most suitable. Work with genetic steriles was abandoned with the advent of cytoplasmic male-sterility sources. Fertility restoration of *G. harknessii* cytoplasm consists of a single dominant gene from *G. harknessii*, but restoration is not complete in some environments.

Horticulture Crops

In highly prized vegetable and ornamental crops and in situations where seed is not required on a massive field scale as in major food crops, F_1 hybrids are commonly found and some crosses may be produced by hand. In other cases, self-incompatibility, male sterility, dioecy, or monoecy may be used to produce F_1 hybrid seed. Practical production of F_1 seed may involve some inherited form of male sterility introduced into one of the parental lines. When seed set is not a requirement, genes for restoration are not required. To increase the proportion of female flowers in a monoecious crop such as cucumber, chemical growth regulators can be used. Hybrid seed is not dependent on male sterility but on disruption of the development of staminate flowers so that most or all functional flowers on a given plant are pistillate.

Most F_1 hybrid vegetables are superior to regular cultivars in vigor, flavor, appearance, insect and disease resistance, and productivity. F_1 hybrid vegetables commonly include tomato, muskmelon, cucumber, cabbage, carrot, onion, cauliflower, watermelon, broccoli, peppers, spinach, squash, and turnip, many of which are normally cross-fertilized.

F_1 hybrids are offered in over 30 species of seed-propagated ornamentals, representing over 24 genera, including marigold, geranium, *Dianthus*, petunia, zinnia, *Impatiens*, snapdragon, and pansy. F_1 seed costs can be from two to three times higher than for non-hybrid seed for begonia to 25 times higher for zinnia.

Advantages of F_1 hybrids include improved vigor, uniformity, pest resistance, and much superior flower types. The disadvantage of the extra cost of F_1 hybrids is nearly always outweighed by superior performance, but this is not the only criterion for hybrid development. F_1 hybrids provide a measure of protection against pirating because seed is produced each season from controlled parents. Protection offered by **Plant Breeders' Rights** makes this reason less compelling. Plant Breeders' Rights is discussed in Chapter 23.

Rice and Cereals

Approximately 6 million hectares (14.8 million ac.) of hybrid rice were grown in the early 1980s (Virmani and Edwards, 1983). Genetic improvement of floral characteristics influencing cross-pollination have resulted in seed yields of 0.45 to 1.5 tonnes/ha (0.2 to 0.7 tons/ac.). Many of the problems outlined in this chapter for wheat and barley also apply to rice.

Hybrid rice is of major importance in China, possibly a result of a good multiplier effect associated with the fact that rice in many countries is transplanted, resulting in a low seeding rate. In addition, a considerable amount of hand labor has been used to produce hybrid rice seed because the extent of natural cross-fertilization in hybrid seed production blocks is low. The panicle may not emerge fully from the boot (Azzini and Rutger, 1982), and to increase seed set, the boot may be removed by hand. High hand-labor requirements could enhance hybrid rice production in those countries where labor is abundant and inexpensive.

In the oat crop, scanty amounts of pollen and florets shielded by drooping glumes and panicles make it difficult to utilize wind for cross-pollination, even if suitable male-sterile forms are present (Figure 22.8).

Figure 22.8   Panicle of oat showing the pendulous nature of florets, often supported on drooping panicles, These features coupled with large outer glumes may inhibit cross-pollination in this naturally self-fertilized crop.

The success of hybrid maize, sorghum, pearl millet, and other seed-producing cross-fertilized crops has raised the notion that some day all crops will be hybrid. After considerable research effort with self-fertilized crops, many seemingly insurmountable problems have been overcome, and hybrids appear to have potential in an increasing number of crops.

Perhaps the oat panicle with its seemingly unsolvable problems symbolizes the advances plant breeders have made in moving toward commercial F_1 hybrid production in almost every major crop. Challenges in hybrid production as difficult as that exhibited by the oat crop have been overcome.

F_1 hybrids in the oat crop may never be developed because of the high research cost compared with the potential commercial value. This does not mean, however, that the problems are insurmountable.

Photo courtesy Ontario Ministry of Agriculture and Food.

Soybeans

The limited number of studies of hybrid soybeans focus on hand-pollinated seed, but hand pollination in soybeans is a difficult task. To determine the degree of heterosis in soybean hybrids, Nelson and Bernard (1984) used four genetic male-sterile female lines, which were crossed by insects to produce 27 hybrid combinations. Five hybrids outyielded the better parent by 13 to 19%, but only one hybrid exceeded the yield of the best pure-line cultivar in the test. The inability to produce large quantities of hybrid seed economically is the major barrier to the use of commercial hybrid soybean cultivars.

Oilseed Rape *(Brassica napus L.)*

Eight single-cross oilseed rape hybrids were produced by hand-emasculation and pollination, then evaluated in a replicated yield trial (Grant and Beversdorf, 1985). Some of the hybrids demonstrated midparent heterosis, but none of the eight exhibited high-parent heterosis. In most cases, the hybrids were intermediate to the parents for flowering date, plant height, lodging resistance, physiological maturity, percent oil, and percent protein.

CONCLUSION

Hybrid development in self-pollinated crops appears to be poised for takeoff after a long and difficult struggle to overcome the barriers to hybrid development (Wilson, 1968). Lehmann (1981) predicted that it is only a matter of time before hybrid barley will be back on a commercial basis since the first hybrid was released in 1968. Hybrid wheat was described by Wilson (1984) as still in the juvenile stage with regard to ultimate potential and impact. An enormous challenge still faces plant breeders.

If F_1 hybrid production comes to dominant self-fertilized crops, what impact will this have on the breeding of pure-line cultivars and on other systems of plant breeding? The breeding of superior pure lines will continue, for these are the parents of F_1 hybrids. Nevertheless, hybrid breeding and conventional breeding programs are at the same time complementary and competitive: better parental lines should produce better hybrids, and better parents or pure-line cultivars should be derived from superior hybrids (Lucken, 1986).

In selected crops such as cotton, the problem of high seed costs may be alleviated by use of F_2 seed. F_2 performance for quantitative characters, such as yield, was satisfactory. Apparently, allelic interactions tend to disappear in the course of reproducing the hybrid populations. Yield loss was minimal in the F_2 generation (Gulyaev and Kyzlasov, 1987).

REFERENCES

Alper, J. 1983. Pollen Suppressors Open New Options for Hybrids. *Biotechnology* 1(1):14.
Azzini, L. E., and J. N. Rutger. 1982. Amount of Outcropping on Different Male Steriles in Rice. *Crop Science* 22:905-907.

Carver, M. F., and R. J. Nash. 1984. The Future for Hybrid Cereals. *Span* 27(2):64-65.

Cook, O. F. 1909. *Suppressed and Intensified Characters in Cotton Hybrids.* United States Department of Agriculture Bureau of Plant Industry, Washington, D.C., Bulletin 147.

Davis, D. D. 1978. Hybrid Cotton: Specific Problems and Potentials. *Advances in Agronomy* 30:129-157.

De Vries, A. P. 1971. Flowering Biology of Wheat, Particularly in View of Seed Production: A Review. *Euphytica* 20:152-170.

————. 1973. Some Aspects of Cross-pollination in Wheat (*Triticum aestivum* L.). 2: Anther Extrusion and Ear and Plant Flowering Pattern and Duration. *Euphytica* 22:445-456.

Duvick, D. N. 1966. Influence of Morphology and Sterility on Breeding Methodology. In *Plant Breeding,* edited by K. J. Frey. Iowa State University Press, Ames, Iowa, pp. 85-138.

Falk, D. E., K. J. Kasha, and E. Reinbergs. 1981. Presowing Selection of Genetic Male Sterile Plants to Facilitate Hybridization in Barley. In *Barley Genetics IV,* Proceedings of the Fourth International Barley Genetics Symposium, Edinburgh University Press, Edinburgh, pp. 778-785.

Foster, C. A., and M. Fothergill. 1981. Breeding F₁ Hybrid Barley. In *Barley Genetics IV,* Proceedings of the Fourth International Barley Genetics Symposium, Edinburgh University Press, Edinburgh, pp. 766-771.

Fukasawa, H. 1953. Studies on Restoration and Substitution of Nucleus of Aegilotricum. I: Appearance of Male-Sterile Durum in Substitution Crosses. *Cytologia* 18:167-175.

Grant, I., and W. D. Beversdorf. 1985. Agronomic Performance of Triazine-Resistant Single-Cross Hybrid Oilseed Rape (*Brassica napus* L.), *Canadian Journal of Plant Science* 65:889-892.

Gulyaev, G., and V. Kyzlasov. 1987. New Approach to Determining the Breeding Value of Hybrids. *Khlopkovodstvo* 6:36-40 (abstract).

Kearney, T. H. 1924. A Hybrid Between Different Species of Cotton. *Journal of Heredity* 15:309-320.

Keydel, F. 1978. The Influence of Pollinator and Various Strip Ratios on the Seed Yield of Hybrid Wheat. *Zeitschrift für Pflanzenzuchtung* 81:319-326.

Kihara, H. 1951. Substitution of Nucleus and its Effects on Genome Manifestations. *Cytologia* 16:177-193.

Lehmann, L. C. 1981. Where is Hybrid Barley Today? In *Barley Genetics IV*, Proceedings of the Fourth International Barley Genetics Symposium, Edinburgh University Press, Edinburgh, pp. 772-777.

Lucken, K. A. 1986. The Breeding and Production of Hybrid Wheat. (In) *Genetic Improvement in Yield of Wheat,* edited by E. L. Smith. Crop Science Society of America Special Publication No. 13. American Society of Agronomy, Madison, Wisconsin, pp. 87-107.

McRae, D. H. 1985. Advances in Chemical Hybridization. In *Plant Breeding Reviews*, edited by J. Janick. vol. 3. Avi Publishing Company, Westport, Connecticut, pp. 169-191.

Meyer, V. G. 1975. Male Sterility from *Gossypium harknessii*. *Journal of Heredity* 66:23-27.

Moore, R. H. 1950. Several Effects of Maleic Hydrazide on Plants. *Science* 112:52-53.

Naylor, A. W. 1950. Observations on the Effects of Maleic Hydrazide on Flowering of Tobacco, Maize and Cocklebur. *Proceedings of the National Academy of Science* 36:230-232.

Nelson, R. L., and R. L. Bernard. 1984. Production and Performance of Hybrid Soybeans. *Crop Science* 24:549-553.

Ramage, R. T. 1963. Chromosome Aberrations and Their Use in Genetics and Breeding Translocations. *Barley Genetics* 1:99-115.

————. 1965. Balanced Tertiary Trisomics for Use in Hybrid Seed Production. *Crop Science* 5:177-178.

————. 1975. Hybrid Barley. In *Barley Genetics III,* Proceedings of the Third International Barley Genetics Symposium, Garching. Verlag Karl Thiessig, Munich, pp. 761-770.

Rutger, J. N., and H. L. Carnahan. 1981. A Fourth Genetic Element to Facilitate Hybrid Cereal Production: A Recessive Tall in Rice. *Crop Science* 21:373-376.

Scholz, F., and G. Kunzel. 1981. Progress and Problems with Hybrid Barley. In *Barley Genetics IV*, Proceedings of the Fourth International Barley Genetics Symposium, Edinburgh University Press, Edinburgh, pp. 758-765.

Srinivasan, K., V. Santhanan, and S. Rajasekaran. 1972. Development of Hybrid Cotton Utilising Male-Sterile Line. *Cotton Development* 2:27-39.

Thompson, R. K. 1971. Barley as a Cross-pollinated Crop. In *Barley Genetics II*, Proceedings of the Second International Barley Genetics Symposium, edited by R. A. Nilan. Washington State University Press, Pullman, pp. 319-322.

Uddin, M. N., F. W. Ellison, L. O'Brien, and B.D.H. Latter. 1992. Heterosis in F_1 Hybrids from Crosses of Adapted Australian Wheats. *Australian Journal of Agricultural Research* 43:907-919.

Virmani, S. S., and I. B. Edwards. 1983. Current Status and Future Prospects for Breeding Hybrid Rice and Wheat. *Advances in Agronomy* 36:145-214.

Waller, G. D. 1972. Evaluating Responses of Honey Bees to Sugar Solutions Using an Artificial Flower Feeder. *Annals of the Entomological Society of America* 65:857-861.

Waller, G. D., J. O. Moffett, G. M. Loper, and J. H. Martin. 1985. An Evaluation of Honey Bee Foraging Activity and Pollination Efficacy for Male-Sterile Cotton. *Crop Science* 25:211-214.

Wilson, J. A. 1968. Problems in Hybrid Wheat Breeding. *Euphytica* 17 (Supplement 1):13-33.

———. 1984. Hybrid Wheat Breeding and Commercial Seed Development. In *Plant Breeding Reviews*, vol. 2. Avi Publishing Company, Westport, Connecticut, pp. 303-319.

Wilson, J. A., and W. M. Ross. 1962. Male Sterility Interaction of the *Triticum aestivum* Nucleus and the *Triticum timopheevi* cytoplasm. *Wheat Information Service* 14:29.

CHAPTER 23

Cultivar Release,
Seed Certification and
Multiplication, and
Proprietary Protection

A sexually produced seed or an asexual propagule takes on a special significance when it represents a cultivar of a particular crop. Considerable trust is placed in a seed or propagule to perform according to the genetic makeup that has been designed so meticulously by the plant breeder. The breeder probably will have only a few plants or possibly a few kilograms of seed, and these must be increased to thousands of plants or tonnes of seed for sale to the producer. To assure crop producers that seed has been produced according to accepted and supervised methods, standards of certification have been established by which **pedigreed seed** can be achieved. It is the responsibility of the producer to use the best seed available. It is the responsibility of the seed grower to maintain the genetic constitution of a particular hybrid or cultivar. This is accomplished by proper management, which includes consideration of isolation and previous crop produced on a field; care in cleaning, seeding, harvesting, and processing equipment to prevent admixtures; and roguing off-types.

Three broad classes of seed can be identified: (1) breeder seed, (2) seed for increase purposes only, and (3) seed of commerce. Each time a crop is propagated, slight genetic changes can occur, and to counteract these changes, it is necessary to rogue or remove off-types. Therefore, there is a limit to the number of times seed for increase can be increased before a return to breeder seed is required.

Seed growers often seek advice about the off-types expected in a specific cultivar. The plant breeder may observe increase plots on seed growers' fields, and thus seed production becomes an extension of plant breeding. Quality seed is recognized as a critical step in crop production and an important component of plant breeding. Over the years, the value of quality seed has become symbolized by the sower of seed. In keeping with the symbol of the sower, the Canadian Seed Growers' Association has adopted the stylized model in Figure 23.1.

The benefits of plant breeding are realized when a new cultivar is produced, licensed, released, and grown commercially. The final tests of merit and acceptability are determined by the harshest critics of all—the producers and consumers who grow, process, and purchase the final product. Popular opinion sometimes can make or break a cultivar, but valid observations and constructive comments can serve as guides for plant breeders and provide direction for future improvement.

Procedures for seed certification and multiplication are well established in many countries, but because of the diversity of crops and environmental conditions, procedures may vary among

473

A.   B.

Figure 23.1 The sower of seed has become a symbol for seed in the modern world. The Mexican papier-mâché statuette in Photo A symbolizes planting as practiced before mechanization. Perhaps it is no accident that the artist has included a mixture of plant species in the seed pouch of the sower to characterize the generally poor quality of seed in an earlier era.

The stylized sower of seed in Photo B has been adopted by the Canadian Seed Growers' Association as their logo, which is used on letterheads, literature, and certification tags.

High production goals, economic considerations, and a growing awareness that quality seed is a sound investment have forced producers to demand and get seed of the highest quality. Systems of seed certification have been established in many countries to provide guidelines for seed multiplication and to assure a high-quality product.

The importance of obtaining proper seed was emphasized with the advent of F_1 hybrids. New hybrid seed must be purchased for each crop because F_2 seed would result in a 15 to 20% yield loss due to the reduction of hybrid vigor from inbreeding. Renewing the seed source of self-fertilizing homogeneous crops has been encouraged and accepted as common practice. As a result, an extensive seed industry has developed to supply the seed needs of horticulture and field crops with both hybrid and conventionally produced seeds.

Seed certification provides growers with assurance that seed production has been conducted under accepted and supervised methods, including removal of off-types. Seed production can be considered an extension of plant breeding.

Photo A courtesy Ontario Ministry of Agriculture and Food; Figure B courtesy Canadian Seed Growers' Association.

regions and crops. All of these procedures cannot be covered in this chapter; however, it is the responsibility of every breeder to become acquainted with and understand pertinent regulations.

The Canadian approach to handling seed certification and multiplication has evolved since 1904 when the Canadian Seed Growers' Association was established as the sole pedigreeing agency for most agricultural crops in Canada, except potatoes and tree seeds. It is a proven and successful system, and many of the principles outlined in this section are based on the Canadian experience (Canadian Seed Growers' Association, 1988). The Canadian system was modeled after the Swedish Seed Association, which was established in 1886.

CULTIVAR LICENSING

When a new hybrid, synthetic, self-fertilized cultivar, or clone of most crops has been tested adequately for performance, adaptation, and quality, and when objective data of superiority can be demonstrated, registration may be considered. This decision rests primarily with the developing organization, either public or private, but must be supported by other plant breeders. Registration comes under the federal government, as outlined in the Seeds Act. Plant breeders, other researchers, and representatives of seed growers, the seed trade, and producers meet annually to consider registration requests and to make recommendations to the government accordingly. When registration has been granted, the new cultivar is given a name or number.

Licensing is based on the following criteria:

- Clear indication of superiority and a place in the agriculture of some specified area or areas.
- Adequate testing at a number of sites over several years to evaluate yield stability as well as merit.
- Genetic stability to ensure that sexually reproduced cultivars will remain unchanged in their essential and distinctive characteristics, at least with a reasonable degree of reliability.
- Uniformity in the sense that any variations are describable.
- A degree of distinctiveness from previous cultivars in one or more identifiable morphological, physiological, or other characteristics.

For self-fertilized crops such as cereal, bean, canaryseed, fava bean, flax, lentil, peanut, pea, and soybean, five pedigreed classes exist: breeder, select, foundation, registered, and certified. For cross-fertilized crops such as maize, forage legumes, and most forage grasses, three pedigreed classes exist: breeder, foundation, and certified. Breeder, select, foundation, and registered seed are principally multiplication classes. The final pedigreed class, certified, is the seed recommended for use by the commercial farmer.

BREEDER SEED

Once the decision to release a cultivar is made, the plant breeder makes a limited seed increase of the new cultivar to produce breeder seed. Production techniques of breeder seed may vary with the crop and preference of the plant breeder (Figure 23.2). An essential feature of

Figure 23.2   Photo A shows an increase plot of breeder seed of the oat crop. Note the high degree of uniformity. There are approximately 300 rows in this plot. Each row originated from a single plant selected to represent the population and evaluated for uniformity in rows 2 m long. Roguing is done on a row basis. Remaining rows are used to establish the long rows in Photo A. Off-type plants, as illustrated by the two plants shown in the foreground, should be removed prior to anthesis. At harvest, seed is bulked. A plant is deemed an off-type if it displays any differences in morphological features, such as growth habit, leaf color, growth rate, heading date, height, panicle characteristics, or disease reaction.

The breeder seedlot in Photo B is an open-pollinated block of pearl millet isolated by distance from other pearl millet plants. Although hybrid pearl millet was released in India in 1965, only about 10% of the total area sown to pearl millet is hybrid. The increase block in Photo B is an important seed source for seed multiplication by growers in the region of adaptation.

Breeder seed plots are subject to inspection to verify isolation and purity and to observe the genetic characteristics of a particular cultivar.

Photo B courtesy International Crops Research Institute for the Semi-Arid Tropics (ICRISAT), India.

breeder seed is that source plants must be maintained so that the genetic constitution is preserved and seed can be renewed as needed.

In cross-fertilized forage crops, breeder seed is constituted from the original blend of seed harvested from two or more individual strains or clones composing the synthetic and designated the Syn 0 generation. The component strains are maintained so that the synthetic cultivar may be reconstituted as needed at regular intervals. This may be accomplished by either (1) storing large quantities of the original seed of each strain from which portions are taken and blended each time the synthetic is reconstituted or (2) increasing the original strains in isolation in the area of their original adaptation. When the synthetic is reconstituted, strains are composited in the same proportions used in the original synthetic.

Rincker, Dean, and May (1984) studied two methods of breeder seed synthesis of the bromegrass cultivar Saratoga. Polycross seed was produced in vegetatively established nurseries at two sites in the states of California and Washington. Two breeder seed lots were synthesized using (1) equal parts of pure, live seed from each of five parent clones, and (2) portions equal to relative seed yields. These lots, plus similar blends produced in New York State, were used to plant four isolated increase blocks under selected conditions in the western part of the United States in order to produce second generation seed harvested at early, optimum, and late maturity.

In general, no consistent trends were found to indicate that method of synthesizing breeder seed or location of production affected the population characteristics as measured by heading date, time of anthesis, plant height, and rhizome spread after two years. Maturity of seed at harvest had no effect on plant characteristics. Based on these results, these workers concluded that even though the parental clones were significantly different, the Saratoga bromegrass synthetic, and perhaps other bromegrass synthetics, would be very stable under differing management and environmental conditions.

Breeder seed for hybrid corn is produced for single-parent materials such as inbred lines, which can be maintained by self-fertilization. Seed of a single cross used in the production of a double cross, a three-way cross, or a top cross becomes foundation single cross. There are no generation limits for breeder or foundation classes.

CIMMYT (1984) suggested four methods for producing breeder seed of an open-pollinated maize cultivar:

1. *Bulk Fertilization.* F_2 bulk seed produced by open-pollinating eight to ten superior families is grown to obtain about 8,000 plants. About 3,000 plants are selected that fit the phenotypic description for the cultivar. The pollen of selected plants is bulked and used to pollinate these plants, which are harvested to obtain the progenitor of breeder seed by selecting about 500 plants with the kernel and ear characteristics of the cultivar. A bulk sample consisting of an equal quantity of seed from each of these ears becomes breeder seed. To obtain the required amount of breeder seed, the number of ears selected as progenitors and the quantity of seed sampled from them can be manipulated.

2. *Isolated Mass Selection Plot.* In an isolated field, F_2 bulk seed is planted, which is produced by combining eight to ten selected families to obtain about 8,000 plants from grid sections of the entire field. Undesired plants are detasseled before pollen is released, and before harvest, selected plants are marked in each grid. At harvest, about 500 ears are selected to serve as progenitors of breeder seed and are obtained by uniformly sampling each of the ears.

3. *Isolated Bulk Planting Converted into Half-Sib Crossing Block.* About 12,000 plants are established from F_2 bulk seed rows from eight to ten selected families grown in isolation. Rows are arbitrarily designated male and female in a ratio of one male to two or three female rows, and all female rows are detasseled prior to pollen release. Off-types in the male rows are rogued out. Prior to harvest, each female row of 16 plants is treated as a grid, desirable plants are identified in each grid, and about 500 ears are harvested as progenitors of breeder seed.

4. *Isolated Half-Sib, Ear-to-Row Crossing Block.* A plot from individually shelled F_2 ears is established in isolation or seed is used from the individual ears saved as progenitors of breeder seed in any of the previous three methods. First, about 500 selected half-sib ears are shelled individually. These are planted as female plants in rows adjacent to male rows in an acceptable ratio, with male rows produced by compositing equal quantities of seed from each half-sib ear. Just before flowering, all female rows are detasseled, as are about 20 to 40% of the plants in the male rows, which may be off-type or not true-to-type. In the second step, about 50% of the female rows that meet the description of the cultivar are selected, using the male rows as checks to select the female rows. From each of the selected female rows, two to four ears are selected to obtain about 500 ears as progenitors of breeder seed.

If a larger quantity of breeder seed is required, additional true-to-type ears from the selected plants in the selected families may be saved in any one of the four systems described.

CONSIDERATIONS IN SEED MULTIPLICATION FROM BREEDER SEED

Seed of each cultivar is sold as a commodity capable of producing a describable and identifiable population. The genetic composition of a cultivar, however, may shift over time for a number of reasons, and to ensure that seed is accurately named and described, national systems of seed certification have been developed. Shifts in genetic composition may occur for the following reasons:

1. A series of minor natural mutations at low frequency with the potential to increase over several generations. **Fatuoids** are a common mutation in some oat cultivars, but usually these can be rogued out.
2. The selective influence of disease, which may reduce the yield of selected plants, notably in heterogeneous and heterozygous forage populations.
3. Natural outcrossing, which may take place even in naturally self-fertilized species. Interaction of the cytoplasm with the environment may induce a low level of male sterility in selected florets that become outcrossed. If the two parents differ genetically, variation will be found in the progeny.
4. Mechanical mixtures that can result if seeding, harvesting, storage, and cleaning equipment are not completely free of foreign seeds.
5. Developmental variation that arises as a differential growth response in diverse environments. A cultivar may respond to different soil, different fertility, and minor environmental variations associated with different photoperiods or elevations. Specific genotypes may be favored or suppressed.
6. Minor genetic variations that may occur during seed production. In cross-fertilized

crops, genetic recombination among heterozygous and heterogeneous progeny may result in genetic shifts. In self-fertilized crops, bulked samples might exhibit some minor genetic variation, although the population appears phenotypically uniform.

Shifts resulting from one or more of these factors have been noted by producers, who describe such cultivar deterioration as "running out." A cultivar resistant to disease was described as having run out when new races of obligate parasites were able to attack. Seed certification programs have served a useful role in educating producers about the temporary nature of major gene or vertical resistance to highly obligate parasites.

In the cotton crop, which cross-fertilizes to some degree, minor genetic variability is a constant feature of seed increase. A system of mass selection may be used to maintain uniformity. Lewis (1970) noted that mass selection procedures for maintaining cotton cultivars are based on the assumption that the population derived from plants that are allowed to reproduce will be equal or superior to the population that might be generated if all plants were permitted to contribute to the ensuing generation. Any system of selection, however, will result in changes in gene frequency in a population, which will vary with selection pressure. Removal of off-types may result in the smallest change of gene frequencies. Both mass selection and off-type removal should tend to increase frequencies of those genes that contribute the most to development of the phenotype and that are regarded as representing the cultivar under maintenance.

In forage crops, the recombination of selected clones or strains to produce a synthetic cultivar is an effective method of utilizing stocks with superior characteristics. Breeder seed (Syn 0 generation) is used to produce F_1 (Syn 1), but this small amount of seed must be increased by random mating to the Syn 3 or Syn 4 generation before it is available to the commercial producer. The maximum performance can be expected in the Syn 1 generation. A decline in subsequent generations is dependent upon yield performance of parental strains or clones, number of parental clones or strains combined, yield of F_1 crosses, and degree of natural cross-fertilization with other cultivars. If natural selection is not a factor, the yield of a synthetic cultivar should not decline beyond the Syn 2 generation. If natural selection is a factor, the cultivar may be maintained in a limited-generation program.

When heterogeneous and heterozygous species are grown outside their region of adaptation, opportunity exists for shifts away from selected characteristics. Bula *et al.* (1965, 1969) noted that when red clover (*Trifolium pratense* L.) was grown in southern locations, a shift occurred toward earliness and susceptibility to winter damage. Taylor, Dade, and Garrison (1966) suggested that early-flowering genotypes, particularly in southern locations, produce the most seed, thus shifting the cultivar toward an earlier type. Differential survival of genotypes due to diseases in diverse environments may be a factor, particularly in old stands. Other factors, such as lack of attention to previous crops grown, volunteer seedlings, inadequate isolation, and seed mixtures in harvesting and cleaning equipment, may be important.

The effect of environment during seed production on seedling vigor of two alfalfa cultivars was reported by Walter and Jensen (1970) (Figure 23.3). Evidence was found of a cultivar x environment interaction.

Sterling, Johnston, and Munroe (1977) reported that barley yields were influenced by seed source, but this may be a reflection of contamination of seed by pathogenic organisms because

Figure 23.3   The seed pods on this alfalfa branch are profuse, and seed set is high. Some regions are notably superior for seed production, but when seed is produced outside of its region of adaptation, the questions of genetic shifts and the impact of environment arise.

The impact that environment has on parental plants and subsequently on the performance of progeny is relatively unexplained. Environment may account for phenomena related to seedling growth and plant vigor, but environment has no effect on the genetic makeup of the progeny, except through selection. During seed maturation, environmental conditions may influence seed dormancy by influencing seed coat permeability and/or proper enzyme or biochemical reactions.

Alfalfa seeds produced under cool temperatures (6° C at night and 21° C during the day, or 43 to 70° F) were heavier and had a higher percentage of hard seeds than those produced under hot conditions of 16° C at night and 32° C during the day (61 to 90° F) (Walter and Jensen, 1970).

In the cultivar Ranger, seed produced under cool temperatures produced larger seedlings. Seedling vigor of Ranger seed increased with decreasing soil moisture. Temperature and soil moisture did not affect the vigor of the two alfalfa cultivars Moapa and Ranger equally. Recognition of a cultivar x environment interaction may help explain why seed characteristics such as germination and seedling vigor vary not only from year to year but also from location to location in the same year.

seed treatment reduced the range of variation in yield among sources. Yield increases of 8% or more were possible through seed treatment.

Cultivar maintenance involves retention of the cultivar without genetic change, which generally is interpreted as deterioration. Some programs are designed to improve the cultivar during seed increases. If a cultivar is to be maintained with the least chance for genetic change, the most effective procedure is storage of breeder seed, which is brought out as needed to start subsequent cycles of seed increase. For popular cultivars, storage of an initially large quantity of seed may not be feasible because storage costs might be too high. In addition, seed aging was shown to cause genetic shifts in a mixture of eight snap beans (*Phaseolus vulgaris*) (Roos, 1984). Allowing the germination of a population to fall to 50% or less, coupled with the use of small sample sizes, can result in elimination of genetic components within five generations, thus significantly reducing the genetic variability within the population.

Other systems involving roguing of off-types are not intended to improve the cultivar but to maintain it. This objective assumes that the removal of off-types will have a minimal effect on gene frequencies.

Seed of perennial grass and forage legumes is most efficiently produced in specialized seed-production areas in western Canada, the western United States, and Israel, often under irrigated conditions. When seed is produced out of the area of adaptation, care must be taken to prevent genetic shifts before the seed is returned to its area of adaptation.

Attention to the following points will help preserve the original identity of a cross-fertilized cultivar (Smith, 1956):

- Minimize genetic changes that may occur in successive seed generations by a periodic return to foundation stock seed.
- Assure that isolation is adequate to minimize cross-fertilization from adjacent fields of other cultivars of the same species.
- Limit the number of seed harvests that are taken from a single field to avoid disappearance of plants with increasing age of stand.
- Attempt to control volunteer stands arising from shattered seed.
- Do not use land planted to the same forage crop species for a number of years preceding the seed crop to avoid contamination with plant survivors.
- Avoid differential seed-producing capacity of plants within the cultivar.

When a synthetic cultivar is produced from asexually propagated clones, clones are maintained so that the synthetic may be reconstituted at regular intervals.

PEDIGREE, OR SEED CERTIFICATION

The purpose of seed certification is to provide guidelines for seed multiplication of various crops. Guidelines have been formulated by federal government seed inspectors, administrators responsible for the Seeds Act, university personnel, regional research and regulatory specialists, representatives of the seed trade, and practical growers. If all the requirements for pedigreed seed production are met, certificates are issued that show the pedigree status (Figure 23.4).

The pedigreed seed system operates on a limited-generation basis. A limited number of generations can be produced before returning to the original breeder seed to maintain the genetic purity of the cultivar.

Three classes of maize seed are recognized in Canada: breeder, foundation, and certified. There is no generation limit for breeder or foundation classes. Certified status may be granted to the first generation of single cross, double cross, three-way cross, top cross, and varietal crosses.

Hybrid seed production has been accepted fully in North America, with an associated elaborate infrastructure of seed production, storage, and distribution (Figure 23.5). The development of an F_1 hybrid seed industry has lagged in some developing countries because of the lack of organized seed production, but considerable progress is being made in several countries (Figure 23.6).

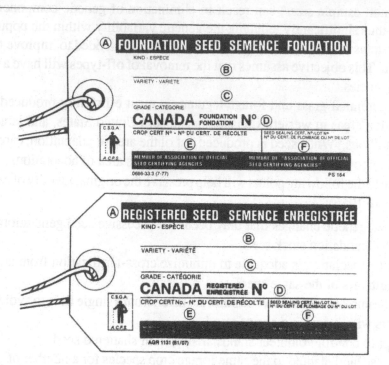

Figure 23.4   The identification tags in this figure are the official Canadian seed-sealing tags placed on each bag of seed. Certificates are issued for crop kinds authorized for sale in bulk. The tags are color-coded and provide the following information:

A. Class of pedigreed sale. Up to five classes exist: breeder, select, foundation (white), registered (purple), and certified (blue).
B. Crop kind; for example, lentils, soybeans, flax.
C. Cultivar (variety).
D. Grade and class of seed (as designated in the Canada Seeds Act).
E. Crop certificate number; for example 89-6000101-21 assigned by the Canadian Seed Growers' Association. The first two digits represent the year in which the crop was produced. The third to ninth digits inclusive represent the identification number of the grower who produced the seed. The tenth digit represents the code for the generation or class issued to the crop that produced the seed. The eleventh digit represents the number of different certificates of the same status issued to the grower.
F. Seed sealing number issued by Canada Agriculture for each distinct lot of seed that has been graded.

Photo courtesy Canadian Seed Growers' Association.

An important reason for seed certification is to assure cultivar identity, genetic purity, and trueness to type (Figure 23.7). Red Turkey wheat, which formed the cornerstone of the hard red winter crop in North American about 1870, had numerous names. Likewise, Mandscheuri barley, which was imported from Russia into North America in 1889, was known by numerous names. Fultz wheat, distributed in 1871 in the United States, was reported under 24 names (Parsons, 1985). An oat cultivar introduced in 1895 as Silvermine was eventually grown under 18 names. Seed certification was established to correct this situation.

A.   B.

Figure 23.5   Some aspects of F_1 hybrid seed production as exemplified by the North American hybrid maize seed industry are illustrated in these photographs. F_1 hybrid seed production is carried out by independent seed growers, producers, processors, and distributors. In the United States, the hybrid maize seed industry produces over 350,000 tons of seed for over 32 million hectares (79 million ac.) annually. The industry is composed of over 500 producers who grow seed on farms ranging in size from 2 ha (5 ac.) to over 400 ha (1,000 ac.). The hybrid maize seed maize industry grew from an infant industry in 1930 to one that supplied seed for 50% of the seeded maize area ten years later. Today it supplies seed for 100% of the seeded area. The effectiveness of this well-established but loosely defined industry is illustrated in sorghum. Hybrid sorghum acreage in the United States increased from no acreage planted to hybrid seed in 1954 to 100% planted to hybrid seed in 1960.

Alternating male and female rows are evident in Photo A. Independent seed growers produce such fields under company supervision using clearly marked seed bags. The seed is produced, dried, shelled, cleaned, sized, treated, and stored. A large refrigerated warehouse is shown in Photo B and is an integral part of the infrastructure needed for commercial exploitation of F_1 hybrid seed. Seed quality, as measured by germination, is preserved, and surplus seed required in case of seedling failure can be carried over under controlled-atmosphere conditions. The need for saving and storing reserve seed to guard against losses resulting from crop failure is well recognized.

Photo A courtesy Pfister Hybrid Corn Co., El Paso, Illinois.

Seed producers must be familiar with procedures for seed production in specific countries, regions, or crops and must take into consideration the following aspects of pedigreed seed production:

1. *Land.* Crops should not be seeded on land where volunteer growth from a previous crop may cause contamination. For alfalfa, birdsfoot trefoil, clovers, crown vetch, and sainfoin, crops planted with breeder seed for foundation status must not be planted on land that in the preceding five crop seasons grew a crop of the same kind. For certified status, the time limit is two years. For cereal crops, the same crop of a different cultivar cannot have been grown the previous year.

2. *Seed requirements.* Seed sown must be eligible to produce in the next generation a class of seed that can be verified by a crop certificate. In cereals, for example, breeder seed

Figure 23.6   Grain sorghum, shown in this photo, commonly is grown under harsh conditions of drought, and hybrids are frequently higher yielding and more stable than ordinary cultivars. Although hybrid sorghums were introduced into India in 1965, only about 10% of the current crop consists of F_1 hybrids. The problem of seed production remains a major constraint in the use of hybrids. A basic concern is how to provide quality seeds to Indian farmers, who traditionally have relied on their own seed for the next crop.

The production of F_1 hybrid seed presents new challenges to traditional farmers. Factors such as plant isolation on small farms, experience, and attention become prohibitively expensive and inefficient. The majority of traditional farmers therefore continue to rely on conventional seed supplies.

At the 1962 FAO (Food and Agriculture Organization) meeting of the Technical Committee on Seed Production, Seed Control and Distribution, the following summary statement was made (House, 1985, p. 155): "It was felt that, in developing countries, seed production and seed distribution should gradually be taken over by non-governmental agencies, such as seed cooperatives, seed growers' associations, and private firms; provided an adequate level of quality control could be maintained."

As a result of action taken in 1960 to develop an organization that would assure the rapid multiplication and distribution of cultivars of dependable, high-quality supplies of seed, India made remarkable progress in the development of a well-organized seed industry. A Seed Law was enacted, and an India Crop Improvement and Certified Seed Producers' Association formed.

---

may be used to produce select, select to produce foundation, foundation to produce registered, and registered to produce certified. A subsequent class of seed can be produced from any previous class. Certified, for example, can be produced from breeder, select, or foundation. More than one generation may be produced from each class. In cross-fertilized forage crops, seed may be multiplied through breeder, foundation, and certified with one generation in each class. Seed production may extend for several years on a single field, however.

3. *Crop inspection.* Each field to be pedigreed must be inspected by a qualified person.

4. *Isolation, roguing, and management.* Roguing to remove off-types must be done at frequent intervals during crop development. Isolation must meet specified guidelines, and adequate weed control must be practiced (Table 23.1).

5. *Record keeping.* Growers are responsible for records that include previous crop(s) grown

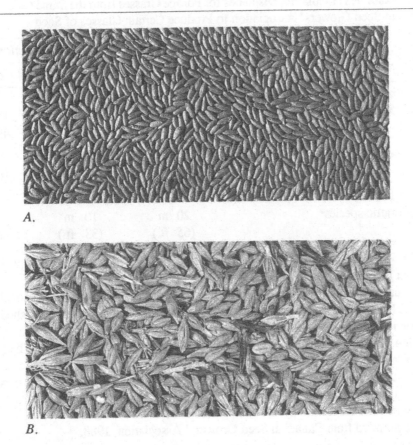

A.

B.

Figure 23.7   A comparison of the two seedlots shown in these photos illustrates the value of pedigreed seed. Photo A is a pedigreed sample of diploid rye, and Photo B is a bin run sample of six-row barley.

In addition to being genetically pure, pedigreed seed is graded and mechanically pure. Harvested seed for which a crop certificate has been issued is not considered pedigreed seed eligible for sale with a cultivar name unless it is processed, inspected, graded, and labeled according to the Seeds Act and Regulations. Impurity tolerances and number of weed seeds are clearly delineated. All crops for pedigreed status must be free of prohibited and primary noxious weeds and have an acceptable level of germination. Careful examination of Photo A will reveal some cracked or damaged seeds and one cotyledon of a soybean seed, but such variations are within acceptable limits of tolerance. In a foundation lot of forage grass seeds, foreign types or other crop kinds cannot exceed 0.1% of the plant population of the inspected crop. For certified status, the limit is 1%.

The classes of pedigreed seed in Canada are designated as follows:

1. *Breeder seed.* This is the seed produced by a plant breeder and is the first supply of a new cultivar. Breeder seed is not for commercial sale.
2. *Select seed.* This is seed produced from breeder seed by experienced growers.
3. *Foundation seed.* In cereals and other self-fertilized crops, and in cross-fertilized forages foundation seed is produced from select seed.
4. *Registered seed.* Normally, registered seed is produced from foundation seed.
5. *Certified seed.* Certified seed is produced from foundation or registered seed and is recommended for commercial use.

Photo courtesy Ontario Ministry of Agriculture and Food.

Table 23.1   Examples of Isolation Distances for Forage Grasses from the Same Crop Kind Required by the Canadian Seed Growers' Association to Produce Certain Classes of Seed

Mode of Fertilization and Field Size	Isolation Distance for Production of		
	Foundation	Registered	Certified
Cross-fertilized species			
2 ha (5 ac.) or less	400 m (1,312 ft.)	300 m (984 ft.)	150 m (492 ft.)
Exceeding 2 ha (5 ac.)	300 m (984 ft.)	100 m (328 ft.)	50 m (164 ft.)
Highly self-fertilized and apomictic species[a]	20 m (65 ft.)	10 m (33 ft.)	5 m (16 ft.)

[a] Slender wheatgrass (*Agropyron trachycaulum*) is a highly self-fertilized species; Kentucky bluegrass (*Poa pratensis* L.) is an apomictic species.

Note: If a crop in an adjacent field is beyond the control of the seed grower, removal of the border of the crop may be feasible in lieu of required isolation. The border must be allowed to shed pollen before being discarded in order to provide a pollen source for plants within the border and to serve as a buffer zone for pollen beyond the border. Border removal is not practical for fields of 2 ha (5 ac.) or less. Removal of 5 m (16 ft.) reduces the isolation distance for fields exceeding 2 ha (5 ac.) by 50% of that shown in this table. Isolation requirements for the certified seed class of alfalfa are based on the size of the certified field and the percentage of the field within the required 50 m (164 ft.) of another alfalfa cultivar. A portion of the field may not be eligible for certified status.

Source: Compiled from Canadian Seed Growers' Association, 1988.

on a field, size of field, crop certificates, off-types rogued out, and general agronomic information.

6. *Clean equipment.* All equipment used in the production, handling, and processing of pedigreed seed, including seeding and harvesting equipment, and seed cleaning and processing equipment, must be cleaned thoroughly before use to avoid contamination.

7. *Age of stand for perennial crops.* The class of seed that may be produced will vary by crop species, the number of classes designated by the plant breeder, and the age of stand. For most perennial crops, there is a specified number of years during which pedigreed seed may be harvested from one planting (Table 23.2).

8. *Grading.* All seed that is sold with a cultivar name must be processed, inspected, graded, and labeled according to the National Seeds Act and Regulations.

9. *Number of cultivars and kinds permitted.* Guidelines exist for the number of cultivars and kinds of crops that one seed grower can produce. For grasses and legumes, only one cultivar is allowed.

Select seed is produced from breeder seed and is grown by seed growers who have completed successfully three years of probation plot production in order to be granted select seed grower status. Select seed is grown in small plots to permit intensive roguing to remove off-types and abnormal plants (Figure 23.8). The select seed plot is the avenue for new growers

Table 23.2   Guidelines for Seed Production of Various Crops in Canada

	Established with Breeder Seed		*Established with Foundation Seed*
	Years for Foundation	*Years for Certified*	*Years for Certified*
Legumes			
Alfalfa	5 plus	3	8
Alsike clover	2 plus	2	4
Birdsfoot trefoil	4 plus	no limit	no limit
Crown and milk vetch	5 plus	3	8
Red clover, double cut	2 plus	1	3
Red clover, single cut	1 plus	1	2
Sainfoin	5 plus	0	5
Sweet clover	1 plus	0	1
White clover	2 plus	2	4
Zigzag clover	5 plus	3	8
Grasses			
Bluegrass, apomictic	4 plus	2	6
Bromegrass (smooth)	4 plus	4	8
Fescue (tall)	3 plus	3	6
Meadow foxtail	3 plus	2	5
Orchardgrass	3 plus	3	6
Reed canarygrass	4 plus	4	8
Ryegrass			
altai	5 plus	5	10
perennial	2 plus	1	3
Timothy	3 plus	2	5
Wheatgrass, slender			
(self-fertilized)	3 plus	2	5
Wheatgrass			
crested	4 plus	4	8
tall	4 plus	2	6

Note: Age of stand has an effect on class of seed produced for perennial forages, as shown here. Note that the class of seed that may be produced varies with crop species. The number of classes and the age of stand are designated by the plant breeder. When forages are grown outside the region of adaptation, the breeder can specify the limitations on size of stand. For calculating age of stand, the first seed crop is the first year in which a seed crop could normally be harvested, without regard for time or method of planting. Certified seed production guidelines must be flexible to meet changing conditions and circumstances. The possibility of hybrid seed may result in major changes. To overcome the limitations of propagating enough sterile plants of alfalfa to plant seed production fields, artificial seeds involving tissue culture and **somatic embryo** encapsulation could be used to clone male-sterile and female-sterile lines of alfalfa. These two lines would be analogous to inbreds in maize, and the seed that would be produced on the male-sterile plants after pollination by the female-sterile plant would be hybrid seed. This seed could be harvested and used directly for the establishment of forage production fields. Careful consideration would have to be given to how long male and female plants could be used to produce hybrid seed. The impact of somaclonal variation would require close attention.

Source: Compiled from Canadian Seed Growers' Association, 1988.

A.   *B.*

Figure 23.8   Photos showing select plots of spring wheat (Photo A) and soybean (Photo B). Both plots are planted to facilitate easy passage through the plot, which helps in effective roguing.

An individual seed grower who has completed successfully three years of probation plot production may be granted select grower status. Such growers may obtain breeder seed for select seed production, and normally five generations of production from breeder seed are allowed. Select seed obtained from another select seed grower may be used only for foundation, registered, or certified seed production. The area of a select plot of one cultivar may not exceed one hectare (2.5 ac.). Production of more than one cultivar of the same crop kind for select status normally is not permitted.

If a probationary seed grower meets all the requirement including isolation, roguing off-types, adequate weed control, and inspection on schedule, seed from a probationary plot is granted foundation status for the first two years. Sufficient seed is retained for second- and third-year probation plots and the balance used to produce registered or certified seed. In the third probationary year, select status may be granted and the seed used for further select or foundation production. A grower who completes the three-year probationary period becomes eligible to produce select seed of any crop having this class of seed.

to become established and recognized. Seed growers take great pride in producing a uniform field (Figure 23.9).

INTERNATIONAL ORGANIZATIONS

To facilitate the international movement of certified seed and to encourage the use of seed of consistently high quality, the Organization for Economic Cooperation and Development seed certification schemes (OECD) was established in 1966. There are five OECD schemes involved with cultivar certification of seed moving in international trade: (1) oilseed and forage seed, (2) cereal seed, (3) vegetable seed, (4) sugarbeet and fodder beet, and (5) maize.

Figure 23.9 This field of rye, a cross-fertilized cereal crop, has met all the requirements for pedigreed status. The cross-fertilized nature of rye means that air currents result in outbreeding, so that when two or more cultivars are grown in close proximity, they are difficult to keep pure for seed increase or testing purposes. Perhaps this is why breeding and selection for improved cultivars has been limited. There are fewer rye cultivars than in wheat or other cereal crops. Diploid and tetraploid forms of rye exist.

Seed set in tetraploid female parents is reduced when pollinated with diploid rye. When a tetraploid cultivar was tested among diploid yield trials in Minnesota, it yielded only about one-half to two-thirds as when grown in isolated plots (Koo, 1958), results comparable to those in Europe. The reason for such a marked reduction in seed set is related to the formation of triploid zygotes, which fail to develop into mature seed because of abortion or shriveled seed development. Tetraploid and diploid rye cultivars must be grown in isolation.

The following guidelines for pedigreed seed production of rye have been laid down by the Canadian Seed Growers' Association. Isolation of 300 m (984 ft.) is required between a pedigreed rye crop and crops of a different rye cultivar or nonpedigreed rye crops, whereas 3 m (10 ft.) of isolation is required for most other crops.

Note the high degree of uniformity exhibited in this field of rye.

The schemes are operated in each participating country by an authority from the government who takes responsibility for arranging approval from the country of origin, seed multiplication of foreign cultivars according to OECD rules and procedures, maintaining records, conducting field inspections, issuing certificates, labeling seed, and other associated duties. The OECD schemes recognize three categories or classes of seed, namely, prebasic or breeder seed, basic, and certified seed. Basic seed of forage cultivars is equivalent to foundation seed and is imported and multiplied to certified seed. Foreign unlicensed cultivars multiplied outside of their country of origin are produced for export purposes only.

In North America, the Association of Official Seed Certifying Agencies (AOSCA) grew out of the International Crop Improvement Association, formed in 1919. The purpose of the AOSCA is to coordinate, standardize, and establish minimum standards for genetic purity and to identify minimum standards for seed quality for all classes of pedigreed seed. Movement of pedigreed seed across country and state boundaries is facilitated because all member agencies recognize AOSCA seed as meeting the genetic standards for the class as labeled. The certification standards established by AOSCA are included in the Federal Seeds Act of the United States.

SEED VIGOR AND PLANT PERFORMANCE

In addition to ensuring the pedigree of the seed, attempts are usually made to measure the quality of the seed. One aspect of quality is the ability of the seed to germinate. Germination tests using standardized procedures are conducted routinely on all pedigreed seed samples and are widely accepted as an index of seed quality. In many crops, such as sunflower (*Helianthus annuus* L.), only the standard germination test is used to evaluate seed quality. Germination of seed under optimum conditions is insufficient to provide information on seed quality and related field performance. Germination tests do not reflect germination vigor, a transitory quality aspect that needs to be addressed in terms of tolerances and expiration date. Germination vigor was viewed by Spain (1976) as a significant criterion.

In any commercial production operation, crop yield potential is only as good as the plant stand. Environmental factors such as cold soil conditions, soil crusting, deep seeding, competition among seedlings and weeds, early seeding, soil organisms, and seedling diseases can affect performance of seed in the soil, so much so that it is surprising that seed germination is so reliable under such a wide range of conditions (Thomas, 1981). Plant stands are dependent on the viability and germination of seed.

Germination may be defined as the emergence and development from the seed embryo of those essential structures that are indicative of the ability to produce a normal plant under favorable conditions. Germination carried out under laboratory conditions may overestimate germination levels found under field conditions.

The concept of seed vigor evolved as seed technologists attempted to identify those seeds having the potential to produce normal, vigorous, healthy seedlings under less than optimum conditions. Those with seed vigor must be distinguished from those that lack vigor because of genetic makeup, seed deterioration, injury, or other causes. Seed vigor was defined by McDonald (1980) as those seed properties that determine the potential for rapid, uniform emergence and the development of normal seedlings under a wide range of field conditions. In attempting to measure and quantify seed vigor, it was recognized that many factors are involved, including environmental and nutritional aspects of the mother plant, genotype of the seed, stage of seed maturity at harvest, seed size and weight, or specific gravity, seed mechanical integrity, seed deterioration and aging, and pathogens or insects attacking the seed (Orzolek and Daum, 1984). These factors may impact on seed quality singly or in combination and make seed vigor evaluation difficult. To ensure seed quality and vigor, a rating of seed vigor shown on the label might prove useful but brings up several problems.

An understanding of seed quality and vigor may lead to seed treatments that enhance seed germination and vigor. Artificial seed development may be contingent upon a better understanding of seed vigor, and in turn, artificial seeds might enhance our understanding of true seed quality and vigor. A review of seed treatments to improve germination and emergence was provided by Heydecker and Coolbear (1977), who listed several treatment categories, as follows:

- Physical treatments to initiate germination before sowing, such as soaking seeds in gibberellic acid, indoleacetic acid, or kinetin.
- Application of nutrients, vitamins, magnetic fields, electrical treatments, and ultrasound treatments to increase tolerance to adverse environmental conditions such as drought, salinity, cold, or heat.
- Mechanical treatments of seed, such as pelleting, intended to facilitate sowing, and precision drilling, or scarifying, seed to eliminate hard seeds in forage crops.

To standardize seed vigor determinations a series of tests are involved. Some of these tests are described briefly in the following section.

SEED VIGOR TESTS

1. Electrical Conductivity of Leachate. Loss of cell membrane integrity by mechanical damage and/or aging and associated loss of cytoplasmic solutes can be measured by the electrical conductivity of seed leachate (Figure 23.10). Estimates of seed germination and vigor can be conducted in 24 hours or less (Hopper and Hinton, 1987).

2. Tetrazolium Test. One of the most widely used tests is the tetrazolium chloride test, based on the reduction of a tetrazolium salt to a red pigment (formazan) by dehydrogenase respiratory enzymes (Smith, 1952). Living tissue is turned red, whereas dead tissue remains colorless. With this test, it is possible to evaluate internal seed injury, insect damage, the destructiveness of frost, and the viability of dormant seed (Moore, 1973).

The tetrazolium test was suggested by Johnston *et al.* (1986) as a possible rapid technique for screening spring wheat genotypes for seedling growth at low temperatures. The formazan extraction method provides a quantitative measure of differences among genotypes. Wheat genotypes could be classified for low temperature seedling growth using as few as five to ten seeds per sample and possibly only two replications. Genetic variability identified by tetrazolium tests may be useful to plant breeders selecting genotypes for seedling vigor under low field temperatures. A four-day imbibition period at 7° C (45° F) appeared to offer the greatest efficiency for genotype screening.

3. Cold Test. The performance of maize seed under cool, wet soil conditions can be predicted in the laboratory with the cold test (Isely, 1950). This test is used routinely and is conducted by placing seed in soil watered to 70% saturation at 10° C (50° F) for seven days, followed by four days at 25° C (77° F). High vigor is indicated when 95% of the seedlings develop normally. For satisfactory field performance, 70 to 85% of the seedlings should develop normally. The cold test was not a consistently reliable predictor of early field emergence. The variability inherent in the test makes comparisons among laboratories difficult (Burris and Navratel, 1979).

A.   B.

Figure 23.10   Obtaining and selecting high-quality seed is extremely important in producing an adequate and uniform stand. Although the soybeans in these photos appear to be of excellent quality, accurate evaluation cannot be based on visual assessment but only on the results of various laboratory tests of germination and vigor conducted prior to planting.

One factor that determines vigor of seed is cellular membrane integrity. Cell membranes provide for compartmentalization of cellular and organelle constituents in viable seeds. Loss of membrane integrity and the subsequent loss of cytoplasmic solutes with electrolytic properties are indicative of rapidly deteriorating seed.

During the late 1970s, a commercial instrument, the Automatic Seed Analyzer, was developed to measure the electrical conductivity of seed leachate. This instrument measures the electrical conductivity of 100 seeds soaked in water, with each seed in an individual cell of the instrument. A negative relationship exists between electrical conductivity and the germination of soybean seedlots with germination values below 20% and above 80% (McDonald and Wilson, 1979).

Soybeans' cellular membrane integrity can be damaged by the beans striking a hard surface with force, resulting in accelerated aging (McDonald and Wilson, 1980). Electrical conductivity tests were reported by Miles and Copeland (1980) to correlate with soybean field emergence.

Photos courtesy Ontario Ministry of Agriculture and Food.

4.   *Accelerated Aging Test.* To predict longevity of seed in storage, seed samples are held at 41° C (106° F) for four to five days at nearly 100% relative humidity, followed by a germination test (Delouche and Baskin, 1973).

5.   *Seedling Growth Rate.* An indication of vigor may be obtained by measuring length or weight of shoot, root, or entire seedling after four or more days of growth in an appropriate germination medium. Methods may vary from placing seed in a rolled moist towel to providing

a six-week growth period in soil in a greenhouse. Short-term tests of four to five days may be influenced unduly by the length of time required for germination. Small seeds may germinate more rapidly than large seeds, thereby giving the impression that small seed has greater vigor. Growth subsequent to germination may be a more meaningful indicator of seedling growth rate.

6. *Speed of Germination Test.* Vigor evaluation for this test is based on the concept that more vigorous seeds germinate faster than less vigorous ones, assuming equal seed size.

7. *Seedling Vigor Classification Test.* An added dimension of standard germination tests is to classify seedlings visually as strong or weak. Valuable information on relative vigor of seed lots can be obtained easily by this test, despite its subjective nature.

8. *Brick-Grit Test.* Sterilized brick grit, which is heavy and acid in nature, is used as a 3 cm (1.18 in.) barrier to seedling emergence. Weak and strong seedlings are differentiated by their ability to penetrate this layer in 14 days.

9. *Mechanical Damage Test.* Internal bruises, fractures, or cracks in the seed coat can reduce seedling vigor and can be evaluated by several methods. External damage, such as breaks in the seed coat, can be detected visually with the aid of a dye or under low-power magnification. Internal damage may be detected by x-rays. The electrical conductivity test, which rapidly measures leachate from 100 individually compartmentalized seeds very accurately and objectively, offers a superior means of measuring mechanical damage.

10. *Seed-Size and Seed-Density Test.* Although seed size is relative and is not directly related to yield, knowledge of seed-size distribution within the lot of seed may be of value. Seed uniformity may be most desirable for equal competition.

In pearl millet, which is often established under stressful environmental conditions associated with uncertain rainfall and without benefit of advanced tillage and seedling technology, seed vigor to increase establishment capability is crucial to success of the crop. Lawan *et al.* (1985) divided pearl millet seed into three density fractions and three size fractions for field and laboratory evaluations. Seedling emergence increased from 40% with small, low-density seed to 62% with large, high-density seed. Results indicated that seed density and seed size were effective criteria of field establishment and seed quality. Density x size interactions implied a compensatory relationship: low-density seeds gave satisfactory emergence, provided they were sufficiently large; small seeds performed satisfactorily, provided they were high in density.

In laboratory tests with soybeans, seed size and seed density were not of sufficient importance to allow for significant seedlot improvement (Hoy and Gamble, 1985).

11. *Mitochondrial Efficiency.* Mitochondrial activity is calculated as the ratio of micromoles of ADP to microatoms of oxygen uptake during acceptor-stimulated respiration. Schneiter *et al.* (1974, 1976) reported significant differences in the ratio of conversion of ADP to ATP coupled with O_2 uptake (ADP:0) among mitochondria isolated from alfalfa seedlings. Significant positive correlations between ADP:0 and forage yield were reported. Mitochondrial efficiency of seeds was measured by germinating seeds in the dark for 96 hours at 21° C (70° F) prior to laboratory analysis.

Percent seedling emergence in the field was compared with laboratory vigor test results for sunflower (Anfinrud and Schneiter, 1984) to determine which laboratory method(s) most accurately evaluated seed-seedling vigor of sunflower. Seed leachate electrical conductivity tests, seedling vigor classification, germination at 10° C (50° F), and accelerated aging tests resulted in significantly higher correlations with percent seedling emergence in the field than did the standard germination test at 21° C (70° F).

PLANT BREEDERS' RIGHTS, OR PROPRIETARY PROTECTION

The initial efforts in plant breeding, described in Chapter 1, were largely a hobby. Over the years, since the discovery of Mendel's laws, increasing costs have made plant breeding less and less attractive and feasible as a hobby. As seed-producing and marketing organizations developed, the profession of plant breeding assumed ever larger dimensions. The demand for the seed of superior cultivars was a powerful stimulant for plant breeders to develop cultivars with further improvements and to multiply and market this seed. In general, the production of a new cultivar that has a desirable combination of characteristics distinguishing it from any other known cultivar and that is homogeneous and stable may require 10 to 15 years. To offset this associated high cost of plant breeding, funds from the sale of products were needed, and for private plant breeders this revenue was the only source of income.

A problem with selling seed of homozygous crops was that the crop producer could use a portion of his harvest as seed for the next crop and sell the remainder to other producers, thus becoming a competitor to the organization that had developed the new cultivar. The result was that developing organizations multiplied seed to obtain large quantities before launching it on the market, thereby delaying the benefits to producers. Plant breeders' rights offers a solution to this problem (Figure 23.11).

Patent legislation, parallel to that for inanimate material, was discarded initially because of the reproductive ability of animate plants, which lack absolute stability and exact repeatability. Cultivars are developed by the chance combinations of genes. Although hybridization may be conducted with known procedures, such chance combinations occur. However, despite the element of chance, superior combinations do result. An urgent need existed to rationalize public and private plant breeding, and movement of seed and other propagating material among countries, and to harmonize as much as possible the seed laws of nations engaging in seed trade.

It took over 60 years to develop protection for the products of plant breeding and to provide a means for reward. It was not until 1961 that the International Convention on the Protection of Plant Varieties (UPOV) was established in Paris to provide protection for crop cultivars. This protection is referred to as plant variety protection or plant breeders' rights. Such legislation occurs on a nation-by-nation basis but is designed to satisfy the minimum demands of the UPOV. Legislation exists in practically every country of the world with a significant trade in seed. In 1988, 18 countries[1] were members of the UPOV (Mastenbroek, 1988; Overton, 1987).

The intent of plant breeders' rights legislation is to grant legal recognition to plant breeders and provide an opportunity to obtain a return from the release of cultivars. Revenue from royalties provides a potential means of defraying the long-term investments required and could act as a stimulus to plant breeding. This potential is achieved, however, only if improved

---

1. Belgium, Canada, Denmark, France, Germany (Federal Republic of), Hungary, Ireland, Israel, Italy, Japan, the Netherlands, New Zealand, South Africa, Spain, Sweden, Switzerland, the United Kingdom, and the United States of America.

A.

B.

Figure 23.11 In self-fertilizing crops, such as the soft white wheat shown in Photo A and including most cereals, peas, beans, and some horticultural crops such as lettuce and tomato, it is easy for the purchaser of seed of a new cultivar to arrange the growing and sale of succeeding generations of seed. In the case of some cross-fertilizing plants, such as rye, oilseed rape, and many grasses and legumes, the situation is similar.

Traditional plant breeding methods involving the hybridization of large numbers of crosses, selection, and possibly backcrossing may produce agronomically superior cultivars that are homogeneous and stable. Such a process can take up to ten or fifteen years. Although superior cultivars may generate enormous income for society, plant breeders have to face the serious problem of covering their breeding costs.

Figure 23.11 continues overleaf

cultivars that truly are superior are produced and sold. Licensing procedures and official tests for recommendation purposes protect the producer from less-than-superior cultivars. In no country does the legislation prevent any individual from reproducing seed for on-the-same-farm use as seed or from selling such seed for food, feed, or any purpose other than for propagation.

The public benefit from plant breeders' rights derives from the increase in plant breeding and associated cultivar improvements.

Plant breeders' rights do not restrict anyone from using a protected cultivar as an initial source of variation for the purpose of developing other cultivars or from marketing such cultivars. National plant breeders' legislation, regardless of membership in UPOV, also includes a clause of similar intent, thereby ensuring that scientists, plant breeders, and developing nations, whether or not they are members of UPOV, may use any protected cultivar for research and evaluation in developing improved cultivars. Free access to protected cultivars for purposes of plant breeding to develop new selections has stimulated plant breeding considerably and has greatly enhanced the distribution of important properties and genes from one breeder to another and from one country to another.

To be eligible for protection, a new cultivar must fulfill specific requirements, which include the following:

1. *Distinguishability*. To avoid problems of confusing one cultivar with another, a cultivar for which protection is sought must be distinct from any other cultivar commonly known at the time of application. Identifying characteristics must be reliably discernible and exactly describable. Such characteristics may be morphological, physiological, pathological, or chemical in nature.

2. *Stability*. A named and protected cultivar should be identical from one production cycle to another, which means that care must be taken in preserving the original inbred parents of a hybrid, the clones of a synthetic cultivar, and breeder seed of a cultivar.

3. *Homogeneity*. Uniformity, or homogeneity, is a requirement, although this may vary among self-fertilized and cross-fertilized crops. Vegetatively reproduced cultivars may have stringent yet easily satisfied requirements. In contrast, a cross-fertilized cultivar reproduced by seed may not be as uniform but must show a degree of uniformity that can be described and maintained. Homozygous self-fertilized crops tend to be uniform, but mutation or outcrossing can produce variation. The requirement of "sufficient uniformity" is subject to interpretation.

4. *Novelty*. To qualify for protection, a new cultivar should be novel. Novelty may be interpreted as superiority in yield, quality, disease resistance, hardiness, and other

---

Figure 23.11 continued

Seed of advanced lines of potential new cultivars, produced at isolated regional sites such as in the test in Photo B, can be secured easily by unscrupulous vendors.

In order to protect the interests of plant breeders and provide due reward for the effort and investment involved, various national plant breeders' rights laws have evolved to give special protection for plant cultivars. Such systems have encouraged private plant breeding, for without some form of protection, self-financed breeding of these plants is hardly possible because the costs are too high.

Photo A courtesy Ontario Ministry of Agriculture and Food.

agronomic traits that make it valuable for producers. The intention of the novelty requirement is to prevent protection being granted to cultivars that had been bred and marketed before plant breeders' rights legislation was enacted.

5. *Denomination*. Each cultivar must be given a name or number. The plant breeder generally makes up a suitable name.

An invention in biotechnology may be patentable but also faces these criteria (Overton, 1987). Advances in biotechnology raise the prospect of generating new or improved types of plants by applying microbiological techniques in genetic manipulation. Such techniques have more in common with biochemical or chemical processes as applied in the pharmaceutical and agrochemical industries than with traditional plant breeding.

In a precedent-setting case in 1980, the Supreme Court of the United States held that microorganisms were inherently patentable by virtue of being living cells and allowed a patent to be granted for a new bacterium (Overton, 1987). In 1985, the U.S. Board of Patent Appeals and Interferences held that plants can be patented also. In contrast, the European Patent Convention in 1978 specifically stated that European patents shall not be granted for plant cultivars, for animals, or for the biological processes in their production. As controversy over intellectual property rights increase, clear definition of procedures become necessary (Belcher and Hawtin, 1991; ASA Special Publication, 1989).

The minimum period of protection prescribed by the UPOV is 15 years, and 18 years for grapes and forest and ornamental trees. Individual member states may establish their own period of protection, which generally varies from 20 to 30 years. After the protection period is over, the cultivar may be multiplied and commercialized without any royalty payment. Historically, the life of a cultivar seldom extends beyond 20 years because the development of new and superior cultivars renders older ones obsolete.

The holder of the right to a cultivar is obliged to provide the market with propagating material in quantities sufficient to meet market demands and at reasonable prices. Plant breeders' rights do not make it legal to withhold seed in order to inflate prices.

All breeders, whatever their background, can enjoy plant breeders' rights protection if their cultivar satisfies all the requirements. This has the beneficial effect of encouraging private enterprise or private citizens to find and develop superior genotypes.

The need for plant breeders to screen a large number of genotypes and the promise of royalties may encourage systems of cooperative breeding between a plant breeder and a grower with a keen eye and a desire to have the opportunity to select individual plants from a segregating population. A segregating F_2 population may be space planted on a cooperating grower's field with specific instructions to select plants with agronomically desirable traits. The involvement of a number of growers, each handling one population, could result in the screening of many crosses for superior segregates.

A farm grower may prove to be more demanding than a plant breeder who sees potential in almost every segregating plant he produces. For example, plants with distinct genetic traits may be saved for genetic studies by a graduate student; an interesting ideotype may be selected for testing under a unique production system; a selection with one or more superior traits but lacking other traits may be saved for backcrossing. Plant breeders love all their plants, and for them discarding is a difficult task. Farm producers who are interested in on-farm performance may exert strong selection pressure, and when large populations are involved, this may prove highly successful. It is possible that plant breeders' rights may open new avenues to plant breeding by encouraging cooperation that allows for large numbers of plants to be screened.

REFERENCES

Anfinrud, M. N., and A. A. Schneiter. 1984. Relationship of Sunflower Germination and Vigor Tests to Field Performance. *Crop Science* 24:341-344.

ASA Special Publication. 1989. *Intellectual Property Rights Associated with Plants.* Crop Science Society of America, American Society of Agronomy, Soil Sciences Society of America, Madison, Wisconsin. 206 pp.

Belcher, B., and G. Hawtin. 1991. *A Patent on Life: Ownership of Plant and Animal Research.* International Development Research Centre, Madison, Wisconsin. 40 pp.

Bula, R. J., R. G. May, C. S. Garrison, C. M. Rincker, and J. G. Dean. 1965. Comparisons of Floral Response of Seed Lots of Dollard Red Clover, *Trifolium pratense* L. *Crop Science* 5:425-428.

―――. 1969. Floral Response, Winter Survival and Leaf Mark Frequency of Advanced Generation Seed Increases of "Dollard" Red Clover, *Trifolium pratense* L. *Crop Science* 9:181-184.

Burris, J. S., and R. J. Navratel. 1979. Relationship Between Laboratory Cold-Test Methods and Field Emergence in Maize Inbreds. *Agronomy Journal* 71:985-988.

Canadian Seed Growers' Association. 1988. *Regulations and Procedures for Pedigreed Crop Production.* Circular 6-88. Ottawa. 65 pp.

CIMMYT. 1984. *Development, Maintenance and Seed Multiplication of Open-pollinated Maize Varieties.* International Maize and Wheat Improvement Center, El Batan, Mexico. 11 pp.

Delouche, J. C., and C. C. Baskin. 1973. Accelerated Aging Techniques for Predicting the Relative Storability of Seed Lots. *Seed Science and Technology* 1:427-452.

Heydecker, W., and P. Coolbear. 1977. Seed Treatments for Improved Performance Survey and Attempted Prognosis. *Seed Science and Technology* 5:535-425.

Hopper, N. W., and H. R. Hinton. 1987. Electrical Conductivity as a Measure of Planting Seed Quality in Cotton. *Agronomy Journal* 79:147-152.

House, L. R. 1985. *A Guide to Sorghum Breeding.* International Crop Research Institute for the Semi-Arid Tropics (ICRISAT), Andra Pradesh, India, pp. 151-165.

Hoy, D. J., and E. E. Gamble. 1985. The Effects of Seed Size and Seed Density on Germination and Vigor in Soybean (*Glycine max* L. Merr.). *Canadian Journal of Plant Science* 65:1-8.

Isely, D. 1950. The Cold-Test for Corn. *Proceedings of the International Seed Testing Association* 16:299-311.

Johnston, W. J., H. A. Yusuf, C. F. Konzak, and J. D. Maguire. 1986. Tetrazolium Chloride Test for Spring Wheat Seedling Vigor at Low Temperatures. *Crop Science* 26:167-169.

Koo, F.K.S. 1958. Deleterious Effects from Interpollination of Diploid and Autotetraploid Winter Rye Varieties. *Agronomy Journal* 50:171-172.

Lawan, M., F. L. Barnett, B. Khaleeq, and R. L. Vanderlip. 1985. Seed Density and Seed Size of Pearl Millet as Related to Field Emergence and Several Seed and Seedling Traits. *Agronomy Journal* 77:567-571.

Lewis, C. F. 1970. Concepts of Varietal Maintenance in Cotton. *Cotton and General Economic Review* 47:272-284.

McDonald, M. B., Jr. 1980. Vigor Test Committee Report. *Association of Seed Analysts Newsletter* 54:37-40.

McDonald, M. B., Jr., and D. O. Wilson. 1979. An Assessment of the Standardization and Ability of the ASA-610 to Rapidly Predict Soybean Germination. *Journal of Seed Technology* 4(2):1-11.

————. 1980. ASA-610 Ability to Detect Changes in Soybean Seed Quality. *Journal of Seed Technology* 5(1):56-66.

Mastenbroek, C. 1988. Plant Breeders' Rights: An Equitable Legal System for New Plant Cultivars. *Experimental Agriculture* 24:15-30.

Miles, D. F., and L. O. Copeland. 1980. The Relationship of Vigor Tests and Field Performance in Soybeans (*Glycine max* L. Merr.). *Agronomy Abstracts*. American Society of Agronomy, Madison, Wisconsin. p. 111.

Moore, R. P. 1973. Tetrazolium Testing for Assessing Seed Quality. In *Seed Ecology*, edited by W. Heydecker. Butterworths, London, pp. 347-366.

Orzolek, M. D., and D. R. Daum. 1984. Effect of Planting Equipment and Techniques on Seed Germination and Emergence: A Review. *Journal of Seed Technology* 9:99-113.

Overton, J. M. 1987. Patents in Biotechnology and Plant Breeding. *Span* 30(1):17.

Parsons, F. G. 1985. The Early History of Seed Certification, 1900-1970. In *The Role of Seed Certification in the Seed Industry*, edited by M. B. McDonald, Jr., and W. D. Pardee. Crop Science Society of America Special Publication 10. American Society of Agronomy, Madison, Wisconsin, pp. 3-7.

Rincker, C. M., J. G. Dean, and R. G. May. 1984. Stability of "Saratoga" Bromegrass Populations from Different Breeder Seed Syntheses, Locations of Seed Production, and Seed Maturities at Harvest. *Crop Science* 24:233-236.

Roos, E. E. 1984. Genetic Shifts in Mixed Bean Populations. I: Storage Effects. *Crop Science* 24:240-244.

Schneiter, A. A., R. G. McDaniel, A. K. Dobrenz, and M. H. Schonhorst. 1974. Relationship of Mitochondrial Efficiency to Forage Yield in Alfalfa. *Crop Science* 14:821-824.

————. 1976. Mitochondrial Efficiency of Individual Alfalfa Plants as Related to Forage Yield. *Agronomy Journal* 68:511-513.

Smith, D. C. 1956. Progress in Grass Breeding. *Advances in Agronomy* 8:127-162.

Smith, F. G. 1952. The Mechanism of Tetrazolium Reaction in Corn Embryos. *Plant Physiology* 27:445-456.

Spain, D. F. 1976. Seed Vigor: A Seed Control Official's Viewpoint. *Journal of Seed Technology* 1(2):13-17.

Sterling, J.D.E., H. W. Johnston, and D. C. Munroe. 1977. Effects of Seed Source and Seed Treatment on Barley Emergence, Yield and Kernel Weight. *Canadian Journal of Plant Science* 57:251-256.

Taylor, N. L., E. Dade, and C. S. Garrison. 1966. Factors Involved in Seed Production of Red Clover Clones and Their Polycross Progenies at Two Diverse Locations. *Crop Science* 6:535-538.

Thomas, T. H. 1981. Seed Treatments and Techniques to Improve Germination. *Scientia Horticulturae* 32:47-59.

Walter, L. E., and E. H. Jensen. 1970. Effect of Environment During Seed Production on Seedling Vigor of Two Alfalfa Varieties. *Crop Science* 10:635-638.

Glossary

A-line The seed-bearing parent line used to produce hybrid seed that is male sterile.

Accession A genetic acquisition.

Acclimatized A physiological adjustment by plants to changed environmental or stress conditions.

Acentric fragment A piece of chromosome lacking a centromere.

Achene A seed with unfused seed coat and fruit wall that do not separate at maturity.

Additive gene effects Gene action in which the effects on a genetic trait are quantitatively enhanced by each additional gene, either an allele at the same locus or genes at different loci. Additive genes contribute to the phenotype without dominance within a locus and without epistasis among loci.

ADP Initials for adenosine-diphosphate, a complex sugar-phosphorus compound formed as a result of energy expenditure and loss of a phosphate group from the energy-rich ATP (adenosine-triphosphate) compounds.

Agamospermy A collective term used to describe essential mechanisms of seed development in apomixis.

Agriculture The science of transforming sunlight energy into products that can be stored and used by humans at a later date. Plants that photosynthesize are the primary converters and animals that consume the products of photosynthesis secondary.

Albino In plants, lacking chlorophyll pigmentation.

Alien germplasm Genes introduced from a wild relative or unadapted species.

Allele Genes that occur at similar loci of homologous chromosomes.

Allogamy Cross-fertilization.

Allogenotypic competition Competition within a segregating population in which different genotypes occur.

Alloploid A polyploid incorporating genetically different chromosome sets from two or more different species.

Allozyme Enzymes produced by genes at several alleles.

Alternation of generations The progression between the diploid and the haploid phase of the life cycle.

Amphidiploid A polyploid whose chromosome complement consists of the entire somatic complements of two species.

Anaphase A stage in the process of mitosis and meiosis where sister chromatids or homologous centromeres are separated by the contraction of spindle fibers.

Androgenesis Development of a pollen grain, the haploid male gamete, into an embryo and plant.

Aneuploid Having a somatic chromosome number that is not an even multiple of the haploid number; for example, n+1 or 2n-1.

Angiosperms Flowering plants whose seeds are enclosed within an ovary.

Anther culture *In vitro* culture of an anther in which the gametic tissue (i.e., microspores) can either produce a haploid callus or regenerate a haploid plant.

Anthesis The period when anthers shed pollen.

Anticodon A three-nucleotide sequence or transfer RNA that complements a messenger RNA codon in the formation of protein.

Apomixis Plant embryo development without fertilization.

Apospory A form of apomixis in which an embryo develops from a cell of the integuments or nucellus in the absence of a haploid embryo sac.

Artificial seed A single somatic embryo in a protective coating that mimics the size, shape, and physiological condition of true seeds.

Artificial selection Elimination of segregating plants by mechanical, chemical, or mechanical human selection.

Asexual reproduction A reproductive process without the union of male and female gametes.

ATP Initials for adenosine-triphosphate, an energy-rich, complex, sugar-phosphorus compound that provides energy for many metabolic reactions.

Autogamous Describing species in which self-fertilization is the norm.

Autogamy Self-fertilization.

Autogenotypic competition Competition within a cultivar where genotypes are identical or very similar.

Autoploids A polyploid arising through multiplication of the complete haploid chromosome set of a species.

B-line A term used in connection with hybrid seed production. The fertile counterpart or maintainer of the A-line. Lacks fertility-restorer genes.

Backcross The cross of a hybrid to one of the parents.

Backcross breeding A plant breeding system whereby repeated backcrosses are made to one of the parents of a cross, accompanied by selection for a specific character or characters.

Biomass Total dry weight produced by a plant.

Biotechnology A technology that modifies or changes the hereditary material of plants or animals such that they possess new or altered functions that may have enhanced value.

Biotypes (1) A physiologic race of a disease organism. (2) Individuals in a population with identical genotypes, which may be heterozygous or homozygous. Distinct races within a species.

Bivalent In meiosis, the unit composed of associated homologous chromosomes during synapsis.

Brand A commercial name applied to seed.

Breed A line having the character type and qualities of its origin.

Breeder seed Seed produced by the originating plant breeder or sponsoring agency, recognized as the initial source of seed for subsequent multiplication and commercial distribution. Mass selection may be used to establish and maintain breeder seed.

Bulbosum method The use of *Hordeum bulbosum* to produce haploids in cereals.

Bulk breeding Also known as the bulk population method. Involves the production of genetically diverse populations of self-fertilized crops in a bulk plot and subjecting them to natural selection and/or mass selection, followed by single-plant selection.

Callus tissue An unorganized, proliferative growth of plant cells *in vitro* on a defined growth medium.

Carcinogens Agents capable of inducing cancer.

Centgener method One of the earliest established pure-line systems of plant breeding based on 100 selected plants.

CentiMorgan Map unit for distance between genes on a chromosome. One centiMorgan is one map unit.

Centric fragment A piece of chromosome with a centromere.

Centromere Constricted area of a chromosome where spindle fibers attach.

Check cultivar A commercial cultivar or experimental strain with well-known characteristics and performance that normally is included for comparison purposes with other selections in all testing procedures.

Chemical hybridizing agent Compound applied to plants prior to anthesis to selectively induce male sterility.

Chiasma (plural chiasmata) Visible association of chromatids in prophase resulting in genetic exchange between bivalents (paired chromosomes).

Chimera An individual containing different types of cells.

Chloroplast Cytoplasmic body containing chlorophyll in which important reactions such as sugar or starch synthesis take place during photosynthesis.

Chromatids During prophase of mitosis and meiosis, two identical units (chromatids) held together at the centromere make up each chromosome. Chromatids become chromosomes when the centromeres divide and the chromatids separate.

Chromatin Readily stained material of the nucleus consisting primarily of nucleoprotein.

Chromosome elimination When crosses between *Hordeum vulgare* and *Hordeum bulbosum* are made, fertilization occurs but the chromosomes of *H. bulbosum* are rejected and a haploid plant is produced.

Chromosomes Structural units contained in the nucleus that carry genes in a linear, constant order. Chromosome number is typically constant in any species.

Cleistogamy Occurs when pollination and fertilization take place in an unopened flower bud, thereby preventing cross-fertilization.

Clonal lines Plants established from asexual propagules.

Clone All individuals asexually derived from an original parent.

Codon A three-nucleotide sequence, or messenger RNA, that signals a particular amino acid in protein synthesis.

Colchicine An alkaloid that induces polyploidy by arresting spindle formation during mitosis.

Coleoptile The protective covering of an embryonic stem growing through the soil following seed germination.

Competition When the immediate supply of a single necessary factor falls below the combined demands of a plant or plants, competition occurs. Competition may result in a retardation of plant growth, and if severe and repeated over a number of generations, it may reduce or eliminate certain genotypes.

Complementary competition Natural selection that sorts out superior genotypes of agricultural value.

Conservation tillage Seed bed preparation systems that have 30% or more of the residue cover on the surface after planting.

Conventional tillage Seed bed preparation involving primary and secondary tillage systems.

Correlation An association between two variables such that as an increase or decrease occurs in one, an increase or decrease occurs in the other. Correlation coefficients may range in value from -1 to +1.

Cross-resistance Resistance associated with a change in one genetic factor that results in resistance to different chemical pesticides that were never applied.

Crossing over Process of reciprocal exchange between chromatids.

Cultigen A cultivated genotype, as contrasted to wild species.

Cultivar A contraction of the words *cultivated variety*. To be classed as such, a cultivar must be distinct from other cultivars, the international equivalent of, and synonym for, variety.

Cytokinesis The division of the cytoplasm into two daughter cells after nuclear division.

Cytoplasmic inheritance Inheritance determined by nonchromosomal genetic factors, often in chloroplasts and/or mitochondria that often display maternal inheritance.

Deductive reasoning Begins with a generalization that infers a factual conclusion.

Deficiency The deletion or absence of a chromosome segment.

Dehiscence The rupture of anthers to shed pollen content.

Diallel cross The crossing of a series of genotypes in all possible combinations.

Dihaploid An organism produced by doubling the haploid chromosome number.

Dioecious Bearing male and female organs on separate plants.

Diploid The state when each cell or nucleus has each chromosome in two copies.

Diploidization In polyploids, stabilizing chromosome pairing to form bivalents similar to that in diploids.

Diplontic selection Selection in an organism having two sets of chromosomes in its somatic cells and a single, haploid set of chromosomes in its gametes.

Directional mutation A genetic change that favors a certain genotype in a population.

Disomic Involving two homologous chromosomes that pair readily.

Disomic polyploid As exemplified by the allopolyploid wheat, exhibits disomic genetics in which regular bivalent pairing occurs only between strictly homologous chromosomes.

DNA Deoxyribonucleic acid, the material of inheritance (genes comprise sequences of DNA). Consists of two strands linked together by base pairing and coiled into a helix, each strand comprising a large number of deoxyribonucleotides attached to each other in single file.

Domestication The modification of plants from independent survival to a form that is dependent on humans for survival. The loss of the seed shattering habit is associated with plant domestication.

Dominant Describes a gene that manifests at a locus over its alternative allele when in the heterozygous condition.

Donor parent The parent from which one or a few genes in a backcross program are transferred to the recurrent parent.

Double fertilization The union of one sperm nucleus from the pollen grain with the egg nucleus in the ovary to form the embryo, along with the union of the other sperm nucleus with the polar nuclei to form the endosperm.

Double helix A structure of DNA consisting of two helixes rotating about a common axis.

Double reduction Pattern of chromatid segregation after crossing over between gene and centromere.

Drift See Genetic drift.

Duplication The occurrence of a chromosome segment twice in a haploid set.

Elite An agronomically superior and high-performing local cultivar.

Emasculation Removal of the male organs (stamens).

Embryo culture *In vitro* development or regeneration of an isolated mature or immature embryo.

Embryogenesis The formation and development of the embryo.

Empirical process Based on observation and experience without a scientific understanding.

Environmental influences The sum total of external conditions that affect growth and development of an organism.

Enzyme electrophoresis Migration of proteins in a starch of polyacrylamide gel from a starting point at the base of the gel and across an electrified field. The amount of migration is dependent on molecular weight, charge, and three-dimensional structure.

Epistasis The interaction between nonallelic genes.

Equational division The second meiotic division that does not reduce chromosome numbers.

EUCARPIA Acronym for the European Association for Research in Plant Breeding.

Euploid Cell having one or more complete genomes.

Evolution Changes in the genetic composition of a population with the passage of each generation.

Evolutionary development The continuous genetic adaptation of a species to the environment, influenced by factors such as selection, hybridization, inbreeding, mutation, and humans.

Exine The outer coat of a pollen grain or spore.

Explant Living tissue taken from a plant and placed on a culture medium.

Facultative apomicts Ovules capable of development without sexual union. Species that can produce both sexual and apomictic seeds.

Fatuoids Spontaneous mutants in oat cultivars that resemble wild oat in that they have a typical basal cavity or sucker mouth, a prominent, twisted, knee-bent awn, and long hairs or bristles on the base of the lemma and on the rachella segment. At maturity, fatuoids shatter their seed.

Fertilization The fusion of male and female gametes to form a zygote.

Gametes Sex cells, which in plants are known as pollen (male) and the ovule (female).

Gametic union The combination of the egg and sperm cells to form a zygote.

Gametophyte The sexual form of a plant having the haploid (n) number of chromosomes that produces male and female gametes in the life cycle as opposed to sporophyte.

Gametophytic generation The haploid (n) phase of angiosperms that produces male and female gametes in the life cycle.

Gene That portion of the DNA molecule containing the instructions for the synthesis of a single protein. The basic unit of inheritance.

Gene banks Units to store and preserve genetic resources of the world.

Genetic drift A shift in gene frequencies associated with small numbers and random sampling. A random change from generation to generation.

Genetic engineering The use of specific DNA endonuclease enzymes to cleave known DNA sequences and the subsequent rearrangement of those DNA sequences to form new DNA combinations.

Genetic erosion The loss of heterogeneity as a result of overuse of narrow germplasm for breeding new cultivars.

Genetic transformation The altering of plants by insertion of one or several genes.

Genome The sum of all the genes carried by a single gamete.

Genotype The genetic constitution of an organism.

Germplasm The collective hereditary materials within a species.

Germplasm resources Any materials of potential use in breeding programs that are not themselves of immediate value as cultivars of the crop under consideration. Includes present and obsolete cultivars, landraces, breeding lines, wild or weed forms, and crop relatives.

Gibberellin A plant hormone associated with cell elongation, seed germination, flowering, and seed dormancy.

Gigas features Enlarged plant features, such as wider and thicker leaves, larger and sturdier stems, and larger flowers and seeds, often resulting from polyploidy.

Haploid The state when each cell or nucleus has one copy of each chromosome.

Hardy Weinberg equilibrium The allelic or gene frequencies reached in one generation by a random mating population given by the expansion of $(p + q)^2 = (p^2 + 2pq + q^2)$ where p and q represent the frequency of a gene.

Harvest index Grain weight as a percentage of total above-ground dry weight at maturity.

Heat units A measure of growing degree days accumulated during the growing season.

Hemizygous The state in which a diploid plant carries a gene(s) in a single dose.

Heritability, broad sense Estimation of heritability from the total genetic variance.

Heritability, narrow sense Estimation of heritability from the additive portion of genetic variance.

Heritability value A measure of observed variability caused by genetic factors as opposed to environmental causes and that is transmitted to the next generation.

Hermaphrodite An individual higher plant in which both sex organs are present in the same flower.

Heterochromatin Parts of the chromosome that stain more darkly during interphase. Heterochromatin contains repetitive DNA that is rarely transcribed.

Heterogeneous Referring to a population that may have individual plants that are homozygous but have a different genetic makeup.

Heterokaryon A cell containing the combined organellar traits from different sources.

Heterosis The superiority of an F_1 hybrid when compared with its parents.

Heterostyly Having ovaries with styles of different lengths.

Hexaploid An individual having six sets of the basic chromosome number.

Homoeologous chromosomes Nonidentical chromosomes that are derived from a common ancestor and that exhibit partial homology.

Homogeneous The condition in which all individuals within a population are similar.

Homologous chromosomes Members of a pair of chromosomes, essentially identical, that synapse during meiosis.

Homomorphy Genetic incompatibility regulated by specific genes.

Horizontal resistance Resistance conditioned by polygenes or quantitative genes.

Hybridization As applied to plants, involves the deliberate act of applying male pollen onto the female stigma to effect fertilization.

Hybrids Cultivars in which the F_1 populations are used to produce the commercial crop. F_1 cultivars may be produced from inbreds, clones, cultivars, or other populations (e.g., F_1 x F_1).

Hyperploid A cell or organism that contains extra chromosome(s) or segments of extra chromosomes.

Ideotype A hypothetical plant described in terms of traits that are thought to enhance genetic yield potential.

Ideotype breeding A method of breeding to enhance genetic yield potential based on modifying individual traits where the breeding goal (phenotype) for each trait is specified.

Inductive reasoning Begins with a specific set of observations and moves to a generalization.

Interaction The tendency of one factor to have different effects at different levels of another factor.

Intercrossing Random mating among cross-pollinated plants.

Intergeneric Involving two or more genera of plants.

Interspecific Involving two or more species of plants.

Introgression The transfer of a small amount of genetic information from one species or genus to another.

Inversion Change in a segment of a chromosome whereby the linear order of its genes are reversed.

Inviable Incapable of living under normal conditions.

In vitro Conducted in a defined environment such as a culture container of glass or plastic, or measurement of biological processes outside of the intact organism.

In vivo Conducted within a living organism.

Isogenic lines Genetically similar lines that differ by one gene only. Also known as isolines.

Isozymes Variations of an enzyme.

Landrace cultivar A cultivar developed and used for crop production before the modern era of plant breeding. It may be regarded as a primitive cultivar that became adapted to a local area.

Landraces Genetically variable traditional cultivars with sufficient genetic integrity to be morphologically identifiable. Landraces differ in adaptation and in their responses to cultural practices.

Line A genotype that has been progeny tested. Several may be blended to form a multiline.

Linkage The association of genetic characters on the same chromosome with different loci on that chromosome.

Locus (plural loci) The location of a gene on a chromosome.

M_1 **generation** Plants grown from seed treated with a mutagenic agent.

Mass selection Selection of plants based on a desired feature and with the progeny of selected plants grown in bulk.

Meiosis Reduction division whereby daughter cells are produced that contain nuclei with one-half the original number of chromosomes.

Metaphase A stage in the process of mitosis and meiosis in which spindle fibers are attached to centromeres and the chromosomes are positioned in the center of the cell.

Micropropagation Asexual *in vitro* production of plants using meristems or other plant parts as starting tissue, which provides an opportunity to produce plants that are disease free and that can be used for cell manipulation.

Micropyle A minute opening in the integuments found at the top of the ovule through which the pollen tube commonly enters.

Microspore culture *In vitro* culture of microspores, which results in the growth of a haploid callus or regeneration of a haploid plant.

Microtubules Components of spindle fibers made up of protein and appearing as hollow cylinders.

Mitochondria (singular mitochondrion) Membrane-bounded organelles in a cell that function in respiration (Krebs cycle) and where transport reactions take place.

Mitosis Nuclear division producing two daughter cells, each containing a nucleus identical to the original.

Model A computerized simulation of scientific events in crop development based on a series of logically assembled, justifiable beliefs about a system.

Monoecious Refers to plants bearing separate male and female flowers.

Monoploids Plants having only a single basic set of chromosomes.

Monosomic Refers to plants deficient in one chromosome of a pair and symbolized as 2n-1.

Multilines Blends or mixtures of cereal cultivars of one species that are generally very similar but differ in specificity of resistance or strains of a disease.

Multisomic A general term describing an organism lacking several chromosomes. A nullisomic is deficient in both members of a pair and is symbolized as 2n-2.

Mutagen An agent capable of causing mutation.

Mutation A sudden and heritable modification in a gene or chromosome.

Natural selection Selection that allows the forces of nature to exert pressure, such as winter kill of nonhardy plants.

Neolithic The last phase of the Stone Age that began 10,000 to 8,000 B.C. and that was marked by the domestication of plants and animals, the first steps in agriculture.

Nick Synchronization of the receptivity of the female organ to the maximum pollen load of the pollinator for cross-fertilization.

Nuclear gene A spherical repository carried in the nucleus containing the chromosomes bearing the genetic information.

Nucleolus A small body in the nucleus consisting of RNA.

Nucleotide A molecule composed of phosphate, sugar, and one of four bases that form the building blocks of DNA and RNA. The four bases act as the letters that form the words, or codons, that together form genes and that in turn form chromosomes.

Nucleus A spherical repository within a cell containing the chromosomes bearing the genetic information.

Null hypothesis An approach in statistical analysis that assumes treatments had no effect rather than trying to demonstrate that treatments had some effect.

Nullisomics Organisms lacking both representatives of a pair of homologous chromosomes.

Obligate apomicts Ovules only capable of developing without sexual union or plants in which sexual reproduction seldom occurs.

Open-pollinated maize In addition to describing a cross-fertilized plant, the term refers to an entire class of corn that stands in contrast to hybrid maize.

Organelle DNA Nucleic acid contained in chloroplasts, mitochondria, golgi structures, and endoplasmic reticulum.

Outbreeding Inducing cross-fertilization in a normally self-fertilized plant.

Panmictic population Plants in which random mating occurs.

Parallel variation Independent but similar evolutionary development from a common ancestor, with the result that many common traits may be found among many distantly related plants.

Parthenocarpy Production of fruit without fertilization.

Parthenogenesis Embryo development from an egg without fertilization and resulting in haploidy.

Pedigree method A selection procedure in a segregating population in which progenies of selected F_2 plants are reselected in succeeding generations until genetic homozygosity is obtained.

Pedigreed seed Seed of a named cultivar that is produced under the supervision of a certification agency to ensure genotype and purity.

Perfect flowers Flowers possessing both male and female parts, i.e., stamens and pistils.

Phenotype Physical or external appearance of an organism without knowledge of its genetic makeup or genotype.

Phenotypic selection Development of a cultivar based on its physical appearance without regard to its genetic makeup.

Photoperiod Length of day to which plants may respond.

Photorespiration A form of respiration stimulated by light, found in C_3 plants, and having no known essential function.

Physiological maturity The point in grain development at which no further increase in dry weight takes place.

Pin flowers Flowers with long styles and short stamens. Associated with heterostyly.

Pistillody The development of flowers bearing pistils but no stamens.

Plant breeders' rights Legal rights of ownership granted by government to the developer of a new cultivar. Rights refer to sale of the cultivar as seed, not to other uses, such as food or feed.

Pleiotropy A genetic situation in which one gene affects more than one trait.

Polycross Random mating among a group of cross-fertilized selected genotypes grown in isolation from other compatible genotypes.

Polycross nursery A group of selected clones or seed lines planted in isolation in a randomized replicated manner to facilitate random interpollination.

Polyembryony The production of two or more embryos from one egg.

Polyploid A plant with other than two genomes per cell. Plants may be monoploid, triploid, tetraploid, hexaploid, octoploid, and so on; includes aneuploids.

Polysomic inheritance Genetic combinations associated with a polyploid.

Polysomic polyploid Exemplified by the tetraploid potato. Exhibits tetrasomic genetics in which meiotic pairing can occur between any two or more of the four homologous chromosomes.

Position effect A phenotypic modification resulting from the arrangement of genes relative to each other.

Prophase The first stage of mitosis and meiosis in which chromosomes become visible and the spindle apparatus develops.

Protandry A mechanism that prevents self-pollination in which stamens mature and shed their pollen before the stigma of the same flower matures.

Protogyny A mechanism to prevent self-pollination in which the stigma matures and becomes pollinated by foreign pollen before the stamens of the same flower shed their pollen.

Protoplast The colloidal mass of protein contained in a cell that constitutes the living mass.

Pseudogamy Apomictic seed set induced by pollination but without fertilization. Incomplete fertilization in which the polar nuclei are fertilized by a sperm for endosperm development and seed viability.

Pure line Genetically pure or homozygous plants resulting from continued inbreeding or self-fertilization. All members of a pure line have descended from a single homozygous individual and may be produced by continued self-fertilization or by doubling of the haploid chromosome number.

Pyramid To incorporate several genes for a specific trait into one plant.

Qualitative traits Traits having precise and defined characteristics.

Quantitative traits Features influenced by a group of genes at different loci that are cumulative in their effect and hence show continuous variation.

Race A group of organisms with common ancestry and common characteristics. A subdivision of a species.

Ratoon The after harvest regrowth of a grain crop, arising from basal buds.

Recessive A gene of an allelic pair that is not expressed when a dominant gene is at the same allele.

Reciprocal recurrent selection A breeding system of repeated cycles of selection following crosses in which the male parent of one cross is the female parent of the second cross.

Recombinant An organism, cell, or plasmid containing new genetic information resulting from the *in vitro* insertion of DNA.

Recombination Formation of new gene combinations as a result of independent assortment or crossing over of the genes when two genetically different parents are hybridized.

Recurrent parent The parent in a backcross breeding program to which successive backcrosses are made.

Recurrent selection A breeding system with the objective of increasing the frequency of favorable genes of a quantitatively inherited characteristic by repeated hybridization and cycles of selection.

Reductional division The first meiotic division that reduces the number of chromosomes to half that of the original cell.

Regression A measure of the rate of change in the dependent variables per unit rate of change in the independent variables.

Restitution type Unreduced product of meiosis.

Restorer line A parent used to overcome cytoplasmic male sterility.

Restriction enzyme An enzyme that severs DNA at a specific nucleotide sequence.

Ribosomes Structures in the cell associated with protein synthesis.

RNA Ribonucleic acid. Consists of a large number of ribonucleotides attached together to form a single file, commonly existing as a single-stranded molecule. RNA is normally copied from DNA by the process of transcription and is responsible for decoding genes.

Roguing The removal of individual plants that deviate in a significant manner from the norm or average type for the cultivar.

S_1, S_2 Symbols for designating first selfed generation, second selfed generation, and so on. The S_0 symbol is used to designate the original selfed plant. S_1 progeny are produced by selfing an S_0 plant, S_2 progeny are produced by selfing an S_1 plant, and so on.

Screen The detection of a particular phenotype arising from the expression of a single gene or combination of genes.

Segregate Offspring resulting from the separation of paternal and maternal chromosomes at meiosis and recombination.

Selection In plant breeding and as a verb, to discriminate deliberately among individuals in the number of offspring contributed to the next generation. As a noun, it is an individual chosen among segregating offspring.

Self-incompatibility A physiological hindrance to self-fertilization controlled by Mendelian factors.

Semigamy Abnormal fertilization in which the male gamete fertilizes the egg but does not fuse with the egg nucleus.

Sib A close relative, brother, or sister.

Somaclonal variation Variation among plants regenerated from tissue culture, which may be epigenetic or genetic as a result of the tissue culture process.

Somatic Refers to a body cell normally containing one set of chromosomes from the male and one set from the female parent.

Somatic embryo A single body cell induced to develop into a whole plant.

Somatic hybridization Combining genetic traits of two plants by union of two body cells, usually by fusing plant protoplasts, often between plant species that are sexually incompatible.

Spindle fibers Microtubules that control chromosome movement during mitosis and meiosis.

Sporophytes Having the diploid (2n) number of chromosomes, resulting in plants with leaves, roots, and stems.

Sporophytic generation The diploid (2n) phase of angiosperms, resulting in a plant with leaves, roots, and stems.

Stock A genotype maintained as a preferred parent in a breeding program but which may not be released as a cultivar.

Strain A group of individuals of common origin, generally denoting a more narrowly defined group than a cultivar.

Sward A stand of forage grasses or legumes.

Synthetic A group of selected plants or cultivars allowed to intercross freely.

Synthetic cultivars Open-pollinated cultivars producing a seed mixture by cross-pollination in all possible hybrid combinations of a number of genotypes selected for good combining ability.

Systemic A chemical absorbed into the entire system or transpiration stream of the plant, thus spreading throughout the plant rather than being localized.

Tapetum Layer of cells encasing the pollen grains.

Telophase The concluding stage of mitosis and meiosis in which chromosomes uncoil, the spindle fibers disintegrate, and cytokinesis occurs.

Tester parent The pollen parent of a top cross, usually an open-pollinated cultivar or double-cross hybrid.

Tetrasomics Plants having one chromosome in four doses in their nuclei.

Thrum flowers Flowers with short styles and long stamens. Associated with heterostyly.

Tiller A secondary stems(s) arising from a basal node and capable of producing reproductive organs.

Top cross An outcross of selections to a common pollen parent. In maize, commonly an inbred x cultivar combination.

Totipotency The genetic potential of a single cell to give rise to a complete organism.

Transcription The initial step in protein synthesis whereby specific DNA segments are copied in the form of RNA.

Transformation Genetic transformation is the transfer and incorporation of DNA, especially recombinant DNA, into a plant cell from another species.

Transgressive segregation Individuals in segregating generations that exceed the parental range for a particular feature.

Translation The conversion of a messenger RNA sequence into a protein.

Translocation A change in the position of a segment of a chromosome either to another location in the same chromosome or to a different chromosome.

Triazine A group of chemicals with herbicidal properties. One such chemical, atrazine, is well known.

Triploid Refers to tissue having three times the reduced primary chromosome number.

Trisomics Plants having a chromosome number of 2n+1.

Type Refers to a phenotype or genotype that is distinctive.

Unconscious selection Nonintentional human selection. Synonymous with automatic selection.

Vegetative propagation Production of a new plant from a portion of another plant such as a stem or a branch. The offspring is identical to the mother plant.

Vernalization Exposure of metabolically active seeds or young plants to specific cool temperatures and short photoperiods, which are prerequisites for the development of reproductive organs.

Vertical disease resistance Resistance conditioned by one or a few qualitative genes.

Viroids Nonencapsulated RNA, capable of autonomous replication and pathogenic to plants; for example, the potato spindle tuber viroid.

Virulent Able to cause disease.

Viruses A group of ultramicroscopic infective agents.

Vybrids The first- and subsequent-generation progenies of crosses of heterozygous facultative apomicts.

Wide cross Hybridization of parents differing greatly in their genes. Parents may belong to the same species or to different species having the same chromosome number. The cross is considered very wide if parents have different chromosome numbers or belong to different genera.

X-rays Penetrating electromagnetic radiations having wavelengths shorter than those of ultraviolet or visible light.

Xenia Changes due to the effect of foreign pollen on visible characters of the endosperm resulting from the phenomenon of double fertilization.

Yield ability The capacity of a genotype to produce a very high yield under ideal conditions.

Yield stability The capacity of a genotype to perform reliably but not necessarily to give a high yield from year to year at a given location.

Zero-tillage A production method in which seed is planted into untilled soil.

Zertation Restricted or reduced growth speed of pollen tubes through genetically different stylar tissue.

Zygote The cell that results from the fusion of the male and female gametes.

About the Book and Authors

In response to the need for an integrated approach to plant breeding, this core text combines theory and practice in a highly illustrative manner. Theoretical chapters cover genetics, segregation ratios, and locating and incorporating parental material with desirable genes. Breeding systems, ranging from the newest methods of dihaploid development to older methods of mass selection, the pedigree system, backcrossing, recurrent selection, and hybrid development, are outlined and illustrated and their strengths and weaknesses discussed.

Representing the combined experience of three authors in self-pollinated, cross-pollinated, and molecular genetics, *Plant Breeding* has applications that extend beyond field crops to include horticulture and forestry. Further, the authors provide an international perspective through their choice of figures, a broad range of examples (including tropical crops), and the inclusion of worldwide references. A comprehensive review of pertinent literature is included to familiarize the student with selected plant breeders and geneticists and to demonstrate various views on controversial issues.

Hundreds of photos and illustrations, each with a comprehensive caption, introduce techniques associated with various crops. The figures complement the text closely and bring the basic principles of plant breeding to life.

Neal C. Stoskopf is professor of crop sciences at the University of Guelph. **Dwight T. Tomes** is research scientist in the Department of Biotechnology Research at Pioneer Hi-Bred International. **B. R. Christie** is research scientist at Agriculture Canada.

Author Index

Subject Index